THE SECRETARY OF TRANSPORTATION
WASHINGTON, DC 20590

AF510043

December 16, 2016

The Honorable Paul D. Ryan
Speaker of the House of Representatives
Washington, DC 20515

Dear Mr. Speaker:

I am pleased to submit the "2015 Status of the Nation's Highways, Bridges, and Transit: Conditions and Performance" report in accordance with the requirements of 23 U.S.C. §503(b)(8) and 49 U.S.C. §308(e). The report provides Congress and other policymakers with an objective appraisal of the physical conditions, operational performance, and financing trends of highways, bridges, and transit systems. The analysis is based both on the current state of these systems and on the projected future state of these systems under a set of alternative future investment scenarios.

This report offers comprehensive data-driven background information to support the development and evaluation of legislative, program, and budget options at all levels of government. It also serves as a primary source of information for national and international news media, transportation associations, and industry. This report consolidates conditions, performance, and financial data provided by States, local governments, and mass transit operators to provide a national-level summary. The future investment scenario analyses are developed specifically for this report and provide national-level projections only. These illustrative scenarios are based on combined levels of Federal, State, local, and private capital investment, and no specific level of Federal spending is implied in any scenario.

I have sent a similar letter to the President of the Senate; the Chairman and Ranking Member of the Senate Committee on Environment and Public Works; the Chairman and Ranking Member of the Senate Committee on Banking, Housing, and Urban Affairs; and the Chairman and Ranking Member of the House Committee on Transportation and Infrastructure.

Sincerely,

Anthony R. Foxx

Enclosure

Table of Contents

List of Exhibits

Highlights

Executive Summaries

Chapter 1

Chapter 2

Chapter 3

Chapter 4

Chapter 5

Main Report

Appendices

Abbreviations

AADT	average annual daily traffic
AADTT	average annual daily truck traffic
AASHTO	American Association of State Highway and Transportation Officials
ACS	American Community Survey
ADA	Americans with Disabilities Act of 1990
ADT	annual daily traffic; average daily traffic
AEO	Annual Energy Outlook
APTA	American Public Transportation Association
ATM	Active Traffic Management
ATRI	American Transportation Research Institute
B/C	benefit-cost
BCA	benefit-cost analysis
BCR	benefit-cost ratio
BIA	U.S. Bureau of Indian Affairs
BLM	Bureau of Land Management
BLS	U.S. Bureau of Labor Statistics
BPI	Bid Price Index
BTS	U.S. Bureau of Transportation Statistics
C&P	Conditions and Performance
CAFE	Corporate Average Fuel Economy
CCSP	Climate Change Scenario Planning Project, Central New Mexico
CFR	Code of Federal Regulations
CMAQ	Congestion Mitigation and Air Quality
CMIP	Coupled Model Intercomparison Project
CO_2	carbon dioxide
CPI	Consumer Price Index
CRFC	Critical Rural Freight Corridor
CSS	Context Sensitive Solutions
CUFC	Critical Urban Freight Corridor
DO	directly operated
DOD	Department of Defense
DOI	Department of the Interior
DOT	Department of Transportation
DR	demand response
DRM	directional route mile
EDC	Every Day Counts
EERPAT	Energy and Emissions Reduction Policy Analysis Tool
EPA	Environmental Protection Agency
FAF	Freight Analysis Framework
FARS	Fatality Analysis Reporting System
FAST Act	Fixing America's Surface Transportation Act
FH	Forest Highways
FHWA	Federal Highway Administration
FLHP	Federal Lands Highway Program
FLMA	Federal Land Management Agency

FLTP	Federal Lands Transportation Program
FLTTP	Federal Lands and Tribal Transportation Program
FMCSA	Federal Motor Carrier Safety Administration
FMEI	Freight Movement Efficiency Index
FPM	freight performance measures
FRA	Federal Railroad Administration
FTA	Federal Transit Administration
FWS	Fish and Wildlife Service
FY	fiscal year
GARVEE	Grant Anticipation Revenue Vehicle
GDP	gross domestic product
GHG	greenhouse gas
GIS	geographic information system
GPS	global positioning system
GRS-IBS	geosynthetic reinforced soil-integrated bridge system
HERS	Highway Economic Requirements System
HOT	high occupancy toll
HOV	high occupancy vehicle
HPMS	Highway Performance Monitoring System
HPMS-AP	HPMS Analytical Process
HR	heavy rail
HSIP	Highway Safety Improvement Program
HSM	Highway Safety Manual
HTF	Highway Trust Fund
ICE	Infrastructure Carbon Estimator
INVEST	Infrastructure Voluntary Evaluation Sustainability Tool
IRI	International Roughness Index
IRR	Indian Reservation Roads
ISTEA	Intermodal Surface Transportation Efficiency Act of 1991
ITE	Institute of Transportation Engineers
ITS	intelligent transportation system
KACTS	Kittery Area Comprehensive Transportation System
LAI	Location Affordability Index
LR	light rail
LRTP	Long Range Transportation Plans
LTPP	Long-Term Pavement Performance
MAP-21	Moving Ahead for Progress in the 21st Century Act
MB	motorbus
MCDM	multi-criteria decision method
MEPDG	Mechanistic Empirical Pavement Design Guide
MFN	Multimodal Freight Network
MIRE	Model Inventory of Roadway Elements
MOVES	Motor Vehicle Emission Simulator
mpg	miles per gallon
MPO	metropolitan planning organization
MR&R	maintenance, repair, and replacement; maintenance, repair, and rehabilitation
MRCOG	Mid Region Council of Governments
MTA	Mass Transit Account
NBE	National Bridge Element
NBI	National Bridge Inventory

NBIAS	National Bridge Investment Analysis System
NCHRP	National Cooperative Highway Research Program
NFN	National Freight Network
NHCCI	National Highway Construction Cost Index
NHFN	National Highway Freight Network
NHPP	National Highway Performance Program
NHS	National Highway System
NHTS	National Household Travel Survey
NHTSA	National Highway Traffic Safety Administration
NPMRDS	National Performance Management Research Data Set
NPRM	Notice of Proposed Rulemaking
NPS	National Park Service
NTD	National Transit Database
OMB	Office of Management and Budget
P3	Public-Private Partnership
PAB	Private Activity Bond
PBES	prefabricated bridge elements and systems
PFN	Primary Freight Network
PHFS	Primary Highway Freight System
PLDR&T	public lands development roads and trails
PLHD	Public Lands Highway Discretionary Program
PM-1	Safety Performance Measures Rule
PM-10	particulate matter of 10 microns in diameter or smaller
PM-2	Pavement and Bridge Performance Measures Rule
PM-3	System Performance Measures Rule
PMT	passenger miles traveled; person miles of travel
PRP	park roads and parkways
PSR	Present Serviceability Rating; Pavement Serviceability Rating
PT	purchased transportation
RAIRS	Rail Accident/Injury Reporting System
Reclamation	Bureau of Reclamation
Recovery	Act American Recovery and Reinvestment Act
ROW	right-of-way
RP	Recommended Practice
RR	Refuge Roads
RTP	Recreational Trails Program
SAFETEA-LU	Safe, Accountable, Flexible, Efficient Transportation Equity Act: A Legacy for Users
SCATS	Sydney Coordinated Adaptive Traffic System
SCOOT	Split Cycle Offset Optimization Technique
SGR	state of good repair
SHSP	Strategic Highway Safety Plan
SP	System Planning
SQC	Synthesis, Quantity, and Condition
SR	State Route
SRTS	Safe Routes to School
STAA	Surface Transportation Assistance Act of 1982
STBG	Surface Transportation Block Grant
STIC	Small Transit Intensive Cities
STP	Surface Transportation Program
STRAHNET	Strategic Highway Network

STRIDE	Southeastern Transportation Research, Innovation, Development and Education Center, University of Florida
STURAA	Surface Transportation and Uniform Relocation Assistance Act of 1987
SUV	sport-utility vehicle
TA	Transit Alternatives
TAP	Transportation Alternatives Program
TDM	Transportation Demand Management
TE	Transportation Enhancement
TEA-21	Transportation Equity Act for the 21st Century
TEAM	Transit Electronic Award Management
TERM	Transit Economic Requirements Model
TIFIA	Transportation Infrastructure and Finance Innovation Act
TMAS	Traffic Monitoring Analysis System
TPM	Transportation Performance Management
TRIP	Transit in the Parks
TSI	Transportation Services Index
U.S.	United States
U.S.C.	United States Code
UPT	unlinked passenger trips
USACE	U.S. Army Corps of Engineers
USAGE	United States Applied General Equilibrium
USDA	U.S. Department of Agriculture
USFS	U.S. Forest Service
UZA	urbanized area
V/SF	volume/service flow
V2I	Vehicle-to-Vehicle
V2V	Vehicle-to-Infrastructure
VAST	Vulnerability Assessment Scoring Tool
VII	Vehicle Infrastructure Integration
VIUS	Vehicle Inventory and Use Survey
VMS	variable message signs
VMT	vehicle miles traveled
VRM	vehicle revenue mile
VSL	value of a statistical life; variable speed limit
VT	vehicle type
VT1	small auto
VT2	medium auto
VT3	four-tire truck
VT4	six-tire truck
VT5a	three- or four-axle truck
VT5b	bus
VT6	four-axle combination truck
VT7	five- or more axle combination truck
WMA	warm mix asphalt

Introduction

This is the 11th in a series of combined documents the U.S. Department of Transportation (DOT) has prepared to satisfy requirements for reporting to Congress on the condition, performance, and future capital investment needs of the Nation's highway and transit systems. This report incorporates highway, bridge, and transit information required by 23 United States Code (U.S.C.) §503(b)(8) and transit system information required by 49 U.S.C. §308(e). Beginning in 1993, the Department combined two separate existing report series that covered highways and transit to form this report series; before then, 11 reports had been issued on the condition and performance of the Nation's highway systems, starting in 1968. Five separate reports on the Nation's transit systems' performance and conditions were issued beginning in 1984.

This *2015 Status of the Nation's Highways, Bridges, and Transit: Conditions and Performance* report to Congress (C&P report) draws primarily on 2012 data. The 2013 C&P report, transmitted on January 14, 2014, was based largely on 2010 data.

In assessing recent trends, many of the exhibits presented in this report present statistics for the 10 years from 2002 to 2012. Other charts and tables cover different periods, depending on data availability and years of significance for particular data series. The prospective analyses presented in this report generally cover the 20-year period ending in 2032.

Report Purpose

This document is intended to provide decision makers with an objective appraisal of the physical conditions, operational performance, and financing mechanisms of highways, bridges, and transit systems based on both their current state and their projected future state under a set of alternative future investment scenarios. This report offers a comprehensive, data-driven background context to support the development and evaluation of legislative, program, and budget options at all levels of government. It also serves as a primary source of information for national and international news media, transportation associations, and industry.

This C&P report consolidates conditions, performance, and financial data provided by States, local governments, and public transit operators to present a national-level summary. Some of the underlying data are available through DOT's regular statistical publications. The future investment scenario analyses are developed specifically for this report and provide projections at the national level only.

Report Organization

This report begins with a Highlights section that summarizes key findings of the overall report, which is followed by an Executive Summary that summarizes the key findings in each individual chapter.

The main body of the report is organized into four major sections. The six chapters in Part I, *Description of Current System*, contain the core retrospective analyses of the report. Chapters 2 through 6 each include separate highway and transit sections discussing each mode in depth. This structure is intended to accommodate report users who might be interested primarily in only one of the two modes. Chapter 1 follows a similar approach, except that the two sections focus on personal travel and freight movement.

The Introduction to Part I provides background information on DOT strategic goals and issues pertaining to transportation performance management, both of which relate closely to the material presented in Part I.

- Chapter 1 discusses selected topics relating to personal travel and highway freight movement.
- Chapter 2 presents information on recent trends in highway and transit system characteristics.
- Chapter 3 describes the current physical conditions of highways, bridges, and transit systems.
- Chapter 4 discusses issues relating to the safety of highways and transit.
- Chapter 5 presents information on various aspects of current system performance for highways and transit, including operational performance, quality of life, and environmental sustainability.
- Chapter 6 discusses highway and transit revenue sources and expenditure patterns for all levels of government.

The four chapters in Part II, *Investment/Performance Analysis*, contain the core prospective analyses of the report, including 20-year future capital investment scenarios. The Introduction to Part II provides critical background information and caveats that should be considered while interpreting the findings presented in Chapters 7 through 10.

- Chapter 7 projects the potential impacts of different levels of future highway, bridge, and transit capital investment on the future performance of various components of the system.
- Chapter 8 describes selected capital investment scenarios in more detail and relates these scenarios to the current levels of capital investment for highways, bridges, and transit.
- Chapter 9 provides supplemental analysis relating to the primary investment scenarios, comparing the findings of the future investment scenarios to findings in previous reports and discussing scenario implications.
- Chapter 10 discusses how changing some of the underlying technical assumptions would affect the future highway and transit investment scenarios.

Part III, *Special Topics*, explores topics related to the primary analyses in the earlier sections of the report.

- Chapter 11 discusses issues pertaining to pedestrian and bicycle transportation.
- Chapter 12 examines the transportation systems serving Federal and Tribal lands.

Part IV, *Recommendations for HPMS Changes*, provides information on the status and planned direction of the Highway Performance Monitoring System (HPMS).

The C&P report contains three technical appendices that describe the investment/performance methodologies used in the report for highways, for bridges, and for transit. A fourth appendix describes an ongoing research effort for *Reimagining the C&P Report in a Performance Management-Based World*.

Highway Data Sources

Highway characteristics and conditions data are derived from HPMS, a cooperative data/analytical effort dating from the late 1970s that involves the Federal Highway Administration (FHWA) and State and local governments. HPMS includes a statistically drawn sample of more than 100,000 highway sections containing data on current physical and operating characteristics and projections of future travel growth on a highway section-by-section basis. All HPMS data are provided to FHWA through State DOTs from existing State or local government databases or transportation plans and programs, including those of metropolitan planning organizations.

FHWA annually collects bridge inventory and inspection data from the States, Federal agencies, and Tribal governments and incorporates the data into the National Bridge Inventory (NBI). NBI contains information from all bridges covered by the National Bridge Inspection Standards (Title 23, Code of Federal Regulations, Part 650, Subpart C) located on public roads throughout the United States and Puerto Rico. Inventory information for each bridge includes descriptive identification data, functional characteristics, structural design types and materials, location, age and service, geometric characteristics, navigation data, and functional classifications; condition information includes inspectors' evaluations of the primary components of a bridge, such as the deck, superstructure, and substructure.

State and local finance data are derived from the financial reports States provide to FHWA in accordance with *A Guide to Reporting Highway Statistics*. These data are the same as those used in compiling the annual Highway Statistics report. Highway safety performance data are drawn from the Fatality Analysis Reporting System.

Highway operational performance data are drawn primarily from the National Performance Management Research Data Set (NPMRDS). This database compiles observed average travel times, date and time, and direction and location for freight, passenger, and other traffic. The data cover the period after the Moving Ahead for Progress in the 21st Century Act (MAP-21) for the National Highway System plus arterials at border crossings. The dataset is made available to States and metropolitan planning organizations (MPOs) monthly to assist them in performance monitoring and target setting. Because NPMRDS data are available only for 2012 onward, historic time series data are drawn from the Texas Transportation Institute's *Urban Mobility Scorecard*.

Under MAP-21, the FHWA was charged with establishing a national tunnel inspection program. In 2015, development began on the National Tunnel Inventory database system, and inventory data were collected for all highway tunnels reported. Concurrently, the FHWA implemented an extensive program to train inspectors nationwide on tunnel inspection and condition evaluation. Complete inventory and condition data for all tunnels will be collected annually, beginning in 2018, and will be available for use in subsequent C&P reports.

Transit Data Sources

Transit data are derived from the National Transit Database (NTD) and transit agency asset inventories. NTD comprises comprehensive data on the revenue sources, capital and operating expenses, basic asset holdings, service levels, annual passenger boardings, and safety data for the more than 800 urban and 1,770 rural transit operators that receive annual funding support through the Federal Transit Administration's (FTA's) Section 5307 (Urbanized Area) and Section 5311 (Rural Area) Formula Programs. Except for fleet vehicle holdings (for which NTD provides data on the composition and age of transit fleets), however, NTD provides no data required to assess the current physical condition of the Nation's transit infrastructure.

To meet this need, FTA collects transit asset inventory data from a sample of the Nation's largest rail and bus transit operators. In direct contrast to the data in NTD and HPMS—which local and State funding grantees must report to FTA and FHWA, respectively, and which are subject to standardized reporting procedures—the transit asset inventory data used to assess current transit conditions are provided to FTA in response to direct requests submitted to grantees and are subject to no reporting requirements. Although asset inventory data are subject to no current reporting requirements or reporting standards, MAP-21 requires that grantees begin submitting this information to NTD. Once rules for collecting these data are formalized in regulation and grantees begin submittals, FTA will have improved data on which to base its forecasts.

In recent practice, data requests primarily have been made to the Nation's 20 to 30 largest transit agencies because they account for roughly 85 percent of the Nation's total transit infrastructure by value. Considering the slow rate of change in asset holdings of transit agencies over time (excluding fleet vehicles and major expansion projects), FTA has requested these data from any given agency only every 3 to 5 years. The asset inventory data collected through these requests document the age, quantity, and replacement costs of the grantees' asset holdings by asset type. The nonvehicle asset holdings of smaller operators have been estimated using a combination of the (1) fleet-size and facility-count data reported to NTD and (2) actual asset age data of a sample of smaller agencies that responded to previous asset inventory requests. This method of obtaining asset data has served FTA well in the past (and the quality of the reported data has improved over time), but the accuracy and comprehensiveness of FTA's estimates of current asset conditions and capital reinvestment needs will benefit from the standardized reporting requirements to be developed through MAP-21.

Multimodal Data Sources

Personal travel data are derived primarily from the National Household Travel Survey, which collects details of travel by all modes for all purposes for each household member. The survey has collected data intermittently since 1969 using a national sample of households in the civilian noninstitutionalized population. The survey was last conducted in 2009. The survey obtains demographic characteristics of households and people and information about all vehicles in the household.

Freight data are primarily derived from the Freight Analysis Framework version 3.4, which includes all freight flows to, from, and within the United States. The framework is built from a variety of datasets, such as the Census Bureau's Commodity Flow survey and HPMS.

Investment/Performance Analytical Procedures

The earliest versions of the reports in this combined series relied exclusively on engineering-based estimates for future investment/performance analysis, which considered only the costs incurred by transportation agencies. This approach failed to consider another critical dimension of transportation programs adequately—the impacts of transportation investments on the costs users of the transportation system incur. Executive Order 12893, Principles for Federal Infrastructure Investments, dated January 1994, directs each executive department and agency with infrastructure responsibilities to base investments on "systematic analysis of expected benefits and costs, including both quantitative and qualitative measures." New approaches have been developed to address the deficiencies in earlier versions of this report and to meet this Executive Order. The analytical tools used in this report now have an added economic component.

The highway investment scenarios presented in this report are developed in part from the Highway Economic Requirements System (HERS), which models highway investment using benefit-cost analysis. The HERS model quantifies user, agency, and societal costs for various types and combinations of capital improvements. HERS considers costs associated with travel time, vehicle operation, safety, routine maintenance, and emissions, including greenhouse gases. Bridge investment scenario estimates are developed from the National Bridge Investment Analysis System (NBIAS) model. Unlike earlier bridge models (and similar to HERS), NBIAS incorporates benefit-cost analysis into the bridge investment/performance evaluation.

The transit investment analysis is based on the Transit Economic Requirements Model (TERM). TERM consolidates older engineering-based evaluation tools and uses a benefit-cost analysis to ensure that investment benefits exceed investment costs. TERM identifies the investments needed to replace and rehabilitate existing assets, improve operating performance, and expand transit systems to address the growth in travel demand.

HERS, NBIAS, and TERM have not yet evolved to the point that they can be used for direct multimodal analysis. Although the three models use benefit-cost analysis, their methods for implementing this analysis are very different. Each model is based on a separate, distinct database. Each model uses data applicable to its specific part of the transportation system and addresses issues unique to each mode. For example, HERS assumes that adding lanes to a highway causes highway user costs to decline, which results in additional highway travel. Under this assumption, some of this increased traffic would be newly generated travel and some could be the result of travel shifting from transit to highways. HERS, however, does not distinguish between different sources of additional highway travel. Similarly, TERM's benefit-cost analysis assumes that some travel shifts from automobile to transit because of transit investments, but the model cannot project the effect of such investments on highways.

In interpreting the findings of this report, recognizing the limitations of these analytical tools and the potential impacts of different assumptions made for the analyses is essential. The technical appendices and the Introduction to Part II contain information critical to contextualizing the future investment scenarios. These issues are also discussed in Q&A boxes presented in Chapters 7 through 10.

Changes to C&P Report Scenarios from 2013 Edition

The 2013 C&P Report presented two versions of each highway and bridge scenario in Chapter 8: (1) a set based on modeled projections of future vehicle miles traveled (VMT) for individual highway sections that States provided to HPMS ("Forecast-Based") and (2) a set based on the historic trend in VMT growth over the past 15 years ("Trend-Based"). This edition of the C&P report reverts to the traditional approach of presenting only one set of highway and bridge scenarios; however, the process used for developing the VMT forecasts for use in the analysis is new. For this edition, a modified version of the "Trend-Based" procedure was applied in which the State-provided VMT projections for individual highway sections were each reduced proportionally to match a national-level VMT forecast developed by the Volpe National Transportation Systems Center for FHWA.

The Low-Growth and High-Growth scenarios for transit presented in the 2013 C&P Report were based on growth in passenger miles traveled (PMT) applied at the urbanized area level. For this edition, both scenarios draw on the trend-based growth rates stratified by FTA region, urbanized area size, and type of transit mode. The Low-Growth scenario assumes an annual PMT growth rate of 0.5 percent less than the 15-year trend, while the High-Growth scenario assumes an annual PMT growth rate of 0.5 percent more than the 15-year trend.

The 2013 C&P Report presented Sustain 2010 Spending scenarios for both highways and transit, which projected the impacts of sustaining spending at base year 2010 levels in constant-dollar terms over 20 years. Because the base year for the current report is 2012, the scenarios have been renamed Sustain 2012 Spending.

The Maintain Conditions and Performance scenario for highways and bridges presented in the 2013 C&P Report used average pavement roughness, average delay per VMT, and the average bridge sufficiency rating as primary indicators. This edition substitutes the percentage of deck area on bridges classified as deficient for the average bridge sufficiency rating in defining this scenario and applies the pavement roughness and delay indicators in a somewhat different manner.

Cautionary Notes on Using This Report

To interpret the analyses presented in this report correctly, understanding the framework in which they were developed and recognizing their limitations are critical. This document is not a statement of Administration policy, and the future investment scenarios presented are illustrative. **The report does not endorse any particular level of future highway, bridge, or transit investment.** It neither addresses how future Federal programs for surface transportation should look, nor identifies the level of future funding for surface transportation that could or should be provided by the Federal, State, or local governments; the private sector; or system users. Making recommendations on such policy issues is beyond the legislative mandate for this report and furthermore would violate its objectivity. Analysts outside FHWA can and do use the statistics presented in the C&P report to draw their own conclusions, but any analysis attempting to use the information presented in this report to determine a target Federal program size would require a series of additional policy and technical assumptions that are well beyond what is reflected here.

The investment scenario results presented in this report are estimates of the performance that **could** be achieved with a given level of funding, not necessarily what **would** be achieved. The analytical tools used in developing these estimates combine engineering and economic procedures that determine deficiencies based on engineering standards while applying benefit-cost analysis procedures to identify potential capital improvements to address deficiencies that might have positive net benefits. The models generally assume that projects are prioritized based on their benefit-cost ratios, but that assumption deviates somewhat from actual patterns of project selection and funding distribution that occur in the real world. Consequently, the level of investment identified as the amount required for achieving a certain performance level should be viewed as illustrative only—not as a projection or prediction of an actual condition and performance outcome likely to result from a given level of national spending.

As in any modeling process, simplifying assumptions have been made to make analysis practical and to report within the limitations of available data. Because operators at the State and local levels primarily make the ultimate decisions concerning highways, bridges, and transit systems, they have a much stronger business case for collecting and retaining detailed data on individual system components. The Federal government collects selected data from States and transit operators to support this report and several other Federal activities, but these data are not sufficiently robust to make definitive recommendations concerning specific transportation investments in specific locations. Improvements are evaluated based on benefit-cost analysis, but not all external costs (such as noise pollution or construction-related loss of wildlife habitat) or external benefits (such as productivity gains that might result from transportation improvements that open markets to competition) are fully considered. Across a broad program of investment projects, such external effects might cancel each other; but, to the extent that they do not, the true "needs" could be either higher or lower than the models predict.

Highlights

This edition of the C&P report is based primarily on data through 2012; consequently, the system conditions and performance measures presented do not reflect the impacts of the Moving Ahead for Progress in the 21st Century Act (MAP-21), which authorized Federal highway and transit funding for Federal Fiscal Years 2013 and 2014. None of the impact of funding authorized under the Fixing America's Surface Transportation Act is reflected.

In assessing recent trends, this report generally focuses on the 10-year period from 2002 to 2012. The prospective analyses generally cover the 20-year period ending in 2032; the investment levels associated with these scenarios are stated in constant 2012 dollars.

Highlights: Highways and Bridges

Extent of the System

- The Nation's road network included 4,109,418 miles of public roadways and 607,380 bridges in 2012. This network carried over 2.987 trillion vehicle miles traveled (VMT), and almost 4.275 trillion person miles traveled (PMT), up from 2.874 trillion VMT and down from 4.667 trillion PMT in 2002.
- The 1,005,378 miles of Federal-aid highways (24 percent of total mileage) carried 2.527 trillion VMT (85 percent of total travel) in 2012.
- Although the 223,257 miles on the National Highway System (NHS) comprise only 5 percent of total mileage, the NHS carried 1.644 trillion VMT in 2012, approximately 55 percent of total travel.

Highway System Terminology

"Federal-aid highways" are roads that generally are eligible for Federal funding assistance under current law. (Note that certain Federal programs do allow the use of Federal funds on other roadways.)

The "National Highway System" (NHS) includes those roads that are most important to interstate travel, economic expansion, and national defense. It includes the entire Interstate System. MAP-21 directed that the NHS system be expanded. Except where noted, the statistics presented in this report reflect the expanded NHS.

- The 47,714 miles on the Interstate System carried 0.736 trillion VMT in 2012, slightly over 1 percent of total mileage and just under 25 percent of total VMT. The Interstate System has grown since 2002, when it consisted of 46,747 miles carrying 0.694 trillion VMT.

Spending on the System

- All levels of government spent a combined $221.3 billion for highway-related purposes in 2012. About 47.5 percent of total highway spending ($105.2 billion) was for capital improvements to highways and bridges; the remainder included expenditures for physical maintenance, highway and traffic services, administration, highway safety, and debt service.

2012 Extent of the Highway System

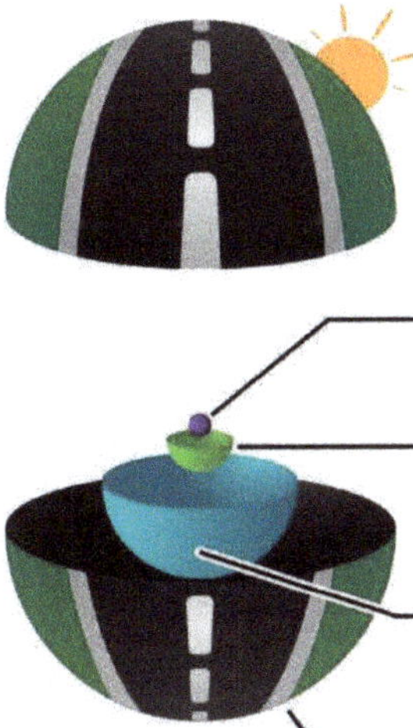

United States Road and Bridge System	EXTENT (Miles and Bridges per System)		VMT (Vehicle Miles Traveled)
Interstate System	1% 47,714 miles	9% 55,959 bridges	25% 0.736 trillion VMT
National Highway System	5% 223,257 miles	19% 117,485 bridges*	55% 1.644 trillion VMT
Federal-aid Highways	24% 1,005,378 miles	53% 321,724 bridges	85% 2.527 trillion VMT
Systemwide (All Roads)	100% 4,109,418 miles	100% 607,380 bridges	100% 2.987 trillion VMT

*Reflects bridge count prior to expansion of the NHS under MAP-21.

You drive 85% of your VMT on Federal-aid Highways.

- In nominal dollar terms, highway spending increased by 62.8 percent (5.0 percent per year) from 2002 to 2012; after adjusting for inflation, this equates to a 28.9-percent increase (2.6 percent per year).
- Highway capital expenditures rose from $68.2 billion in 2002 to $105.2 billion in 2012, a 54.3-percent (4.4 percent per year) increase in nominal dollar terms; after adjusting for inflation, this equates to a 23.5-percent (2.1 percent per year) increase in constant-dollar terms.

> **Constant-Dollar Conversions for Highway Expenditures**
>
> This report uses the Federal Highway Administration's National Highway Construction Cost Index and its predecessor, the Composite Bid Price Index, for inflation adjustments to highway capital expenditures and the Consumer Price Index for adjustments to other types of highway expenditures.

- The portion of total highway capital spending funded by the Federal government decreased from 46.1 percent in 2002 to 43.1 percent in 2012. Federally funded highway capital outlay grew by 3.7 percent per year over this period, compared to a 5.0-percent annual increase in capital spending funded by State and local governments.
- The composition of highway capital spending shifted from 2002 to 2012. The percentage of highway capital spending directed toward system rehabilitation rose from 53.1 percent in

> **Highway Capital Spending Terminology**
>
> This report splits highway capital spending into three broad categories. "System rehabilitation" includes resurfacing, rehabilitation, or reconstruction of existing highway lanes and bridges. "System expansion" includes the construction of new highways and bridges and the addition of lanes to existing highways. "System enhancement" includes safety enhancements, traffic control facilities, and environmental enhancements.

2002 to 59.0 percent in 2012. Over the same period, the percentage directed toward system enhancement rose from 11.1 percent to 15.1 percent, while the percentage directed toward system expansion fell from 35.8 percent to 25.8 percent.

Conditions and Performance of the System

The data systems and performance metrics for different aspects of system conditions and performance are at different stages of development. Progress is being made on measuring the impact of transportation investments on livability. Several resources and tools, such as the Location Affordability Portal, Sustainable Communities Indicator Catalog, Infrastructure Voluntary Evaluation Sustainability Tool, and the Community Vision Metrics Web Tool have been developed to measure the impact of transportation investments on quality of life.

Bridge Conditions Have Improved

- Based directly on bridge counts, the share of bridges classified as structurally deficient has improved, dropping from 14.2 percent in 2002 to 11.0 percent in 2012. The share of NHS bridges classified as structurally deficient also improved over this period, dropping from 5.9 percent to 4.5 percent.
- Weighted by deck area, the share of bridges classified as structurally deficient improved, declining from 10.4 percent in 2002 to 8.2 percent in 2012. The deck area-weighted share of structurally deficient NHS bridges dropped from 8.6 percent to 7.1 percent over this period.

Bridge Geometry Has Slightly Improved

- Based directly on bridge counts, the share of bridges classified as functionally obsolete declined from 15.4 percent in 2002 to 14.0 percent in 2012. The share of NHS bridges classified as functionally obsolete also improved over this period, dropping from 17.2 percent to 16.2 percent. Functional obsolescence tends to be a more significant problem on larger bridges carrying more traffic, such as those located on the NHS.
- Weighted by deck area, the share of bridges classified as functionally obsolete improved slightly, dropping from 20.4 percent in 2002 to 20.1 percent in 2012. The deck area-weighted share of functionally obsolete NHS bridges dropped slightly from 21.1 percent to 21.0 percent over this period.

FHWA Bridge Classifications

Bridges are considered "structurally deficient" if (1) significant load-carrying elements are found to be in poor or worse-than-poor condition due to deterioration or damage, or (2) the adequacy of the waterway opening the bridge provides is determined to be insufficient to the point of causing intolerable traffic interruptions due to high water. That a bridge is structurally deficient does not mean it is unsafe.

Functional obsolescence in general is a function of the geometrics (e.g., broad roadway width, load carrying capacity, clearances, approach roadway alignment) of the bridge in relation to the geometrics required by current design standards. The magnitude of such deficiencies determines whether a bridge is classified as "functionally obsolete."

These classifications are often weighted by bridge deck area, recognizing that bridges are not all the same size and, in general, larger bridges are more costly to rehabilitate or replace to address deficiencies. They are also sometimes weighted by annual daily traffic, recognizing the more heavily traveled bridges have a greater impact on total highway user costs.

2002–2012 Highway System Trends

Highway Safety Improved Overall, but Nonmotorist Fatalities Rose

- The annual number of highway fatalities was reduced by 21.4 percent from 2002 to 2012, dropping from 43,005 to 33,782. The fatality rate per 100 million VMT declined from 1.51 in 2002 to 1.14 in 2012. (Since 2012, the number of highway fatalities has risen to 35,092 in 2015; the fatality rate per 100 million VMT was 1.08 in 2015).
- The number of traffic-related injuries decreased by more than 19 percent, from 2.9 million in 2002 to 2.4 million in 2012. The injury rate per 100 million VMT declined from 102 in 2002 to 80 in 2012.
- Fatalities related to roadway departure decreased by 31.0 percent from 2002 to 2012, but roadway departure remains a factor in over half of all highway fatalities. Intersection-related fatalities decreased by 21.5 percent from 2002 to 2012, but over one-fifth of highway fatalities in 2012 occurred at intersections.

- In 2012, roadway departure, intersection, and pedestrian fatalities accounted for 52.2 percent, 21.7 percent, and 14.1 percent, respectively, of the 33,561 fatalities.
- From 2002 to 2012, the number of nonmotorists killed by motor vehicles increased by 1.1 percent, from 5,630 to 5,692. Since 2009, the number of pedestrians and pedacylists (such as bicyclists) killed by motor vehicle crashes has each increased by approximately 15.6 percent.

Pavement Condition Trends Have Been Mixed

- In general, pavement condition trends over the past decade have been better on the NHS (the 5 percent of total system mileage that carries 25 percent of total system VMT) than on Federal-aid highways (the 24 percent of system mileage that carries 85 percent of total system VMT, including the NHS).
- The share of Federal-aid highway VMT on pavements with "good" ride quality rose from 43.8 percent in 2002 to 44.9 percent in 2012. The share of mileage with good ride quality declined from 46.6 percent to 36.4 percent over this same period, however, indicating that conditions have worsened on roads with lower travel volumes.
- The share of Federal-aid highway pavements with "poor" ride quality rose from 2002 to 2012, as measured on both a VMT-weighted basis (rising from 14.7 percent to 16.7 percent) and a mileage basis (rising from 12.6 percent to 19.7 percent). Although this trend is exaggerated due to changes in data reporting instructions beginning in 2010, the data clearly show that more of the Nation's pavements have deteriorated to the point that they are adding to vehicle operating costs and reducing driver comfort.
- The share of VMT on NHS pavements with good ride quality rose from 50 percent in 2002 to 57.1 percent in 2012. This gain is even more impressive considering the significant expansion of the NHS under MAP-21, as pavement conditions on the additions to the NHS were not as good as those on the pre-expansion NHS. When adjusted for the NHS expansion, the share of VMT on NHS pavements with good ride quality improved by an average of more than 2 percentage points per year. The share rose from 50 percent in 2002 to 60 percent in 2010 based on the pre-expansion NHS and from an estimated 54.7 percent in 2010 to 57.1 percent in 2012 based on the post-expansion NHS.

Pavement Condition Terminology

This report uses the International Roughness Index (IRI) as a proxy for overall pavement condition. Pavements with an IRI value less than 95 inches per mile are considered to have "good" ride quality. Pavements with an IRI value greater than 170 inches per mile are considered to have "poor" ride quality. Pavements that fall between these two ranges are considered "fair"; the term "acceptable" combines the "good" and "fair" categories.

Operational Performance Has Slowly Worsened

- The Texas Transportation Institute estimates that the average commuter experienced a total of 41 hours of delay resulting from congestion in 2012, up from 39 hours in 2002. Total delay experienced by all travelers combined rose from 5.6 billion hours in 2002 to 6.7 billion hours in 2012, an all-time high.

- The combined cost of wasted time and wasted fuel caused by congestion rose from an estimated $124 billion in 2002 to $154 billion in 2012. Although these costs had declined during the recent recession, by 2012, they had reverted to their pre-recession peak in 2007.

Future Capital Investment Scenarios – Systemwide

The scenarios that follow pertain to spending by all levels of government combined for the 20-year period from 2012 to 2032 (reflecting the impacts of spending from 2013 through 2032); the funding levels associated with all of these analyses are stated in constant 2012 dollars. The results below apply to the overall road system; separate results based on applying the scenario criteria separately to the Interstate System, the NHS, and Federal-aid highways, are presented in the body of this report.

Sustain 2012 Spending Scenario

- The Sustain 2012 Spending scenario assumes that capital spending by all levels of government is sustained in constant-dollar terms at the 2012 level ($105.2 billion systemwide) through 2032. At this level of investment, average pavement roughness on Federal-aid highways would be projected to improve by 4.5 percent, while average delay per VMT improves by 13.4 percent. The share of bridges classified as structurally deficient would be projected to improve, declining from 8.2 percent in 2012 to 2.9 percent in 2032.

> **Highway Investment/Performance Analyses**
>
> To provide an estimate of the costs that might be required to maintain or improve system performance, this report includes a series of investment/performance analyses that examine the potential impacts of alternative levels of future combined investment by all levels of government on highways and bridges for different subsets of the overall system. Rather than assuming an immediate jump to a higher (or lower) investment level, each analysis assumes that spending will grow by a uniform annual rate of increase (or decrease) in constant-dollar terms using combined highway capital spending by all levels of government in 2012 as the starting point.
>
> Drawing on these investment/performance analyses, a series of illustrative scenarios was selected for more detailed exploration and presentation. The scenario criteria were applied separately to the Interstate System, the NHS, all Federal-aid highways, and the overall road system.

Maintain Conditions and Performance Scenario

- The Maintain Conditions and Performance scenario assumes that capital investment gradually changes in constant-dollar terms over 20 years to the point at which selected measures of future conditions and performance in 2032 are maintained at 2012 levels. The average annual level of investment associated with this scenario is $89.9 billion, 14.6 percent less than actual capital spending by all levels of government in 2012.

Improve Conditions and Performance Scenario

- The Improve Conditions and Performance scenario assumes that capital investment gradually rises to the point at which all potential highway and bridge investments that are estimated to be cost-beneficial (i.e., those with a benefit-cost ratio [BCR] of 1.0 or higher) could be funded by 2032. The average annual level of systemwide investment associated with this scenario is $142.5 billion, 35.5 percent higher than actual 2012 spending.

- As of 2012, the United States had an estimated $836 billion of unmet capital investment needs for highways and bridges that would be cost-beneficial to address. The Improve Conditions and Performance scenario would eliminate this backlog, while addressing other needs as they arise over 20 years through 2032. Eliminating this backlog would require increasing highway capital spending by 2.81 percent per year faster than the rate of inflation.
- Under the Improve Conditions and Performance scenario, average pavement roughness on Federal-aid highways is projected to improve by 14.0 percent, while average delay per VMT is projected to improve by 16.5 percent. The share of bridges classified as structurally deficient is projected to improve, declining from 8.2 percent in 2012 to 1.9 percent in 2032.
- The State of Good Repair benchmark represents the subset of this scenario that is directed toward addressing deficiencies in the physical condition of existing highway and bridge assets. The average annual investment level associated with this benchmark is $85.3 billion. This level of investment would not eliminate all poor pavement or structurally deficient bridges because, in some cases, addressing such deficiencies until after they arise would not be cost-beneficial. Therefore, at the end of any given year, some portion of the pavement and bridge population would remain deficient.

2012–2032 Future Highway Capital Investment Scenarios

	SPENDING NEEDED (Billions of Dollars)	AVERAGE ROUGHNESS	AVERAGE DELAYS
Maintain Conditions and Performance at 2012 Levels	$89.9	NO CHANGE 0.0%	IMPROVE 12.2%
Sustain Spending at 2012 Level by All Levels of Government, Adjusted for Inflation	$105.2	IMPROVE 4.5%	IMPROVE 13.4%
Improve Conditions and Performance (BCR 1.0 or Higher)	$142.5	IMPROVE 14.0%	IMPROVE 16.5%

Highlights: Transit

Extent of the System

- Of the transit agencies that submitted data to the National Transit Database in 2012, 800 provided service to urbanized areas and 1,703 provided service to rural areas. Urban agencies operated 661 bus systems, 629 demand-response systems, 18 heavy rail systems, 29 commuter rail systems, 25 light rail systems, 17 streetcar systems, and 4 hybrid rail systems. Additionally, 74 transit vanpool systems, 23 ferryboat systems, 5 trolleybus systems, 8 monorail and automated guideway systems, 3 inclined plane systems, 1 cable car system, and 1 Público were in operation.
- Bus and heavy rail modes continue to be the largest segments of the industry, providing 50 percent and 36 percent of all transit trips, respectively. Commuter rail supports a relatively high share of passenger miles (20.0 percent). Although light rail is the fastest-growing rail mode (with passenger miles growing at 5.7 percent per year from 2002 to 2012), it still provides only 4.0 percent of transit passenger miles. Vanpool growth during this period was 10.7 percent per year, but vanpools still accounted for only 2.0 percent of all transit passenger miles.
- Urban transit operators reported 10.4 billion unlinked passenger trips on 4.0 billion vehicle revenue miles. Rural transit operators reported an additional 124 million unlinked passenger trips and 558 million vehicle revenue miles.

Bus, Rail, and Demand Response: Transit Modes

Public transportation is provided by several different types of vehicles that are used in different operational modes. The most common is fixed-route bus service, which uses different sizes of rubber-tired buses that run on scheduled routes. Commuter bus service is similar, but uses over-the-road buses and runs longer distances between stops. Bus rapid transit is high-frequency bus service that emulates light rail service. Públicos and jitneys are small, owner-operated buses or vans that operate on less-formal schedules along regular routes.

Larger urban areas often are served by one or more varieties of fixed-guideway (rail) service. These include heavy rail (often running in subway tunnels), which is characterized primarily by third-rail electric power and exclusive dedicated guideway. Extended urban areas might have commuter rail, which often shares track with freight trains and usually uses overhead electric power (but might also use diesel power). Light rail systems are common in large and medium-sized urban areas; they feature overhead electric power and run on track that is partially or entirely on city streets shared with pedestrian and automobile traffic. Streetcars are small light rail systems, typically with only one or two cars per train that usually run in mixed traffic. Hybrid rail, previously reported as light rail and commuter rail, is a mode with shared characteristics of these two modes. The average station density (stations per track mileage) for hybrid rail is greater than for commuter rail and lower than for light rail, and unlike commuter rail, it has smaller peak-to-base ratio. Cable cars, trolley buses, monorail, and automated guideway systems are less-common rail variants.

Demand-response transit service is usually provided by vans, taxicabs, or small buses dispatched to pick up passengers upon request. This mode is primarily used to provide paratransit service as required by the Americans with Disabilities Act. Demand-response transit does not follow a fixed schedule or route.

2012 Extent of the Transit System

	Number of Transit Agencies (Providers of Transit Service)	Unlinked Passenger Trips (Billions) (Number of Boardings)	Vehicle Revenue Miles (Billions) (Mileage in Revenue Service)
Rural	1,703	0.1	0.6
Urban	800	10.4	4.0

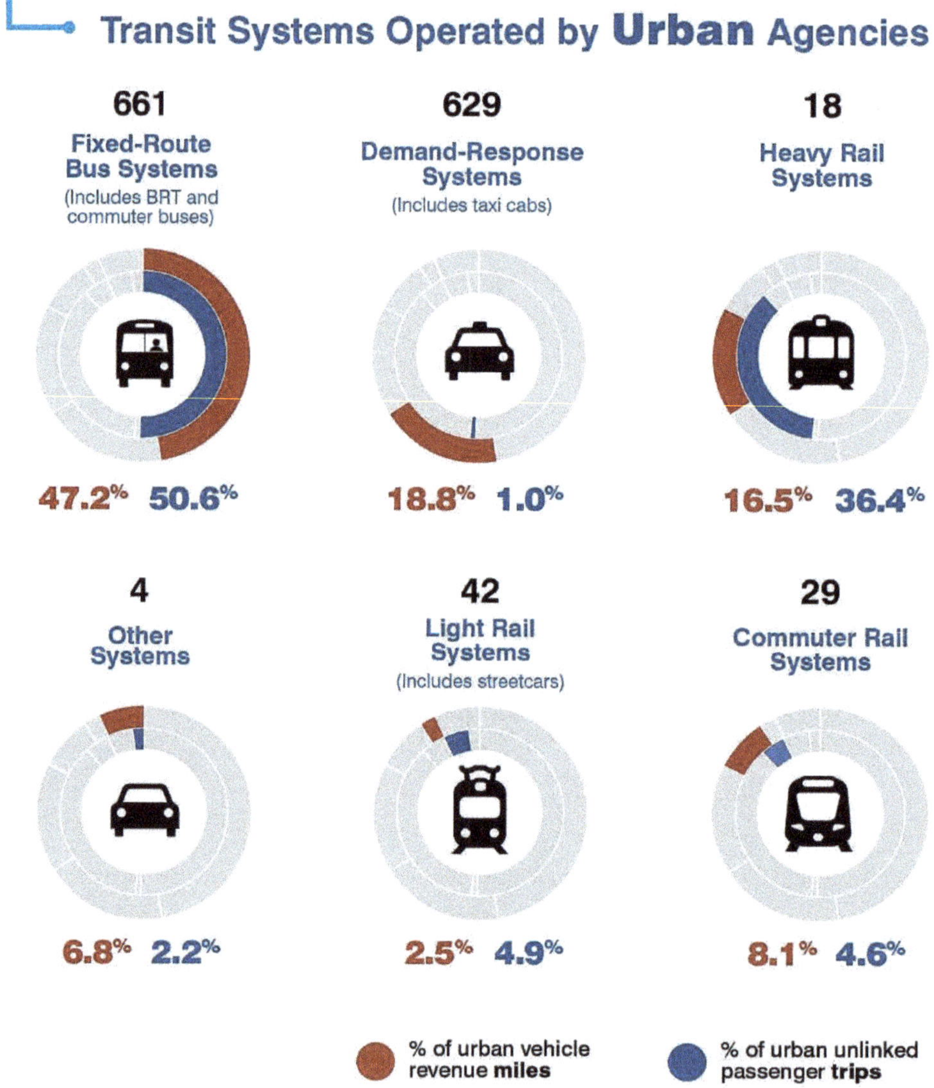

Spending on the System

- All levels of government spent a combined $58 billion to provide public transportation and maintain transit infrastructure. Of this, 26.7 percent was system-generated revenue, most of which came from passenger fares. The Federal government was the source of 19 percent of revenues, while the remaining funds came from State and local sources.

- Public transit agencies spent $16.9 billion on capital investments in 2012. Annually authorized Federal funding comprised 36 percent of these capital expenditures. Funds from the Federal American Recovery and Reinvestment Act provided another 9 percent.

- Federal funding is primarily targeted toward capital assistance; however, Federal funding for operating expenses at public transportation agencies has increased from 19 percent of all Federal funding in 2002 to 35 percent in 2012. Virtually all of the increase is due to the 2004 change that made "preventive maintenance" eligible for reimbursement from section 5307 grant funds. Meanwhile, farebox recovery ratios, representing the share of operating expenses that come from passenger fares, have remained close to the 2000 value of 35.5 percent throughout this period.

Federal Transit Funding Urban and Rural

Federal Transit Administration (FTA) Urbanized Area Formula Funds are apportioned to urbanized areas (UZAs), as defined by the Census Bureau. UZAs in this report were defined by the 2000 census. Data from the 2010 census was used starting in the 2013 apportionment. Each UZA has a designated recipient for the Federal funds, usually a metropolitan planning organization or large transit agency, which then reallocates those funds in its area according to local policy. In small urban and rural areas, FTA apportions funds to the State, which allocates them according to State policy. Indian Tribes are apportioned their formula funds directly, once they are obligated in a grant. All funds then become available, on a reimbursement basis, through application to FTA.

Conditions and Performance of the System

Transit Remains Safe despite High Increase in Fatalities in 2012

- The number of fatalities from 2002 to 2011 (excluding suicides and commuter rail) remained stable, hovering around 150 fatalities per year. In 2012, however, fatalities significantly increased to 202 fatalities. In 2012, one in four transit-related fatalities was classified as a suicide. In 2002, the rate was just 1 in 13. The rate of suicides on transit facilities has increased every year since 2005.

Some Aspects of System Performance Have Improved

- From 2002 to 2012, transit agencies have provided substantially more service. The annual rate of growth in route miles ranged from 0.3 percent per year for heavy rail to 6.2 percent per year for light rail. This growth has resulted in 32 percent more route miles available to the public.

- From 2004 to 2012, the number of annual service miles per vehicle (vehicle productivity) remained unchanged and the average number of miles between breakdowns (mean distance between failures) increased by 24 percent.

- Growth in service offered was nearly in accordance with growth in service consumed. In spite of steady growth in route miles and revenue miles, average vehicle occupancy levels did not decrease. Passenger miles traveled grew at a 1.6-percent annual pace, while the number of trips grew 1.3 percent annually. This growth rate is significantly higher than the annual growth rate in the U.S. population during this period (0.93 percent), which suggests that transit has been able to attract riders who previously used other modes of travel. Increased availability of transit service has undoubtedly been a factor in this success.

2002–2012 Transit Trends in Urban Areas

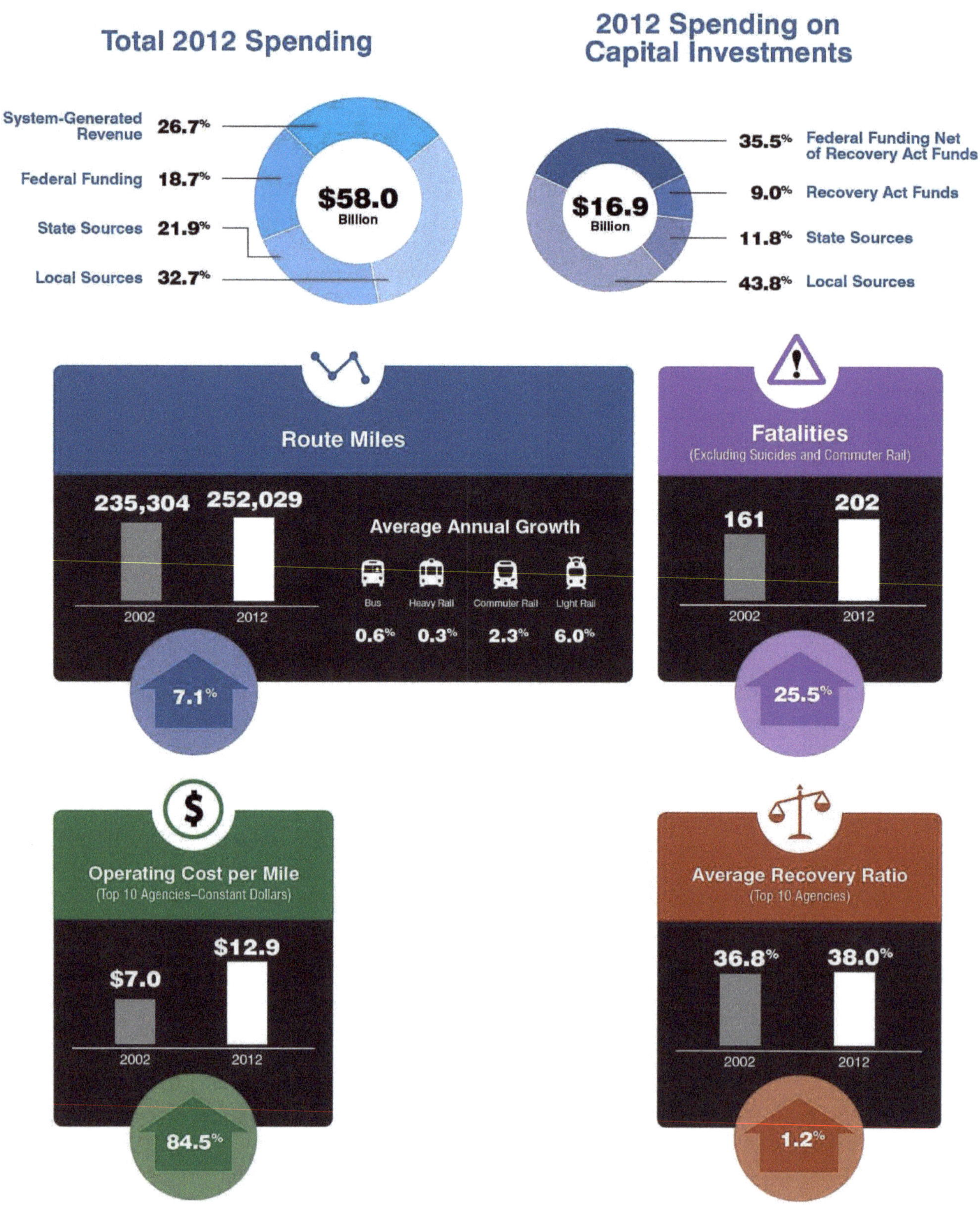

Unlinked Passenger Trips, Passenger Miles, Route Miles, and Revenue Miles

Unlinked passenger trips (UPT), also called boardings, count every time a person gets on an in-service transit vehicle. Each transfer to a new vehicle or route is considered another unlinked trip, so a person's commute to work could count as more than one trip if that person transferred between routes.

Passenger miles traveled (PMT) simply count how many miles a person travels. UPT and PMT are both commonly used measures of transit service consumed.

Directional route miles (DRM) measure the number of miles of transit route available to customers. They are directional because each direction counts separately; thus, a 1-mile-out and 1-mile-back bus route would be 2 DRM. Vehicle revenue miles count the miles of revenue service and are typically much greater than DRM because many trips are taken over each route (and each DRM). These measures are commonly used to describe the transit service provided.

Future Capital Investment Scenarios – Systemwide

As in the highway discussion, the transit investment scenarios that follow pertain to spending by all levels of government combined for the 20-year period from 2012 to 2032 (reflecting the impacts of spending from 2013 through 2033); the funding levels associated with all analyses are stated in constant 2012 dollars. Unlike the highway scenarios, these transit scenarios assume an immediate jump to a higher (or lower) investment level that is maintained in constant-dollar terms throughout the analysis period.

Included in this section for comparison purposes is an assessment of the investment level needed to replace all assets that are currently past their useful life or that will be over the forecast period. This investment level would be necessary to achieve and maintain a state of good repair but would not address any increases in demand during that period. Although not realistic, this scenario does provide a benchmark for infrastructure preservation.

Sustain 2012 Spending Scenario

- The Sustain 2012 Spending scenario assumes that capital spending by all levels of government is sustained in constant-dollar terms at the 2012 level ($16.8 billion systemwide), including Recovery Act funds, through 2032. Assuming that the current split between expansion and preservation investments is maintained, this scenario will allow enough expansion to meet medium growth expectations but will fall far short of meeting system preservation needs. By 2032, this scenario would result in roughly $122 billion in deferred system preservation projects.

Low-Growth Scenario

- The Low-Growth scenario assumes that transit ridership will grow at an annual rate of 1.3 percent between 2012 and 2032. During that period, this scenario also attempts to pay down the current $89.8-billion system preservation backlog. The annualized cost of this scenario is $22.9 billion. In 2012, all levels of government spent a combined $16.8 billion for transit capital improvements.

High-Growth Scenario

- The High-Growth scenario assumes that transit ridership will grow at an annual rate of 2.2 percent between 2012 and 2032. This scenario also attempts to pay down the current $89.8-billion system preservation backlog (subject to the same cost-benefit constraint). The annualized cost of this scenario is $26.4 billion.

State of Good Repair – Expansion vs. Preservation

As used in this report, the term "state of good repair" means that all transit capital assets are within their average service life. This general construct enables FTA to estimate system preservation needs. The analysis examines the age of all transit assets and adds the value of those that are past the age at which that type of asset is usually replaced to a total reinvestment needs estimate. Some assets can continue to provide reliable service well past the average replacement age and others will not; over the large number of assets nationally, the differences average out. Some assets will need require replacement, and some will be refurbished. Both types of cost are included in the reinvestment total. State of good repair is a measure of system preservation needs, which failure to meet will increase operating costs and poor service.

Expansion needs are treated separately in this analysis. They result from the need to add vehicles and route miles to accommodate more riders. Estimates of future demand are inherently speculative. Failure to meet expansion needs results in crowded vehicles and represents a lost opportunity to provide the benefits of transit to a wider customer base.

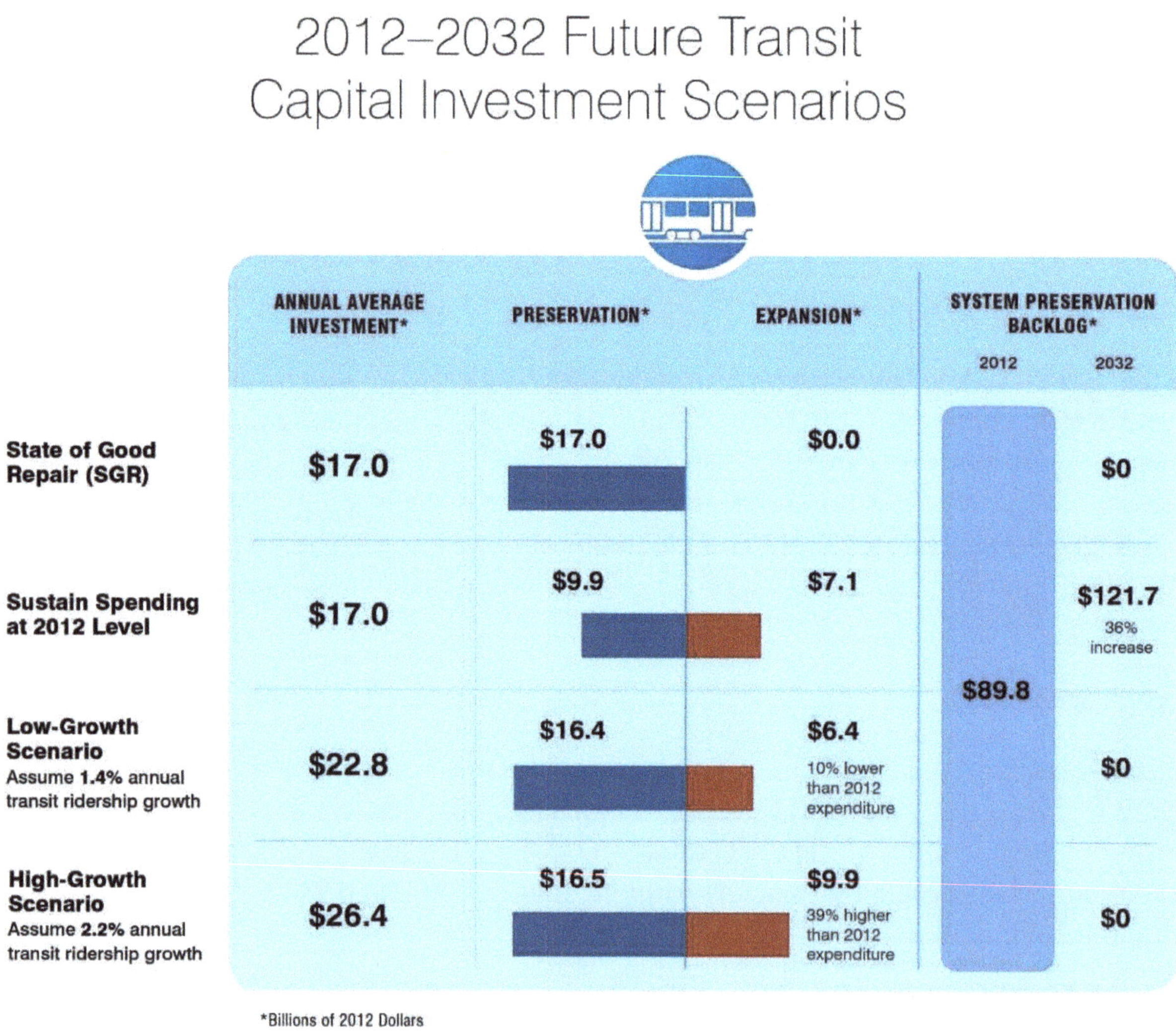

Executive Summary

PART I

Description of Current System

The U.S. Department of Transportation's (DOT's) *Transportation for a New Generation, a Strategic Plan for Fiscal Years 2014–18* presents five strategic goals:

- Safety – Improve public health and safety by reducing transportation-related fatalities, injuries, and crashes.
- State of Good Repair – Ensure that the United States proactively maintains critical transportation infrastructure in a state of good repair.
- Economic Competitiveness – Promote transportation policies and investments that bring lasting and equitable economic benefits to the Nation and its citizens.
- Quality of Life in Communities – Foster quality of life in communities by integrating transportation policies, plans, and investments with coordinated housing and economic development policies to increase transportation choices and access to transportation services for all.
- Environmental Sustainability – Advance environmentally sustainable policies and investments that reduce carbon and other harmful emissions from transportation sources.

Part I Structure

Chapter 1 outlines the trends in travel behavior of households and businesses and describes the freight transportation system. Chapter 2 describes the extent and use of highways, bridges, and transit systems. Chapter 3 addresses issues relating to the State of Good Repair Goal, while Chapter 4 relates to the Safety Goal. Chapter 5 covers topics relating to the Economic Competitiveness Goal, the Quality of Life in Communities Goal, and the Environmental Sustainability Goal. Chapter 6 provides data on highway and transit finance.

Transportation Performance Management

A recurring theme in Part I of this C&P report is the coming impact of changes under the Moving Ahead for Progress in the 21st Century Act (MAP-21). The cornerstone of the MAP-21 program transformation is the transition to a performance and outcome-based program. Performance measures will be established through a set of rulemakings; grant recipients will set performance targets based on these measures, and will periodically report on their progress toward meeting these targets. FHWA is implementing the MAP-21 requirements through six interrelated rulemakings:

- Statewide and Metropolitan/ Nonmetropolitan Planning Rule
- Safety Performance Measures Rule (PM-1)
- Highway Safety Improvement Program (HSIP) Rule
- Pavement and Bridge Performance Measures Rule (PM-2)
- Asset Management Plan Rule
- System Performance Measures Rule (PM-3) (includes measures for freight movement and the Congestion Mitigation and Air Quality program).

Executive Summary

CHAPTER 1

Personal Travel

The Nation is becoming more populated, more diverse, and more urban. Three states (Texas, Florida, and California) are projected to account for nearly half of all national population growth through 2030, with the remainder concentrated in other States in the South and West (Census, 2010).

Overall, migration in the United States has slowed over the past few years, but the southern region showed gains from 2005 to 2010. New immigrants are especially likely to settle in California, New York, Texas, Florida, Illinois, and New Jersey.

Regional Migration and Growth

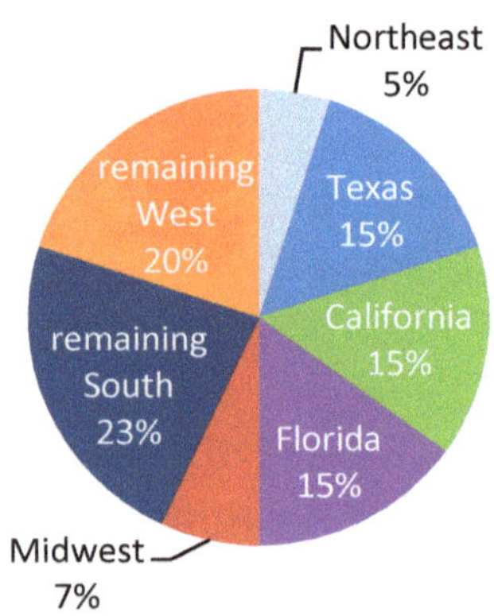

Expanding metropolitan areas, or "megaregions," have also spawned a network of metropolitan centers. Experts believe 75 percent of the U.S. population will reside in megaregions by 2050. The 11 emerging megaregions are the Northeast, Florida, Piedmont Atlantic, Gulf Coast, Great Lakes, Texas Triangle, Arizona Sun Corridor, Front Range, Cascadia, Northern California, and Southern California. Both freight travel and passenger travel are expected to increase within the megaregions, putting further demands on the transportation system.

Young adults (20–34 years old) are gravitating to new suburban development near the fringes of metropolitan areas. In contrast, the share of young adults living in established suburbs declined in recent years.

Percentage Point Change in Young Adult Population (Ages 20–34 Years) by Neighborhood Type, 2000–2010

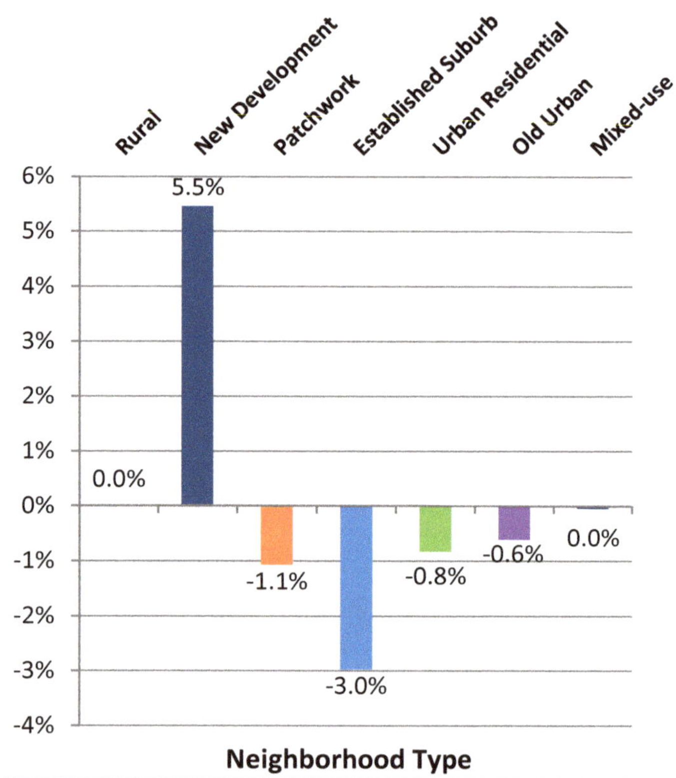

Low-income families are moving from city centers to suburban locations. In 1970, suburbs housed 25 percent of the poor; by 2010 the proportion was 33 percent.

Executive Summary

CHAPTER 1

Freight Movement

Freight transportation is vital to the U.S. economy and the day-to-day needs of citizens. This sector has gained increasing attention under both MAP-21 and the Fixing America's Surface Transportation Act (FAST Act). The success of freight planning across the United States, however, faces significant and varied challenges.

Current freight demands are straining existing system capacity, while freight movement across the United States is expected to increase. In 2012, the U.S. transportation system handled a record amount of freight: 54 million tons of product valued at $48 billion was shipped daily to 118.7 million U.S. households, 7.4 million business establishments, and 89,004 government institutions.

Freight transport and passenger transport are very different. Freight travel patterns tend to change more rapidly than passenger travel patterns in response to short-term economic fluctuations. Improvements targeted at general passenger travel are less likely to aid the flow of freight than are improvements targeted at freight demand.

Trucks transport most U.S. freight, accounting for 67.0 percent of freight tonnage and 64.1 percent of freight value in 2012. Trucking handles most of the lower-valued bulk tonnage, which includes agricultural products, local gasoline delivery, and municipal solid waste pickup, but is also critical for moving high value freight coming off of air cargo or other modes. Trucks are usually the main mode for freight trips of less than 500 miles. As gas prices fluctuate, this threshold will vary.

Virtually all carriers and freight facilities (such as railroad lines and some port terminals) are privately owned but use public highways and airways. This mix of ownership requires complex coordination by a variety of private and public stakeholders. The private sector owns $1.173 trillion in transportation equipment and $739 billion in transportation infrastructure, while the public sector maintains $686 billion in transportation equipment and $3.343 trillion in highway infrastructure.

Although most goods move short distances, usually less than 250 miles, freight often moves over multiple jurisdictions. As half the weight and two-thirds the value of freight products cross State or international boundaries, the benefits of this freight transportation might not accrue to the communities through which it travels. Both the interregional nature of many freight movements and the varying levels of support or opposition in local jurisdictions for freight-generating development can complicate the freight planning process. Assessing and measuring the impacts of full freight trips across multiple jurisdictions is often challenging for State and local transportation planners.

Freight transport—which accounted for 25.5 percent of all gasoline, diesel, and other fuels consumed by motor vehicles in 2013—can create safety and congestion challenges. Hazardous material transport is dangerous; hazardous material incidents across all modes totaled 15,433 in 2012. Increasing freight levels are adding to congestion in urban areas and on intercity routes.

Executive Summary

CHAPTER 2

System Characteristics: Highways

Although the Federal government provides significant financial support for the Nation's highways and bridges, it owns relatively few of these facilities. In 2012, State and local governments owned 96.3 percent of the Nation's 4,103,418 public road miles and 98.3 percent of the Nation's 607,380 bridges. These roads and bridges carried more than 2.9 trillion VMT.

Highway and Bridge Ownership by Level of Government

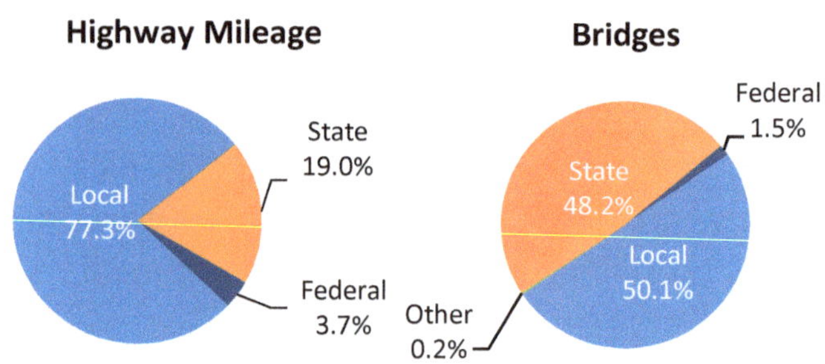

The Nation's roadway system is a vast network connecting places and people within and across national borders. The network facilitates movement of vehicles serving everything from long-distance freight needs to neighborhood travel. To accommodate the Nation's diverse travel needs, different types of roads are constructed to serve two primary purposes: access and mobility.

Roads are categorized into functional classifications to establish their purpose and to determine whether they are eligible for Federal-aid highway funding assistance. In general, public roads that are functionally classified higher than rural minor collector, rural local, or urban local are eligible for Federal-aid highway funding.

Nearly 73 percent of the public road mileage is located in rural settings with populations less than 5,000. Although only 27.1 percent of the public road mileage is located in urban areas with populations of 5,000 or more, the densely populated areas contribute 67.2 percent of VMT.

Similar to the breakdown of public road mileage, 73.6 percent of the Nation's bridges are in rural areas, while 26.4 are in urban areas.

Percentages of Highway Miles, Vehicle Miles Traveled, and Bridges by Functional System, 2012

Functional System	Highway Miles	Highway VMT	Bridges
Rural Areas (4,999 or less in population)			
Interstate	0.7%	8.2%	4.1%
Other Freeway and Expressway[1]	0.1%	0.7%	
Other Principal Arterial[1]	2.2%	6.8%	
Other Principal Arterial[1]			6.0%
Minor Arterial	3.3%	5.0%	6.4%
Major Collector	10.3%	5.9%	15.3%
Minor Collector	6.4%	1.8%	7.9%
Local	49.9%	4.4%	33.8%
Subtotal Rural Areas	**72.9%**	**32.8%**	**73.6%**
Urban Areas (5,000 or more in population)			
Interstate	0.5%	16.5%	5.1%
Other Freeway and Expressway	0.2%	7.5%	3.3%
Other Principal Arterial	1.6%	15.4%	4.6%
Minor Arterial	2.6%	12.5%	4.7%
Collector[1]			3.4%
Major Collector[1]	2.8%	5.9%	
Minor Collector[1]	0.0%	0.1%	
Local	19.4%	9.3%	5.4%
Subtotal Urban Areas	**27.1%**	**67.2%**	**26.4%**
Total	**100.0%**	**100.0%**	**100.0%**

[1] Less functional system detail is available for bridges than for highways. Bridges on rural Other Freeway and Expressway are included under the rural Other Principal Arterial category. Bridges on urban Major Collector and urban Minor Collector are combined into a single urban Collector category.

Executive Summary

CHAPTER 2

System Characteristics: Transit

Most transit systems in the United States report to the National Transit Database (NTD). In 2012, 822 systems served 497 urbanized areas, which have populations greater than 50,000. In rural areas, 1,637 systems were operating. Thus, the total number of transit systems reporting to NTD in 2012 was 2,264.

Modes. Transit is provided through 18 distinct modes in two major categories: rail and nonrail. Rail modes include heavy rail, light rail, streetcar, commuter rail, and other less common modes that run on fixed tracks. Nonrail modes include bus, commuter bus, bus rapid transit, demand response, vanpools, other less common rubber-tire modes, ferryboats, and aerial tramways. This edition of the C&P report includes four new modes: commuter bus and bus rapid transit (previously reported as fixed-route bus); hybrid rail, which shares characteristics of light rail and commuter rail; and demand-response taxi (previously reported as demand response).

Urbanized Areas, Population Density, and Demand. Based on the 2010 census, the average population density of the United States is 82.4 people per square mile. The average population density of all 498 urbanized areas combined is 2,510 people per square mile. The average density for the 50 most-populated areas is 3,132 people per square mile. The chart shows the relationship between ridership and urbanized area density for the top 50 areas in 2012.

National Transit Assets

- Of the 162,830 vehicles in urban and rural areas, most are buses, cutaways (short buses), vans, and rail vehicles (passenger cars).
- Rail systems operate on 12,617 miles of track, and fixed-route bus systems operate over 252,800 route miles.
- Urban and rural areas have 3,281 stations and 1,720 maintenance facilities

2012 Urbanized Area Density vs. Ridership (Top 50 Areas in Population)

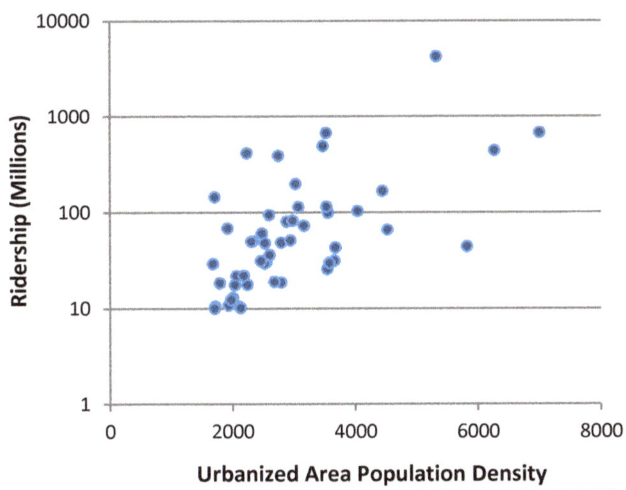

ADA Compliance. The Americans with Disabilities Act of 1990 (ADA) prohibits discrimination and ensures equal opportunity and access for persons with disabilities. The act requires transit agencies to provide accessible vehicles (e.g., with lifts) and stations with barriers on platforms, ramps, elevators and other elements. The level of ADA compliance is high for the national fleet, but lower for rail stations, particularly for old heavy rail systems built before the passing of the ADA.

Executive Summary

CHAPTER 3

System Conditions: Highways

Pavement and bridge conditions directly affect vehicle operating costs because deteriorating pavement and bridge decks increase wear and tear on vehicles and repair costs. Poor pavement also can affect travel time costs if road conditions force drivers to reduce speed and can increase the frequency of crash rates. Poor bridge conditions can force trucks to detour to alternative routes, leading to increased travel time and delays and introducing increased freight impacts to other communities.

The Highway Performance Monitoring System (HPMS) collects data on pavement ride quality on Federal-aid highways. Between 2002 and 2012, the percentage of Federal-aid highway mileage classified as acceptable decreased from 87.4 percent to 80.3 percent. When weighted by VMT during the same period, however, acceptable ride quality decreased from 85.3 percent to 83.3 percent.

Pavement Ride Quality on Federal-Aid Highways, 2002–2012

	2002	2008	2012
By Mileage			
Acceptable (Good + Fair)	87.4%	84.2%	80.3%
Poor	12.6%	15.8%	19.7%
Weighted By VMT			
Acceptable (Good + Fair)	85.3%	85.4%	83.3%
Poor	14.7%	14.6%	16.7%

Although States were instructed to collect pavement data differently in 2009, the variance between mileage and VMT data suggests that ride quality on less-traveled Federal-aid highways has significantly declined since 2002.

Bridges are a vital component of the Nation's highway system. One term used to classify bridges is "structurally deficient." Structural deficiencies are characterized by deteriorated conditions of primary bridge components and potentially reduced load carrying capacity and waterway adequacy, but do not imply safety concerns.

The National Bridge Inventory (NBI) includes data for bridge conditions. The total number of bridges reported in the NBI increased by 16,137 between 2002 and 2012, but the number of bridges classified as structurally deficient decreased by 17,282. During that time, the share of bridges classified as structurally deficient decreased from 14.2 percent to 11.0 percent.

Structurally Deficient Bridges (Systemwide), 2002–2012

	2002	2008	2012
Count			
Total Bridges	591,243	601,506	607,380
Structurally Deficient	84,031	72,883	66,749
Percent Structurally Deficient			
By Bridge Count	14.2%	12.1%	11.0%
Weighted by Deck Area	10.4%	9.3%	8.2%
Weighted by Traffic	8.0%	7.2%	5.9%

As part of its ongoing efforts to encourage the integration of Transportation Performance Management principles into project selection decisions and to implement related provisions in MAP-21, FHWA has proposed moving to "Good" and "Poor" classifications to measure pavement and bridge conditions. Future C&P reports will integrate the final measures that emerge after a rulemaking process is completed.

Executive Summary

CHAPTER 3

System Conditions: Transit

Transit asset infrastructure in the C&P report includes five major asset groups.

Major Asset Categories

Asset Category	Components
Guideway Elements	Tracks, ties, switches, ballasts, tunnels, elevated structures, bus guideways
Maintenance Facilities	Bus and rail maintenance buildings, bus and rail maintenance equipment, storage yards
Stations	Rail and bus stations, platforms, walkaways, shelters
Systems	Train control, electrification, communications, revenue collection, utilities, electrification, signals and train stops, centralized vehicle/train control, substations
Vehicles	Large buses, heavy rail, light rail, commuter rail passenger cars, nonrevenue vehicles, vehicle replacement parts

Condition Rating. FTA uses a capital investment needs tool, TERM, to measure the condition of transit assets. The model uses a numeric scale that ranges from 1 to 5.

Definition of Transit Asset Conditions

Rating	Condition	Description
Excellent	4.8–5.0	No visible defects, near-new condition
Good	4.0–4.7	Some slight defective or deteriorated components
Adequate	3.0–3.9	Moderately defective or deteriorated components
Marginal	2.0–2.9	Defective or deteriorated components in need of replacement
Poor	1.0–1.9	Seriously damaged components in need of immediate repair

The replacement value of the Nation's transit assets was $847.5 billion in 2012, 49 percent of which was guideway elements.

The relatively large proportion of guideway elements and systems assets that are rated below condition 2.0 (poor), and the magnitude of the $140-billion investment required to replace them represent major challenges to the rail transit industry.

2012 Asset Categories Rated Below Condition 2.0 (Poor)

Asset Category	Percentage in Poor Condition
Guideway Elements	31.4
Systems	15.1
Facilities	4.8
Vehicles	4.0
Stations	2.1

State of Good Repair (SGR). An asset is deemed in a state of good repair if its condition rating is 2.5 or higher. An agency mode is in SGR if all its assets are rated 2.5 or higher.

Trends in Urban Bus and Rail Transit Fleet not in SGR. The average condition rating for bus and rail fleets did not change much between 2002 and 2012, ranging between 3.0 and 3.3 for buses, and remaining constant for rail at 3.5. The percentage of the bus fleet not in SGR also did not change much, ranging between 10 and 12 percent. For rail, the percentage decreased from 4.6 to 2.8 percent.

2012 Transit Assets not in SGR (percent)

Category	Bus	Rail	All
Guideway Elements	5.5	35.1	34.6
Systems	17.2	17.1	17.1
Stations	11.5	37.8	37.5
Facilities	7.4	24.3	15.3
Vehicles	9.8	2.8	7.2

Executive Summary

CHAPTER 4

Safety: Highways

In 2012, 31,006 fatal crashes took place on roads in the United States; in the same year, approximately 1.63 million nonfatal injury crashes and 3.95 million property damage-only crashes occurred.

The number of fatalities related to the operation of motor vehicles dropped by 21.4 percent from 2002 to 2012, from 43,005 to 33,782. Over the same period, the fatality rate per 100 million VMT dropped from 1.51 to 1.14. Relative to recent years, the number of fatalities in 2012 was up from a low of 32,479, while the fatality rate was up slightly from a low of 1.10, reached in 2011.

Annual Highway Fatality Rates, 2002–2012

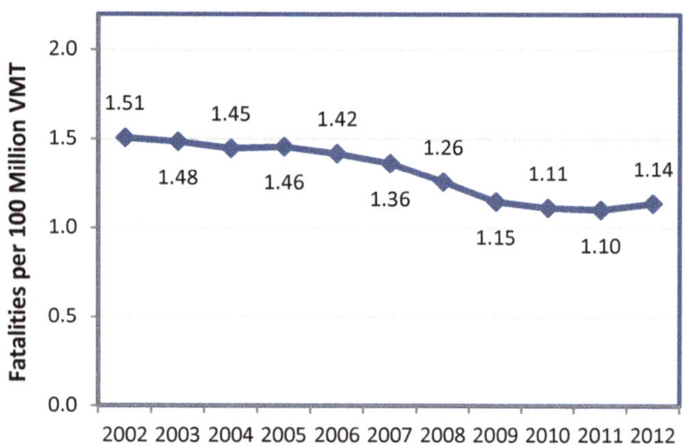

The DOT strategic goal on safety is "Improve public health and safety by reducing transportation-related fatalities and injuries for all users, working toward no fatalities across all modes of travel." In support of this goal, FHWA oversees the Highway Safety Improvement Program (HSIP), which requires a data-driven, strategic approach to improving highway safety on all public roads that focuses on performance. Use of HSIP funds is driven by a Strategic Highway Safety Plan (SHSP), which each State develops in cooperation with a broad range of multidisciplinary stakeholders.

When it occurs, a crash is generally the result of numerous contributing factors. Roadway, vehicle, driver, passenger, and nonoccupant factors all have an impact on the safety of the Nation's highway system. FHWA focuses on infrastructure design and operation to address roadway factors. Based on analyses of crash data, FHWA has established three focus areas: roadway departures, intersections, and pedestrian crashes. In 2012, roadway departure, intersection, and pedestrian fatalities accounted for 52.2 percent, 21.7 percent, and 14.1 percent, respectively, of all crash fatalities. That these three categories overlap one another should be noted, such as when a roadway departure crash results in a pedestrian fatality.

Highway Fatalities by Crash Type, 2002–2012

Crash Type	2002	2012	Percent Change
Roadway Departure-Related	25,415	17,532	-31.0%
Intersection-Related	9,273	7,279	-21.5%
Pedestrian-Related	4,851	4,743	-2.2%

Although progress has been made from 2002 to 2012 in reducing these types of fatalities, pedestrian fatalities have increased since 2009. Nonmotorist fatalities (including pedestrians, bicyclists, etc.) increased to 5,692 in 2012, up from 5,630 in 2002 and 4,888 in 2009.

Executive Summary

CHAPTER 4

Safety: Transit

Rates of injuries and fatalities on public transportation generally are lower than for other modes of transportation. Nonetheless, serious incidents do occur, and the potential for catastrophic events remains. Several transit agencies have had major accidents in recent years. The National Transportation Safety Board has investigated several of these accidents and has issued reports identifying the factors that contributed to them. Since 2004, the National Transportation Safety Board has reported on 9 transit accidents that, collectively, resulted in 15 fatalities, 297 injuries, and more than $30 million in property damage.

Most injuries and fatalities in transit result from collisions, and most victims are not passenger or patrons. They are pedestrians, automobile drivers, bicyclists, or trespassers. Patrons are individuals in stations who are waiting to board or just got off transit vehicles. In 2012, of the 265 fatalities, only 4 percent were passengers. In 2002, of the 175 fatalities, the share of passenger fatalities was similarly small, 6 percent. The most striking change over that time has been the increase in the percentage of suicides, from 8 percent in 2002 to 23 percent in 2012.

The Federal Transit Administration reports the rate of fatalities per 100 million passenger miles traveled. This rate did not change significantly between 2002 and 2012 for bus and heavy rail, the two largest modes and the ones with the most fatalities. The rate for light rail, which includes streetcars, is more volatile and has increased over the past 10 years due to a significant increase in service and number of systems.

Fatalities by Type of Person, 2012 and 2002

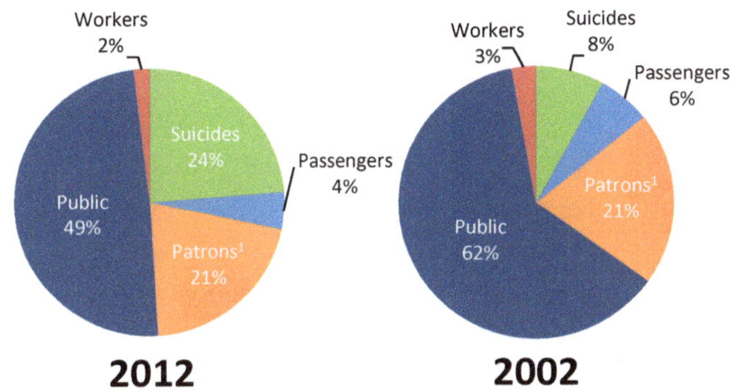

[1] *Includes individuals waiting for or leaving transit at stations; in mezzanines; on stairs, escalators, or elevators; in parking lots; or at other transit-controlled property.*

Annual Transit Fatality Rates per 100 Million Passenger Miles Traveled for Fixed-Route Bus, Heavy Rail, and Light Rail, 2002–2012

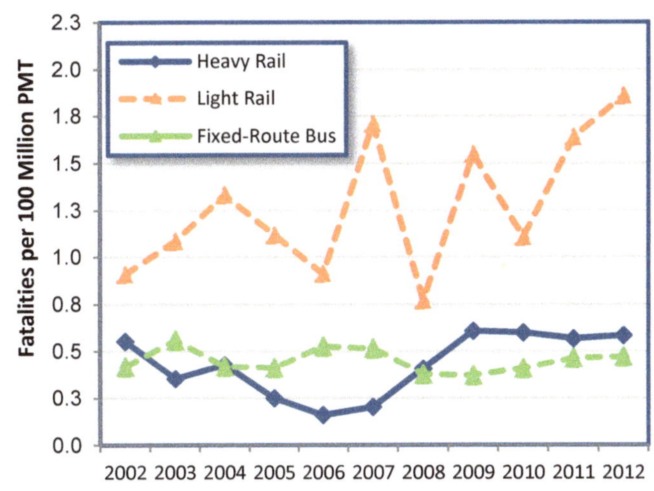

Executive Summary

CHAPTER 5

System Performance: Highways

This chapter relates to three goals presented in DOT's FY 2014–18 Strategic Plan: (1) Economic Competitiveness, (2) Quality of Life in Communities, and (3) Environmental Sustainability.

Economic Competiveness: Congestion harms the U.S. economy and wastes time, fuel, and money. The Texas Transportation Institute's Urban Mobility Scorecard estimates that in 2012, on average, each commuter was delayed 41 hours due to congestion. Congestion wastes 6.7 billion hours and 3 billion gallons of fuel for the Nation as a whole, at a total cost of $154.2 billion in 2012.

The Travel Time Index is calculated as the ratio of travel time required to make a trip during the congested peak period to travel time for the same trip during the off-peak period in noncongested conditions. Based on this measure, congestion rose from 2002 to 2008 before dropping briefly during the recent recession; by 2012, it had risen close to the levels observed before the recession.

Travel Time Index for Urbanized Areas, 2002–2012

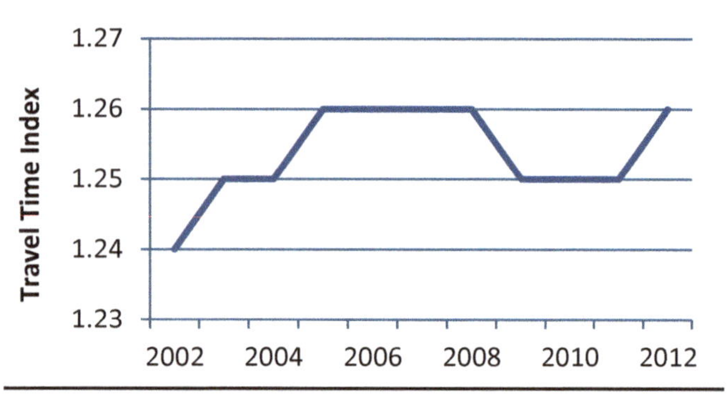

Congestion can be recurring or nonrecurring; most travelers are less tolerant of unexpected delays than they are of everyday congestion. One relatively new source of data for measuring congestion and reliability is the National Performance Management Research Data Set (NPMRDS), which supports the Freight Performance Measures and Urban Congestion Report programs. FHWA measures freight highway congestion using truck probe data from global positioning system equipment. Truck probe data in the NPMRDS also measure corridor-level travel time reliability.

Quality of Life: Fostering livable communities is a continued goal of DOT. Progress is being made on measuring the impact of investments on increasing transportation choices and access to transportation services. Relevant tools such as the Location Affordability Portal, Sustainable Communities Indicator Catalog, Infrastructure Voluntary Evaluation Sustainability Tool, and Community Vision Metrics Web Tool are being used to support quality of life goals in communities.

Environmental Sustainability: Preparing for climate change is critical for protecting the integrity of the transportation system. One-fourth of the greenhouse gas emissions causing climate change in the United States is derived from the transportation sector. FHWA is partnering with State DOTs, MPOs, and Federal Land Management Agencies to develop strategies to reduce greenhouse gas emissions from transportation sources and build climate resilient transportation systems.

Executive Summary

CHAPTER 5

System Performance: Transit

The transit industry has largely succeeded in meeting the demand for its services in communities across the country. Transit data from the end of the past decade show steady increases in service provided and consumed, commensurate with the growth of the urbanized population.

Between 2002 and 2012, the geographic coverage of transit significantly increased. New and extended commuter modes such as vanpools and commuter rail have reached areas previously accessible only by automobile. Light rail systems served 33 communities in 2012, compared with 22 in 2000. This light rail growth has increased the number of revenue service hours by 15 percent, the number of unlinked trips by 14 percent, and the number of passenger miles by 20 percent. The relatively larger increase in passenger miles is due to longer average trip lengths, which could be due to an expansion of service to outlying suburbs.

In 2004–2012, maintenance performance improved (vehicle revenue miles between mechanical failures). Over these 6 years, the average number of miles between failures increased by 21 percent.

Mean Distance Between Failures, Directly Operated Service, 2004–2012

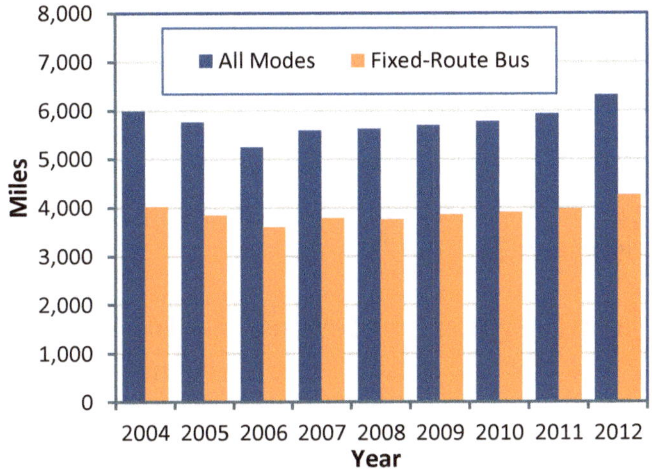

Transit ridership is greatly affected by employment. Transit ridership increased significantly from July 2006 to January 2009 and then plummeted following the 2009 economic crisis. Relatively low fuel prices and employment levels characterized this crisis. Employment, fuel prices, and ridership all increased from 2010 to 2012.

Transit Ridership versus Employment, 2006–2012

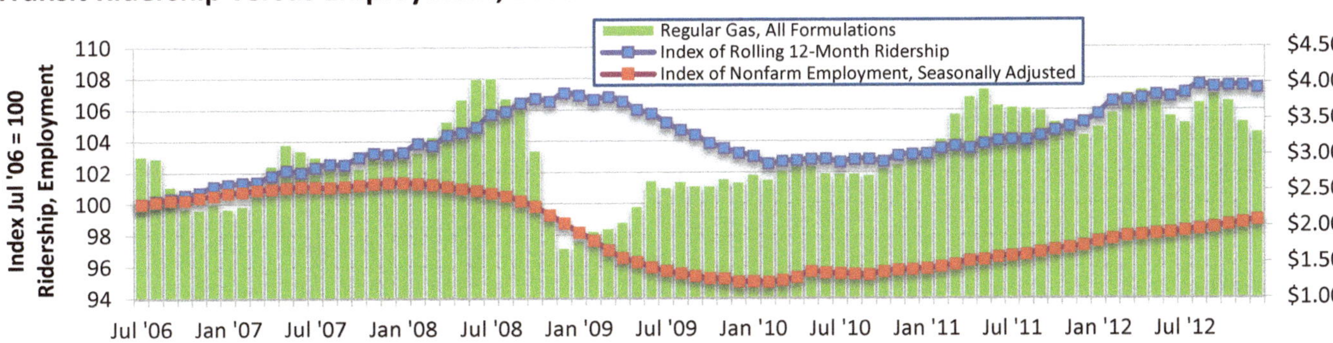

Sources: National Transit Database, U.S. Energy Information Administration's Gas Pump Data History, and Bureau of Labor Statistics' Employment Data.

Executive Summary

CHAPTER 6

Finance: Highways

Combined expenditures for highways by all levels of government totaled $221.3 billion in 2012, with the Federal government funding $47.4 billion [including $3.0 billion authorized by the American Recovery and Reinvestment Act of 2009 (Recovery Act)], States $105.8 billion, and local governments $68.1 billion. Most of the Federal funding was in the form of grants to State and local governments; direct Federal expenditures for federally owned roads, highway research, and program administration totaled $3.2 billion.

Highway capital spending totaled $105.2 billion, or 47.5 percent of total highway spending in 2012. Spending on maintenance totaled $35.1 billion, $12.9 billion was for highway and traffic services, $16 billion was for administrative costs (including planning and research), $17.8 billion was spent on highway patrol and safety, $11.6 billion was for interest on debt, and $22.6 billion was used to retire debt.

Highway Expenditure by Type, 2012

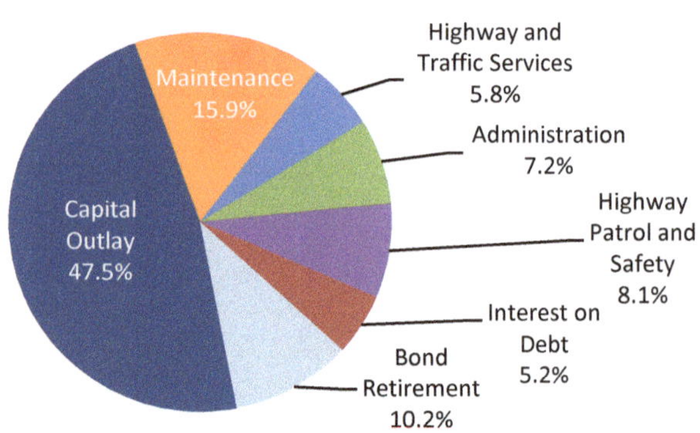

Total highway spending increased 62.8 percent from 2002 to 2012, averaging 5.0 percent per year. (In inflation-adjusted constant-dollar terms, highway spending grew by 2.6 percent per year). Expenditures funded by local governments grew by 7.2 percent per year, outpacing annual increases at the State and Federal levels of 4.4 percent and 3.7 percent, respectively. Over this period, the share of total highway expenditures funded by the Federal government dropped from 24.1 percent to 21.4 percent, while the federally funded share of highway capital spending dropped from 46.1 percent to 43.1 percent.

Combined revenues generated for use on highways by all levels of government totaled $216.6 billion in 2012 (the difference between expenditures and receipts is the amount drawn from reserves). In 2012, $105.2 billion (48.6 percent) of total highway revenues came from highway user charges—including motor-fuel taxes, motor-vehicle fees, and tolls. Other major sources for highways included general fund appropriations of $44.1 billion (20.4 percent) and bond proceeds of $21.3 billion (9.8 percent). All other sources, such as property taxes, other taxes and fees, investment income, and other receipts, totaled $46.0 billion (21.3 percent).

Revenue Sources for Highways, 2012

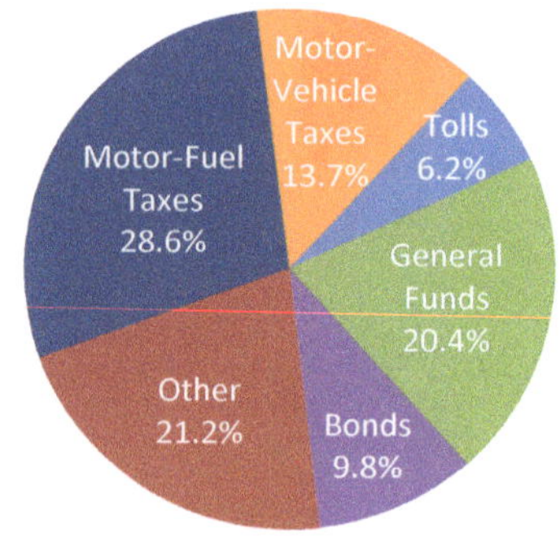

Executive Summary

CHAPTER 6

Finance: Transit

In 2012, $58.0 billion was generated from all sources to finance urban transit. Transit funding comes from public funds that Federal, State, and local governments allocate and system-generated revenues that transit agencies earn from the provision of transit services. Of the funds generated in 2012, **73.3 percent came from public sources and 26.7 percent came from system-generated funds** (passenger fares and other system-generated revenue sources). The Federal share was $10.9 billion (25.5 percent of total public funding and 18.7 percent of all funding).

Guideway assets use the largest share of capital—36 percent ($6 billion)—for expansion and rehabilitation projects.

2012 Urban Capital Expenditure by Asset Category

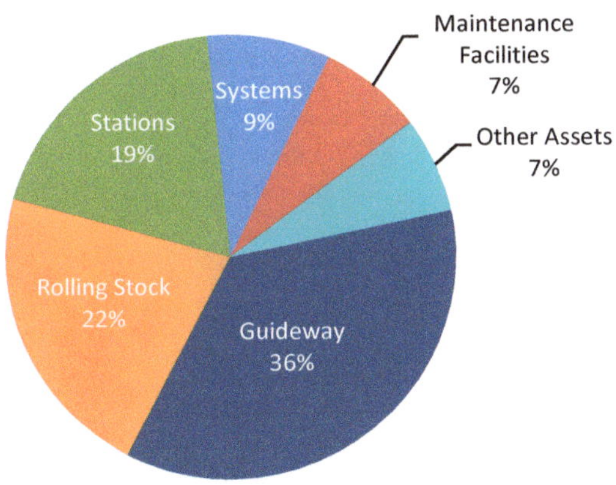

The American Recovery and Reinvestment Act of 2009 (Recovery Act) nearly doubled the total funding for urban transit.

Urban Recovery Act Funding Awards Compared to Other FTA Fund Awards

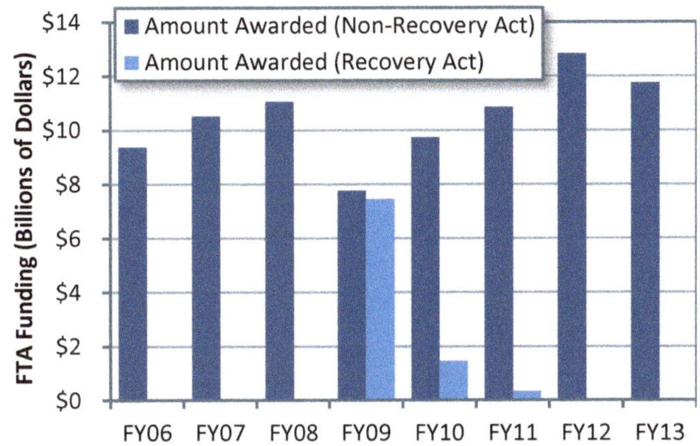

From 2002 to 2012, for the top 10 transit agencies, fringe benefits increased at the highest rate of any operating cost category on a per-mile basis. Over this period, fringe benefits increased at an annual compound average rate of 2.3 percent. Meanwhile, salaries and wages decreased by 1 percent.

Salaries and Wages and Fringe Benefits, Average Cost per Mile—Urbanized Areas Over 1 Million, 2002–2012 (Constant Dollars)

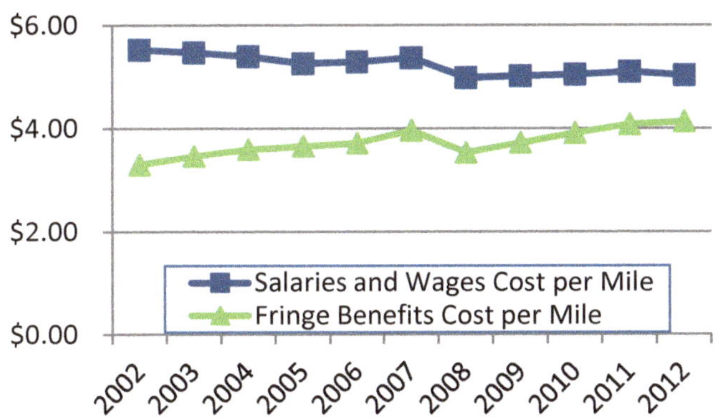

Executive Summary

PART II

Investment/Performance Analysis

The methods and assumptions used to analyze future highway, bridge, and transit investment scenarios for this report are continuously evolving to incorporate new analytical methods, new data and evidence, and changes in transportation planning objectives. Estimates of future requirements for highway investment, as reported in the *1968 National Highway Needs Report* to Congress, began as a combined "wish list" of State highway "needs." Over time, simulation models were developed that applied engineering standards to identify system deficiencies and the investments necessary to remedy these deficiencies. The current generation of analytical tools applied in this report each combine economic analysis with engineering principles in their identification of future capital investment needs.

The economic approach to transportation investment decision-making entails analyzing and comparing benefits and costs. Investments that yield benefits with values exceeding their costs increase societal welfare and are thus considered "economically efficient." To be reliable, such analyses must adequately consider the range of possible benefits and costs and the range of possible investment alternatives. A comprehensive benefit-cost analysis of a transportation investment would consider all potentially significant impacts on society and value them in monetary terms to the extent feasible. For some types of impacts, monetary valuation is facilitated by the existence of observable market prices; such prices are generally available for inputs to the costs of providing transportation infrastructure, such as prices for concrete to build highways or prices for new buses. For some other types of impacts for which market prices are not available, monetary values can be inferred from behavior or expressed preferences. In this category are savings in non-business travel time and reductions in the risk of crash-related fatalities or other injury. For still other impacts, monetary valuation might not be possible because of difficulties with reliably estimating the impact of the improvement, placing a monetary value on that impact, or both. Even when possible, reliable monetary valuation might require time and effort that would be out of proportion to the likely importance of the impact concerned and the inherent uncertainty in the estimates. Benefit-cost analyses of transportation investments therefore typically omit valuing certain impacts that nevertheless could be of interest.

The Highway Economic Requirements System (HERS), National Bridge Investment Analysis System (NBIAS), and Transit Economic Requirements System (TERM) used to develop the analyses presented in Chapters 7 through 10 each omit various types of investment impacts from their benefit-cost analyses. To some extent, such omissions reflect the national scope of their primary databases: Such broad geographic coverage requires some sacrifice of detail to stay within feasible budgets for data collection. The analyses do not consider, for example, environmental impacts of increased water runoff from highway pavements, barrier effects of highways on human and animal populations, and health benefits from the additional walking activity when travelers use transit rather than cars.

Although HERS, NBIAS, and TERM all use benefit-cost analysis, their methods for implementing this analysis differ significantly. These highway, bridge, and transit models each rely on separate databases, making use of the specific data available for each mode of the transportation system and addressing issues unique to that mode. The three models have not yet evolved to the point where direct multimodal analysis is possible; currently the models provide no direct way to analyze the impact that a given level of highway investment in a particular location would have on the transit investment in that vicinity (or vice versa).

Chapter 7 analyzes the projected impacts of alternative levels of future investment on measures of physical condition, operational performance, and benefits to system users. Each alternative pertains to investment from 2013 through 2032, which is presented as an annual average level of investment and as the annual rates of increase or decrease in investment that would produce that annual average. Both the level and rate of growth in investment are measured using constant 2012 dollars.

Chapter 8 examines several scenarios distilled from the investment alternatives considered in Chapter 7. Some of the scenarios are oriented around maintaining aspects of system condition and performance or achieving a specified minimum level of performance, while others link to broader measures of system user benefits. **The scenarios included in this chapter are intended to be illustrative and do not represent comprehensive alternative transportation policies; the U.S. Department of Transportation (DOT) does not endorse any of these scenarios as a target level of investment.**

Chapter 9 explores some of the implications of the scenarios presented in Chapter 8 and contains some additional policy-oriented analyses addressing issues not covered in Chapters 7 and 8. As part of this analysis, highway traffic projections from previous editions of the C&P report are compared with actual outcomes to elucidate the value and limitations of the projections presented in this edition. Chapter 9 also discusses the revised method for estimating transit travel growth rates introduced in this report.

The three investment analysis models used in this report are deterministic, not probabilistic: They provide a single projected value of total investment for a given scenario rather than a range of likely values. As a result, the element of uncertainty in these projections is amenable only to general characterizations based on the characteristics of the projection process; estimates of confidence intervals cannot be developed.

Chapter 10 presents sensitivity analyses that explore the impacts on scenario projections of varying some of the key assumptions. The investment scenario projections in this report are developed using models that evaluate current system condition and operational performance and make 20-year projections based on assumptions about future travel growth and a variety of engineering and economic variables. The accuracy of these projections depends, in large part, on the realism of these assumptions.

Because the future rate of growth in transit travel is uncertain, Chapter 7 considers alternative high and low values for this parameter. Chapter 10 likewise varies the assumed rate of growth in highway travel and the values assumed for the discount rate, the value of travel time savings, and other assumed parameters.

Potential Capital Investment Impacts: Highways

NBIAS evaluates rehabilitation and replacement needs for all bridges. HERS evaluates needs associated with pavement resurfacing or reconstruction and widening needs, including those associated with bridges. HERS analyses are limited to Federal-aid highways, as HPMS does not collect data for other roads.

All levels of government combined spent $105.2 billion on highway capital outlay in 2012, including $16.4 billion for the types of bridge improvements modeled in NBIAS, and $57.4 billion on Federal-aid highways for the types of capital improvements modeled in HERS. The remaining $31.4 billion was divided about evenly between spending on non-Federal-aid highways on the same types of improvements that are modeled in HERS and on types of capital improvements covered by neither model.

The rate of future travel growth can significantly influence the projected future conditions and performance of the highway system. For the HERS and NBIAS analyses presented in this report, future travel volumes were tied to a 20-year national-level forecast averaging 1.04-percent growth per year. As this rate of growth was lower on average than the State projections reflected in HPMS and NBI, future traffic levels for individual highway sections and bridges were proportionally reduced so that the overall rate of growth would match the nationwide forecast.

Sustaining HERS-modeled investment at $57.4 billion in constant-dollar terms over 20 years is projected to result in a 4.5-percent decrease in average pavement roughness per VMT and a 13.4-percent decrease in average delay by 2032 relative to 2012. HERS projects that constant-dollar spending growth of 2.53 percent per year would suffice to finance all cost-beneficial capital improvements on Federal-aid highways by 2032. This spending would translate into an average annual investment level of $75.4 billion and result in a 14.0-percent decrease in average pavement roughness and a 16.5-percent reduction in average delay per VMT.

Projected Change in 2032 Highway Conditions and Performance Measures Compared with 2012 Levels for Various Levels of Investment

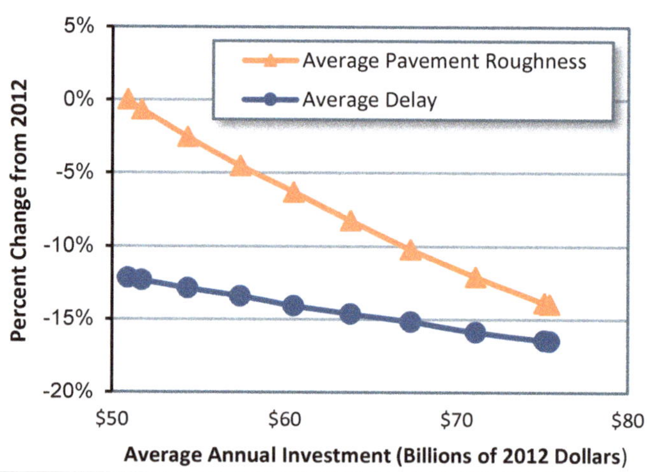

Sustaining NBIAS-modeled investment at $16.4 billion in constant-dollar terms is projected to result in a reduction in the deck area-weighted share of structurally deficient bridges from 8.2 percent in 2012 to 2.9 percent in 2032. NBIAS projects that constant-dollar spending growth of 3.72 percent per year would suffice to eliminate the backlog of cost-beneficial capital improvements to bridges by 2032.

Executive Summary

CHAPTER 7

Potential Capital Investment Impacts: Transit

The current level of investment in preservation and replacement of existing assets is insufficient to prevent the SGR backlog from growing by 2032. The backlog currently stands at $89.9 billion. Maintaining the current annual investment in preservation ($9.8 billion) over the next 20 years would result in a backlog that is 36 percent higher, $122.2 billion. The size of the projected 2032 backlog is sensitive to small variations in annual funding levels. For example, an average reduction of 2.5 percent in capital invested per year over 20 years would result in a backlog that is 81 percent higher, $162.5 billion.

Further, the one-time infusion of Recovery Act funds that began in 2009 might have inflated the assumed average annual investment over 20 years.

The condition of assets is determined using decay curves for each asset type and past maintenance and rehabilitation investments and utilization levels. The TERM assesses the aggregate average physical condition of all existing assets nationwide as of 2032. Nevertheless, even when overall conditions improve due to additional expenditures, some individual assets still will deteriorate.

The aggregate average condition rating of all nationwide transit assets in 2012 was 3.5, that is, in the middle of the range for "adequate" condition.

The current investment level of $9.8 billion per year is not sufficient to maintain the current aggregate average condition rating (3.5) in 2032. Instead, by 2032, the condition would decrease to 3.1, near the lower bound of the "adequate" range. Even higher levels of annual investment, however, still would result in average condition of existing assets below current levels. This outcome is due, in part, to assets for which the useful lives are much longer than the 20-year period of analysis. These assets include several new light-rail systems.

Transit capital investment in 2012 was at $17.1 billion. Preservation accounted for 58.5 percent, and expansion 41.5 percent. Urbanized areas with populations over 1 million accounted for 88 percent of all capital investment.

Impact of Expansion Investments on Transit Ridership – Basic Assumption: Maintain vehicle occupancy rates at current levels over the next two decades. The current level of investment in expansion ($7.1 billion) supports roughly 3.6 billion additional trips by 2032, which corresponds to an annual growth in ridership of 1.5 percent.

The current level of investment in expansion is insufficient to support forecasted ridership growth, leading to increased crowding on systems that currently experience high utilization. Increased crowding, in turn, could lead to increased dwell times, reduced operating speeds, and increased vehicle wear. Our demand forecasts predict an average annual rate of increase in ridership of 1.7 percent per year over 20 years: 2012–2032. An annual investment of $8.0 billion is required to keep the average rate of vehicle occupancy constant over this period.

Executive Summary

CHAPTER 8

Selected Capital Investment Scenarios: Highways

This report presents a set of illustrative 20-year capital investment scenarios based on simulations developed using HERS and NBIAS, with scaling factors applied to account for types of capital spending that are not currently modeled. The scenario criteria were applied separately to the Interstate System, the NHS, Federal-aid highways, and the highway system as a whole.

The Sustain 2012 Spending scenario assumes that capital spending is sustained in constant-dollar terms at the 2012 level of $105.2 billion from 2013 through 2032. (In other words, spending would rise by exactly the rate of inflation during that period.)

The Maintain Conditions and Performance scenario assumes that capital investment gradually changes in constant-dollar terms over 20 years to the point at which selected measures of highway and bridge performance in 2032 are maintained at their 2012 levels. For the highway system as a whole, the average annual investment level associated with this scenario is $89.9 billion; this suggests that sustaining spending at the 2012 level of 105.2 billion should result in improved overall conditions and performance. Such is not the case for Interstate highways, as the $24.1-billion average annual investment level needed for the Maintain Conditions and Performance scenario is more than the $20.5 billion of capital spending on Interstate highways in 2012.

The Improve Conditions and Performance scenario assumes that capital investment gradually rises in constant-dollar terms to the point at which all cost-beneficial investments could be implemented by 2032. This scenario can viewed as an "investment ceiling," above which it would not be cost-beneficial to invest. Of the $142.5 billion average annual investment level under the Improve Conditions and Performance scenario, $85.3 billion (59.8 percent) would be directed toward improving the physical condition of existing infrastructure assets (system rehabilitation); this portion is identified as the State of Good Repair benchmark. This scenario also includes $35.7 billion (25.1 percent) directed toward system expansion and $21.5 billion (15.1 percent) for system enhancement.

Average Annual Cost by Investment Scenario (Billions of 2012 Dollars)

System Subset	Sustain 2012 Spending	Maintain C&P	Improve C&P
Interstate	$20.5	$24.1	$31.8
NHS	$54.6	$51.7	$72.9
Federal-aid Highways	$79.0	$69.3	$107.9
All Roads	$105.2	$89.9	$142.5

In addition to addressing future highway and bridge needs as they arise over 20 years, the Improve Conditions and Performance scenario would address the estimated $830.0-billion existing backlog of cost-beneficial highway and bridge investments as of 2012. Approximately $156.8 billion (18.8 percent) of the total backlog is for the Interstate System, $394.9 billion (47.2 percent) is for the NHS, and $644.8 billion (77.1 percent) is for Federal-aid highways.

Executive Summary

CHAPTER 8

Selected Capital Investment Scenarios: Transit

Chapter 8 explores the consequences of three distinct investment scenarios: maintaining current levels, meeting low-growth ridership levels, and meeting high-growth ridership levels. It also includes the SGR Benchmark.

Average Annual Investment 2013 through 2032

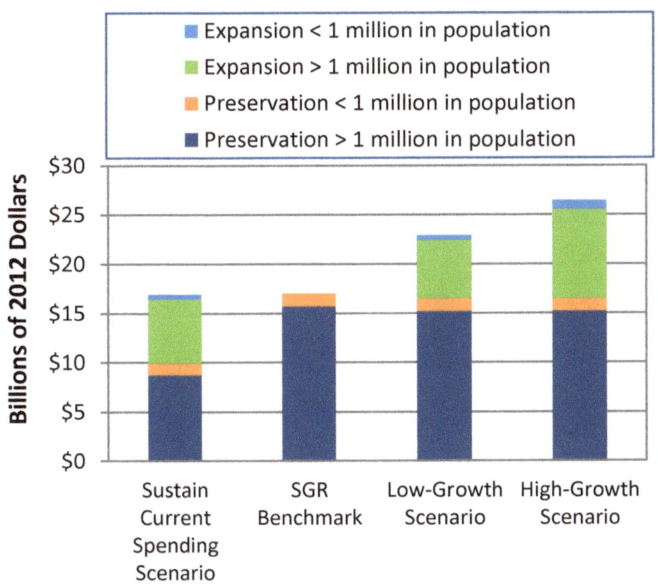

- **Sustain 2012 Spending Scenario:** Total spending under this scenario is well below that of the other needs-based scenarios, indicating that sustaining recent spending levels is insufficient to attain the investment objectives of the SGR Benchmark and the Low-Growth and High-Growth scenarios. This projection suggests future increases in the size of the SGR backlog and deterioration in the quality of service and likely increase in the number of transit riders per peak vehicle—including an increased incidence of crowding—in the absence of increased expenditures.

- **SGR Benchmark:** The level of expenditures required to attain and maintain an SGR over the upcoming 20 years—which covers preservation needs but excludes any expenditures on expansion investments—is 8.6 percent higher than that currently expended on asset preservation and expansion combined.

- **Low- and High-Growth Scenarios:** The level of investment to address expected preservation and expansion needs is estimated to be about 46 to 69 percent higher than the Nation's transit operators currently expend.

Projected Total Ridership Per Year

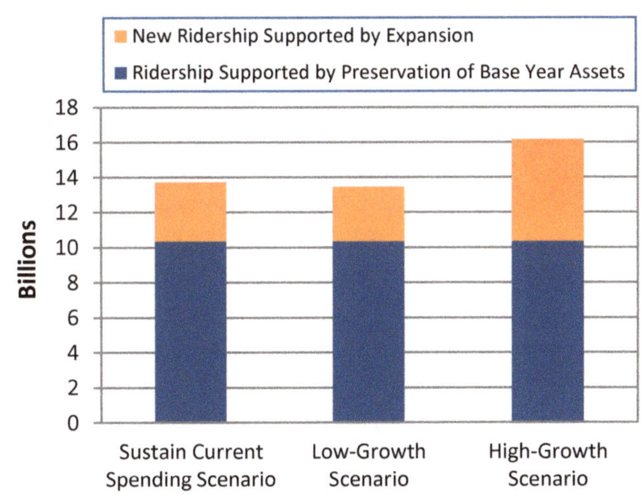

The share of transit assets exceeding their usual lives will increase if the level of investment over the next 20 years is maintained at 2012 levels. The proportion of existing asset categories exceeding their useful lives will undergo a near-continuous increase across each of these categories.

Executive Summary

CHAPTER 9

Supplemental Scenario Analysis: Highways

The 2013 C&P Report presented two values for each scenario based on alternative assumptions about future VMT growth. The average annual investment level for the Maintain Conditions and Performance scenario ranged from $65.3 to $86.3 billion in 2010 dollars. Adjusting these values for inflation shifts this range to $69.3 to $91.6 billion in 2012 dollars. The comparable amount for this scenario presented in Chapter 8 of this edition is $89.9 billion in 2012 dollars, approximately 1.9 percent lower than the high end of the adjusted 2013 C&P Report range.

The 2013 C&P Report estimated an average annual investment range of $123.7 to $145.9 billion for the Improve Conditions and Performance scenario in 2010 dollars; adjusting for inflation increases this range to $131.3 to $154.9 billion in 2012 dollars. The comparable amount for the Improve Conditions and Performance scenario presented in Chapter 8 of this edition is $142.5 billion, toward the middle of the adjusted 2013 C&P Report range.

The names and definitions of the highway scenarios presented in the C&P report have varied over time, but each edition has generally included one primary scenario oriented toward maintaining the overall state of the system and one oriented toward improving the overall state of the system. Starting with the 1997 C&P Report, the "gap" between base-year spending and the average annual investment level for the primary "Maintain" and "Improve" scenarios has varied, rising as high as 34.2 percent and 121.9 percent, respectively, in the 2008 C&P Report (comparing needs in 2006 dollars with actual spending in 2006). These larger gaps coincided with a 43.3-percent increase in construction costs between 2004 and 2006. For the current 2015 C&P Report, the gap associated with the Improve Conditions and Performance scenario has fallen to 35.5 percent, while the gap with the Maintain Conditions and Performance scenario is negative (-14.6 percent).

Gap Between Average Annual Investment Scenarios and Base Year Spending as Identified in the 1997 to 2015 C&P Reports

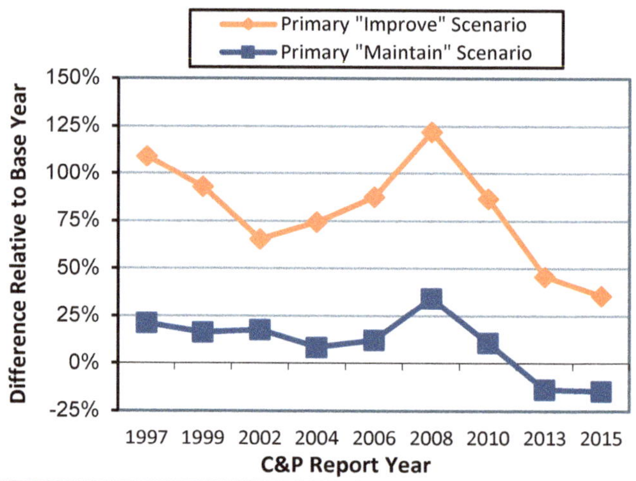

The decision to base the 2015 C&P Report scenarios on a national-level VMT forecast model rather than higher State-provided forecasts was driven in part by a review of the accuracy of the VMT forecasts used in previous C&P reports. States have tended to underpredict future VMT during periods when actual VMT was growing rapidly and to overpredict at times when actual VMT growth was slowing or declining.

Executive Summary

CHAPTER 9

Supplemental Scenario Analysis: Transit

New technologies have had an impact on transit investment needs. As with most industries, the existing stock of assets used to support transit service is subject to ongoing technological change and improvement. Such change and improvements tend to increase investment costs. For example, by 2032, alternative, cleaner fuels are expected to propel more than 70 percent of the national bus fleet. These vehicles are more expensive to purchase and operate.

A cleaner, more fuel-efficient national bus fleet will not affect capital investment needs significantly over the next 20 years. The chart below adds the estimated additional funding required to support a national transit bus fleet composed of more than 70 percent alternative fuel vehicles in 2032 to the forecasted capital investment needs under the Low-Growth scenario.

Impact of Shift to Vehicles Using Hybrid and Alternative Fuels on Investment Needs: Low-Growth Scenario

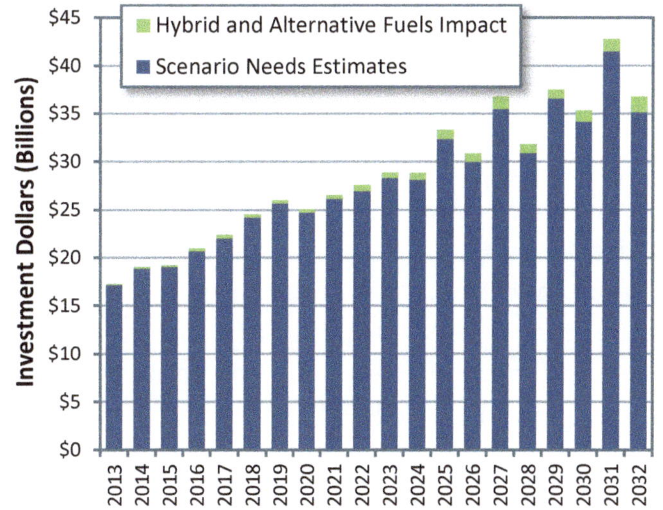

If the levels of investment in preservation and expansion are kept the same as in 2012, the average condition of all assets will decline from the middle of the adequate range (3.5 in 2012) to 3.1, near the upper bound of the marginal range (2.0–2.9), in 2032.

Asset Condition Forecast for All Existing and Expansion Transit Assets

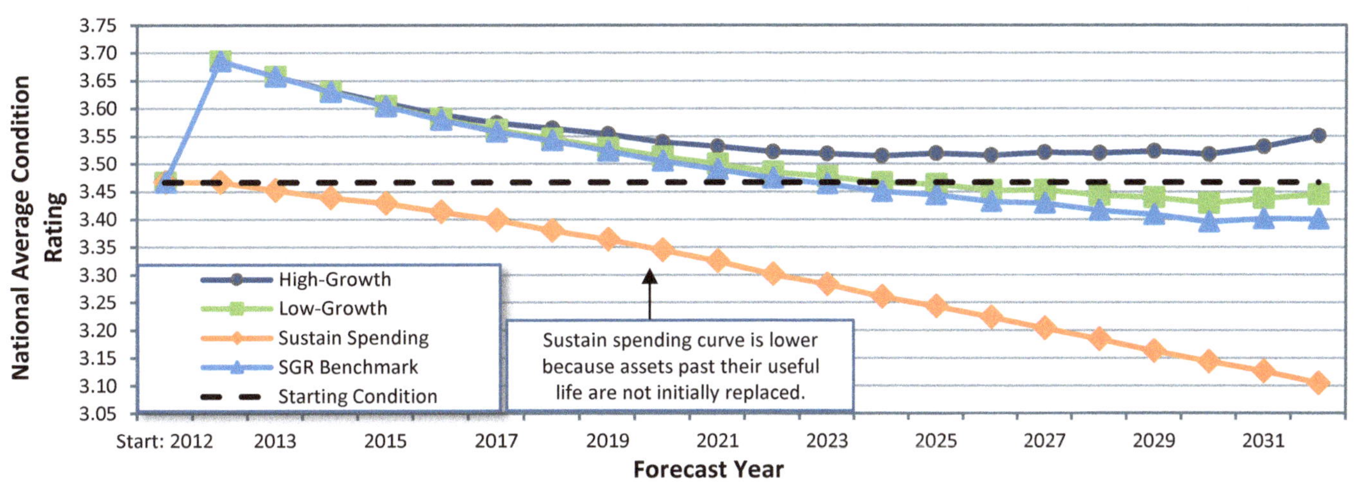

Executive Summary

CHAPTER 10

Sensitivity Analysis: Highways

Sound practice in modeling includes analyzing the sensitivity of key results to changes in assumptions. Chapter 10 demonstrates how the baseline scenarios presented in Chapter 8 would be affected by changing some HERS and NBIAS parameters.

If VMT per capita were to remain constant from 2012 to 2032, VMT would grow by 0.74 percent per year (based on U.S. Census population projections), rather than the 1.04-percent annual rate assumed in the baseline analyses. This assumption would reduce the average annual investment level under the Improve Conditions and Performance scenario to $129.0 billion. If travel increases at the annual rates States project in the HPMS (1.41 percent) and the NBI (1.46 percent), the cost of this scenario would increase to $159.8 billion.

The valuation of travel time savings assumed in the baseline scenarios is linked to average hourly income; personal travel is valued at 50 percent of income, while business travel is valued at 100 percent. Alternative tests were run reducing these shares to 35 percent and 80 percent, respectively, and increasing them to 60 percent and 120 percent. Applying a lower value of time reduces the benefits associated with travel time savings and reduces the average annual investment level under the Improve Conditions and Performance scenario from $142.5 billion to $134.6 billion, as some potential projects would no longer qualify as cost beneficial. Assuming a higher value of time increases the annual cost of this scenario to $147.7 billion.

The baseline scenarios assume the value of a statistical life is $9.1 million when computing safety-related benefits. Reducing this value to $5.2 million would reduce the annual cost of the Improve Conditions and Performance scenario to $138.3 billion; increasing the value to $12.9 million would increase the annual cost to $144.2 billion.

Benefit-cost analyses use a discount rate that scales down benefits and costs arising later in the future relative to those arising sooner. The baseline scenarios assume a 7-percent rate; changing this to 3 percent would increase the average annual investment level under the Improve Conditions and Performance scenario to $171.5 billion.

Impact of Alternative Assumptions on Highway Scenario Average Annual Investment Levels

Parameter Change	Maintain C&P	Improve C&P
	(Billions of 2012 Dollars)	
Baseline	**$89.9**	**$142.5**
Slower Growth in VMT	$81.3	$129.0
Faster Growth in VMT	$101.1	$159.8
Lower Value of Time	$84.7	$134.6
Higher Value of Time	$92.7	$147.7
Lower Value of Statistical Life	$89.0	$138.3
Higher Value of Statistical Life	$90.6	$144.2
3 Percent Discount Rate	$88.0	$171.5

The impacts of alternative assumptions on the Maintain Conditions and Performance scenario are generally smaller and are linked to either the models' distribution of spending among different capital improvement types or to reduced VMT.

Executive Summary

CHAPTER 10

Sensitivity Analysis: Transit

TERM relies on several key input values, variations of which can significantly influence the projections of capital needs for the scenarios considered in the C&P report: the SGR benchmark and Low-Growth and High-Growth scenarios.

Impact of alternative replacement condition thresholds on transit preservation needs—Baseline: Assets are replaced at a condition rating of 2.5. Analysis suggests that each scenario is sensitive to changes in the replacement condition threshold. The sensitivity increases disproportionately with higher replacement condition thresholds. For example, reducing the condition threshold to 2.0 tends to reduce the SGR backlog by $0.7 billion (4 percent). In contrast, increasing the threshold to 3.0 increases preservation needs by more than $1 billion (6 percent). Note that selecting a higher replacement condition results in asset replacement at an earlier age, which in turn results in more replacements over the 20-year forecast.

Impact of increase in capital costs on transit investment estimates. The asset costs used in TERM are based on actual prices agencies paid for capital purchases as reported to the FTA in the Transit Electronic Award Management system and in special surveys. Asset prices in the current version of TERM were converted from the dollar-year replacement costs in which assets were reported to FTA by local agencies (which vary by agency and asset) to 2012 dollars using the RSMeans© construction cost index. Any increase in capital costs without a similar increase in transit benefits results in lower benefit-cost ratios and failure of some investments to pass this test. The analysis shows that a 25-percent increase in capital costs for the Low-Growth and High-Growth scenarios would yield a roughly 13-percent and 15-percent increase, respectively, in capital needs that pass TERM's benefit-cost test.

Impact of alternative value of time rates on transit investment estimates. The most significant source of transit investment benefits as assessed by TERM's benefit-cost analysis is the net cost savings to users of transit services, a key component of which is the value of travel time savings. The current hourly rate based on U.S. Department of Transportation guidance is $12.50. Increasing this rate results in higher benefits, which results in including projects that failed the benefit-cost test at the standard rate. Decreasing the rate has the opposite effect. Doubling the rate (to $25.00) results in increases of 4.2 percent and 6.5 percent in capital needs for the Low-Growth and High-Growth scenarios, respectively. Reducing the rate by half (to $6.25) results in decreases of 12 percent and 13 percent, respectively.

Impact of discount rate. TERM's benefit-cost test is responsive to the discount rate used to calculate the present value of investment costs and benefits. TERM's analysis uses a rate of 7 percent in accordance with Office of Management and Budget guidance. The analysis using a rate of 3 percent (57 percent smaller) leads to an increase of 17.2-percent investment needs in the High-Growth scenario, but no change in the Low-Growth scenario.

Executive Summary

CHAPTER 11

Pedestrian and Bicycle Transportation

DOT is committed to making walking and bicycling safer and more comfortable transportation options for everyone. The 1994 National Bicycling and Walking Study set a goal to double the percentage of trips made by bicycling and walking from 7.9 percent to 15.8 percent. This share had risen to 11.5 percent by 2009, just shy of halfway toward reaching this goal.

Bicycle and Pedestrian Travel Trends as Percentage of All Trips, 1994 and 2009

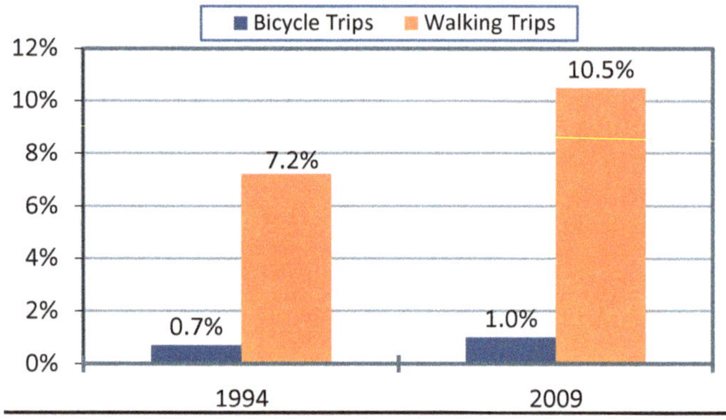

The 1994 study also set a goal of reducing pedestrian and bicycle injuries and fatalities by 10 percent. This goal has been exceeded, but recent trends indicate some reversal of progress, as injuries and fatalities have risen since 2009.

Federal funding for pedestrian and bicycle transportation has increased significantly, from $113 million in 1994 to a peak level of $1.2 billion in 2009; funding for 2014 was $820 million.

Federal policies and guidance supporting the inclusion of pedestrian and bicycle transportation in routine transportation planning, design, and construction have advanced multimodal planning and project development at all levels. Hundreds of communities, MPOs, and State Departments of Transportation (DOTs) have adopted Complete Streets policies, which require the formal consideration of all modes of travel throughout the project planning and development process. States and communities now routinely accommodate people with disabilities when developing pedestrian facilities and pedestrian access routes.

Context Sensitive Solutions, a collaborative, interdisciplinary, and holistic approach to the development of transportation projects, has become increasingly accepted by a broad range of stakeholders in all phases of program delivery, including long-range planning, programming, environmental studies, design, construction, operations, and maintenance.

The field of pedestrian and bicycle transportation engineering and planning has evolved, enabling practitioners at all levels to become more effective in improving safety and mobility for pedestrians and bicyclists. Professional organizations such as the Association of Pedestrian and Bicycle Professionals and pedestrian and bicycle advocacy organizations have played a key role in this process. Information-sharing resources such as the Pedestrian and Bicycle Information Center have been established, and professional training programs, guidebooks, and other educational resources have been developed.

Executive Summary

CHAPTER 12

Transportation Serving Federal and Tribal Lands

The Federal government holds title to approximately 30 percent (650 million acres) of the total land area of the United States. Additionally, on behalf of Tribal governments, approximately 55 million acres of land is held in trust. Federal lands have many uses, including the facilitation of national defense, recreation, grazing, timber and mineral extraction, energy generation, watershed management, fish and wildlife management, and wilderness maintenance.

More than 450,000 miles of Federal roads provide access to Federal lands, creating opportunities for recreational travel and tourism, protection and enhancement of resources, and sustained economic development in both rural and urban areas. Annual visits to Federal lands total nearly 1 billion, and are expected to rise as the population increases, posing a challenge to Federal land management agencies in fulfilling their missions of providing visitor enjoyment while conserving precious resources. Accommodating growing traffic volumes and demands for visitor parking will require innovation and creative solutions.

Roads Serving Federal Lands[1]

Federal Agency	Public Paved Road Miles	Paved Road Condition			Public Unpaved Road Miles	Public Bridges		Backlog of Deferred Maintenance[2]
		Good	Fair	Poor		Total	Structurally Deficient	
Forest Service	9,500	42%	55%	3%	362,500	4,200	11%	$2.9 billion[3]
National Park Service	5,500	59%	29%	12%	4,100	1,442	3%	$6 billion
Bureau of Land Management	500	65%	20%	15%	600	835	3%	$350 million
Fish and Wildlife Service	400	60%	25%	15%	5,200	281	7%	$1 billion
Bureau of Reclamation	762	65%	25%	10%	1,253	331	12%	N/A
Bureau of Indian Affairs	8,800	N/A	N/A	N/A	20,400	929	15%	N/A
Tribal Governments	3,300	N/A	N/A	N/A	10,200	N/A	N/A	N/A
Military Installations	27,900	N/A	N/A	N/A	N/A	1,418	26%	N/A
U.S. Army Corps of Engineers	5,247	56%	30%	14%	2,549	416	6.20%	N/A

[1] Data shown are not for a consistent year, but instead reflect the latest available information as of late 2014 when these data were obtained from the FLMAs. Road condition categories are based on definitions of each, which are not fully consistent. Structural deficiencies are classified using a uniform definition consistent with that presented in Chapter 3.
[2] Backlog includes only transportation-related amounts.
[3] Value is for passenger car roads only.

part I

Description of Current System

Introduction

In 2014, the U.S. Department of Transportation (DOT) developed *Transportation for a New Generation, a Strategic Plan for Fiscal Years 2014–18*, outlining the objectives and performance goals for the Nation's transportation system. The plan includes five strategic goals:

- **Safety** – Improve public health and safety by reducing transportation-related fatalities, injuries, and crashes.
- **State of Good Repair** – Ensure that the United States proactively maintains critical transportation infrastructure in a state of good repair.
- **Economic Competitiveness** – Promote transportation policies and investments that bring lasting and equitable economic benefits to the Nation and its citizens.
- **Quality of Life in Communities** – Foster quality of life in communities by integrating transportation policies, plans, and investments with coordinated housing and economic development policies to increase transportation choices and access to transportation services for all.
- **Environmental Sustainability** – Advance environmentally sustainable policies and investments that reduce carbon and other harmful emissions from transportation sources.

Each goal relates to a different aspect of system conditions and performance and thus relates directly to the types of information presented in the C&P report.

Part I Chapters

Part I of this 2015 C&P Report includes six chapters, each of which describes the current system from a different perspective:

- Chapter 1, **Household Travel and Freight Movement**, outlines the trends in travel behavior of households and businesses. The household travel discussion relies heavily on the results of the 2009 National Household Travel Survey and delves into topics relating to traveler demographics, travel geography, and emerging travel trends. The freight section describes the freight transportation system, freight demand, and challenges facing the movement of freight.
- Chapter 2, **System Characteristics**, describes the extent and use of highways, bridges, and transit systems. Highway and bridge data are presented for system subsets based on functional classification and Federal system designation, while transit data are presented for different types of modes and assets.
- Chapter 3, **System Conditions**, presents data on the physical condition of the Nation's highway, bridge, and transit assets. This chapter relates directly to DOT's State of Good Repair goal.

- Chapter 4, **Safety**, relates directly to DOT's Safety goal. The highway section presents national-level statistics on safety performance, focusing on the most common roadway factors that contribute to fatal and serious injury crashes. The transit section summarizes safety and security data by mode and type of transit service.
- Chapter 5, **System Performance,** covers a range of topics relating to three separate DOT goals: Economic Competitiveness, Quality of Life in Communities, and Environmental Sustainability.
- Chapter 6, **Finance**, provides detailed data on the revenue collected and expended by different levels of governments to fund transportation construction and operations throughout the United States.

Transportation Performance Management

In addition to the DOT goals referenced above, a recurring theme in Part I of the C&P report is the impact of changes under the Moving Ahead for Progress in the 21st Century legislation (MAP-21), in particular those changes pertaining to transportation performance management.

What is Transportation Performance Management?

The Federal Highway Administration (FHWA) defines Transportation Performance Management (TPM) as a strategic approach that uses system information to make investment and policy decisions to achieve national performance goals. FHWA is working with States and metropolitan planning organizations to transition toward and implement a performance-based approach to carrying out the Federal-aid highway program. This transition supports MAP-21, which integrates performance into many Federal transportation programs. In short, TPM

- is systematically applied in a regular ongoing process;
- provides key information to help decision makers, enabling them to understand the consequences of investment decisions across multiple markets;
- improves communications among decision makers, stakeholders, and the traveling public; and
- ensures targets and measures are developed in cooperative partnerships and is based on data and objective information.

National Goals – Federal-Aid Program [23 United States Code §150(b)]

The cornerstone of MAP-21's highway program transformation is the transition to a performance- and outcome-based program. States will invest resources in projects to achieve individual targets that collectively will make progress toward national goals. FHWA is collaborating with State and local agencies across the country to focus on the national goals MAP-21 established, regardless of resource limitations.

The national performance goals for Federal highway programs as established in MAP-21 are as follows:

- Safety – To achieve a significant reduction in traffic fatalities and serious injuries on all public roads.
- Infrastructure Condition – To maintain the highway infrastructure asset system in a state of good repair.
- Congestion Reduction – To achieve a significant reduction in congestion on the National Highway System.
- System Reliability – To improve the efficiency of the surface transportation system.
- Freight Movement and Economic Vitality – To improve the national freight network, strengthen the ability of rural communities to access national and international trade markets, and support regional economic development.
- Environmental Sustainability – To enhance the performance of the transportation system while protecting and enhancing the natural environment.
- Reduced Project Delivery Delays – To reduce project costs, promote jobs and the economy, and expedite the movement of people and goods by accelerating project completion through eliminating delays in the project development and delivery process, including reducing regulatory burdens and improving agencies' work practices.

Transportation Performance Management Elements

The Federal Highway Administration has organized the performance-related provisions within MAP-21 into six TPM elements to communicate the efforts underway for implementing these requirements more effectively. These six TPM elements are listed below.

National Goals	MAP-21 establishment of goals or program purpose to focus the Federal-aid highway program into specific areas of performance.
Measures	The establishment of measures by FHWA to assess performance/condition to carry out performance-based Federal-aid highway programs.
Targets	Establishment of targets by recipients of Federal-aid highway funding for each of the measures to document expectations of future performance.
Plans	Development of strategic or tactical plans, or both, by recipients of Federal funding to identify strategies and investments that will address performance needs.
Reports	Development of reports by recipients of Federal funding that would document progress toward the achievement of targets, including the effectiveness of Federal-aid highway investments.
Accountability and Transparency	Requirements developed by FHWA for recipients of Federal funding to use in achieving or making significant progress toward achieving targets established for performance.

Summary of MAP-21 Performance Requirements

The MAP-21 legislation integrates performance into many Federal transportation programs and contains several performance elements. FHWA will help coordinate the alignment of MAP-21 requirements and provide guidance and resources. Listed below is more information regarding the performance requirements for the National Highway Performance Program, the Highway Safety Improvement Program, the Congestion Mitigation and Air Quality Improvement Program, and Freight Movement, as established in MAP-21.

- National Highway Performance Program
  http://www.fhwa.dot.gov/tpm/about/nhpp.cfm
- Highway Safety Improvement Program
  http://www.fhwa.dot.gov/tpm/about/hsip.cfm
- Congestion Mitigation and Air Quality Improvement Program
  http://www.fhwa.dot.gov/tpm/about/cmaq.cfm
- Freight Movement
  http://www.fhwa.dot.gov/tpm/about/freight.cfm
- Implementation Schedule
  http://www.fhwa.dot.gov/tpm/about/schedule.cfm

Implementation of MAP-21 Performance Requirements

FHWA is implementing the MAP-21 performance requirements through six interrelated rulemakings:

- A Final Rule on **Statewide and Metropolitan/Non-metropolitan Transportation Planning** published May 27, 2016 to implement a performance-based planning process at the State and metropolitan levels. The Final Rule defines coordination in the selection of targets, linking planning and programming to performance targets.
- A Final Rule for **Safety Performance Management Measures (PM-1)** published March 15, 2016 with an effective date of April 14, 2016 defines fatalities and serious injuries measures, along with target establishment, progress assessment, and reporting requirements. The Final Rule discusses the implementation of MAP-21 performance requirements.
- A Final Rule for **Highway Safety Improvement Program (HSIP)** published March 15, 2016 integrates performance measures, targets, and reporting requirements into the HSIP. The Final Rule contains three major policy changes: Strategic Highway Safety Plan (SHSP) Updates, HSIP Report Content and Schedule, and the Subset of the Model Inventory of Roadway Elements (MIRE).
- An NPRM for a **Pavement and Bridge Performance Measures Rule (PM-2)** published January 5, 2015 proposes and defines pavement and bridge condition performance measures, along with minimum condition standards, target establishment, progress assessment, and reporting requirements.

- An NPRM for an **Asset Management Plan Rule** published February 20, 2015 proposes and defines the contents and development process for an asset management plan. The NPRM also proposes minimum standards for pavement and bridge management systems.
- An NPRM for **System Performance Measures Rule (PM-3)** published April 22, 2016 proposes and defines performance measures to assess performance of the Interstate System, non-Interstate National Highway System, freight movement on the Interstate System, CMAQ traffic congestion, and on-road mobile emissions. The NPRM will also summarize all three MAP-21 highway performance measure proposed rules.

The Fixing America's Surface Transportation Act (FAST Act) continues MAP-21's overall performance management approach, but includes some clarifications and adjustments to certain provisions pertaining to individual programs. All of the Department's performance management rulemakings will comply with the performance management provisions of the FAST Act.

chapter 1

Personal Travel and Freight Movement

Personal Travel

The movement of people constitutes the vast majority of travel within the Nation's transportation system. Estimates from the Highway Performance Monitoring System show that 91 percent of the miles traveled in the United States are by passenger vehicles (see *Exhibit 1-1*).

Exhibit 1-1 Vehicle Miles Traveled by Type of Travel

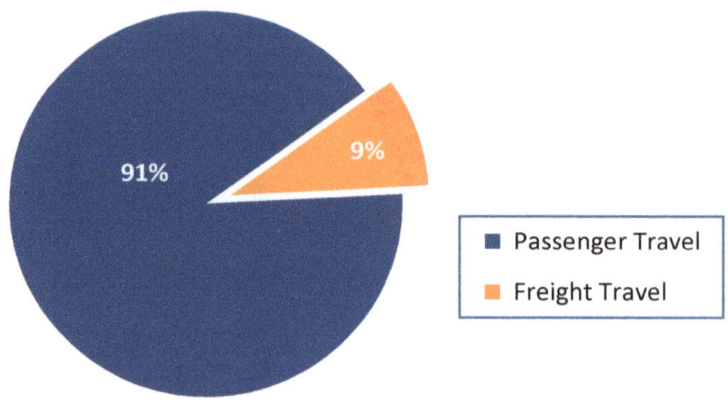

Source: Highway Performance Monitoring System.

Population changes, in both demographics and geographic location, historically have had significant impacts on the size and distribution of travel demand. The growth of the suburbs, and women entering the workforce are examples of influences demographic shifts can have on increasing travel.

Today, the U.S. population is undergoing changes in age distribution, racial and ethnic composition, migration, and immigration that affect the way we travel. In addition, advancements in information communication technologies, global positioning systems (GPS), sensors, and automation are affecting the personal travel experience.

Travel Trends

The past decade has experienced a shift in trends of vehicle ownership, vehicle miles traveled (VMT), and licensing rates, especially among teens and young adults. The number of registered vehicles rose from 111.2 million to 235.1 million between 1970 and 2013, but the number of cars per person peaked in 2008. Approximately 8.7 percent of households have no vehicle.

According to the Highway Performance Monitoring System, total VMT peaked in 2008 and per capita VMT peaked in 2004. Both have since decreased and leveled off (see *Exhibit 1-2*).

Licensing rates for teens and young adults have declined since the 1980s. In 1978, 46 percent of all 16-year-olds (1.9 million) and 97 percent of all 24-year-olds (3.8 million) had driver's licenses. In 2012, the rates dropped to 28.2 percent for 16-year-olds (1.2 million) and 83.4 percent for 24-year-olds (3.6 million). This downward trend, which began well before the December 2007 to June 2009 recession, suggests that other factors could explain declining rates, such as the effects of graduated licensing and youth attitudes about driving (see *Exhibit 1-3*).

Exhibit 1-2 Per Capita Vehicle Miles Traveled, 1987–2012

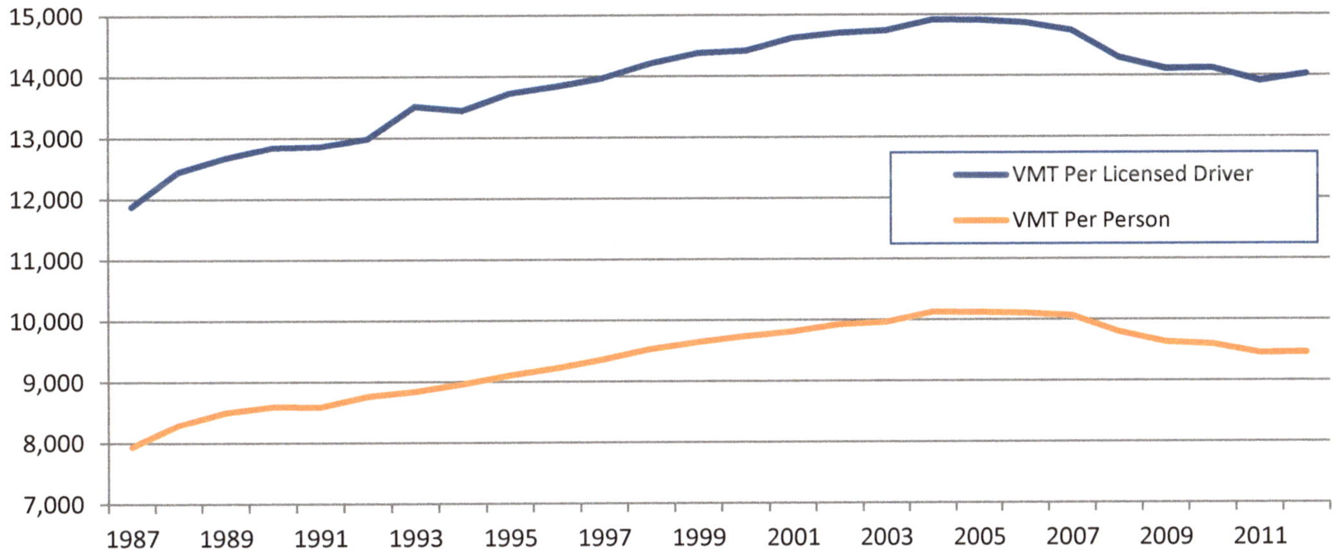

Source: Highway Performance Monitoring System.

Exhibit 1-3 Licensing Percentages for 16- and 24-year olds, 1978–2012

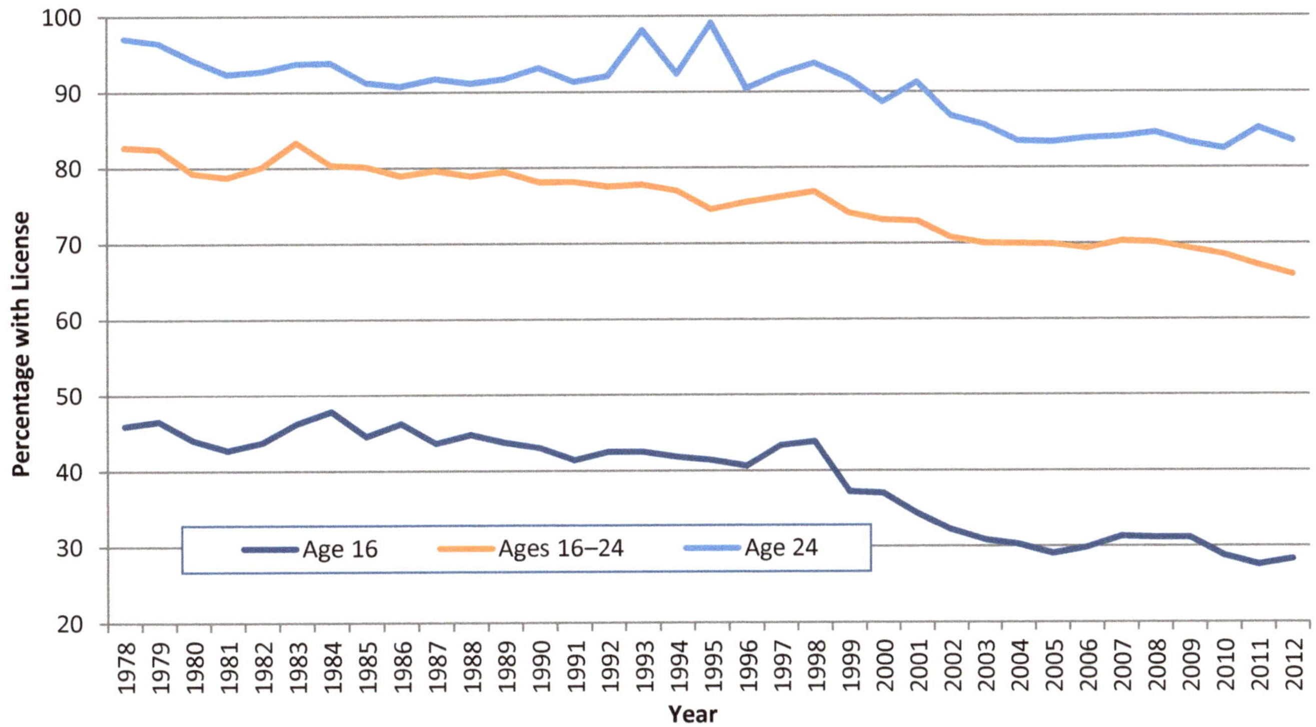

Source: Highway Statistics, Table DL-20.

Person Miles Traveled and Mode Share

Based on data available from the National Household Travel Survey (NHTS), person miles of travel (PMT), which includes vehicle and nonvehicle travel, declined less between 2001 and 2009 than VMT. Average vehicle occupancy decreased over this period but was more than offset by increases in nonvehicle travel. Nonvehicle miles include travel by modes other than personal vehicle (with the exception of car pools), including all public transportation, bike, walk, ferry, and air travel (see Chapter 11 for information on bicycle and pedestrian travel). While trips in private vehicles declined from 2001 to 2009, transit trips increased (see *Exhibits 1-4* and *1-5*).

Exhibit 1-4 Per Capita Daily Passenger Miles Traveled and Vehicle Miles Traveled, 1977–2009

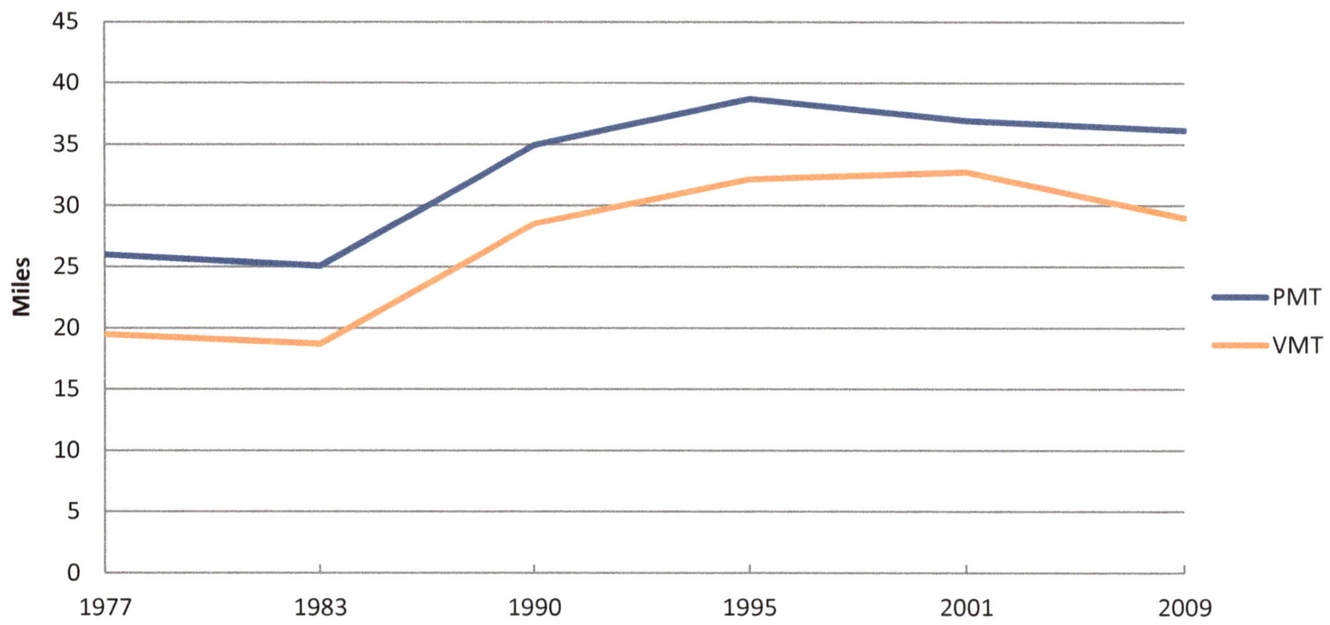

Source: National Household Travel Survey.

Exhibit 1-5 Percentage Change in Trips by Travel Mode, 1990–2009

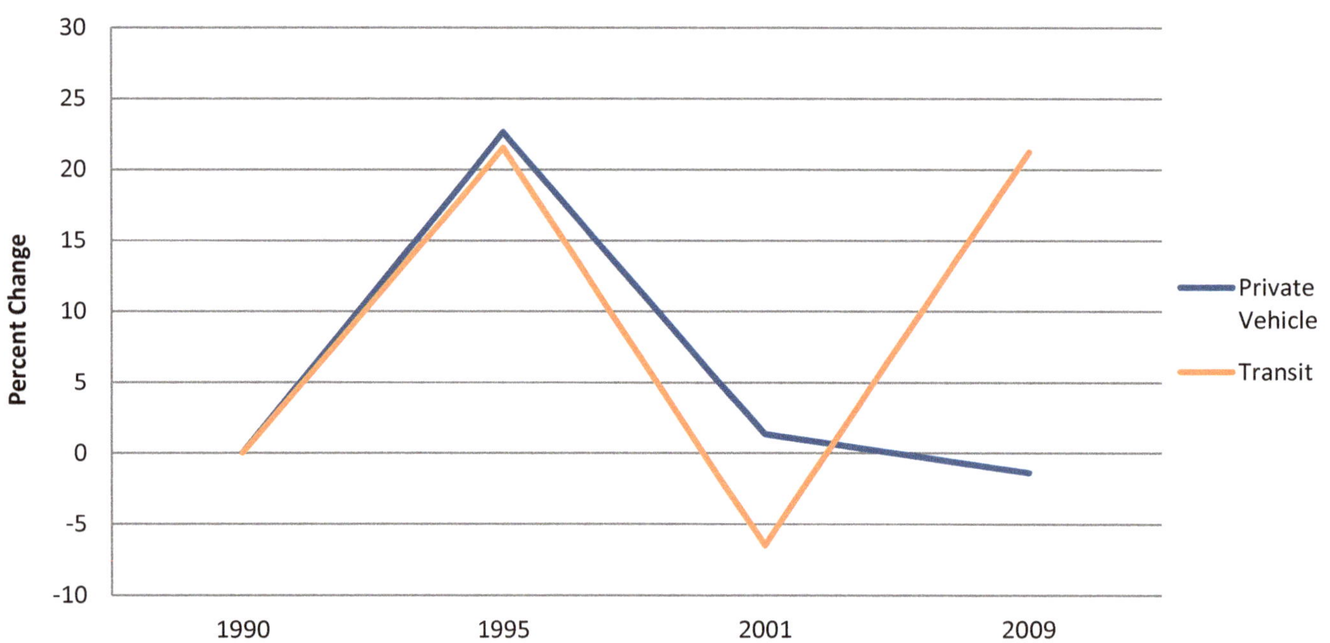

Source: National Household Travel Survey.

Where a person lives influences his or her mode choices. In densely populated areas, where public transportation is easily accessible, a large percentage of the public is likely to use the services. In New York City, for example, more than 50 percent of the population commutes to work using public transit (based on the metropolitan statistical area defined by the U.S. Census Bureau). As private vehicle travel has decreased in the past decade, use of public transit has increased; this shift has occurred, however, primarily in metropolitan statistical areas with a population of 500,000 and higher. Population density also plays a role in the use of other travel modes: The highest percentages of households without a vehicle occur within areas having population density of 10,000 or more per square mile. In addition to location, income influences mode choice. Lower-income households are less likely to own a vehicle. Individuals living in lower-income households and new immigrants are more likely to carpool, take transit, walk, or bike. As these populations continue to migrate from city centers to suburban locations, local area transportation agencies should consider the need for travel options in these areas.

Trip Purpose

Understanding why a person makes a trip provides transportation planners and policy makers the knowledge to better understand and anticipate travel volumes and demand and the needs of different traveling populations. For instance, aging populations might engage in more social and recreational travel, which can contribute to congestion during off-peak hours. Younger populations might rely on public transit to get to school or work places, emphasizing the need for reliable services to these destinations.

Q&A **For what years are NHTS data available?**

Data from the NHTS and comparable predecessor surveys are available for 1969, 1977, 1983, 1990, 1995, 2001, and 2009. The next NHTS will be conducted in 2016.

The NHTS is the only national data source that asks the American public why they took a given trip. Purposes of trips are classified into several categories: getting to or from work, shopping, running family or personal errands, and making social or recreational trips.

From 2001 to 2009, travel for all trip purposes declined or remained stagnant. Travel to work showed a 10-percent decrease in miles and a 7-percent decrease in number of trips. American household travel for family or personal errands decreased by 13.9 percent and the length of trips for such errands dropped by 10 percent compared to 2001. In addition, daily PMT for social and recreational purposes declined by 9.5 percent between 2001 and 2009 (see *Exhibit 1-6*).

Of all trip purposes, discretion is greatest for traveling for family and personal errands and social/recreational trips. As non-work travel comprises a large percentage of daily travel, further research might be needed to examine whether the reductions from 2001 to 2009 were due primarily to economic reasons or to demographic and lifestyle changes among the American public.

Exhibit 1-6 Average Annual Person Miles and Person Trips per Household by Trip Purpose[1]

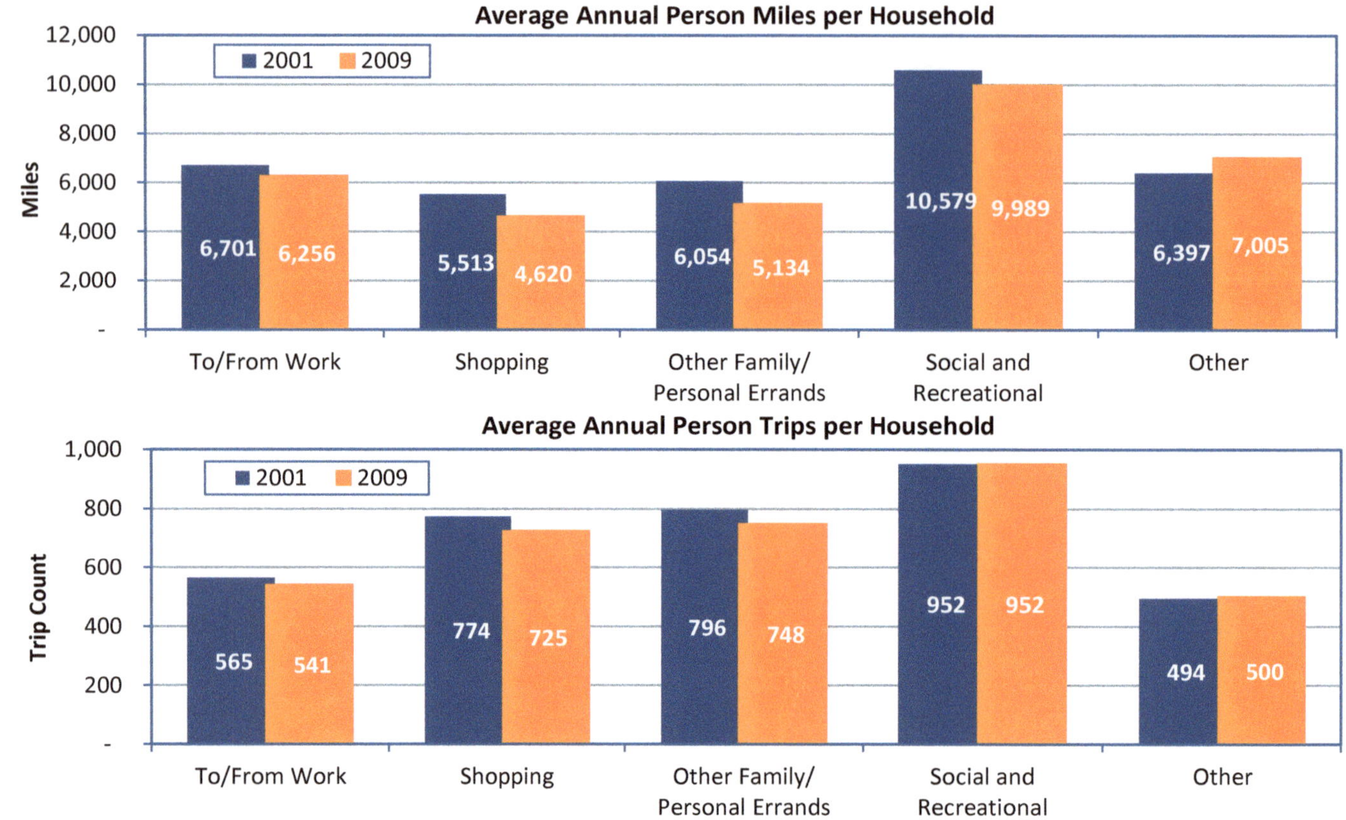

[1] The travel of children aged 0–4 years old is excluded from 2001 NHTS data to make it compatible with other years.

Sources: 2001 and 2009 NHTS.

Overall Population Trends

Between 1970 and 2013, the population of the United States increased by 53 percent, from 203 million to 316 million. By 2050, the U.S. population is projected to be just under 400 million. The annual rate of growth has declined in recent years to approximately 0.7 percent in 2013, the lowest growth rate since 1937. The rate is expected to continue to drop and by 2050 is projected to be 0.5 percent. Even though the growth rate is declining, the overall increase in population will result in an increase in total travel and an increased amount of freight being moved even if the miles and freight per person stabilizes or declines.

Aging Population

For the coming decades, the older population of the United States is expected to experience considerable growth. Most researchers expect the "baby boom" generation to enjoy increased longevity and to drive more miles than today's older adults. Currently, the number of people aged 65 years and older is approximately 40 million, and they account for 13 percent of the population. By 2050, this age group will double and account for 20 percent of the population, with the most significant increases occurring in the 85 years and older age group (see *Exhibit 1-7*).

Ex

1960

Male Age Female

6 5 4 3 2 1 0 6 5 4 3 2 1 0

Percentage of total population Percentage of total population

1990

Male Age Female

6 5 4 3 2 1 0 6 5 4 3 2 1 0

Percentage of total population Percentage of total population

2020

Male Age Female

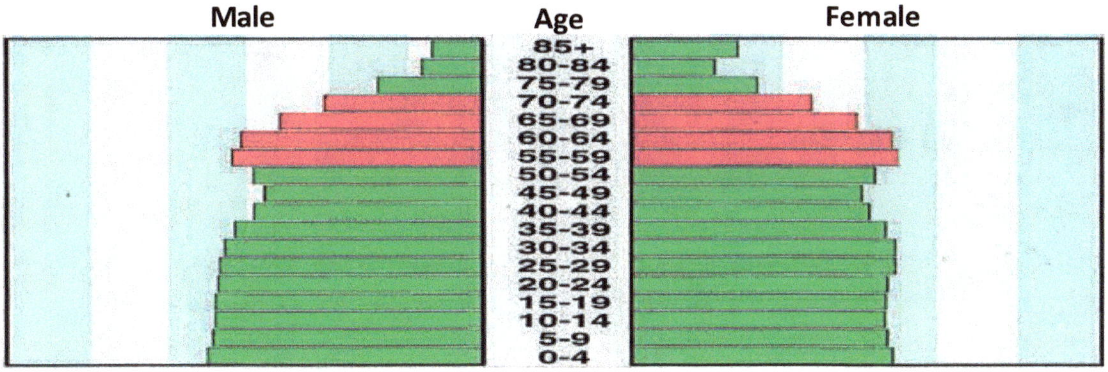

6 5 4 3 2 1 0 6 5 4 3 2 1 0

Percentage of total population Percentage of total population

[1] Pi

Sou

As the U.S. population ages, the percentage of older people continuing to work and drive will increase. Although some of the baby boom generation will retire, some will switch to part-time work, second careers, or volunteer activities. At the same time, the increase in the aging population also will result in an increased number of nondrivers requiring alternative means of mobility.

Daily patterns of travel shift with age. The proportion of travel for shopping, recreation, and other purposes (including medical appointments and to visit friends) increases as people age. More than 60 percent of daily travel by older adults occurs between 9 a.m. and 4 p.m., compared to young adults whose travel peaks during three distinct periods: morning (7 a.m. to 8 a.m.), midday (12 noon to 1 p.m.), and after work (5 p.m. to 6 p.m.). The growth in the older population could add significantly to midday travel.

Diversity

Not only is the population aging, it is also becoming more ethnically diverse. Minority groups, which made up 12 percent of the total population in 1970, have increased to 27 percent in recent years. In 2050, this percentage is expected to increase to 50 percent, with no single race or ethnic group having the majority (see *Exhibit 1-8*).

Exhibit 1-8 Racial and Ethnic Composition of the United States, 1970–2050

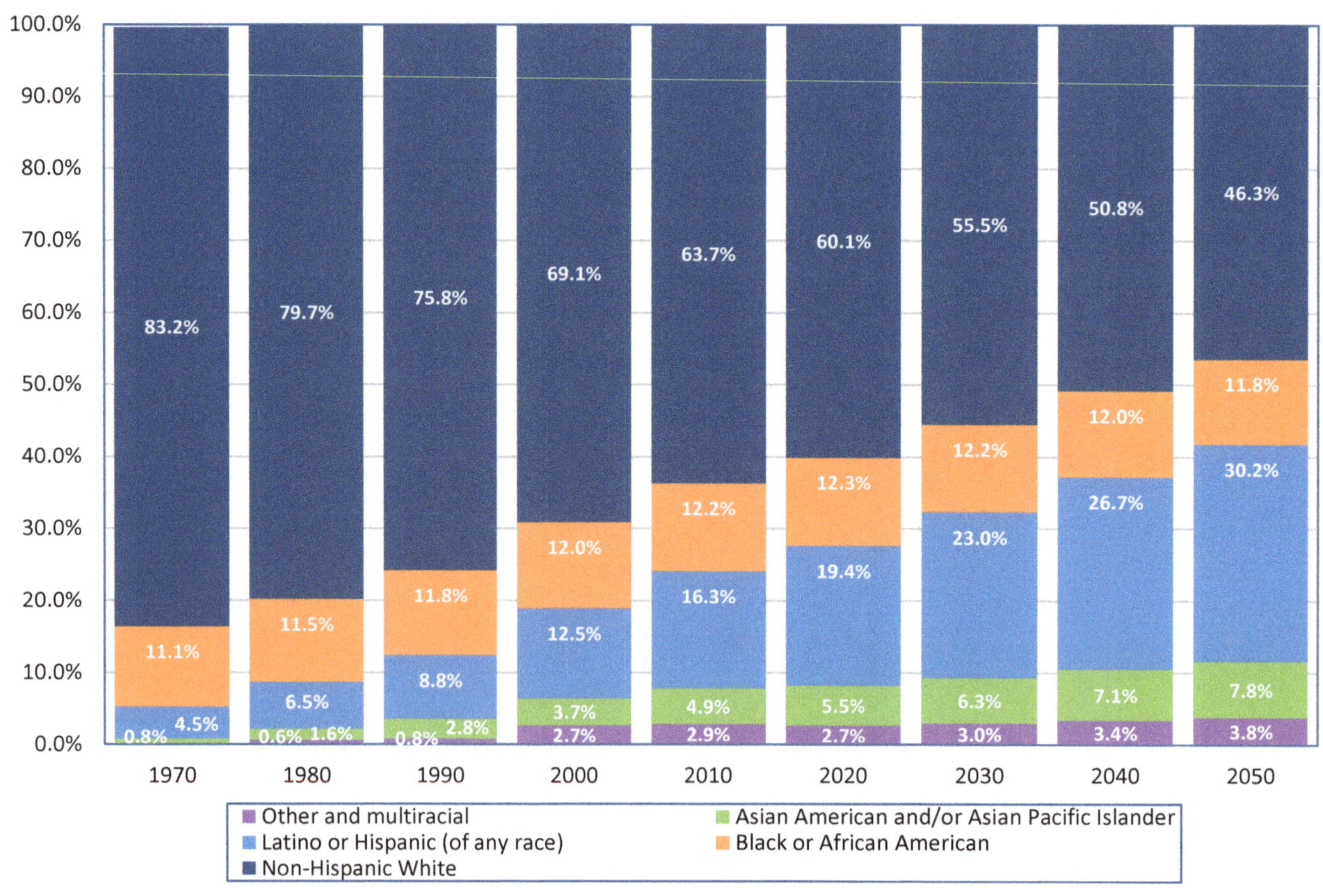

Sources: Data for 1970 and 1980 obtained from Statistical Abstract of the United States; data for 1990, 2000, and 2010 obtained from the U.S. Census Bureau; data for 2020 through 2050 came from the U.S. Census Bureau Population Projections by Race and Ethnicity (2008).

Immigration has a significant impact on national, regional, and local transportation needs. New immigrants travel differently than long-term residents because they are less likely to own a vehicle and more likely to depend on other modes of travel such as carpooling, public transit, walking, and bicycling.[1] Because of this, immigration places a different set of demands on the transportation system, and, for the first time since 1920, immigrants comprise more than 12 percent of the U.S. population. Although the immigrant population (even at the projected 2050 levels) is not substantial enough to influence national VMT and PMT projections significantly, regions in the United States with high levels of immigration will experience a significant shift in travel demand and forecasting assumptions.

Trends in Household Size

Historically, the number of households has increased much faster than the total population. Although the overall population of the United States increased by 270 percent from 1900 to 2000, the total number of households grew from 16 million in 1900 to more than 105 million in 2000, an increase of 561 percent.

Between 1960 and 2013, total households increased from roughly 53 million to more than 122 million, while the average household size decreased from 3.33 to 2.54 persons per household. The major reason for this rapid household growth is changing household structure.

Nationally, household size has been steadily declining for decades, trending toward smaller families and more single-person households. Average household size has decreased since 1970, when the average household had 3.11 persons. In 2010, it had 2.63, a slight increase from 2000. Even though the proportion of nuclear family households (those with married parents and at least one child under 18 years of age) has declined, the proportions of all other household types have increased: families without children at home, single-person households, and households of unrelated persons. Smaller household size increases the number of trips for some and reduces mobility for others due to limits in transportation access (e.g., shared vehicles, carpooling).

Income and Labor

Income

Between 2000 and 2009, U.S. transportation costs increased at nearly twice the rate of incomes. In the United States, personal income has been the most important predictor of personal travel demand, although these trends have begun to diverge over the past 10 years. Other economic factors, including employment levels (and the resulting commuting patterns) and transportation costs such as transit prices, fuel prices, and automobile prices (including vehicle operating costs), also affect transportation decisions and vehicle utilization rates. Due to this direct and longstanding influence on personal travel, economic and employment trends are an important consideration in transportation policy.

On average, U.S. citizens spend 20 percent of their income on travel-related expenditures, and transportation is the second largest expenditure for households after housing, although this varies by location and income. Households located in more diffuse, automobile-dependent regions are estimated to spend 25 percent of their income on transportation, while households located closer to employment and other amenities spend an estimated 9 percent of household income on transportation.

Poverty

The current national poverty rate is close to 15 percent and by 2020 is expected to decline to around 14 percent. Despite this decline, analyses suggest many years will pass before it returns to rates experienced before the December 2007 to June 2009 recession.

Both unemployment and poverty rates respond to changes in workforce location and geographic influences. For several years, the geography of poverty has been shifting from urban to suburban, presenting new transportation-related challenges. The economic gap between high- and low-income working families is growing. In 2011, 46.2 million people were poor—22.5 percent of them working poor—and they comprised 7 percent of the total work force. The working poor spend a much higher portion of their income on commuting and use public transit, carpooling, biking, and walking more frequently than higher income workers do.

Unemployment and Labor Force Participation

Although related to economic factors, workforce composition and distribution affect travel-related trends in different ways. The availability and location of employment have a significant impact on where people live and how they commute to work. Commuting to work constitutes approximately 16 percent of all person trips and 19 percent of all PMT. For roadway travel, commuting comprises 28 percent of household VMT and for transit systems, 39 percent of all transit PMT. Although the average commuting time has remained relatively consistent at 25 minutes, typical commuting patterns are becoming more complex and are likely to involve more trip chains that incorporate stops along the way that are not related to work.

In 2012, the unemployment rate was 8.1 percent. The rate has declined since the December 2007 to June 2009 recession but is still higher than it has been since the 1990s. The rate, however, is expected to continue to decline. Furthermore, after a long-term increase, the overall labor force participation rate has declined in recent years. Although a sharp rise in participation occurred among individuals aged 55 and older, the largest drop was in 16-to 24-year-olds and especially in teenagers. The driving factors for this drop include increased rates of school enrollment, a slower than average labor market recovery, and higher competition for available jobs (from older workers and recent immigrants). Looking forward, the rate of labor force participation is expected to continue decreasing through at least 2022 (see *Exhibit 1-9*).

Exhibit 1-9 Labor Force Participation Rates, 1947–2012 and Projected Rates for 2022[1]

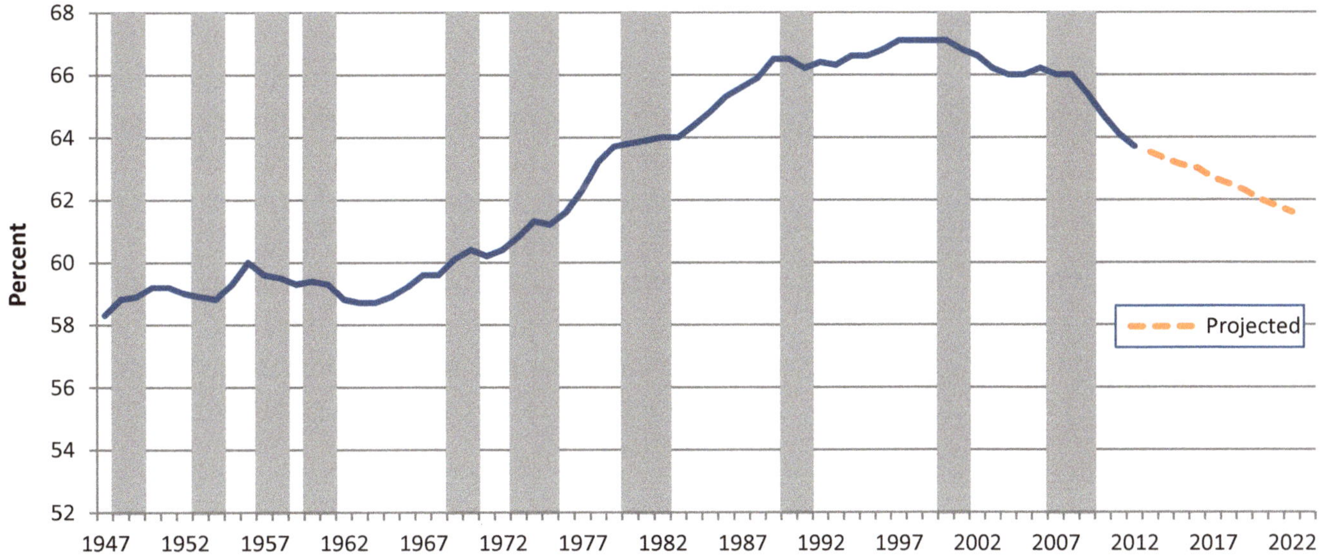

[1] Shaded regions represent recessions as designated by the National Bureau of Economic Research. Turning points are quarterly.

Source: U.S. Bureau of Labor Statistics.

Geography

Where people live, work, and recreate and how they travel are intimately related to their place of residence. For example, living in a dense urban area, where origins and destinations are closer together and public transit is available, slows automobile speeds, increases parking costs, and provides more opportunities to travel by other modes.

The distribution of the population significantly influences the amount of freight, business, and personal travel in a given area. As metropolitan regions continue to expand, maintaining and improving our Nation's transportation system is critical.

Regional Migration

Census projections through 2030 show that population growth will continue to be sharply skewed geographically, with almost half the national growth occurring in three States: Texas, Florida, and California. These three States account for most of the growth that is occurring in the South and West, contributing to a projected 88 percent in growth. In comparison, about half the States in the Nation have shown slow or stagnant growth (see *Exhibit 1-10*). Such a sharply skewed population distribution will make defining an equitable national transportation program difficult to achieve. Additionally, increases in both freight and passenger travel are likely to continue to concentrate in specific geographic areas, adding to system performance and maintenance issues.

Exhibit 1-10 Regional Migration and Growth

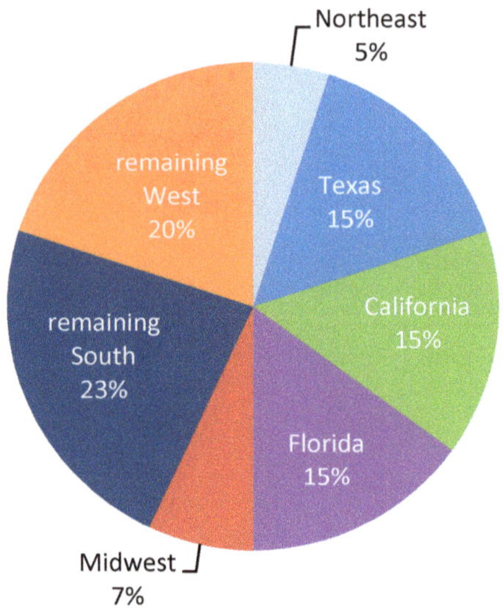

Source: U.S. Census Bureau.

Population change is due to both migration and immigration. Migration has slowed in the United States over the past few years; in fact, the 5-year move rate for 2010 was the lowest in history at 35.4 percent (U.S. Census Bureau, *Geographic Mobility: 2005–2010*). People in their 20s have the highest move rate (65.5 percent), followed by the unemployed (47.7 percent). The South was the only region to show a significant gain due to migration from 2000 to 2010; the Northeast and Midwest lost 2.6 million in population to the southern and western United States.

Knowing the distribution of new immigrants across our Nation's cities and States helps us better understand the transportation needs of people in the geographic locations where immigration levels are very high. New immigrants are still more likely to take up residence in the "Big Six" immigrant magnet States: California, New York, Texas, Florida, Illinois, and New Jersey. These States accounted for 65 percent of the foreign-born population in 2012.

Another notable trend is the dispersion of new immigrant populations to other States, including Georgia, North Carolina, Arizona, Washington, and Virginia. Over the past 5 years, Georgia has experienced a 38-percent increase in its foreign-born population. The State of Washington has experienced a 24-percent increase in the number of foreign-born residents since 2000.

By 2050, the immigrant population in the United States is projected to reach 68 million, 16.2 percent of the total population. High-series projections from the census estimate the immigrant population to reach 114 million in 2050.

Increased Urbanicity

The Bureau of the Census does not provide detailed metropolitan population growth projections, but substantial evidence from the past 100 years, and certainly over the past 50 years, indicates what the direction might be.

Most of the Nation's growth has occurred in suburbs. Today, the suburbs contain more than half the national population. Just over half of young people (ages 20–34) live in suburban areas. Youth are more commonly taking up residence in suburban neighborhoods, particularly in new developments. These developments are often located on the fringes of metropolitan areas, where access to public transportation can also be sparse (see *Exhibit 1-11*). Additionally, low-income families are moving from city centers to more suburban locations. In 1970, suburbs housed 25 percent of the poor, and by 2010, it was 33 percent.

Exhibit 1-11 Percentage Point Change in Young Adult Population (Ages 20–34 Years) by Neighborhood Type, 2000–2010

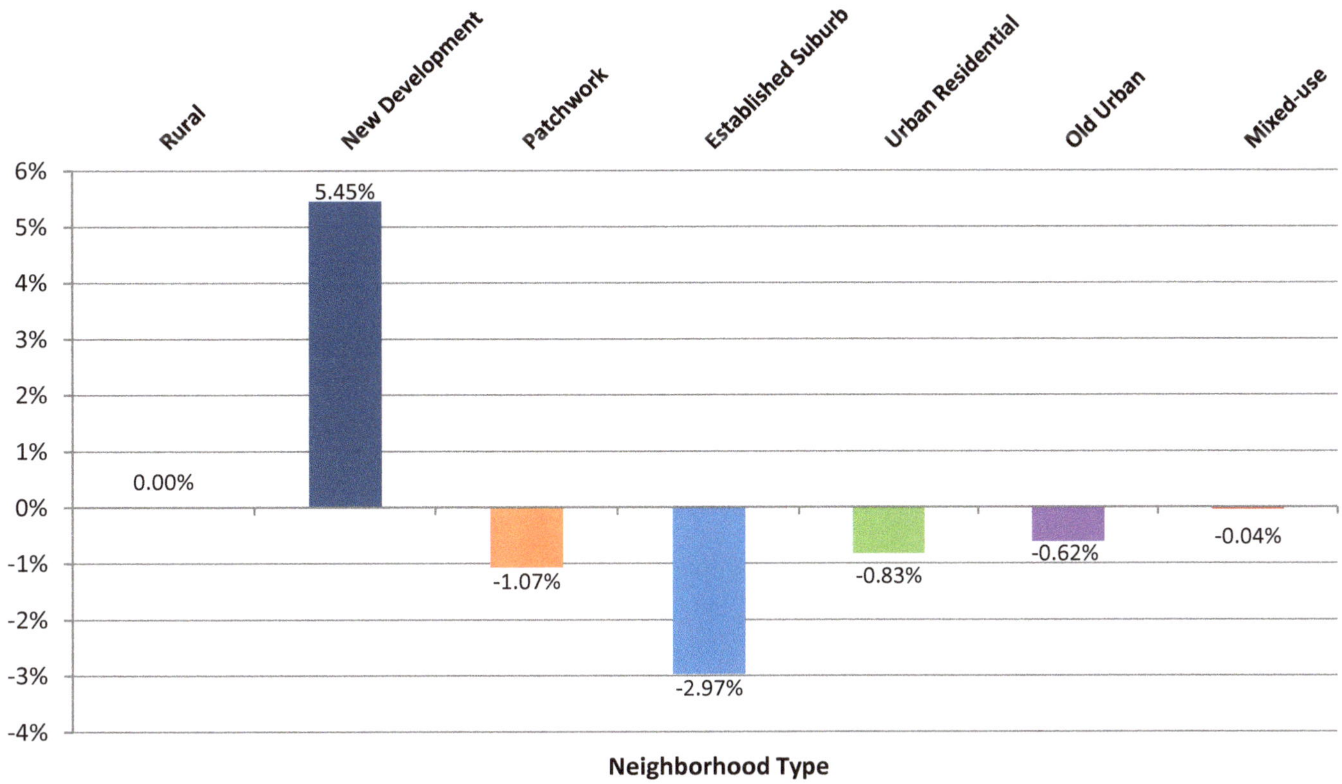

Source: Typecasting Neighborhoods and Travelers: Analyzing the Geography of Travel Behavior Among Teens and Young Adults in the U.S., Institute of Transportation Studies, UCLA Luskin School of Public Affairs, 2015.

A large part of the "suburban" growth has occurred because rural areas are becoming incorporated into neighboring metropolitan areas. Currently, the rural population is declining in terms of both the percentage of U.S. population and in actual population size. More than 40 rural counties were classified as metropolitan in the 2000 census.

Across most communities in the country, travel differences between suburban and urban residential areas are surprisingly small; significant differences occur, however, in rural areas and in old urban neighborhoods. In rural areas, VMT is significantly higher, which reflects the need for residents to travel longer distances to destinations using a vehicle.

In old urban neighborhoods, residents make fewer trips, travel fewer miles, are less likely to have a license, and are more likely to walk or take public transit. Old urban areas tend to have more households without vehicles and better access to transit services, compared to other neighborhoods. Nearly three-fourths of old urban neighborhoods are located in New York and Los Angeles.

Growth in Megaregions

Between 2000 and 2010, urban population growth (12.7 percent) outpaced overall national growth (9.7 percent). More than 50 percent of U.S. population growth through 2050 is projected to occur in the Nation's largest metropolitan areas.

As cities grow in size and population, a network across metropolitan centers is created with environmental, economic, and infrastructural relationships. These "megaregions" often cross county and State lines—linked by both transportation and communication networks. Many experts believe that by 2050, 75 percent of the U.S. population will live in megaregions and more than half the Nation's population growth will occur there.

At present, 11 emerging megaregions have been identified. They include the Northeast, Florida, Piedmont Atlantic, Gulf Coast, Great Lakes, Texas Triangle, Arizona Sun Corridor, Front Range, Cascadia, Northern California, and Southern California.

Megaregions span State and political boundaries, which will require substantial coordination among several entities in planning future transportation systems. Regional economies are likely to become so interwoven that the transportation system will have to endure dramatic increases in both freight and passenger travel. The northeastern corridor, for instance, is expected to experience a 50-percent growth in rail traffic by 2050 (see *Exhibit 1-12*).

Exhibit 1-12 Emerging Megaregions of the United States

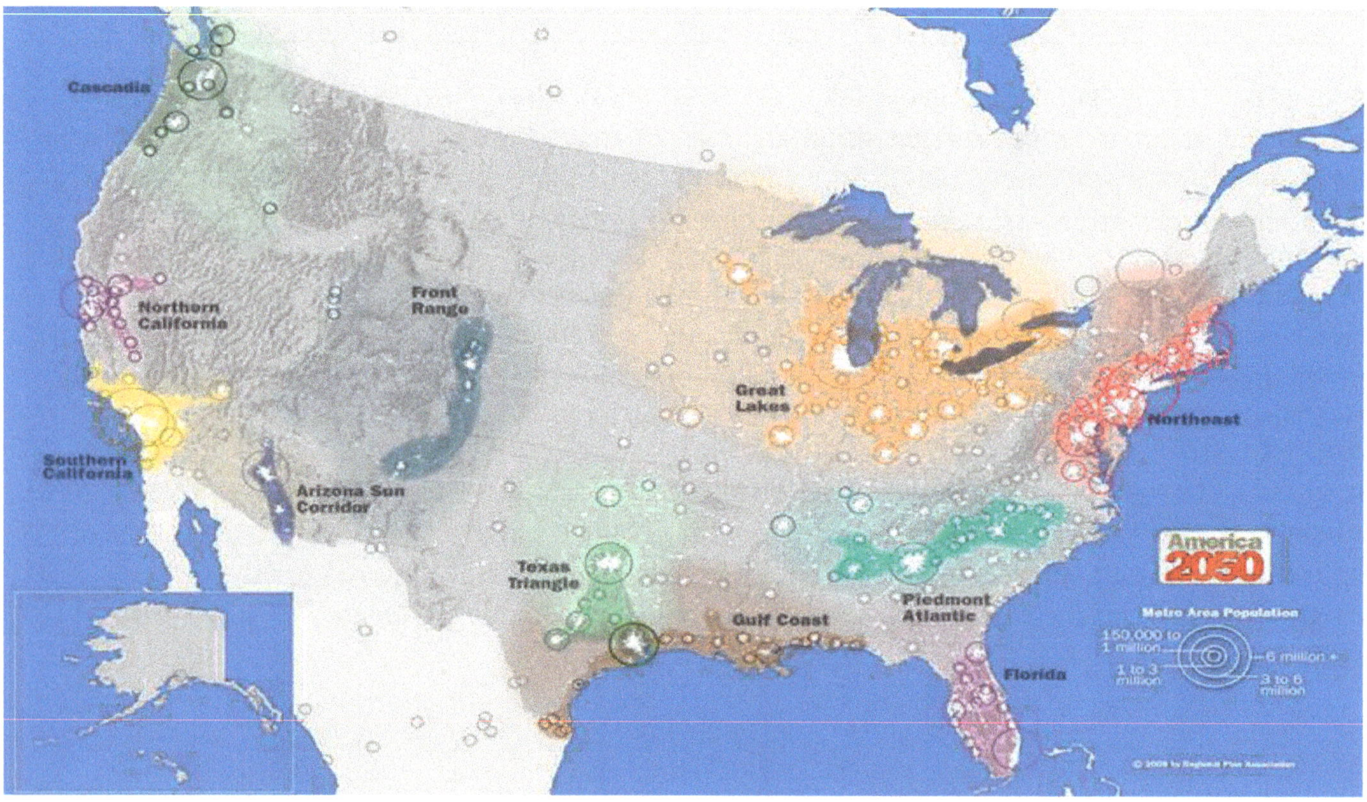

Source: *http://www.america2050.org/maps/.*

Technological Trends and Travel Impacts

Over the past two decades, advancements in Information Communication Technologies have not only changed the way we communicate, but also the way we travel. Our lifestyles and travel decisions are influenced by our need for time. New technologies have enabled us to do things faster, with greater accuracy and more options.

We live in the age of the "informed traveler." Our understanding and use of the transportation system has grown with the availability of travel apps and real-time data. We now know how best to reach our destination, how long the trip will take, and how to pay for it easily, if required. If we need a vehicle, we can find one online and complete the transaction in minutes. We can know ahead of time that traffic is unbearable and change our travel plans to save time.

This section discusses a few technological advancements that have influenced our travel decisions and the way we travel, including GPS use in travel data collection, payment systems, the rise of the sharing economy organizations and businesses, and telecommuting.

Using GPS and Smartphones to Collect Personal Travel Data

Over the past several years, interest has grown in advancing data collection techniques to help reduce survey burden and increase data accuracy. Emerging technologies to collect travel data, such as GPS and smartphones, are becoming more widely used as a means not only to collect location data, but also to obtain additional information about the personal trip-making experience.

Made possible by sensors, smartphones can detect motion, speed, and location via cellular network or wireless fidelity (Wi-Fi), and proximities to nearby objects. Most phones today are equipped with accelerometers that measure linear speeds. The raw data are fed through software programmed with a trip-detection algorithm that can determine whether a person is traveling by car, bike, taking public transit, or walking.

In addition to mode detection, GPS or other network-based services can use location data to identify frequently visited locations, including arrival and departure times. Land use information obtained from GIS maps can be used to determine most frequently traveled routes and details about the transportation system infrastructure, such as the availability of sidewalks.

All these data can be collected passively, requiring nothing from the participant except to carry the smartphone and keep it charged. When more detail is needed about a person's trip-making experience, such as the purpose of the trip, reasons for route selection, or travel party size, a follow-up survey can be programmed into a phone application that prompts the user to answer questions about trips taken.

Since the first FHWA-sponsored study in Lexington, Kentucky in 1996, GPS has been used to collect data on travel behavior in other areas of the country, including Portland, Oregon; Chicago, Illinois; San Francisco, California; Austin, Texas; and South Florida. For the NHTS, several States will be integrating GPS technologies into add-on survey work as part of the national collection.

Integrating GPS and smartphone technologies into travel behavior surveys is quickly becoming a more common practice among metropolitan planning organizations and State Departments of Transportation; however, hurdles to overcome remain, such as the cost of acquiring and distributing smartphones and how best to extend phone battery life during the survey period.

Electronic Payment Systems

Advancements in the area of payment systems have made paying for transportation more convenient. These advancements are important to State Departments of Transportation and local transit agencies that are facing increasing pressures to reduce operating costs and increase revenues, in addition to improving customer convenience and quality of service. Integrated transportation payment systems lead to greater efficiencies, provided these systems are secure, preserve privacy, and do not lead to fraud.

Technologies for integrated transportation payment systems include the use of magnetic stripe cards, "smart cards," and electronic toll collection transponders and systems that enable the user to make a payment electronically. Value can be added to the card or device, and the cost of a trip is deducted for each trip. Where available, the card or device can act as a pass that allows the user unlimited access for a certain period of time, typically a month. The card or device can also contain client information.

The advent of electronic tolling has helped alleviate congestion surrounding toll facilities by allowing for ease of payment. Electronic tolling that involves variable pricing (pricing that is based on the time of day, level of service, or other factors) might shift traffic away from peak hours. A New Jersey study found a small, but statistically significant, shift in car traffic to prepeak hours in the morning (5 a.m. to 6 a.m.) and afternoon (3 p.m. to 4 p.m.), especially among younger drivers and those who come from lower income households.[2] Additionally, traffic could be diverted to alternative non-tolled routes.

Integrated transportation payment systems commonly used in public transportation include magnetic stripe cards and smart cards. Magnetic stripe cards are inexpensive, reliable, and have a high customer acceptance. These cards, which can store value over time, can be used throughout multiple transit networks. Trip origins and destination information can be recorded on the cards.

Smart cards are made of plastic, similar to a credit card and contain microprocessors and memory chips with wireless communication capabilities.[3] Smart cards, sometimes called integrated circuit cards, are similar to magnetic stripe cards but store the information on an embedded microcomputer chip rather than the stripe. Smart cards have been used in a range of applications, including toll and parking payments, Internet access, and mobile commerce.

These technologies make the use of public transportation more convenient, possibly encouraging ridership. Riders no longer need to carry correct change to ride the bus or keep track of multiple train tickets. Increasingly, riders who transfer between modes of public transportation no longer need to carry a transfer pass, as transfer passes are often automatically loaded onto smart cards.

Innovations in smartphone technology and in the financial service industry have provided additional convenience to paying for transportation. Paying for services by mobile phone is already a common practice abroad and is becoming a more frequent method of payment in the United States. The American Public Transportation Association Universal Transit Farecard Standards is one example of a program that actively promotes the Contactless Fare Media System Standard[4] for use in contactless fare systems throughout North America. Additionally, exciting developments are occurring in the contactless payments industry that will simplify the future of transit fare collection. These include using contactless bankcards, mobile devices, and identification credentials to pay for transit fare.

The added convenience that integrated transportation payment systems bring to transportation users and the industry is not the only benefit of using these systems. The wealth of data that can be acquired from trip tracking provides planning agencies and other transportation professionals with the information necessary to understand the needs and value of existing systems more completely.

Sharing Economy

The rise of sharing economy organizations and businesses is having a marked influence on the way people travel. In certain areas, travelers are more often choosing to forgo car ownership and rely on other means of travel, among them, car and ridesharing services.

Increasing use of the Internet and mobile phones and the development of enabling technologies, such as social media platforms, open data sources, and phone applications, have created a market that makes goods and services easier to share and more accessible to a larger audience. Forbes estimates that the revenue flowing through the share economy will surpass $3.5 billion in 2013, with growth exceeding 25 percent.[5]

In the share economy, owners make money from underused assets, or the value of unused time that goods and services remain idle. Peer-to-peer car sharing, for instance, is a service that offers car owners the opportunity to rent out their personal vehicle. Mobile phone applications help facilitate the transaction by making the process quick and easy. Car owners who rent their vehicles using peer-to-peer services such as Relay Rides reportedly can make an average of $250 per month, and some make more than $1000.[6]

Many carsharing organizations have received startup grants and incentives from Federal, State, and local sources. FHWA's CMAQ (Congestion Mitigation and Air Quality) program provides funding to States for eligible activities that reduce VMT or encourage the use of alternative fuels. The State of California, for instance, has enacted legislative initiatives to support carsharing by reducing barriers that owners might face when sharing their vehicles and has worked with local governments to provide exclusive use of onstreet parking for carsharing vehicles.[7]

The continuing growth of the sharing economy could affect personal travel. Studies have shown that carsharing programs have mixed effects on VMT. In some cases, households have slight increases in VMT, but households that lose one or all vehicles show substantial reductions in VMT.

These households learn to adapt to a new travel lifestyle that leads to modal shifts facilitated by car sharing.[8] Carsharing vehicles typically are also newer and more fuel efficient than the average privately owned vehicles, which helps decrease emissions, even if miles are not reduced (see *Exhibit 1-13*).[9]

Ridesharing service companies are connecting drivers to riders and coordinate rides in minutes using mobile phone, GPS technology, and online payment systems. Unlike traditional taxis, these companies boast faster and cheaper service, without the need for hailing.

As the public continues to adapt to new technologies and ways of travel, modal shifts are likely to increase, primarily among nonwork trips (as commute and short trips are typically traveled by walking, biking, and public transit use[10]). The future of the traveling public will be influenced by sharing economy practices and the effect they could have on increased modal options.

Exhibit 1-13 Impact of Carshare Membership on Household Vehicle Miles Traveled, Gasoline Consumption, and Greenhouse Gas Emissions

Study	Study Location	Study Year(s)	Difference in VMT	Difference in Emissions	Difference in Gasoline Consumption
			After vs. before joining carsharing		
Martin and Shaheen, 2011a	Multiple cities	Varies–2008	-26.9%	-34.5%	N/A
Cervero et al., 2007	San Francisco Bay area	2001–2003	Not significant	N/A	-36.5%
		2001–2005	-32.9%	N/A	-59.5%
			Carsharing members (or pre-members) vs. non-members		
Cervero et al., 2007	San Francisco Bay area	2001 (pre-launch)	-33.1%	N/A	-65.1%
		2003	-66.4%	N/A	89.9%
		2005	-68.2%	N/A	90.3%

Source: Boarnet, Handy, Lovejoy, Impacts of Carsharing on Passenger Vehicle Use and Greenhouse Gas Emissions, California Air Resources Board, 2013.

Telecommuting

Someone who telecommutes or teleworks is most often working from home or a place other than their usual worksite. Over the past two decades, advancements in information communication technologies, increased access to broadband services, and changes in workplace policies have made the ability to telework a possibility for those whose jobs are telework eligible.

In 2010, 13.4 million people worked at least one day at home per week, an increase in more than 4 million people (35 percent) in the past decade.[11] In the past three decades, the number of teleworkers has almost tripled, from just 2 million in 1980 to 6 million in 2010.[12] As of 2009, approximately 2.3 percent of the total workforce telecommutes at least one day a week (see *Exhibit 1-14*).

Exhibit 1-14 Teleworking Population, 2005 and 2009

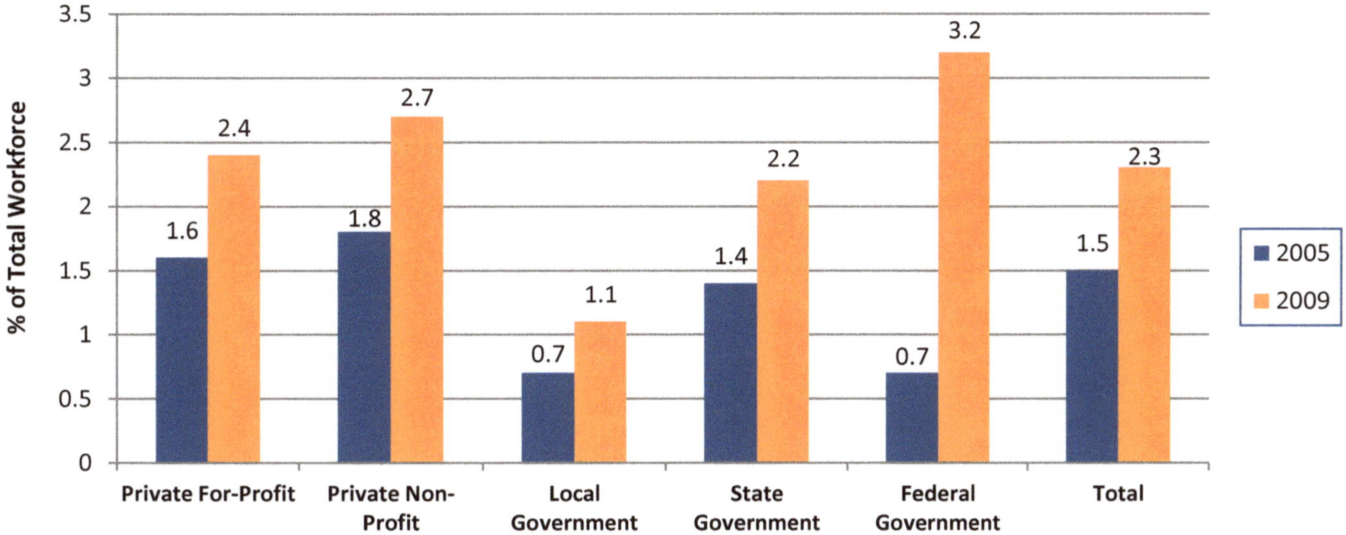

Source: 2005 and 2009 American Community Survey Collected by Telework Research Network.

Telework Relative to Transit

In certain urban areas of the country (primarily in the West and Southwest), teleworkers outnumber transit commuters.

Source: Potential Impacts of Increased Telecommuting on Passenger Travel Demand, National Surface Transportation Policy and Revenue Study Commission, January 2007

Teleworkers are more likely to be self-employed or work in the private sector. Common occupations associated with home-based telework include jobs in the business and finance fields. The number of telework-eligible positions in the computer, scientific, and engineering fields is growing. More than 50 percent of teleworkers have a bachelor's degree or higher. The largest growth in teleworking by census region in the United States has occurred in the South and West, where overall worker growth was also greater (see *Exhibit 1-15).*

The opportunity to telework benefits workers and companies alike. It adds flexibility to the workday, which helps families and individuals better manage the responsibilities of daily life without being confined to a workplace location. It also saves workers time from the daily commute, which for some, might add up to several hours a week. Companies and public agencies have found that giving employees opportunities to telework can increase productivity and increase employee retention.[13] Teleworking has become an important part of workplace efforts to help employees balance work/life issues.

Exhibit 1-15 Top 10 U.S. Metropolitan Statistical Areas Ranked by Percentage of Workers 16 Years Old and Older Who Worked from Home

Rank	Metropolitan Statistical Area	Percent
1	Boulder, CO	10.9
2	Medford, OR	8.4
3	Santa Fe, NM	8.3
4	Kingston, NY	8.1
5	Santa Rosa-Petaluma, CA	7.9
6	Mankato-North Mankato, MN	7.7
7	Prescott, AZ	7.6
8	St. Cloud, MN	7.6
9	Athens-Clarke County, GA	7.5
10	Austin-Round Rock-San Marcos, TX	7.3

Source: American Community Survey, 2010.

From the transportation perspective, telecommuting has become a component of many

Transportation Demand Management (TDM) programs within State Departments of Transportation and metropolitan planning organizations to relieve congestion at the local/regional level during commute times. In urbanized areas, it can improve air quality by helping reduce emissions associated with traffic congestion.

Federal dollars from the CMAQ program can be used by State Departments of Transportation and metropolitan planning organizations to support telework programs that help reduce emissions. Private institutions have also provided funding to implement telework programs as part of local initiatives to reduce congestion and improve air quality. The Clean Air Campaign in Atlanta, Georgia,[14] for example, has trained thousands of teleworkers and has worked with almost 300 companies to institute telework policies.

On a nationwide scale, the impact of telecommuting on total congestion is difficult to evaluate due to a myriad of factors associated with personal travel decisions. Some might argue that, although telecommuting can reduce peak-hour trip making or VMT, it has no effect on total trip making or total VMT,[2] as teleworkers could travel to other destinations throughout the workday. Various studies have shown, however, that increases in trip making by teleworkers are related more so to individual differences in workers, and are not a result of the act of teleworking. For instance, high-income teleworkers still show more trip making during the workday than low-income teleworkers.[15] Because of the diversity of individuals who make up the workforce and differences in land uses and transportation systems from place to place, the success of telework programs in reducing congestion are best evaluated at the local or regional level.

Looking Forward

New technologies will continue to affect how people travel by increasing our knowledge of the personal trip experience and through increased system efficiencies that influence how we use our time. Changes to the transportation system are inevitable, as vehicle automation features to improve safety and trip reliability continue to gain a predominant place in the car market. In the future, travelers might no longer need to think as much about how to get there, but what to do along the way.

---

[1] Blumenberg, Evelyn, Moving In and Moving Around: *Immigrants, Travel Behavior, and Implications for Transport Policy*, Institute of Transportation Studies, UCLA School of Public Affairs, 2008.

[2] Holguin-Veras, J., K. Ozbat, and A.de Cerreño. *Evaluation Study of Port Authority of New York and New Jersey's Time of Day Pricing*. March 2005. NJ Department of Transportation. FHWA/NJ-2005-005.

[3] Multisystems, Inc. *Fare Policies, Structures, and Technologies: Update.* 2003. TCRP Report 94. http://onlinepubs.trb.org/onlinepubs/tcrp/tcrp_rpt_94.pdf.

[4] This standard defines the data format used on Proximity Integrated Circuit Cards (PICC). This part provides a consistent and uniform method for storing, retrieving, and updating data from contactless fare media used in transit applications. It also references related international standards that define the physical, electrical, and communications aspects of PICCs. Source: APTA. "Universal Transit Fare System Task Force." APTA Standards Development Program Website. Available at: http://www.aptastandards.com/StandardsPrograms/UTFSStandards/TaskForce/tabid/82/language/en-US/Default.aspx, as of 27 July 2011.

[5] Geron, T. *Airbnb and the Unstoppable Rise of the Share Economy*. Forbes. January 23, 2013.

[6] *The rise of the sharing economy*, http://www.economist.com/news/leaders/21573104-internet-everything-hire-rise-sharing-economy, March 9, 2013.

[7] California, Assembly Bill 1871, 2011.
[8] Martin, E.W., and S. Shaheen, *Greenhouse Gas Emissions Impacts of Carsharing in North America*, Mineta Transportation Institute, 2011.
[9] Boarnet, M.G., S. Handy, and K. Lovejoy, *Impacts of Carsharing on Passenger Vehicle Use and Greenhouse Gas Emissions*, California Air Resources Board, 2013.
[10] Martin, E.W., and S. Shaheen, *Greenhouse Gas Emissions Impacts of Carsharing in North America*, Mineta Transportation Institute, 2011.
[11] U.S. Census Bureau, *Journey to Work*, 2010
[12] *Commuting in America 2013: The National Report on Commuting Patterns and Trends*, Brief 6. Job Dynamics, AASHTO, January 2015
[13] *Commuting in America 2013: The National Report on Commuting Patterns and Trends*, Brief 6. Job Dynamics, AASHTO, January 2015, pg. 5.
[14] http://www.cleanaircampaign.org/About-Us.
[15] Sylvia Y. He, Lingqian Hu, *Telecommuting, Income and Out-of-home Activities*, Travel Behaviour and Society, 2015

Freight Movement

The economy of the United States depends on freight transportation to link businesses with suppliers and markets throughout the Nation and the world. Freight affects nearly every American business and household in some way. American farms and mines use inexpensive transportation to compete against their counterparts around the world. Domestic manufacturers rely on remote sources of raw materials to produce goods. Wholesalers and retailers depend on fast and reliable transportation to obtain inexpensive or specialized goods. In the expanding world of e-commerce, households and small businesses increasingly depend on freight transportation to deliver purchases directly to them. Service providers, public utilities, construction companies, and government agencies rely on freight transportation to obtain needed equipment and supplies from distant sources.

The U.S. economy requires effective freight transportation to operate at minimum cost and respond quickly to demands for goods. As the economy grows over the next several decades, the demand for goods and the volume of freight transportation activity will increase. Current volumes of freight are straining the capacity of the transportation system to deliver goods quickly, reliably, and cheaply. Anticipated growth of freight could overwhelm the system's ability to meet the needs of the American economy unless public agencies and private industry work together to improve the system's performance.

Freight Transportation System

The Bureau of Transportation Statistics' (BTS) publication, *Freight Facts and Figures 2015*, shows the U.S. freight transportation system handled a record amount of freight in 2012. About 54 million tons of freight worth more than $48 billion was transported each day across all modes of transportation to meet the logistical needs of the Nation's 118.7 million households, 7.4 million business establishments, and 89,004 government units. This system includes nearly 10.7 million single-unit and combination trucks, more than 1.3 million locomotives and rail cars, and over 40,000 marine vessels. The system operates on almost 450,000 miles of Interstate, other limited-access, and arterial highways; nearly 140,000 miles of railroads; 11,000 miles of inland waterways and the Great Lakes-St. Lawrence Seaway system; and almost 1.75 million miles of petroleum and natural gas pipelines. The U.S. Army Corps of Engineers' *Waterborne Commerce of the United States 2012* identifies 133 ports that handle more than 1 million tons of freight per year.

The freight transportation system is more than equipment and facilities. As reported in *Freight Facts and Figures 2013*, freight employment at for-hire transportation establishments currently is over 4.4 million workers in the United States. Truck transportation businesses comprise the single largest freight transportation occupation, employing more than 1.3 million workers. Other freight transportation and freight transportation-related occupations include rail and water vehicle operations, pipeline operations, equipment manufacturing, infrastructure construction and maintenance, and secondary support services.

Freight Transportation Demand

Freight movements in the United States take a variety of forms, from the shipment of farm products across town to the shipment of electronic devices across the world. These goods move to, from, and within the United States via the Nation's highways, railroads, waterways, airplanes, and pipelines, sometimes using a combination of two or more of the aforementioned modes to complete the trip. Due to the country's well-developed roadway network and the transport connectivity and flexibility the network provides, most freight moved to, from, and within the United States is transported by truck. *Exhibit 1-16* shows a breakdown of freight movements by mode, measured by both tonnage and value of shipment.

Exhibit 1-16 Goods Movement by Mode, 2012

Mode	Tons (Millions)	Percentage	Value (Billions of Dollars)	Percentage
Truck	13,182	67.0%	11,130	64.1%
Rail	2,018	10.3%	551	3.2%
Water	975	5.0%	339	2.0%
Air; Air and Truck	15	<0.1%	1,182	6.8%
Multiple Modes and Mail	1,588	8.1%	3,023	17.4%
Pipeline	1,546	7.9%	768	4.4%
Other/Unknown	338	1.7%	359	2.1%
Total[1]	**19,662**	**100%**	**17,352**	**100%**

[1] Numbers may not add to totals due to rounding. The data are provisional estimates that are based on selected modal and economic trend data. All truck, rail, water, and pipeline movements that involve more than one mode, including exports and imports that change mode at international gateways, are included in multiple modes and mail to avoid double counting. As a consequence, rail and water totals in this table are less than other published sources. In addition, it should be noted that raw tonnage statistics does not take into account the distance these goods were moved. To use one example, a shipment, such as a shipping container, that is transported 2 miles by truck and 2,000 miles by rail is treated the same when measured by tonnage.

Sources: U.S. Department of Transportation, Federal Highway Administration, Office of Freight Management and Operations, Freight Analysis Framework, version 3.4, 2014.

Exhibit 1-17 shows a map containing the tonnage information presented in the table in *Exhibit 1-16* for truck, rail, and inland water shipments, plotted to the U.S. freight transportation network. *Exhibit 1-18* shows the same information as in *Exhibit 1-17*, but includes only long-haul truck shipments on the National Highway System (NHS).

Much of the freight moved on the U.S. transportation system is transported by for-hire carriers—third-party carriers that serve a variety of customers. The Bureau of Transportation Statistics' Freight Transportation Services Index measures the output of services provided by for-hire transportation industries. According to the Bureau, this freight index correlates strongly with U.S. economic activity and helps illustrate the relationship between freight transportation and long-term changes in the U.S. economy. *Exhibit 1-19* shows the annual Freight Transportation Services Index figures for recent years.

Exhibit 1-17 Tonnage on Highways, Railroads, and Waterways, 2010

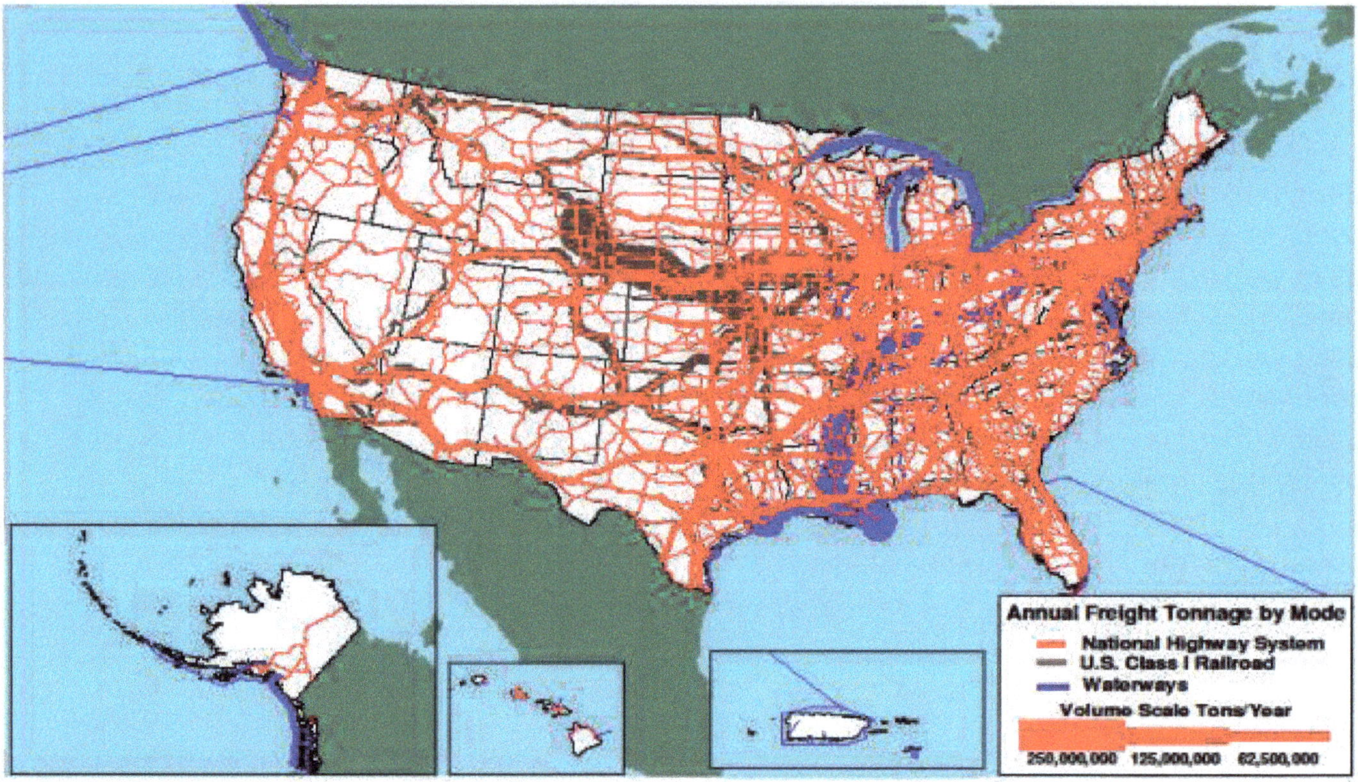

Sources: Highways—Federal Highway Administration, Freight Analysis Framework, Version 3.4, 2013; Rail—Surface Transportation Board, Annual Carload Waybill Sample, Federal Railroad Administration, rail freight flow assignments (2013); Waterways—U.S. Army Corps of Engineers (USACE), Annual Vessel Operating Activity, Tennessee Valley Authority, Lock Performance Monitoring System data for USACE, USACE Institute for Water Resources, Waterborne Foreign Trade Data, USACE water flow assignments (2013).

Freight Statistics

Many of the freight statistics in this section are derived from the Freight Analysis Framework (FAF) version 3 (FAF^3). FAF includes all freight flows to, from, and within the United States. FAF estimates are recalibrated every 5 years, primarily with data from the Commodity Flow Survey, and are updated annually with provisional estimates. The Commodity Flow Survey, conducted every 5 years by the Census Bureau and DOT's Bureau of Transportation Statistics, measures approximately two-thirds of the tonnage covered by the FAF. FAF^3 incorporates data from the 2007 Commodity Flow Survey.

Statistics on trucking activity are primarily from FHWA's Highway Performance Monitoring System and the Census Bureau's Vehicle Inventory and Use Survey. This survey links truck size and weight, miles traveled, energy consumed, economic activity served, commodities carried, and other characteristics of significant public interest, but was discontinued after 2002. See www.ops.fhwa.dot.gov/freight/freight_analysis/faf for additional information. Efforts are underway to restart the Vehicle Inventory and Use Survey collection.

Freight movements are expected to increase over the next few decades as U.S. and global populations grow and consumer spending power increases both nationally and globally. More people and greater spending power will boost the production and consumption demand for many types of goods. All freight transportation modes are expected to experience increased volumes, although the amount of expected growth will vary from mode to mode, as *Exhibit 1-20* shows.

Exhibit 1-18 Average Daily Long-Haul Freight Truck Traffic on the National Highway System, 2011[1]

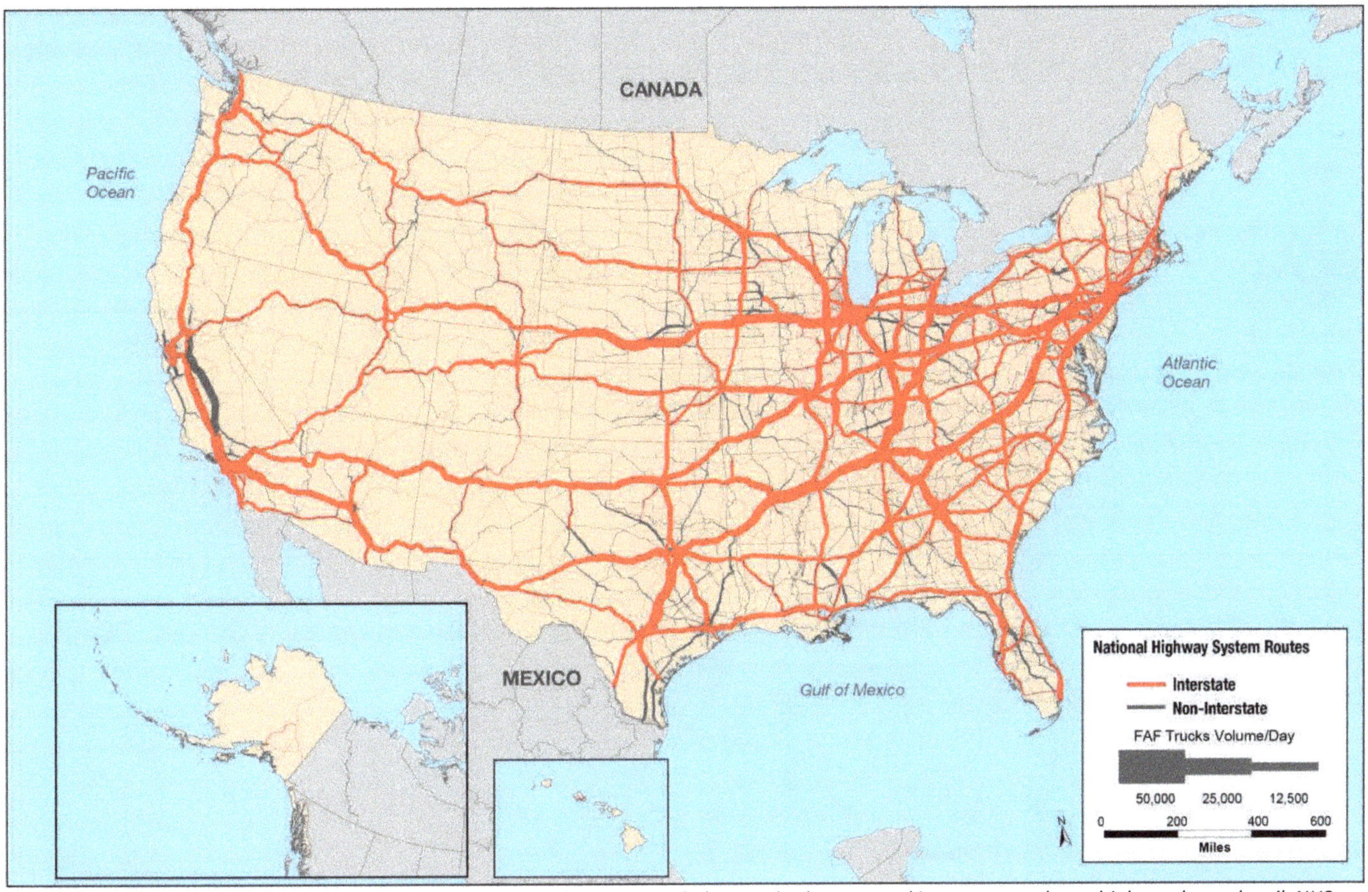

[1] *Long-haul freight trucks typically serve locations at least 50 miles apart, excluding trucks that are used in movements by multiple modes and mail. NHS mileage as of 2011, prior to MAP-21 system expansion.*

Sources: U.S. Department of Transportation, Federal Highway Administration, Office of Freight Management and Operations, Freight Analysis Framework, Version 3.4, 2013.

Even though the annual volume increases are modest for all modes, the cumulative increase over 30 years for each mode is significant. This increased volume will strain the entire freight transportation network, most notably the highway network. *Exhibit 1-21* displays a map of the 2040 truck tonnage information shown in *Exhibit 1-20* plotted on the NHS.

Truck volume on many key truck routes of the NHS is expected to increase significantly between 2012 and 2040. These projected traffic increases would have major implications for highway congestion and freight movement efficiency, especially near large urban areas along or near major truck corridors.

Exhibit 1-19 Annual Freight Transportation Services Index Values, 2000–2014

Year	Freight TSI[1]
2000	100
2005	112.4
2006	111.5
2007	110.1
2008	108.8
2009	98.3
2010	106.4
2011	110.8
2012	112.1
2013	116.2
2014	120.4

[1] The TSI is indexed such that the Year 2000 TSI equals 100.0.

Sources: U.S. Department of Transportation, Office of the Assistant Secretary for Research and Technology, Bureau of Transportation Statistics.

Exhibit 1-20 Weight of Shipments by Transportation Mode[1]

Mode	Weight of Shipments (Millions of Tons) 2007	2012	2040 Projected	Compound Annual Growth, 2010–2040
Truck	12,778	13,182	18,786	1.3%
Rail	1,900	2,018	2,770	1.1%
Water	950	975	1,070	0.3%
Air; Air and Truck	13	15	53	4.6%
Multiple Modes and Mail[2]	1,429	1,588	3,575	2.9%
Pipeline	1,493	1,546	1,740	0.4%
Other/Unknown	316	338	526	1.6%
Total	**18,879**	**19,662**	**28,520**	**1.3%**

[1] Data do not include imports and exports that pass through the United States from a foreign origin to a foreign destination by any mode. Numbers may not add to total due to rounding.

[2] In this table, Multiple Modes and Mail includes export and import shipments that move domestically by a different mode than the mode used between the port and foreign location.

Sources: U.S. Department of Transportation, Federal Highway Administration, Office of Freight Management and Operations, Freight Analysis Framework, version 3.4, 2014.

Exhibit 1-21 Average Daily Long-Haul Freight Truck Traffic on the National Highway System, 2040[1]

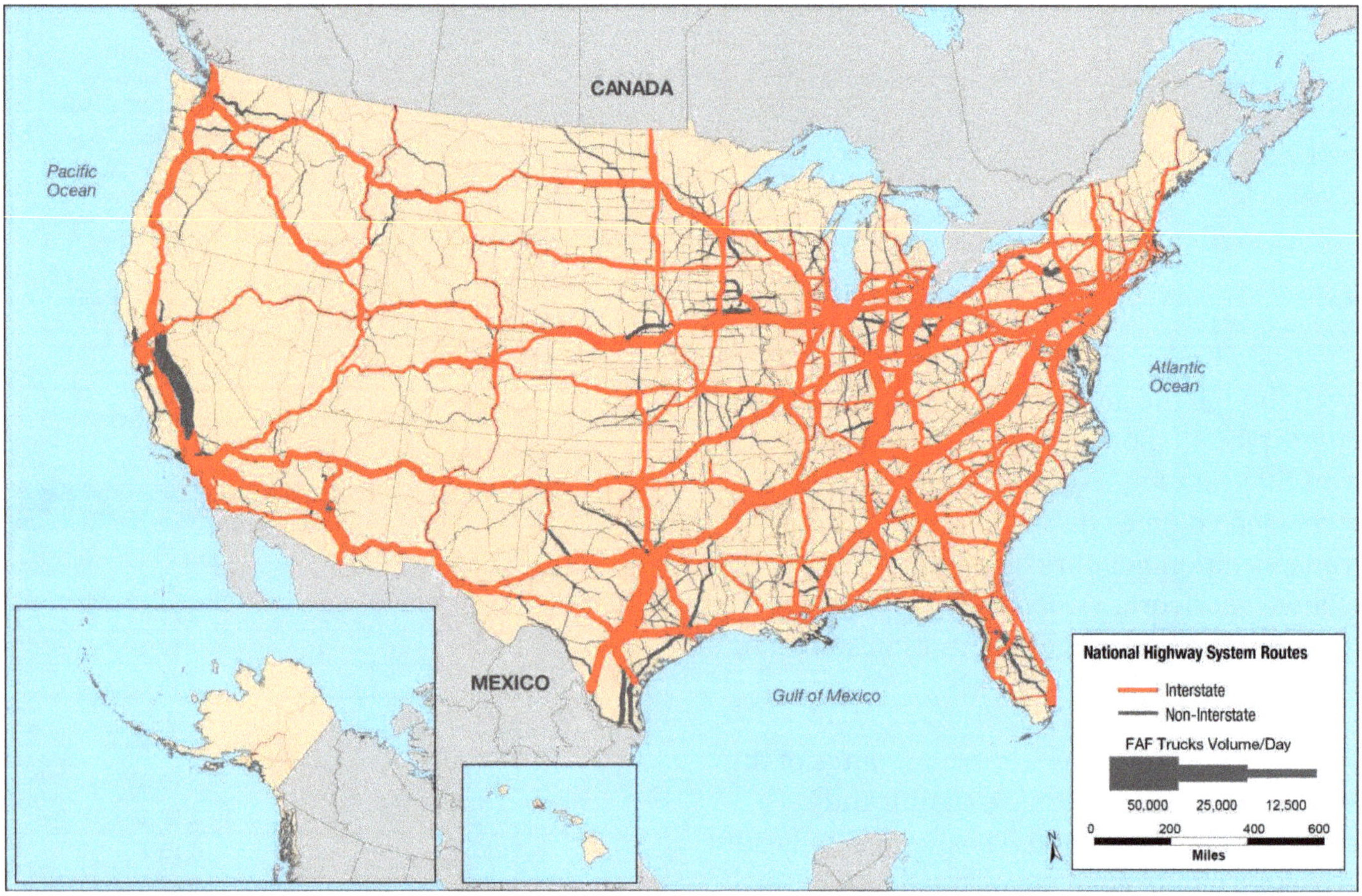

[1] Long-haul freight trucks typically serve locations at least 50 miles apart, excluding trucks that are used in movements by multiple modes and mail. NHS mileage as of 2011, prior to MAP-21 system expansion.

Sources: U.S. Department of Transportation, Federal Highway Administration, Office of Freight Management and Operations, Freight Analysis Framework, Version 3.4, 2013.

The differing freight volume and freight growth characteristics of the various freight transportation modes is related in large part to each mode's operating characteristics. These operating characteristics are key to determining how certain types of goods are transported. The routes, facilities, volumes, and service demands differ between higher-value, time-sensitive goods moving at high velocities and lower-value, cost-sensitive goods moving in bulk shipments, as shown in *Exhibit 1-22*.

Exhibit 1-22 The Spectrum of Freight Moved, 2007

	Commodity Type	
Parameter	**High Value/Time Sensitive**	**Bulk**
Top Three Commodity Classes	Machinery	Gravel
	Electronics	Cereal Grains
	Motorized Vehicles	Non-metallic mineral products
Share of Total Tons	13%	65%
Share of Total Value	58%	16%
Key Performance Variables	Reliability	Reliability
	Speed	Cost
	Flexibility	
Share of Tons by Domestic Mode	87% Truck	71% Truck
	5% Multiple Modes and Mail	12% Rail
	4% Rail	9% Pipeline
		4% Multiple Modes and Mail
		3% Water
Share of Value by Domestic Mode	70% Truck	71% Truck
	16% Multiple Modes and Mail	12% Pipeline
	10% Air	7% Multiple Modes and Mail
	2% Rail	6% Rail
		2% Water

Sources: U.S. Department of Transportation, Federal Highway Administration and Bureau of Transportation Statistics, Freight Analysis Framework, version 3.6, 2015.

Although trucking typically is considered a "faster" mode and handles a very high volume of high-value, time-sensitive goods, it also handles a significant share of lower-valued bulk tonnage. This share includes movement of agricultural products from farms, local distribution of gasoline, and pickup of municipal solid waste that cannot be handled readily by other transportation modes. The length of haul for activities such as these is typically very short.

Most trucking activity involves moving freight, and truck movements are a significant component of overall highway traffic. Three-fourths of vehicle miles traveled (VMT) by trucks larger than pickups and vans involves carrying freight, which encompasses a wide variety of products ranging from electronics to sand and gravel. Much of the rest of the large-truck VMT comprises empty backhauls of truck trailers or shipping containers. An increasing number of highways are carrying both a high volume and a high percentage of trucks. In 2011, for example, single-unit and combination trucks comprised more than 25 percent of the total average annual daily traffic on over 14,500 miles of NHS routes. On average, those routes accommodated at least 8,500 trucks per day.

Exhibit 1-23 presents a map identifying those major truck routes on the NHS, showing the routes that handle more than 8,500 trucks per day or experience daily traffic composed of at least 25 percent truck traffic.

Exhibit 1-23 Major Truck Routes on the National Highway System, 2011

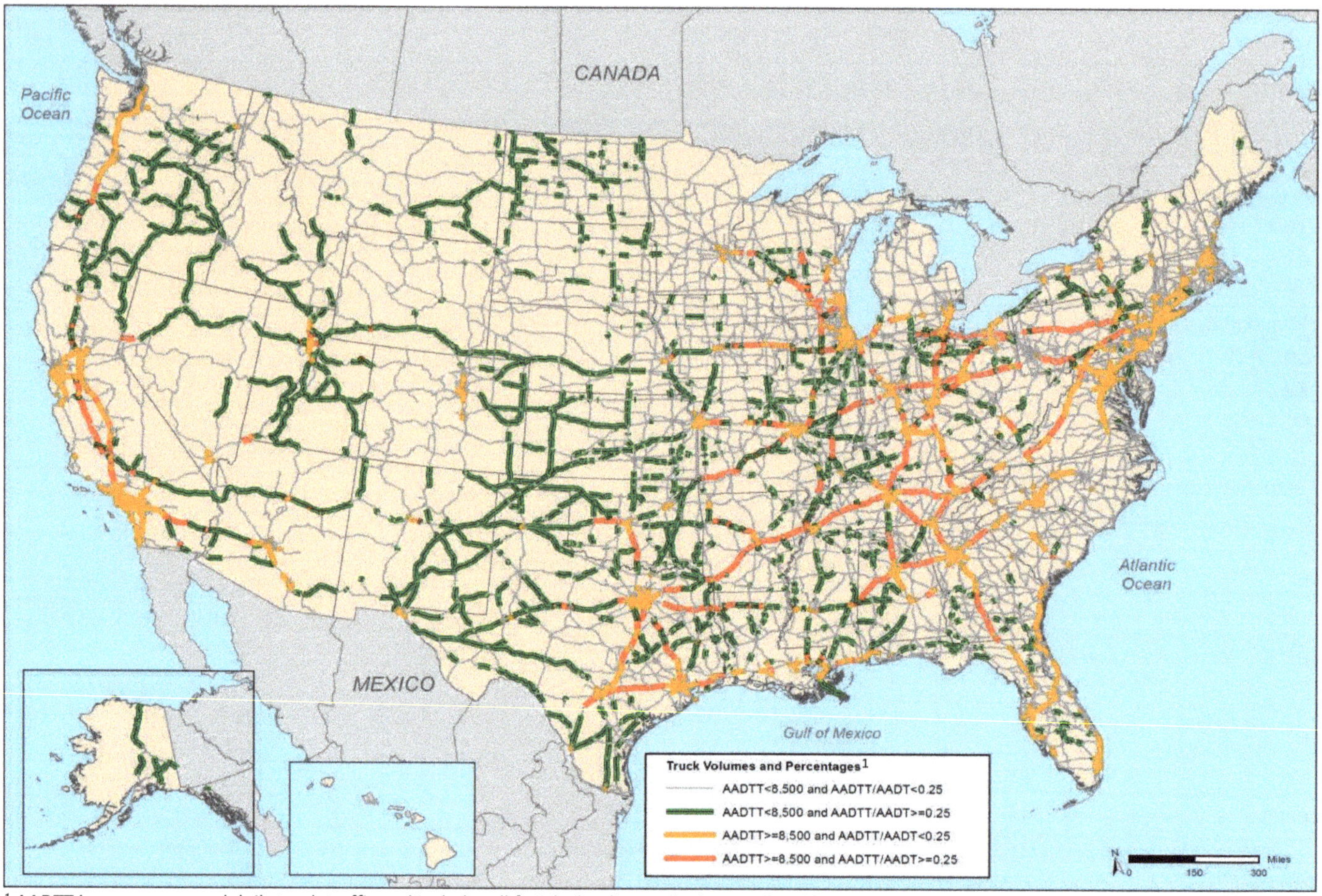

[1] AADTT is average annual daily truck traffic and includes all freight-hauling and other trucks with six or more tires. AADT is average annual daily traffic and includes all motor vehicles. NHS mileage as of 2011, prior to MAP-21 system expansion.

Sources: U.S. Department of Transportation, Federal Highway Administration, Office of Freight Management and Operations, Freight Analysis Framework, Version 3.4, 2013.

Although many freight movements are long-distance shipments to domestic or international locations, a larger percentage of shipments, particularly those by truck, are transported shorter distances. Approximately half of all trucks larger than pickups and vans operate locally—within 50 miles of home—and these short-haul trucks account for about 30 percent of truck VMT. By contrast, only 10 percent of trucks larger than pickups and vans operate more than 200 miles away from home, but these trucks account for more than 30 percent of truck VMT. Long-distance truck travel also accounts for nearly all freight ton miles and a large share of truck VMT. More information is shown in *Exhibit 1-24*.

Exhibit 1-24 Trucks and Truck Miles by Range of Operations[1]

Location	Number of Trucks (percent)	Truck Miles (percent)
Off the Road	3.3%	1.6%
50 Miles or Less	53.3%	29.3%
51 to 100 Miles	12.4%	13.2%
101 to 200 Miles	4.4%	8.1%
201 to 500 Miles	4.2%	12.1%
501 Miles or More	5.3%	18.4%
Not Reported	13.0%	17.3%
Not Applicable	4.1%	0.1%
Total	**100%**	**100%**

[1] Includes trucks registered to companies and individuals in the United States except pickups, minivans, other light vans, and sport utility vehicles. Numbers may not add to total due to rounding.

Sources: U.S. Department of Commerce, Census Bureau, 2002 Vehicle Inventory and Use Survey: United States, EC02TV-US, Table 3a (Washington, DC: 2004), available at http://www.census.gov/prod/ec02/ec02tv-us.pdf as of March 13, 2015.

Many U.S. freight movements are part of international trade between the United States and other countries. Canada and Mexico, which according to the U.S. Census Bureau are the United States' largest and third-largest trading partners, respectively, account for a significant portion of these international freight movements, including all freight movements on land surface modes. *Exhibits 1-25* and *1-26* show U.S.-Canada trade volumes by value and tonnage, respectively, for trucks, railroads, and all modes combined (including non-land surface modes).

Exhibits 1-27 and *1-28* show U.S.-Mexico trade volumes by value and tonnage, respectively, for trucks, railroads, and all modes combined (including nonland surface modes).

Exhibit 1-25 Total U.S.-Canada Trade Value by Transportation Mode, 2000–2014[1]

Year	Total Trade Value (Millions of U.S. Dollars)		
	Truck	Rail	All Modes
2000	$257,642	$62,646	$409,779
2005	$294,917	$79,928	$499,291
2006	$314,202	$85,736	$533,673
2007	$324,747	$91,459	$561,548
2008	$319,946	$93,194	$596,470
2009	$247,757	$61,032	$429,587
2010	$299,886	$82,999	$526,893
2011	$334,012	$94,797	$596,616
2012	$344,919	$103,050	$616,913
2013	$348,332	$105,409	$634,162
2014	$353,955	$104,155	$658,188

[1] The monetary values shown are not adjusted for inflation.

Sources: U.S. Department of Transportation, Bureau of Transportation Statistics, North American Transborder Freight Data, available at www.bts.gov/transborder as of April 10, 2015.

Exhibit 1-26 Total U.S.-Canada Trade Tonnage by Transportation Mode, 2000–2013

Year	Total Trade Tonnage (Thousands of Metric Tons)		
	Truck	Rail	All Modes
2000	N/A	N/A	364,230.00
2005	133,679.40	98,775.90	414,328.40
2006	130,752.80	102,453.70	420,589.40
2007	116,995.90	105,099.80	414,405.50
2008	110,337.00	98,011.60	406,014.30
2009	92,542.00	72,107.00	333,343.30
2010	104,138.60	87,933.70	371,862.20
2011	106,410.30	91,875.90	387,757.20
2012	107,216.10	97,625.80	409,211.30
2013	108,764.80	104,701.30	426,797.60

Sources: U.S. Department of Transportation, Bureau of Transportation Statistics, North American Transborder Freight Data, available at www.bts.gov/transborder as of April 10, 2015.

Exhibit 1-27 Total U.S.-Mexico Trade Value by Transportation Mode, 2000–2014[1]

Year	Total Trade Value (Millions of U.S. Dollars)		
	Truck	Rail	All Modes
2000	$171,058	$31,552	$247,275
2005	$195,609	$36,530	$290,247
2006	$219,455	$43,135	$332,426
2007	$230,084	$46,400	$347,340
2008	$234,488	$47,230	$367,453
2009	$207,195	$34,591	$305,525
2010	$260,331	$48,144	$393,650
2011	$295,522	$57,270	$461,162
2012	$323,170	$64,399	$493,500
2013	$335,351	$69,851	$506,608
2014	$360,668	$73,690	$534,484

[1] The monetary values shown are not adjusted for inflation.

Sources: U.S. Department of Transportation, Bureau of Transportation Statistics, North American Transborder Freight Data, available at www.bts.gov/transborder as of April 10, 2015.

Exhibit 1-28 Total U.S.-Mexico Trade Tonnage by Transportation Mode, 2000–2013

Year	Total Trade Tonnage (Thousands of Metric Tons)		
	Truck	Rail	All Modes
2000	N/A	N/A	161,888.00
2005	47,630.90	17,369.00	190,116.20
2006	49,254.90	17,879.40	195,741.40
2007	56,918.80	35,060.10	212,331.70
2008	54,944.10	35,801.30	200,337.10
2009	48,254.60	26,251.60	172,558.20
2010	65,703.40	33,762.30	214,598.30
2011	82,115.70	36,980.50	242,456.30
2012	70,736.10	41,889.00	228,823.70
2013	69,426.30	38,446.90	222,606.10

Sources: U.S. Department of Transportation, Bureau of Transportation Statistics, North American Transborder Freight Data, available at www.bts.gov/transborder as of April 10, 2015.

Freight Challenges

The challenges of moving the Nation's freight cheaply and reliably on an increasingly constrained infrastructure without affecting safety or degrading the environment are substantial, and traditional strategies to support passenger travel might not apply. The freight transportation challenge differs from that of urban commuting and other passenger travel in several ways:

- Freight often moves long distances through localities and responds to distant economic demands, while most passenger travel occurs locally. Freight movement often creates local problems without local benefits. Local residents and elected officials are also less likely to have direct experience in freight transportation operations, making it more difficult for such improvements to be seen as a priority.
- Freight movement fluctuates more, and more quickly, than passenger travel does. Although both passenger travel and freight respond to long-term demographic changes, freight responds more quickly than passenger travel to short-term economic fluctuations. Fluctuations can be national or local. The addition or loss of even a single major business can dramatically change the level of freight activity in a locality.
- Freight movement is heterogeneous compared with passenger travel. Patterns of passenger travel tend to be similar across metropolitan areas and among large economic and social strata. Freight transportation demands differ radically in terms of the types of freight vehicles used and the locations they serve. For example, farms, mines, manufacturing plants, commercial retail shopping centers, grocery stores, and online retail sales all have significantly different locations, shipment frequencies, and general shipment needs. These differences occur not only *between* freight transportation modes but also *within* freight transportation modes. As

one example, the operating characteristics of long-haul tractor-trailers serving one location per shipment load distinctly differ from those of shorter-haul tractor-trailers and large single-unit trucks serving multiple locations per shipment load. Both are distinctly different from parcel carriers that use smaller, single-unit trucks and serve many locations per shipment load. Solutions aimed at "average" conditions are less likely to succeed because the freight demands of economic sectors vary widely.

- To the extent that freight movement is concentrated in different corridors or locations than passenger travel, transportation system investments targeted solely at improving general traffic conditions may be less likely to specifically aid the flow of freight.
- The reliable movement of freight depends on all modes working together such that the multimodal freight system functions smoothly and without costly delays. Bottlenecks on one mode of transportation can affect the performance of freight throughout the network.

Local public action to support the economic benefits of freight transportation is difficult to marshal because freight traffic and the benefits of serving that traffic rarely stay within a single political jurisdiction. One-half the weight and two-thirds the value of all freight movements cross a State or international boundary. Additionally, locations desirable from a developer's standpoint for industrial and commercial development are often highly sensitive to non-transportation considerations such as local zoning, tax rates, and development incentives. Such considerations can pit adjacent municipalities or counties against one another and undermine comprehensive freight transportation planning efforts. Federal legislation established metropolitan planning organizations in the 1960s to coordinate transportation planning and investment across State and local lines within urban areas. Both the interregional nature of many freight movements and the varying levels of support or opposition in local jurisdictions for freight-generating development, however, complicate the metropolitan area transportation planning process. Creative and ad hoc arrangements often are required through pooled-fund studies and multi-State coalitions to plan and invest in freight corridors that span regions and even the continent, but few institutional arrangements coordinate this activity. One example of a more established multi-State arrangement is the I-95 Corridor Coalition. Additional information about this coalition and similar groups can be found at www.ops.fhwa.dot.gov/freight/corridor_coal.htm.

The growing needs of freight transportation can bring into focus conflicts between national, State, and local interests. Many longer-haul truck, train, and other freight movements create negative impacts, such as increased noise and dirt, and provide only limited benefits to localities. Those transits, however, can greatly influence national freight movement and regional economies.

Beyond the challenges of intergovernmental coordination, freight transportation raises additional issues involving the relationships between public and private sectors. Virtually all carriers and many freight facilities are privately owned. *Freight Facts and Figures 2015* shows that the private sector owns $1.173 trillion in transportation equipment plus $739 billion in transportation structures. In comparison, public agencies own $686 billion in transportation equipment and $3.343 trillion in highways. The private sector owns virtually all freight railroad facilities and services, and trucks owned by the private sector operate over public highways. Likewise, air cargo services that the private sector owns operate in public airways and primarily at public airports.

Challenges for Freight Transportation: Congestion

Congestion affects economic productivity in several ways. American businesses require more operators and equipment to deliver goods when shipping takes longer, more inventory when deliveries are unreliable or disrupted in some way, and more distribution centers to reach markets quickly when traffic is slow. Likewise, sluggish traffic on the ground and in the air affects both businesses and households, reducing the number of workers and job sites within easy reach of any location. The growth in freight is a major contributor to congestion in urban areas and on intercity routes, and congestion affects the timeliness and reliability of freight transportation. Long-distance freight movements are often a significant contributor to local congestion, and local congestion typically impedes freight to the detriment of local and distant economic activity.

Growing freight demand increases recurring congestion at freight bottlenecks, places where freight and passenger service conflict with one another, and where room for local pickup and delivery is insufficient. Congested freight hubs include international gateways such as water ports, airports, and border crossings, and major domestic terminals and transfer points such as distribution center hubs in large metropolitan areas and rail yards in major railroad centers such as Chicago, Kansas City, and Dallas/Fort Worth. In many cases, inadequate connections between a freight hub and the nearby highway network create congestion chokepoints. Bottlenecks on intercity corridors between freight hubs are caused by converging traffic at highway intersections and railroad junctions, steep grades on highways and rail lines, lane reductions on highways and single-track portions of railroads, and locks and constrained channels on waterways.

Congestion also is caused by restrictions on freight movement, such as the lack of space for trucks in dense urban areas and limited delivery and pickup times at ports, terminals, and shipper loading docks. The Off-Hours Delivery Project in New York City found that, for a large percentage of urban deliveries (between 40 percent and 78 percent), receivers—the stores and businesses receiving freight shipments—decide when the deliveries are made. The result is that many freight deliveries cannot be shifted readily to lower congestion times.[1] The same study also determined, however, that freight deliveries made during off-hour periods were 30 percent cheaper than deliveries during regular business hours.[2] Limitations on delivery times place significant demands on highway rest areas when large numbers of trucks park outside major metropolitan areas waiting for their destination to open and accept their shipments. The Jason's Law Truck Parking Study mandated in MAP-21 was completed in August 2015 by FHWA, and it examined truck-parking needs throughout the United States. The study highlighted the need for additional truck parking facilities and recommended incorporating truck parking analyses into freight planning at the State and regional level, as well as increased regional coordination by Freight Stakeholder Advisory Groups.

Bottlenecks cause recurring, predictable congestion in various locations having high transportation volume. Additionally, less predictable, nonrecurring congestion can also create challenges for freight movements across all modes, especially those that are time sensitive. Sources of nonrecurring delay include incidents, weather, work zones, and other disruptions. In some cases, disruptions not only cause nonrecurring congestion, but also cause freight diversions. According to the Port of New York-New Jersey, Superstorm Sandy forced a diversion of 57 ships, 9,000 vehicles, and 15,000 shipping containers from the port to other East Coast ports.[3] The Port of Hampton Roads in southeastern Virginia alone handled more than 8,000 of the diverted containers.[4] The Virginia Port Authority, trucking companies, CSX, Norfolk Southern, and U.S. Customs and Border Protection all needed to coordinate to handle the unexpected volume of shipments and ensure the shipments were transported to the proper locations.

Chapter 5 includes a broader discussion of system performance, including congestion's impacts on system performance.

[1] *U.S. Department of Transportation, Integrative Freight Demand Management in the New York City Metropolitan Area, September 30, 2010, page 27* http://transp.rpi.edu/~usdotp/OHD_FINAL_REPORT.pdf.

[2] *U.S. Department of Transportation, Integrative Freight Demand Management in the New York City Metropolitan Area, September 30, 2010, pages 6–7* http://transp.rpi.edu/~usdotp/OHD_FINAL_REPORT.pdf.

[3] *Southworth, F., et al. NCFRP Report 30, Making U.S. Ports Resilient as Part of Extended Intermodal Supply Chains, Transportation Research Board, 2014, page 50* http://onlinepubs.trb.org/onlinepubs/ncfrp/ncfrp_rpt_030.pdf.

[4] *Southworth, F., et al. NCFRP Report 30, Making U.S. Ports Resilient as Part of Extended Intermodal Supply Chains, Transportation Research Board, 2014, page 52* http://onlinepubs.trb.org/onlinepubs/ncfrp/ncfrp_rpt_030.pdf.

Challenges for Freight Transportation: Safety, Energy, and the Environment

Freight transportation is not simply an issue of throughput and congestion. The growth in freight movement has heightened public concerns about safety, energy consumption, and the environment.

Highways and railroads account for nearly all fatalities and injuries involving freight transportation. Most of these fatalities involve people who are not part of the freight transportation industry, such as trespassers at railroad facilities and occupants of other vehicles killed in crashes involving large trucks. The BTS *Freight Facts and Figures 2015* shows that, of the 32,719 highway fatalities in 2013, 2.1 percent were occupants of large trucks and 10.0 percent were others killed in crashes involving large trucks (the remaining 87.9 percent of fatalities were attributed to other types of personal and commercial vehicles). Chapter 5 of *Freight Facts and Figures 2015* discusses highway safety in more detail.

According to *Freight Facts and Figures 2015*, single-unit and combination trucks accounted for 25.5 percent of all gasoline, diesel, and other fuels consumed by motor vehicles in 2013. Fuel consumption by trucks resulted in 78 percent of the 388.3 million metric tons of carbon dioxide equivalent generated by freight transportation, and freight accounted for 21.9 percent of transportation's contribution to the emissions of this major greenhouse gas. Trucks and other heavy vehicles that operate on the U.S. highway system are also a major contributor to air quality problems related to nitrogen oxide and PM-10 (particulate matter 10 microns in diameter or smaller). Absolute freight truck-related emissions, however, have declined significantly with the increased use of ultralow-sulfur diesel; nitrogen oxide emissions from freight trucks fell 44.6 percent between 2005 and 2012, and PM-10 emissions declined 44.8 percent during the same period.

Environmental issues involving freight transportation go well beyond emissions. Disposal of dredge spoil—the mud and silt that must be removed to deepen water channels for commercial vessels—is a major challenge associated with allowing larger ships to berth. Land use and water quality concerns due to various factors such as soil contamination are raised against all types of freight facilities, and invasive species can spread through freight movement.

Incidents involving hazardous materials exacerbate public concern and cause serious disruption. *Freight Facts and Figures 2015* shows that, of the 15,433 transportation incidents in 2012 involving hazardous materials, highways accounted for 13,241 accidents (85.8 percent of hazardous material transportation incidents), air accounted for 1,293 accidents (9.5 percent of incidents), rail accounted for 662 accidents (4.3 percent of incidents), and water accounted for 70 accidents (0.5 percent of incidents). The railcar fire in the Howard Street tunnel in the city of Baltimore in 2001 illustrates both the perceived and real problems of transporting hazardous materials. This incident, which occurred on tracks near a major league baseball stadium at game time during the evening rush hour, forced the evacuation of thousands of people and closed businesses in much of downtown Baltimore. A vital railroad link between the Northeast and the South and a local rail transit line and all east-west arterial streets through downtown were closed for an extended period. More recent hazardous material incidents, such as the multiple petroleum-shipping train derailments that have occurred in different parts of the country, although not as widely disruptive to the U.S. transportation system as the 2001 Baltimore accident was, also have created significant short-term disruptions on freight transportation movements, negative environmental impacts, and in extreme cases, human fatalities.

Privately owned ships operate over public waterways and at both public and private port facilities. Most pipelines are privately owned but are significantly controlled by public regulation. In the public sector, State or local governments own virtually all truck routes, and regional or local authorities typically own airports and harbors. Air and water navigation is typically handled at the Federal level, and safety is regulated by all levels of government. Because of this mixed ownership and management, most solutions to freight problems require coordinated action by a wide variety of public and private-sector organizations and companies. Financial, planning, and other institutional mechanisms for developing and implementing joint efforts have been limited, inhibiting effective measures to improve the performance and minimize the public costs of the freight transportation system.

Freight challenges are not new. Their ongoing importance and increased complexity, however, warrant creative solutions by all parties having a stake in the vitality of the American economy.

National Freight Policy

The 2012 passage of the Moving Ahead for Progress in the 21st Century (MAP-21) transportation reauthorization created a formal U.S. policy to improve the condition and performance of the national freight network. This network is critical for ensuring the United States remains competitive in the global economy and achieves various goals to improve the Nation's freight movement (Section 1115). MAP-21 has greatly increased the visibility and emphasis on freight transportation at the Federal level. MAP-21 required the designation of a primary freight network and creation of a national freight strategic plan, a freight conditions and performance report, and new or refined transportation investment and planning tools to evaluate freight-related and non-freight related projects. All these provisions, and others in MAP-21, such as prioritizing projects to improve freight movement (Section 1116), encouraging States to establish freight advisory committees (Section 1117), encouraging States to develop State freight plans (Section 1118), and requiring creation of freight performance measures and performance targets that the States will use to assess freight movement on the Interstate system (Section 1203), have increased the focus on addressing and improving freight transportation at the Federal, State, and regional/metropolitan level. Many States and metropolitan planning organizations were already engaged in formal or informal freight transportation planning efforts before MAP-21 was passed. The current reauthorization has helped formalize these efforts, however, both in States and metropolitan planning organizations that have already been actively engaged in freight planning and where freight planning efforts have been limited, irregular, or nonexistent.

MAP-21 also indirectly encouraged other initiatives intended to promote better understanding of freight activities and to address freight challenges at all levels of government and in the private sector. DOT has created a Freight Policy Council involving the Office of the Secretary and all of DOT's freight-related modal administrations to coordinate the implementation of MAP-21 freight provisions, including the National Freight Strategic Plan and the National Freight Network. DOT also created a National Freight Advisory Committee composed of various public and private-sector representatives to provide advice and recommendations to the DOT Secretary on many of the MAP-21 freight-related provisions.

The Fixing America's Surface Transportation (FAST) Act established a national policy of maintaining and improving the condition and performance of the National Highway Freight Network (i.e., Network) to identify the highest priority highway freight routes and ensure these routes are appropriately improved and maintained to enhance the United States' ability to compete in the global economy. The FAST Act established a national highway freight program and specified goals associated with this national policy. Those goals include investing in infrastructure improvements and improving the safety, security, efficiency, productivity, resiliency, and reliability of the Network. The FAST Act requires DOT to develop a National Freight Strategic Plan that identifies and assesses the conditions of the Network and forecasts its future needs. It also required DOT to establish a National Multimodal Freight Network that will assist in strategically directing resources toward improved system performance and the prioritization of Federal investments.

chapter 2

System Characteristics

Highway and Bridge System Characteristics

The Nation's extensive network of roadways facilitates movement of people and goods, promotes the growth of the American economy, affords access to national and international markets, and supports national defense by providing the means for rapid deployment of military forces and their support systems. The network's bridges allow for the unimpeded movement of traffic over barriers created by geographical features such as rivers.

This chapter explores the characteristics of the Nation's roadways and bridges in terms of ownership, purpose, and usage. Information is presented for the National Highway System (NHS), including its Interstate Highway System component, and for the overall highway system. Separate statistics also are presented for Federal-aid highways, which include roadways and bridges that are generally eligible for Federal assistance under current law. Subsequent sections within this chapter explore the characteristics of bridges and transit systems.

Road statistics reported in this section draw on data collected from States through the Highway Performance Monitoring System (HPMS). The terms highways, roadways, and roads are generally used interchangeably in this section and elsewhere in the report. Roadways within a community with a population of 5,000 or more are classified as urban, while roadways in areas outside urban boundaries are classified as rural.

Bridge statistics reported in this section draw on data collected from States through the National Bridge Inventory (NBI). This information details physical characteristics, traffic loads, and the evaluation of the condition of each bridge longer than 20 feet (6.1 meters). As of December 2012, NBI contained records for 607,380 bridges. Data for input to NBI are collected regularly as set forth in the National Bridge Inspection Standards.

Tunnels

Under MAP-21, FHWA was charged with establishing a national tunnel inspection program. In 2015, development began on the National Tunnel Inventory database system, and inventory data were collected for all highway tunnels reported. Concurrently, FHWA implemented an extensive program to train inspectors nationwide on tunnel inspection and condition evaluation.

The 2015 preliminary inventory included 473 tunnels. Of these, 271 (57.3 percent) are on the National Highway System. States own 304 (64.3 percent) of the tunnels, 83 (17.5 percent) are owned by Local governments, 77 (16.3 percent) are owned by Federal agencies, and 9 (1.9 percent) are owned by others. Further information can be found at https://www.fhwa.dot.gov/bridge/inspection/tunnel/.

Complete inventory and condition data for all tunnels will be collected annually, beginning in 2018, and will be available for use in subsequent C&P reports.

System History

Before the 20th century, most Americans lived in rural communities or small cities. Railways and waterways were the leading methods of transporting goods and services because the technology was the cheapest. Most of the Nation's paved roads were located in urban centers that did not connect to other urban centers.

As technology progressed, difficulties in transporting agricultural goods to and between population centers continued. The Department of Agriculture established the Office of Road Inquiry in 1893 to determine ways to expand the national road network. The Office of Road Inquiry was moved to the Department of Commerce and renamed the Bureau of Public Roads in 1918 as the road network continued to grow. The agency's mission however, was to collect information on road construction and maintenance. The Federal role on road construction was confined to creating military roads and trails in remote or frontier areas. States were constructing privately operated toll roads.

Although the need for an interstate network to facilitate economic development and national unity had been identified throughout American history, construction of the system did not begin until the 1950s. The 1956 Federal-Aid Highway Act transformed highway financing by expanding the Federal role. Federal user fees based on the amount of gasoline purchased were deposited into the Highway Trust Fund to help fast track the construction of the Dwight D. Eisenhower National System of Interstate and Defense Highways. The Interstate System accelerated interstate and regional commerce, enhanced the country's competitiveness in international markets, increased personal mobility, facilitated military transportation, and furthered metropolitan development throughout the United States. President Eisenhower wrote in his memoir, "More than any single action by the government since the end of the war, this one would change the face of America. Its impact on the American economy ... was beyond calculation."

Roads and Bridges by Ownership

State and local governments own the vast majority of public roads and bridges. As shown in *Exhibit 2-1*, local governments own 77.3 percent of the Nation's public road mileage and 50.1 percent of all bridges. State governments own 19.0 percent of public road mileage and 48.2 percent of the Nation's bridges. State and local governments' owning most of the Nation's surface transportation infrastructure is attributed to the construction of lower-volume routes that feed into a larger network eligible for Federal funding. With a match of 20 percent or less, State and local governments leverage Federal assistance to construct larger transportation projects that aid efficient movement throughout the Nation. Although these larger projects are constructed with Federal funding, State and local governments assume ownership responsibilities for maintaining the facilities and keeping them safe for public use.

Q&A

Who owns the Federal-aid highway components?

In addition to the Interstate System and National Highway System, federally assisted highway mileage is found on other routes. Based on mileage, State highway agencies own the vast majority of the Interstate and National Highway systems; State highway agencies own 94.1 percent of the Interstate System and 88.1 percent of the National Highway System. In contrast, the Federal government owns only 0.2 percent of the 47,432 Interstate System mileage and 0.2 percent of the 222,946 National Highway System mileage. Local levels of government own the remaining mileage.

State highway agencies own 55.9 percent of the 1,001,874 miles of Federal-aid highways, while the Federal government owns only 0.6 percent of those miles. Local government agencies tend to own Federal-aid highway mileage that is not part of the Interstate and National Highway system.

Source: Highway Statistics HM-16 2012

Exhibit 2-1 Highway and Bridge Ownership by Level of Government

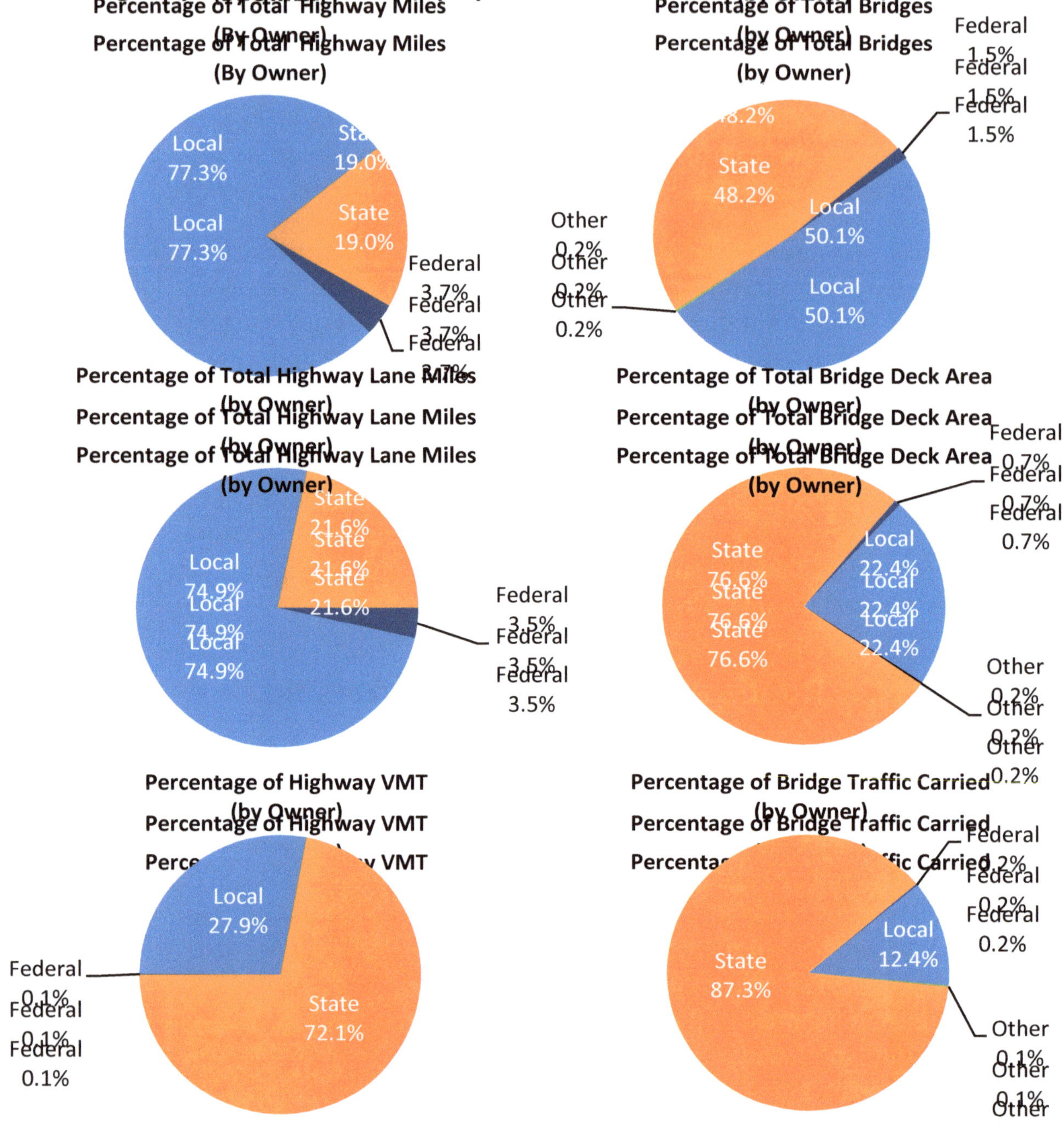

Sources: Highway Performance Monitoring System; National Bridge Inventory.

Federally owned facilities are generally found in National Parks and National Forests, on Indian reservations, and on military bases. Similar to State and local governments' assuming ownership of facilities during construction, federally owned facilities are the responsibility of agencies such as the Department of the Interior and Department of Defense.

The data presented throughout this chapter do not reflect privately owned facilities or facilities not available for public use.

Roads and Bridges by Federal System

The Nation's road network is diversely constructed to fit the needs of its surrounding environment. For example, roads in an urban setting will often have multiple lanes on a facility to support high levels of demand, while a rural setting will have fewer lanes supporting lower traffic levels. Highway mileage measures road distances from one point to another while lane mileage accounts for the number of lanes actually constructed. As shown in *Exhibit 2-2*, highway mileage and its accompanying lane mileage have increased steadily between 2002 and 2012. With population growth expected throughout the Nation, State and local governments are adding and increasing capacity throughout the road network. As this construction continues, the number of bridges cataloged in NBI has increased 0.3 percent between 2002 and 2012.

Exhibit 2-2 Highway Miles, Lane Miles, Vehicle Miles Traveled, Passenger Miles Traveled, and Bridges, 2002–2012

	2002	2004	2006	2008	2010	2012	Annual Rate of Change 2012/2002
Highway Miles	3,981,670	3,997,462	4,032,011	4,059,352	4,083,768	4,109,418	0.3%
Lane Miles	8,327,108	8,372,270	8,460,352	8,518,776	8,616,206	8,641,051	0.4%
VMT (millions)	2,874,455	2,981,998	3,033,957	2,992,779	2,985,095	2,987,403	0.4%
PMT (millions)[1]	4,667,038	4,844,452	4,929,366	4,900,171	4,244,833	4,274,877	-0.9%
Bridges	586,930	591,707	594,101	601,506	604,493	607,380	0.3%

[1] Values for 2002, 2004, 2006, and 2008 were based on a vehicle occupancy rate of approximately 1.63 based on data from the 2001 National Household Travel Survey (NHTS). Values for 2010 and 2012 were based on a vehicle occupancy rate of approximately 1.42 based on data from the 2009 NHTS. PMT data exclude Puerto Rico.

Sources: Highway Performance Monitoring System; Highway Statistics, Table VM-1, various years; National Bridge Inventory.

Vehicle miles traveled (VMT) measures the distance each vehicle traverses the Nation's road network in a year. Passenger miles traveled (PMT) weights travel by the number of occupants in a vehicle. As shown in *Exhibit 2-2,* total highway VMT grew by 0.4 percent between 2002 and 2012. Annual PMT, however, has decreased 0.9 percent during this period, due to a reduction in average vehicle occupancy and an increase in drivers driving alone. The change in vehicle occupancy was measured in the 2009 National Household Travel Survey, and the new PMT value was used from 2010 on.

Q&A Has VMT changed since 2012?

VMT on the Nation's roads is increasing. In 2013, VMT increased 0.6 percent. VMT increased 1.7 percent in 2014.

FHWA forecasts continual VMT growth based on long-term economic and demographics indicators. These indicators include national economic growth, disposable income growth, population growth, and declining global oil prices. Based on these economic indicators, all types of vehicles are expected to experience an increase in VMT.

Source: FHWA Traffic Volume Trends and FHWA Forecasts of Vehicle Miles Traveled

Exhibit 2-3 shows annual VMT growth rates between 1992 and 2012. An examination of recent trends shows VMT growth has fluctuated between 2006 and 2012. The negative growth rates can be attributed partially to the period of economic contraction from December 2007 to June 2009 identified by the National Bureau of Economic Research. Now that the economy has stabilized, Americans are beginning to travel more often. Of note, however, is that VMT growth had been trending downward: Annual VMT growth rate last exceeded 3 percent in 1997 and has been less than 1 percent every year since 2004.

Exhibit 2-3 Annual Growth Rates in Vehicle Miles Traveled, 1992–2012

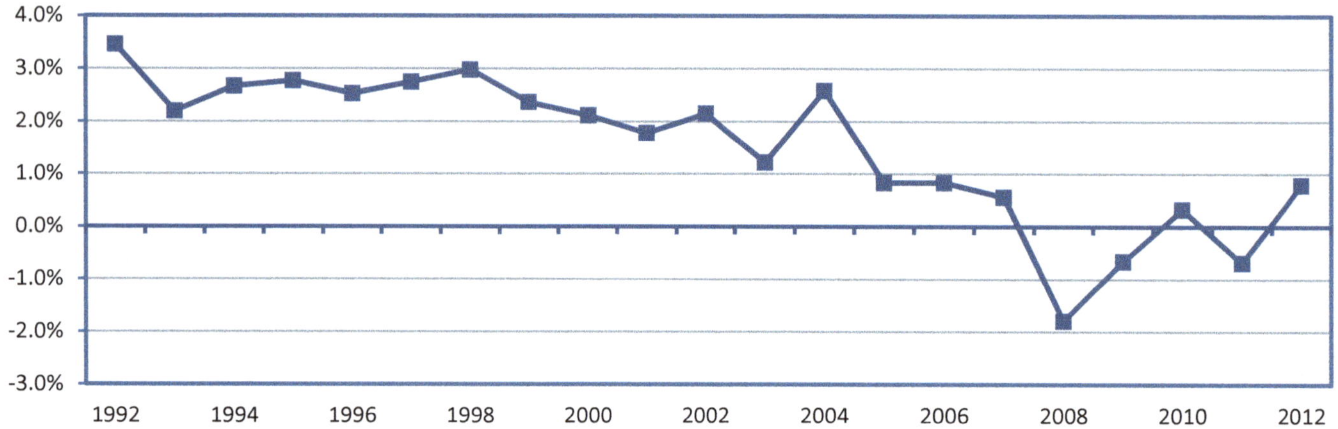

Source: Highway Statistics 2013, Table VM-202.

Federal-Aid Highways

The mileage eligible for Federal-aid highway assistance is much smaller than the total road mileage throughout the Nation. Federal-aid highway assistance mileage, however, consists of longer routes that cross multiple States and facilitate higher traffic volumes at increased speeds. Conversely, non-Federal-aid highway mileage generally consists of shorter and smaller roads that eventually feed into the larger facilities that are eligible for Federal assistance. A discussion on roads eligible for Federal-aid highway assistance is presented later in this section.

As shown in *Exhibit 2-4*, Federal-aid highways comprised approximately 1.0 million miles in 2012 and facilitated more than 2.5 trillion VMT. Federal-aid highway VMT was similarly affected by the economic impacts of 2007, as shown by comparing total VMT in *Exhibit 2-2*. This impact occurred primarily because most of the Nation's VMT occurs on Federal-aid highways.

Between 2002 and 2012, highway mileage, lane mileage, VMT, and the number of bridges have increased slightly.

Exhibit 2-4 Federal-Aid Highway Miles, Lane Miles, Vehicle Miles Traveled, and Number of Bridges, 2002–2012

	2002	2004	2006	2008	2010	2012	Annual Rate of Change 2012/2002
Highway Miles	959,125	971,036	984,093	994,358	1,007,777	1,005,378	0.5%
Lane Miles	2,282,024	2,319,417	2,364,514	2,388,809	2,451,140	2,433,012	0.6%
VMT (millions)	2,430,698	2,531,629	2,573,956	2,534,490	2,525,455	2,526,558	0.4%
Bridges	305,609	307,840	312,062	316,012	319,108	321,724	0.5%

Sources: Highway Performance Monitoring System; National Bridge Inventory.

National Highway System

With the Interstate System essentially complete, the National Highway System Designation Act of 1995 revised the Federal-aid highway program for the post-Interstate System era. The legislation authorized designation of an NHS that would give priority for Federal resources to roads most important for interstate travel, economic expansion, and national defense; that connect with other modes of transportation; and that are essential to the Nation's role in the international marketplace.

The NHS was designed to be a dynamic system capable of changing in response to future travel and trade demands. States must cooperate with local and regional officials in proposing modifications. In metropolitan areas, local and regional officials must act through metropolitan planning organizations and the State transportation department when proposing modifications. Numerous such modifications are proposed and approved each year.

The NHS has five components. The first, the Interstate System, is the core of the NHS and includes the most traveled routes. The second component includes other principal arterials deemed most important for commerce and trade. The third is the Strategic Highway Network (STRAHNET), which consists of highways important to military mobilization. The fourth is the system of STRAHNET connectors that provide access between major military installations and routes that are part of STRAHNET. The final component consists of intermodal connectors. These roads provide access between major intermodal passenger and freight facilities and the other four subsystems that comprise the NHS.

As shown in *Exhibit 2-5*, only 5.4 percent of the Nation's highway mileage and 8.9 percent of the Nation's lane mileage are located on the NHS. Approximately 55.0 percent of the Nation's VMT, however, occurs on the NHS. The NHS is crucial to truck traffic, which carries cargo long distances, often across multiple State lines. Approximately 83.1 percent of truck VMT occurred on the NHS.

Exhibit 2-5 Share of Highway Miles, Lane Miles, Vehicle Miles Traveled, and Truck Vehicle Miles Traveled On and Off the National Highway System, 2012[1]

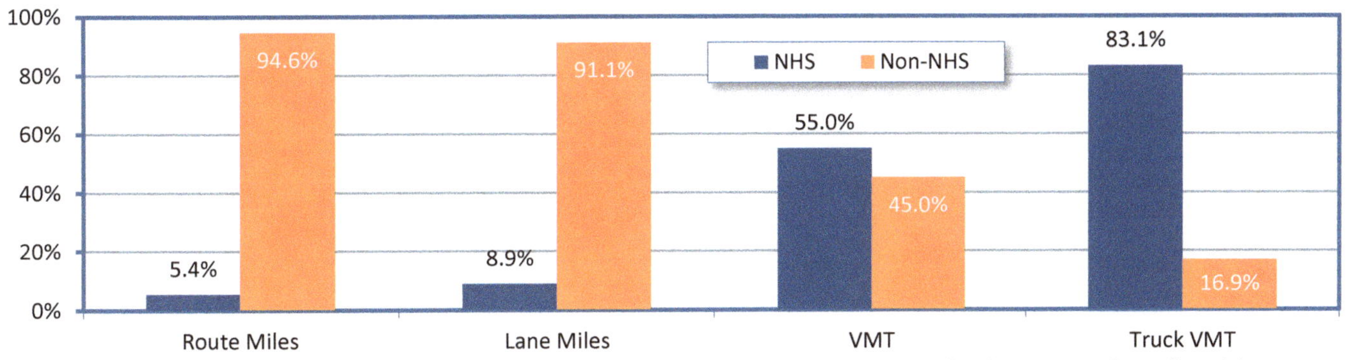

[1] *Data reflect the expansion of the NHS required by MAP-21. (Bridge data are not shown as the 2012 National Bridge Inventory data still used the pre-MAP-21 version of the NHS.)*

Source: Highway Performance Monitoring System.

In view of the importance of the NHS for truck traffic and freight, State DOTs often will design such highways to accommodate trucks at higher volumes and speeds in the safest and most efficient ways possible. Additionally, NHS highways often are constructed with stronger, more robust materials that enable them to withstand the heavier loads trucks convey.

> **MAP-21 Expansion of the NHS**
>
> The Moving Ahead for Progress in the 21st Century Act of 2012 (MAP-21) modified the scope and extent of the NHS to include some additional principal arterial and related connector mileage not previously designated as part of the NHS.
>
> The expansion of the NHS to include all principal arterials increased its size from 4.0 percent of the Nation's roadway miles to 5.4 percent. The NHS share of total lane mileage increased from 6.6 percent to 8.9 percent. The share of total VMT carried by the NHS increased from 43.9 percent to 55.0 percent; for truck VMT, the share carried by the NHS increased from 75.1 percent to 83.1 percent.

Interstate System

With the strong support of President Eisenhower, the Federal-Aid Highway Act of 1956 declared the completion of the "National System of Interstate and Defense Highways" was essential to the national interest. The Act committed the Nation to completing the Interstate System within the Federal-State partnership of the Federal-aid highway program, with the State responsible for construction according to approved standards. The Act also resolved the challenging issue of how to pay for construction by establishing the Highway Trust Fund to ensure that revenue from highway user taxes, such as the motor fuels tax, would be dedicated to the Interstate System and other Federal-aid highway and bridge projects.

As shown in *Exhibit 2-6*, small additions to the Interstate System have occurred between 2002 and 2012 at a rate of 0.2 percent. Lane mileage has also increased by 0.4 percent during this period, suggesting that Interstate capacity has increased slightly.

Exhibit 2-6 Interstate Highway Miles, Lane Miles, Vehicle Miles Traveled, and Numbers of Bridges, 2002–2012

	2002	2004	2006	2008	2010	2012	Annual Rate of Change 2012/2002
Highway Miles	46,747	46,836	46,892	47,019	47,182	47,714	0.2%
Lane Miles	210,896	212,029	213,542	214,880	217,165	220,124	0.4%
VMT (millions)	693,941	727,163	741,002	725,213	731,095	735,914	0.6%
Bridges	55,234	55,315	55,270	55,626	55,339	55,959	0.1%

Sources: Highway Performance Monitoring System; National Bridge Inventory.

Freight System

Freight in America travels over an extensive network of highways, railroads, waterways, pipelines, and airways: 985,000 miles of Federal-aid highways, 141,000 miles of railroads, 11,000 miles of inland waterways, and 1.6 million miles of pipelines. The Nation has more than 19,000 airports, with approximately 540 serving commercial operations, and more than 5,000 coastal, Great Lakes, and inland waterway facilities moving cargo. Although specific commodities are likely to be moved on a particular mode or series of modes, a complex multimodal system is required to meet fully the growing volume of bulk and high-velocity, high-value goods in the United States.

The U.S. freight highway transportation system is, in the broadest sense, composed of all Federal, State, local (county or municipal), and private roads that facilitate the movement of freight-hauling trucks or commercial vehicles. The National Network, however, is the system of roadways officially designated to accommodate commercial freight-hauling vehicles. The National Network was authorized by the Surface Transportation Assistance Act of 1982, and specified in the U.S. Code of Federal Regulations. 23 CFR 658 is the requirement that States allow conventional combinations on the Interstate System and those portions of the Federal-aid Primary System serving to link principal cities and densely developed portions of the States on high volume routes utilized extensively by large vehicles for interstate commerce. Conventional combinations are tractors with one semitrailer up to 48 feet in length or with one 28-foot semitrailer and one 28-foot trailer up to 102 inches wide. Currently, most States allow conventional combination trucks with single trailers up to 53 feet in length to operate without permits on their portions of the National Network (see *Exhibit 2-7*).

Exhibit 2-7 National Network for Conventional Combination Trucks, 2013[1,2]

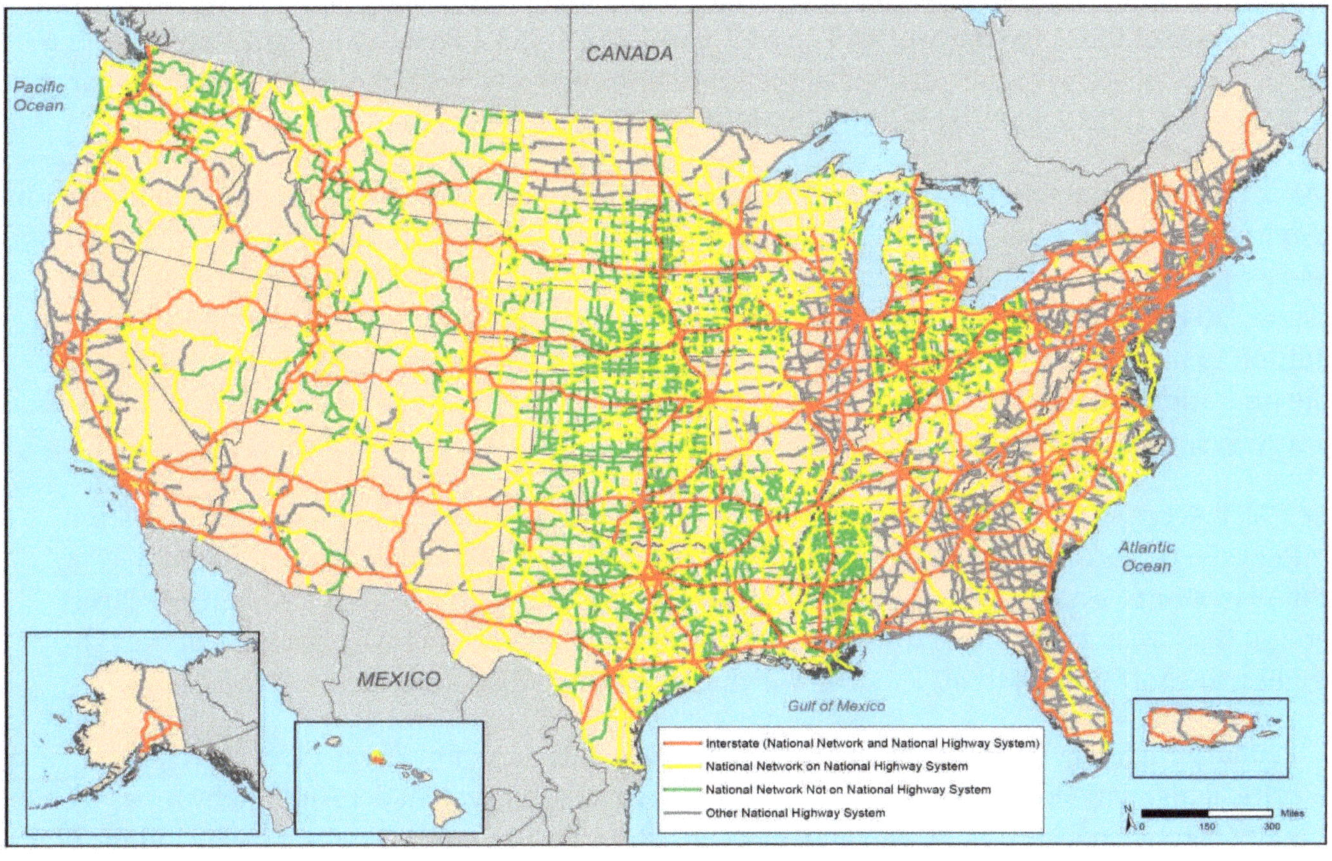

[1] This map should not be interpreted as the official National Network and should not be used for truck size and weight enforcement purposes. The National Network and the 65,000 miles of highways beyond the NHS, and the NHS encompasses about 50,000 miles of highways that are not part of the National Network.National Highway System (NHS) are approximately 200,000 miles in length, but the National Network includes 65,000 miles of highways beyond the NHS, and the NHS encompasses about 50,000 miles of highways that are not part of the National Network.

[2] "Other NHS" refers to NHS mileage that is not included on the National Network. Conventional combination trucks are tractors with one semitrailer up to 48 feet in length or with one 28-foot semitrailer and one 28-foot trailer. Conventional combination trucks can be up to 102 inches wide.

Source: U.S. Department of Transportation, Federal Highway Administration, Office of Freight Management and Operations, Freight Analysis Framework, version 3.4, 2013 (http://ops.fhwa.dot.gov/Freight/freight_analysis/nat_freight_stats/nnnhs2013.htm).

The National Network has remained generally unchanged since its designation in 1982. The network is essential for supporting interstate commerce by maintaining truck access to major industrial centers and freight generators. The National Network differs in extent and purpose from the NHS, which was created more than a decade later by the National Highway System Designation Act of 1995 and modified in 2012 by MAP-21.

The National Network and the NHS share more than 114,000 miles. The National Network includes 65,000 miles of highway not on the NHS, and the NHS includes 50,000 miles not on the National Network. Both the National Network and the NHS were created to support interstate commerce. The National Network protects interstate commerce by ensuring that all States allow certain truck configurations to travel on the system, while the NHS supports long-distance interstate travel such as connecting routes between principal metropolitan areas and industrial centers important to national defense and the national economy.

MAP-21 outlined the requirements for new freight routes and the creation and definition of a highway-focused National Freight Network (NFN). The NFN was intended to include the most important urban, rural, and intercity routes for commercial truck movements. This network overlapped portions of both the National Network and the NHS and includes mileage that is not part of either of those two networks. The NFN consisted of (1) a Primary Freight Network (PFN) that DOT designates, (2) the portions of the Interstate Highway System not selected to be part of the PFN, and (3) Critical Rural Freight Corridors that States designate.

MAP-21 mandated the PFN include no more than 27,000 centerline miles of existing roadways and be defined based on eight factors specified in the legislation. DOT found that these factors did not yield a network representative of the most critical highway elements of the national freight system. DOT had reservations about the limitations of the NFN, and particularly the PFN. In addition to the challenges associated with creating an interconnected PFN that met the 27,000-mile limitation, the MAP-21 NFN provisions did not allow nonhighway modes, such as railroads, waterways, and pipelines, to be included in the NFN.

The FAST Act repealed both the Primary Freight Network and National Freight Network from MAP-21. To replace and improve upon those networks, the FAST Act directed the FHWA Administrator to establish a National Highway Freight Network (NHFN) to strategically direct Federal resources and policies toward improved performance of highway portions of the U.S. freight transportation system. The NHFN includes the following subsystems of roadways:

- Primary Highway Freight System (PHFS): This is a network of highways identified as the most critical highway portions of the U.S. freight transportation system determined by measurable and objective national data. The network consists of 41,518 centerlines miles, including 37,436 centerline miles of Interstate and 4,082 centerline miles of non-Interstate roads.
- Other Interstate portions not on the PHFS: These highways consist of the remaining portion of Interstate roads not included in the PHFS. These routes provide important continuity and access to freight transportation facilities. These portions amount to an estimated 9,511 centerline miles of Interstate, nationwide, and will fluctuate with additions and deletions to the Interstate Highway System.

- Critical Rural Freight Corridors (CRFCs): These are public roads not in an urbanized area that provide access and connection to the PHFS and the Interstate with other important ports, public transportation facilities, or other intermodal freight facilities. These roadways will be identified by State Departments of Transportation.
- Critical Urban Freight Corridors (CUFCs): These are public roads in urbanized areas that provide access and connection to the PHFS and the Interstate with other ports, public transportation facilities, or other intermodal transportation facilities. These roadways will be identified by either State Departments of Transportation or Metropolitan Planning Organizations (MPOs), depending on the population of MPOs' urbanized areas.

After the initial designation, FHWA must redesignate the PHFS every 5 years, with up to 3 percent growth each time.

The FAST Act requires DOT to develop, in consultation with a range of stakeholders, a National Freight Strategic Plan and to update this plan every 5 years. The FAST ACT directed DOT to establish an interim National Multimodal Freight Network to include the NHFN, freight rail systems of Class I railroads, the Great Lakes, the St. Lawrence Seaway, inland and intracoastal waterways, ports and airports that meet specified criteria, and other strategic freight assets. DOT must designate a National Multimodal Freight Network and must redesignate this network every 5 years with input from a wide range of stakeholders.

Roads and Bridges by Purpose

The Nation's roadway system is a vast network that connects places and people within and across national borders. The network serves movements from long-distance freight needs to neighborhood travel. Because of the diverse needs for vehicular travel, the network is categorized under the Highway Functional Classification System. Each functional classification defines the role an element of the network plays in serving travel needs.

As shown in *Exhibit 2-8*, roadways serve two primary travel needs: access and mobility. The two concepts are illustrated on both far ends of the exhibit. Access roads enable many roadway users to enter the system at any given time. Access roads can be found in the urban setting

Q&A How are arterials defined?

Arterials serve the longest distances with the fewest access points. Because they have the longest distance between other routes, arterials facilitate the highest speed limits. Several functional classifications are included in the arterial category.

Interstates are the highest classification of arterials facilitating the highest level of mobility. Interstates are relatively easy to locate due to their official designation by the Secretary of Transportation.

Other Freeways and Expressways are similar to Interstates in that they have directional travel lanes usually separated by a physical barrier. Access and egress points are limited primarily to on- and off-ramps at grade-separated interchanges.

Other Principal Arterials can serve specific land parcels directly and have at-grade intersections with other roadways that are managed by traffic devices.

Minor Arterials, the lowest of arterial classifications, provide service for trips of moderate length and offer connectivity to the higher arterial classifications.

Source: Highway Functional Classification Concepts, Criteria and Procedures 2013

next to office buildings or suburban neighborhoods that have a high concentration of residences. Many vehicles entering the network from multiple directions create higher points of friction. Friction points can occur when a vehicle decelerates or stops so another car can enter a roadway. Access streets have lower speeds and more traffic control devices to accommodate traffic traveling shorter distances. Mobility roads allow many users to travel in the same direction on the network. These roads are found in interstate travel or around urban centers to move vehicles quickly. These roads can facilitate higher speed limits over longer distances because fewer opportunities for entry and exit to the road are available.

Any normal trip on the roadway system could use roads that serve different purposes. For example, a traveler can leave a suburban home located on a local street and use an arterial Interstate to commute to an urban office located on a local street. For this commuter to transition from an accessible road to a mobility road, a collector road must be used. *Exhibit 2-8* depicts collectors as a bridge between local roads and arterials.

Exhibit 2-8 Functional Classifications

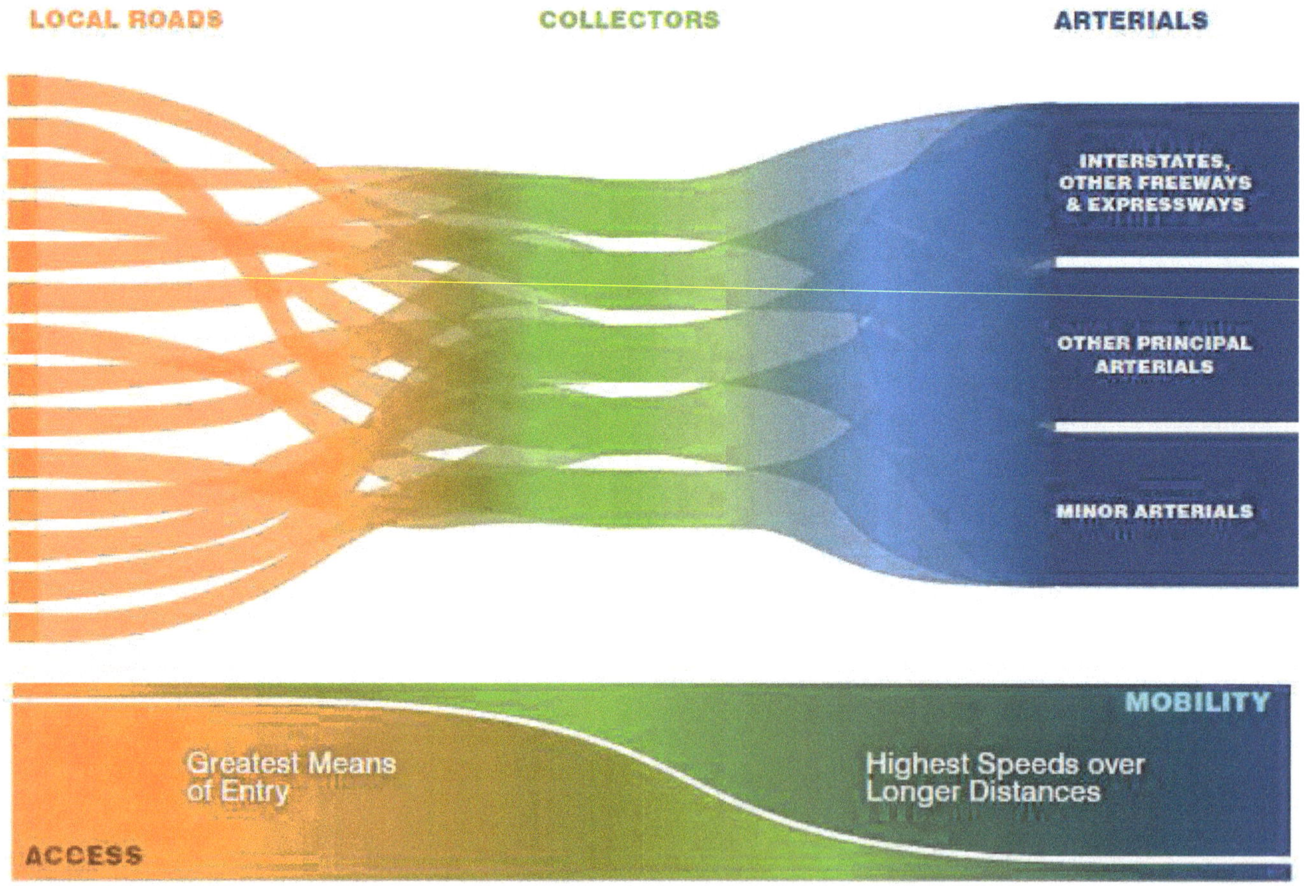

Source: FHWA Functional Classification Guidelines.

Exhibit 2-9 presents a formal hierarchy of road functional classifications. The functional classifications establish which roads are eligible for Federal-aid highway funding. Although the functional classification definitions do not change for each setting, roads are divided into rural and urban classifications.

The hierarchy continues the access and mobility concepts with collector roads bridging the two. Arterials include both principal and minor arterials. Interstates, other principal arterials, and other freeways and expressways are a component of principal arterials. Within the collector classification, roads are divided into major or minor collectors. All other roads are considered local.

Exhibit 2-9 Highway Functional Classification System Hierarchy

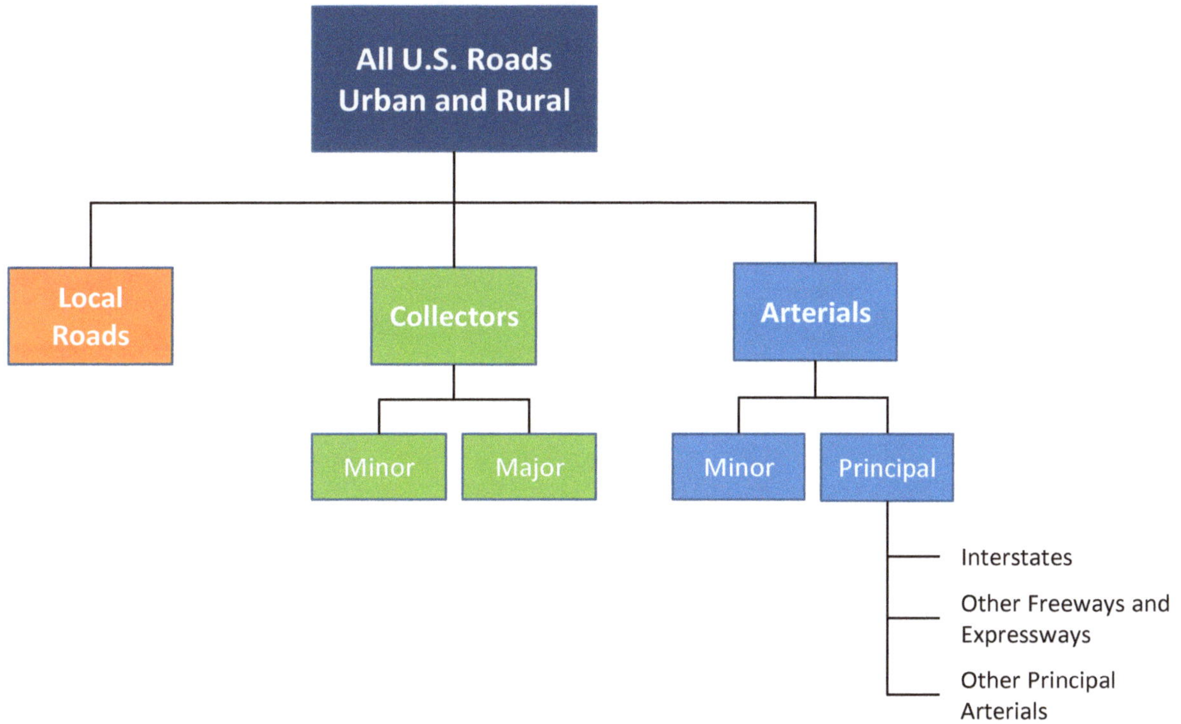

Source: FHWA Functional Classification Guidelines.

Public roads that are functionally classified higher than rural minor collector, rural local, or urban local are eligible for Federal-aid highway assistance. Although bridges follow the hierarchy scheme, they differ in several ways because NBI tracks bridges, while HPMS tracks highways. NBI makes no distinction between urban major and urban minor collectors as HPMS does. Important to note is that MAP-21 allows Federal-aid highway funding to be used on bridges that are not on the Federal-aid highways. States may use funding from their Surface Transportation Program apportionments to fund bridge projects not on Federal-aid highways.

> ## Q&A How are collectors defined?
>
> Collectors serve the critical roles of gathering traffic from local roads and funneling vehicles into the arterial network. Although subtly different, two classifications are included in the collector category.
>
> **Major Collectors** are longer, have fewer points of access, have higher speed limits, and can have more travel lanes.
>
> **Minor Collectors** is the classification for collectors not classified as major collectors. One distinction between the two classifications is that minor collectors are focused more on access than on mobility.
>
> Source: Highway Functional Classification Concepts, Criteria and Procedures 2013

System Characteristics

As stated earlier in this section, local governments own most of the Nation's highway mileage and bridges, due to the large amount of mileage classified as local roads that feed into larger facilities. Local governments tend to own shorter and less traveled roads. As shown in *Exhibit 2-10*, the highest share of the 2012 highway mileage was classified as local, with 49.9 percent in rural areas and 19.4 in urban areas. The share of 2012 VMT on roads classified as local, however, was only 4.4 percent in rural areas and 9.3 percent in urban areas.

Exhibit 2-10 Percentages of Highway Miles, Lane Miles, Vehicle Miles Traveled, Bridges, Bridge Deck Area, and Bridge Traffic by Functional System, 2012

Functional System	Highway Miles	Highway Lane Miles	Highway VMT	Bridges	Bridge Deck Area	Bridge Traffic Volume
Rural Areas (4,999 or less in population)						
Interstate	0.7%	1.4%	8.2%	4.1%	6.9%	8.9%
Other Freeway and Expressway	0.1%	0.2%	0.7%			
Other Principal Arterial	2.2%	2.8%	6.8%			
Other Principal Arterial[1]				6.0%	8.9%	5.8%
Minor Arterial	3.3%	3.3%	5.0%	6.4%	6.1%	3.2%
Major Collector	10.3%	9.8%	5.9%	15.3%	9.1%	3.1%
Minor Collector	6.4%	6.1%	1.8%	7.9%	3.2%	0.8%
Local	49.9%	47.3%	4.4%	33.8%	9.4%	1.4%
Subtotal Rural Areas	**72.9%**	**70.9%**	**32.8%**	**73.6%**	**43.6%**	**23.3%**
Urban Areas (5,000 or more in population)						
Interstate	0.5%	1.1%	16.5%	5.1%	19.4%	35.8%
Other Freeway and Expressway	0.2%	0.7%	7.5%	3.3%	10.8%	16.4%
Other Principal Arterial	1.6%	2.7%	15.4%	4.6%	11.4%	11.9%
Minor Arterial	2.6%	3.3%	12.5%	4.7%	7.5%	7.3%
Collector[1]				3.4%	3.5%	2.8%
Major Collector	2.8%	2.9%	5.9%			
Minor Collector	0.0%	0.1%	0.1%			
Local	19.4%	18.3%	9.3%	5.4%	3.8%	2.4%
Subtotal Urbanized Areas	**27.1%**	**29.1%**	**67.2%**	**26.4%**	**56.4%**	**76.7%**
Total	**100.0%**	**100.0%**	**100.0%**	**100.0%**	**100.0%**	**100.0%**

[1] Highway data reflects revised HPMS functional classifications. Bridge data still uses the previous classifications, so that rural Other Freeway and Expressway is included as part of the rural Other Principal Arterial category, and urban Major Collector and urban Minor Collector are combined into a single urban Collector category.

Sources: Highway Performance Monitoring System; National Bridge Inventory.

Exhibit 2-10 also details the breakdown of travel occurring in rural and urban settings. Urban areas with populations greater than 5,000 have a higher share of VMT and lower highway mileage because urban settings tend to be more consolidated environments. With higher population concentrations, more vehicles use the highway mileage in urban areas. Alternatively, rural areas have a higher share of the highway mileage to provide connectivity between areas with lower population density.

Q&A How are local roads defined?

Local Roads are any road not classified as an arterial or collector. They are not intended for use in long-distance travel, except at the origination or termination of a trip. Local roads are often designed to discourage through traffic.

Source: Highway Functional Classification Concepts, Criteria and Procedures 2013

Although Interstate highway mileage comprises only 1.2 percent of the Nation's highway mileage, it receives the Nation's highest share of VMT by classification at 24.7 percent. Interstate bridges also receive the highest share of bridge traffic volume by classification with 44.7 percent.

As shown in *Exhibit 2-11,* the Nation's public highways comprised nearly 4.11 million miles in 2012, up from 3.98 million miles in 2002. Total mileage in urban areas grew by an average annual rate of 2.2 percent between 2002 and 2012. Highway miles in rural areas, however, decreased at an average annual rate of 0.3 percent during the same period.

Exhibit 2-11 Highway Route Miles by Functional System, 2002–2012

Functional System	2002	2004	2006	2008	2010	2012	Annual Rate of Change 2012/2002
Rural Areas (less than 5,000 in population)							
Interstate	33,107	31,477	30,615	30,227	30,260	30,564	-0.8%
Other Freeway & Expressway[1]					3,299	4,395	
Other Principal Arterial[1]					92,131	91,462	
Other Principal Arterial[1]	98,945	95,998	95,009	95,002			-0.3%
Minor Arterial	137,855	135,683	135,589	135,256	135,681	135,328	-0.2%
Major Collector	431,754	420,293	419,289	418,473	418,848	419,353	-0.3%
Minor Collector	271,371	268,088	262,966	262,852	263,271	262,435	-0.3%
Local	2,106,725	2,051,902	2,046,796	2,038,517	2,036,990	2,039,276	-0.3%
Subtotal Rural Areas	**3,079,757**	**3,003,441**	**2,990,264**	**2,980,327**	**2,980,480**	**2,982,813**	**-0.3%**
Urban Areas (5,000 or more in population)							
Interstate	13,640	15,359	16,277	16,789	16,922	17,150	2.3%
Other Freeway and Expressway	9,377	10,305	10,817	11,401	11,371	11,521	2.1%
Other Principal Arterial	53,680	60,088	63,180	64,948	65,505	65,593	2.0%
Minor Arterial	90,922	98,447	103,678	107,182	108,375	109,337	1.9%
Collector[1]	89,846	103,387	109,639	115,087			3.0%
Major Collector[1]					115,538	116,943	
Minor Collector[1]					3,303	3,588	
Local	644,449	706,436	738,156	763,618	782,273	802,473	2.2%
Subtotal Urban Areas	**901,913**	**994,021**	**1,041,747**	**1,079,025**	**1,103,288**	**1,126,605**	**2.2%**
Total Highway Route Miles	**3,981,670**	**3,997,462**	**4,032,011**	**4,059,352**	**4,083,768**	**4,109,418**	**0.3%**

[1] Starting in 2010, the HPMS data reflect revised functional classifications. Rural Other Freeway and Expressway has been split from the rural Other Principal Arterial category, and urban Collector has been split into urban Major Collector and urban Minor Collector. The annual rate of change was computed based on the older combined categories.

Source: Highway Performance Monitoring System.

In addition to the construction of new roads, two factors have continued to contribute to the increase in urban highway mileage. First, based on the decennial census, more people are living in urban areas, and thus urban boundaries have expanded. This expansion has resulted in the reclassification of some mileage from rural to urban. States have implemented these boundary changes in their HPMS data reporting gradually. As a result, the impact of the census-based changes on these statistics is not confined to a single year. Second, greater focus has been placed on Federal agencies to provide a more complete reporting of federally owned mileage.

Exhibit 2-12 details lane mileage by functional system and population size. Lane mileage represents the length of the roadway multiplied by the number of lanes on that roadway. Because 72.9 percent of the Nation's highway mileage is located in rural areas, lane mileage is also higher in rural areas. Local roads in urban and rural settings also continue to have the highest share of the Nation's lane mileage. Lane mileage in urban areas increased 2.3 percent between 2002 and 2012, while lane mileage in rural areas decreased 0.3 percent during the same period.

Exhibit 2-12 Highway Lane Miles by Functional System, 2002–2012

	Highway Lane Miles						Annual Rate of Change
Functional System	**2002**	**2004**	**2006**	**2008**	**2010**	**2012**	**2012/2002**
Rural Areas (less than 5,000 in population)							
Interstate	135,032	128,012	124,506	122,956	123,762	124,927	-0.8%
Other Freeway and Expressway[1]					11,907	16,593	
Other Principal Arterial[1]					243,065	240,639	
Other Principal Arterial[1]	256,458	249,480	248,334	250,153			0.03%
Minor Arterial	288,391	283,173	282,397	281,071	287,761	281,660	-0.2%
Major Collector	868,977	845,513	843,262	841,353	857,091	842,722	-0.3%
Minor Collector	542,739	536,177	525,932	525,705	526,540	524,870	-0.3%
Local	4,213,448	4,103,804	4,093,592	4,077,032	4,073,980	4,078,552	-0.3%
Subtotal Rural Areas	**6,305,044**	**6,146,159**	**6,118,023**	**6,098,270**	**6,124,107**	**6,109,963**	**-0.3%**
Urban Areas (5,000 or more in population)							
Interstate	75,864	84,016	89,036	91,924	93,403	95,197	2.3%
Other Freeway and Expressway	43,467	47,770	50,205	53,073	53,231	54,160	2.2%
Other Principal Arterial	188,525	210,506	221,622	228,792	235,127	234,469	2.2%
Minor Arterial	233,194	250,769	269,912	274,225	285,954	283,608	2.0%
Collector[1]	192,115	220,177	235,240	245,262			3.0%
Major Collector[1]					252,435	250,760	
Minor Collector[1]					7,404	7,948	
Local	1,288,898	1,412,872	1,476,314	1,527,230	1,564,546	1,604,946	2.2%
Subtotal Urban Areas	**2,022,064**	**2,226,111**	**2,342,329**	**2,420,506**	**2,492,099**	**2,531,088**	**2.3%**
Total Highway Lane Miles	**8,327,108**	**8,372,270**	**8,460,352**	**8,518,776**	**8,616,206**	**8,641,051**	**0.4%**

[1] Starting in 2010, the HPMS data reflects revised functional classifications. Rural Other Freeway and Expressway has been split from the rural Other Principal Arterial category, and urban Collector has been split into urban Major Collector and urban Minor Collector. The annual rate of change was computed based on the older combined categories.

Source: Highway Performance Monitoring System.

Highway Travel by Functional Classification

With regard to VMT and individual functional classifications, rural and urban areas also differ. *Exhibit 2-13* details VMT trends by functional classification between 2002 and 2012. Urban area

VMT increased 1.4 percent in that span, while rural area VMT decreased 1.5 percent. Interstate with Other Freeway and Expressway in urban areas had the biggest increase of VMT share with 1.7 percent. Major collectors in rural areas had the greatest decrease of VMT share at 2.0 percent. VMT in 2012 was more than 2.98 trillion, a 0.4-percent increase from the 2.87 trillion VMT in 2002.

Exhibit 2-13 Vehicle Miles Traveled by Functional System, 2002–2012

	Annual Travel Distance (Millions of Miles)						Annual Rate of Change
Functional System	**2002**	**2004**	**2006**	**2008**	**2010**	**2012**	**2012/2002**
Rural Areas (less than 5,000 in population)							
Interstate	281,461	267,397	258,324	243,693	246,109	246,334	-1.3%
Other Freeway & Expressway[1]					19,603	20,146	
Other Principal Arterial[1]					205,961	203,310	
Other Principal Arterial[1]	258,009	241,282	232,224	222,555			-1.4%
Minor Arterial	177,139	169,168	162,889	152,246	151,307	148,956	-1.7%
Major Collector	214,463	200,926	193,423	186,275	176,301	175,838	-2.0%
Minor Collector	62,144	60,278	58,229	55,164	53,339	53,215	-1.5%
Local	139,892	132,474	133,378	131,796	132,827	130,124	-0.7%
Subtotal Rural Areas	**1,133,107**	**1,071,524**	**1,038,467**	**991,729**	**985,447**	**977,923**	**-1.5%**
Urban Areas (5,000 or more in population)							
Interstate	412,481	459,767	482,677	481,520	482,726	489,580	1.7%
Other Freeway and Expressway	190,641	209,084	218,411	223,837	221,902	225,098	1.7%
Other Principal Arterial	410,926	453,868	470,423	465,965	460,753	460,302	1.1%
Minor Arterial	341,958	365,807	380,069	380,734	378,048	374,915	0.9%
Collector[1]	143,621	164,330	175,516	177,665			-0.7%
Major Collector[1]					178,909	177,217	
Minor Collector[1]					3,837	4,476	
Local	241,721	257,617	268,394	271,329	273,474	277,892	1.4%
Subtotal Urban Areas	**1,741,348**	**1,910,473**	**1,995,489**	**2,001,050**	**1,999,648**	**2,009,480**	**1.4%**
Total VMT	**2,874,455**	**2,981,998**	**3,033,957**	**2,992,779**	**2,985,095**	**2,987,403**	**0.4%**

[1] Starting in 2010, the HPMS data reflects revised functional classifications. Rural Other Freeway and Expressway has been split from the rural Other Principal Arterial category, and urban Collector has been split into urban Major Collector and urban Minor Collector. The annual rate of change was computed based on the older combined categories.

Source: Highway Performance Monitoring System.

Exhibit 2-14 provides an analysis of the types of vehicles comprising the Nation's VMT between 2008 and 2012. Three types of vehicles are identified: passenger vehicles, which include motorcycles, buses, and light trucks (two-axle, four-tire models); single-unit trucks having six or more tires; and combination trucks, including trailers and semitrailers. Passenger vehicle travel accounted for 91.0 percent of total VMT in 2012; combination trucks accounted for 5.5 percent of VMT during this period; and single-unit trucks accounted for the remaining 3.5 percent. The share of truck travel on the rural interstates is considerably higher; in 2012, single-unit and combination trucks together accounted for 23.6 percent of total VMT on the rural Interstates.

Passenger vehicle travel grew at an average annual rate of 0.3 percent from 2008 to 2012. Over the same period, combination truck traffic declined by 2.9 percent per year, and single-unit truck traffic declined by 4.6 percent per year. The decrease in combination truck traffic occurred mostly in urban areas; single-unit truck traffic decreased in both rural and urban areas, but the change was more pronounced in urban areas. Direct comparisons over a longer period cannot be made due to significant revisions to the methodology for estimating vehicle distribution implemented in 2007.

Exhibit 2-14 Highway Travel by Functional System and Vehicle Type, 2008–2012[1,2]

Functional System Vehicle Type	2008	2010	2012	Annual Rate of Change 2012/2008
Rural				
Interstate				
Passenger Vehicles	181,278	185,212	187,932	0.9%
Single-Unit Trucks	11,970	11,206	9,249	-6.2%
Combination Trucks	49,973	49,229	48,691	-0.6%
Other Arterial				
Passenger Vehicles	322,288	324,467	325,071	0.2%
Single-Unit Trucks	20,176	18,922	17,194	-3.9%
Combination Trucks	31,771	33,023	29,689	-1.7%
Other Rural				
Passenger Vehicles	335,206	327,748	326,522	-0.7%
Single-Unit Trucks	19,286	18,059	17,961	-1.8%
Combination Trucks	16,287	16,281	14,316	-3.2%
Total Rural				
Passenger Vehicles	838,772	837,428	839,525	0.0%
Single-Unit Trucks	51,431	48,188	44,404	-3.6%
Combination Trucks	98,031	98,532	92,696	-1.4%
Urban				
Interstate				
Passenger Vehicles	423,699	427,395	434,394	0.6%
Single-Unit Trucks	16,752	14,485	14,539	-3.5%
Combination Trucks	35,663	35,812	35,614	-0.03%
Other Urban				
Passenger Vehicles	1,403,376	1,415,087	1,426,578	0.4%
Single-Unit Trucks	58,672	48,001	46,018	-5.9%
Combination Trucks	50,131	41,567	35,047	-8.6%
Total Urban				
Passenger Vehicles	1,827,075	1,842,482	1,860,972	0.5%
Single-Unit Trucks	75,423	62,486	60,557	-5.3%
Combination Trucks	85,794	77,379	70,662	-4.7%
Total				
Passenger Vehicles	**2,665,848**	**2,679,910**	**2,700,497**	0.3%
Single-Unit Trucks	**126,855**	**110,674**	**104,961**	-4.6%
Combination Trucks	**183,826**	**175,911**	**163,358**	-2.9%

[1] Data do not include Puerto Rico.

[2] The procedures used to develop estimates of travel by vehicle type have been significantly revised; the data available do not support direct comparisons prior to 2007.

Source: Highway Statistics, various years, Table VM-1.

Bridges by Functional Classification

The Nation's bridges help travelers traverse what would be geographical challenges. Bridges help provide travelers a more direct route to their destination. These direct routes help move passengers and goods efficiently, benefiting the Nation's economic productivity and output.

Exhibit 2-15 presents the number of bridges by functional classification between 2002 and 2012. These bridges are identified by NBI and are at least 20 feet long. The number of bridges increased 0.3 percent from 591,243 to 607,380. Less than three-quarters of the Nation's bridges are located in rural areas with most classified as local. The annual rate of change of bridge numbers in rural areas between 2002 and 2012 decreased 0.2 percent. Bridges in urban areas have increased 1.7 percent in the same period, with the largest increase occurring on urban collectors (3.1 percent).

Exhibit 2-15 Number of Bridges by Functional System, 2002–2012

Functional System	2002	2004	2006	2008	2010	2012	Annual Rate of Change 2012/2002
Rural							
Interstate	27,310	27,648	26,633	25,997	25,223	25,201	-0.8%
Other Principal Arterial	35,215	36,258	35,766	35,594	36,084	36,460	0.3%
Minor Arterial	39,571	40,197	39,521	39,079	39,048	39,123	-0.1%
Major Collector	94,766	94,079	93,609	93,118	93,059	92,875	-0.2%
Minor Collector	49,309	49,391	48,639	48,242	47,866	47,922	-0.3%
Local	209,358	208,641	207,130	205,959	205,609	205,192	-0.2%
Subtotal Rural	455,529	456,214	451,298	447,989	446,889	446,773	-0.2%
Urban							
Interstate	27,924	27,667	28,637	29,629	30,116	30,758	1.0%
Other Freeway and Expressway	16,843	17,112	17,988	19,168	19,791	20,139	1.8%
Other Principal Arterial	24,301	24,529	26,051	26,934	27,373	28,141	1.5%
Minor Arterial	24,510	24,802	26,239	27,561	28,103	28,437	1.5%
Collectors	15,169	15,548	17,618	18,932	20,311	20,590	3.1%
Local	26,592	27,940	29,508	31,183	31,877	32,540	2.0%
Subtotal Urban	135,339	137,598	146,041	153,407	157,571	160,605	1.7%
Unclassified	375	288	222	110	33	2	
Total	**591,243**	**594,100**	**597,561**	**601,506**	**604,493**	**607,380**	**0.3%**

Source: National Bridge Inventory.

NHS by Functional Classification

As noted earlier in this section, most of the Nation's road mileage is located outside the NHS and on highways other than Federal-aid highways. As shown in *Exhibit 2-16*, 5.4 percent of the Nation's road mileage is on the NHS, while only 8.9 percent of the Nation's lane mileage is located on the NHS. Of the Nation's VMT, however, 55.0 percent occurs on the NHS.

The highest share of VMT on the NHS occurs on urban area Interstate facilities and urban area other principal arterials. This observation suggests that a substantial portion of the Nation's VMT occurs during morning and afternoon commutes to urban centers. In rural areas, the highest share of VMT also occurs on Interstate and other principal arterials systems.

The NHS encompasses all of the Interstate System and almost all of the facilities classified as other freeway and expressway, and other principal arterial. Local road mileage and other mileage classified lower than principal arterial represent NHS intermodal connectors.

Exhibit 2-16 Highway Route Miles, Lane Miles, and Vehicle Miles Traveled on the National Highway System Compared with All Roads, by Functional System, 2012[1]

	Route Miles		Lane Miles		VMT (Millions)	
Functional System	**Total on NHS**	**Percent on NHS**	**Total on NHS**	**Percent on NHS**	**Total on NHS**	**Percent on NHS**
Rural NHS						
Interstate	30,564	100.0%	124,927	100.0%	246,334	100.0%
Other Freeway and Expressway	4,284	97.5%	16,547	99.7%	20,115	99.8%
Other Principal Arterial	91,181	99.7%	239,899	99.7%	202,580	99.6%
Minor Arterial	2,630	1.9%	6,426	2.3%	4,839	3.2%
Major Collector	662	0.2%	1,439	0.2%	1,055	0.6%
Minor Collector	5	0.002%	9	0.002%	2	0.004%
Local	38	0.002%	77	0.002%	15	0.01%
Subtotal Rural NHS	**129,364**	**4.3%**	**389,324**	**6.4%**	**474,940**	**48.6%**
Urban NHS						
Interstate	17,149	100.0%	95,194	100.0%	489,580	100.0%
Other Freeway and Expressway	11,404	99.0%	53,665	99.1%	223,353	99.2%
Other Principal Arterial	63,407	96.7%	227,208	96.9%	448,105	97.4%
Minor Arterial	1,439	1.3%	4,541	1.6%	7,086	1.9%
Major Collector	384	0.3%	990	0.4%	1,018	0.6%
Minor Collector	9	0.3%	20	0.3%	9	0.2%
Local	101	0.01%	242	0.02%	137	0.05%
Subtotal Urban NHS	**93,893**	**8.3%**	**381,860**	**15.1%**	**1,169,288**	**58.2%**
Total NHS	**223,257**	**5.4%**	**771,184**	**8.9%**	**1,644,228**	**55.0%**

[1] Data reflect the expansion of the NHS required by MAP-21.

Source: Highway Performance Monitoring System.

Transit System Characteristics

System History

The first transit systems in the United States date to the late 19th century. These systems were privately owned, for-profit businesses that were instrumental in defining the urban communities of that time. By the postwar period, competition from the private automobile was preventing transit businesses from operating at a profit. As transit businesses started to fail, local, State, and national government leaders began to realize the importance of sustaining transit services. In 1964, Congress passed the Urban Mass Transportation Act, establishing the Urban Mass Transit Agency to administer Federal funding for transit systems. The Act changed the character of the industry by specifying that Federal funds for transit be given to public agencies rather than private firms; this funding shift accelerated the transition from private to public ownership and operation of transit systems. The Act also required local governments to contribute matching funds as a condition for receiving Federal aid for transit services—setting the stage for the multilevel governmental partnerships that characterize today's transit industry.

State government involvement in the provision of transit services is usually through financial support and performance oversight. Some States, however, have undertaken outright ownership and operation of transit services. Connecticut, Georgia, Louisiana, Maryland, Washington, the U.S. Virgin Islands, and Puerto Rico directly own and operate transit systems. Pennsylvania contracts for transit services. New Jersey Transit, a statewide company, and numerous private fixed-route bus systems operate the State's transit services. New Jersey Transit provides buses to private bus systems but is not involved with their operations or oversight.

In 1962, Congress passed legislation requiring the formation of metropolitan planning organizations (MPOs) for urbanized areas with populations greater than 50,000. MPOs are composed of State and local officials who work to address transportation planning needs of urbanized areas at a regional level. Twenty-nine years later, the Intermodal Surface Transportation Efficiency Act of 1991 (ISTEA) made MPO coordination a prerequisite for Federal funding of many transit projects.

In addition, the ISTEA reauthorization made several other changes to transportation law, including changing the name of the Urban Mass Transportation Administration to the Federal Transit Administration (FTA). On the urban side, ISTEA increased transit formula grant funding to all agencies and initiated the use of a formula to allocate capital funds, rather than determine funding allocation on a discretionary project basis. The Act also increased the flexibility in using Highway Trust funds between transit and highway projects.

The Transportation Equity Act for the 21st Century (TEA-21) was passed in 1998 and over the next 6 years increased transit funding by 70 percent. Part of this additional funding was to offset the increased costs of implementing service for persons with disabilities under the Americans with Disabilities Act of 1990 (ADA). The ADA required public transit services to be open to the public without discrimination and to meet all other requirements of the Act. The ADA also further increased the flexibility in the use of Federal funds.

The Safe, Accountable, Flexible, Efficient Transportation Equity Act: A Legacy for Users (SAFETEA-LU) was enacted in 2005. This Act created some new programs—especially for smaller transit providers—and new program definitions. Within the urban formula program, a new formula allocation was added for Small Transit Intensive Cities (STIC). In the new starts program, a Small Starts program was created, encouraging cost-effective alternative approaches to transit projects such as bus rapid transit, rather than more expensive rail systems. In the rural (other than urbanized area) program, funding was greatly increased for rural transit providers, intercity fixed-route bus transportation became eligible for rural funds, and funds were made available for Native American Tribal transit. SAFETEA-LU also made funding available for parks and public lands. SAFETEA-LU extension acts were continued until July 2012.

On July 6, 2012, Congress passed the new Moving Ahead for Progress in the 21st Century (MAP-21) reauthorization act, covering Fiscal Years 2013 and 2014. MAP-21 is the current law. The law retained the basic structure of the urban formula program, but increased the STIC formula funding and allowed certain smaller systems (100 fixed-route buses or fewer) in large urban systems to use some formula funds for operating expense. MAP-21 also added a new factor: the number of low-income individuals. The Act gave FTA safety oversight authority and set aside funds for FTA to create an office for administering a safety oversight program for public transit. Funds for the rural program are to be allocated as in the past, but a service factor—vehicle revenue miles—and a factor for low-income individuals were added to the formula allocation factors. Funds for Tribal transit were increased, and some funds were distributed by a new formula based in part on vehicle revenue miles. The most dramatic change, however, was the elimination of the Fixed-Guideway Modernization capital program and the creation of the new, formula-based State of Good Repair program in its place. The State of Good Repair program would dedicate capital funds to the repair, upgrading, and modernization of the Nation's transit fixed-guideway infrastructure. This fixed-guideway infrastructure would include the Nation's rail transit systems, high-intensity motor bus systems operating on HOV (high occupancy vehicle) lanes, ferries, and bus rapid transit systems. The Act requires transit agencies to develop a capital asset report that inventories their capital assets and evaluates the condition of those assets.

System Infrastructure

Urban Transit Agencies

State and local transit agencies have evolved into several different institutional models. A transit provider can be a unit of a regional transportation agency; operated directly by the State, county, or city government; or an independent agency with an elected or appointed Board of Governors. Transit operators can provide service directly with their own equipment or they might purchase transit services through an agreement with a contractor.

In 2012, 725 reporters in urbanized areas submitted data to the National Transit Database (NTD). Five agencies were consolidated entities reporting on behalf of 80 transit providers. Thus, the total number of urban providers was 800. Of the 725 reporters, 697 were public agencies, including 369 city, county, or local government transportation units or departments, 250 independent public authorities or agencies for transit service, and 8 State Departments of Transportation (DOTs). The remaining 28 agencies were either private operators or independent agencies (e.g., for-profit organizations, nonprofit organizations, and Indian Tribes).

All transit providers that receive funds from FTA must report to NTD. In the past, small systems operating fewer than nine vehicles could request a reporting exemption; however, all small systems are now required to submit a simplified report to NTD each year. This small-system reporting waiver was granted to 213 agencies having fewer than 30 vehicles in maximum service and not operating fixed-guideway service.

Of the 512 agencies that reported providing service on 1,282 separate modal networks, all but 107 operated more than one mode. In 2012, an additional 1,703 transit operators were serving rural areas. Some agencies that do not have a reporting requirement to NTD will choose to submit a report because doing so can help their region receive additional Federal transit funding.

Urbanized Areas (UZA) with Population over 1 Million in 2010 Census

UZA Rank	UZA Name	2010 Population	2012 Unlinked Transit Trips (in Thousands)
1	New York-Newark, NY-NJ-CT	18,351,295	4,181,730
2	Los Angeles-Long Beach-Anaheim, CA	12,150,996	671,381
3	Chicago, IL-IN	8,608,208	663,752
4	Miami, FL	5,502,379	166,350
5	Philadelphia, PA-NJ-DE-MD	5,441,567	386,746
6	Dallas-Fort Worth-Arlington, TX	5,121,892	79,377
7	Houston, TX	4,944,332	81,381
8	Washington, DC-VA-MD	4,586,770	485,448
9	Atlanta, GA	4,515,419	144,090
10	Boston, MA-NH-RI	4,181,019	409,749
11	Detroit, MI	3,734,090	47,954
12	Phoenix-Mesa, AZ	3,629,114	72,195
13	San Francisco-Oakland, CA	3,281,212	435,867
14	Seattle, WA	3,059,393	196,767
15	San Diego, CA	2,956,746	102,851
16	Minneapolis-St. Paul, MN-WI	2,650,890	93,864
17	Tampa-St. Petersburg, FL	2,441,770	30,025
18	Denver-Aurora, CO	2,374,203	98,716
19	Baltimore, MD	2,203,663	112,927
20	St. Louis, MO-IL	2,150,706	49,559
21	San Juan, PR	2,148,346	59,964
22	Riverside-San Bernardino, CA	1,932,666	25,342

Urbanized Areas (UZA) with Population over 1 Million in 2010 Census (continued)

UZA Rank	UZA Name	2010 Population	2012 Unlinked Transit Trips (in Thousands)
23	Las Vegas-Henderson, NV	1,886,011	65,145
24	Portland, OR-WA	1,849,898	114,196
25	Cleveland, OH	1,780,673	49,139
26	San Antonio, TX	1,758,210	50,804
27	Pittsburgh, PA	1,733,853	67,770
28	Sacramento, CA	1,723,634	30,971
29	San Jose, CA	1,664,496	43,487
30	Cincinnati, OH-KY-IN	1,624,827	21,479
31	Kansas City, MO-KS	1,519,417	17,453
32	Orlando, FL	1,510,516	29,250
33	Indianapolis, IN	1,487,483	10,328
34	Virginia Beach, VA	1,439,666	18,460
35	Milwaukee, WI	1,376,476	47,423
36	Columbus, OH	1,368,035	18,763
37	Austin, TX	1,362,416	35,660
38	Charlotte, NC-SC	1,249,442	28,794
39	Providence, RI-MA	1,190,956	21,611
40	Jacksonville, FL	1,065,219	12,706
41	Memphis, TN-MS-AR	1,060,061	10,035
42	Salt Lake City-West Valley City, UT	1,021,243	42,366
Total		**135,639,208**	**9,331,875**

The Nation's fixed-route bus and demand-response systems are much more extensive than the Nation's rail transit system. Bus fixed-route service includes three distinct modes: regular fixed-route bus, commuter bus, and bus rapid transit.

In 2012, 661 agencies reported fixed-route bus service, including 619 regular bus systems, 67 commuter bus systems, and 10 bus rapid transit systems. Some agencies operate more than one type of fixed-route bus, and so the sum of the three types does not equal the number of agencies operating these systems.

Transit agencies reported 629 demand-response systems (not including demand-response taxi) in urban areas, 18 heavy rail systems, 29 commuter rail systems, 4 hybrid rail systems, 25 light rail systems, and 17 street car systems (some of which are not yet in service).

The number of fixed-route bus systems is greater than the number of demand-response systems because in some urban areas a single, consolidated entity operates paratransit service, while more than one agency provides fixed-route service.

Although every major urbanized area in the United States has fixed-route bus and demand-response systems, 35 urbanized areas were served by at least one of the three primary rail modes, including 20 by commuter rail, 22 by light rail, and 12 by heavy rail. *Exhibit 2-17* depicts the number of passenger cars for each rail mode by urbanized area.

Exhibit 2-17 Rail Modes Serving Urbanized Areas

Uza Rank	Urbanized Area	Commuter Rail Vehicles	Heavy Rail Vehicles	Light Rail Vehicles	Streetcar Vehicles	Hybrid Rail Vehicles	Total Rail Vehicles
1	New York-Newark, NY-NJ-CT	3,441	5,598	56	-	15	9,110
2	Los Angeles-Long Beach-Anaheim, CA	172	70	140	-	-	382
3	Chicago, IL-IN	1,114	1,070	-	-	-	2,184
4	Miami, FL	40	76	-	-	-	116
5	Philadelphia, PA-NJ-DE-MD	347	369	-	126	-	842
6	Dallas-Fort Worth-Arlington, TX	23	-	100	-	-	123
7	Houston, TX	-	-	18	-	-	18
8	Washington, DC-VA-MD	87	868	-	-	-	955
9	Atlanta, GA	-	182	-	-	-	182
10	Boston, MA-NH-RI	416	336	144	-	-	896
12	Phoenix-Mesa, AZ	-	-	26	-	-	26
13	San Francisco-Oakland, CA	100	534	131	24	-	789
14	Seattle, WA	56	-	26	5	-	87
15	San Diego, CA	24	-	95	-	8	127
16	Minneapolis-St. Paul, MN-WI	22	-	27	-	-	49
17	Tampa-St. Petersburg, FL	-	-	-	3	-	3
18	Denver-Aurora, CO	-	-	102	-	-	102
19	Baltimore, MD	132	54	38	-	-	224
20	St. Louis, MO-IL	-	-	58	-	-	58
21	San Juan, PR	-	32	-	-	-	32
24	Portland, OR-WA	-	-	104	7	4	115
25	Cleveland, OH	-	20	13	-	-	33
27	Pittsburgh, PA	-	-	56	-	-	56
28	Sacramento, CA	-	-	61	-	-	61
29	San Jose, CA	-	-	55	-	-	55
34	Virginia Beach, VA	-	-	7	-	-	7
37	Austin, TX	-	-	-	-	4	4
38	Charlotte, NC-SC	-	-	14	-	-	14
41	Memphis, TN-MS-AR	-	-	-	10	-	10
42	Salt Lake City-West Valley City, UT	36	-	82	-	-	118
44	Nashville-Davidson, TN	7	-	-	-	-	7
46	Buffalo, NY	-	-	23	-	-	23
47	Hartford, CT	28	-	-	-	-	28
49	New Orleans, LA	-	-	-	21	-	21
56	Albuquerque, NM	25	-	-	-	-	25
88	Little Rock, AR	-	-	-	3	-	3
102	Stockton, CA	18	-	-	-	-	18
104	Denton-Lewisville, TX	8	-	-	-	-	8
177	Portland, ME	14	-	-	-	-	14
256	Kenosha, WI-IL	-	-	-	1	-	1
	Grand Total	**6,110**	**9,209**	**1,376**	**200**	**31**	**16,926**

Source: National Transit Database.

In addition to fixed-route bus systems, demand-response systems, and rail modes, 74 publicly operated transit vanpool systems, 23 ferryboat systems, 5 trolleybus systems, 8 monorail/automated guideway systems, 3 inclined plane systems, 1 cable car system, and 1 Público were operating in urbanized areas of the United States and its territories.

The transit statistics presented in this report include those for the San Francisco Cable Car, Seattle Monorail, Roosevelt Island Aerial Tramway in New York, and Alaska Railroad (a long-distance passenger rail system included as public transportation by statutory exemption).

Transit Fleet

Exhibit 2-18 provides an overview of the Nation's 199,639 transit vehicles in 2012 by type of vehicle and size of urbanized area. Although some types of vehicles are specific to certain modes, many vehicles—particularly small buses and vans—are used by several different transit modes. For example, vans are used to provide vanpool, demand-response, Público, or fixed-route bus services. The limited classification options for vehicle type in NTD can make classifying smaller vehicles difficult.

Exhibit 2-18 Transit Active Fleet by Vehicle Type, 2012[1,2]

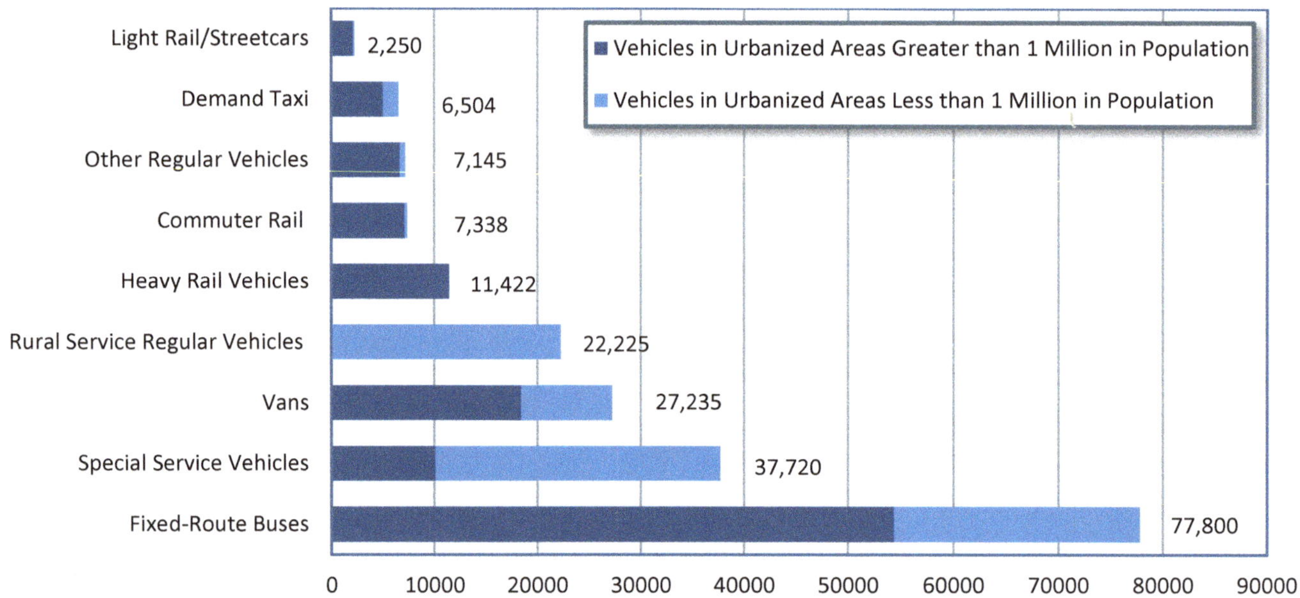

[1] *Vehicle types: "Demand Taxi" includes taxicab sedan, taxicab station wagon, and taxicab vans. "Other Regular Vehicles" includes aerial tramway vehicles, Alaska railroad vehicles, automated guideway vehicles, automobiles, cable cars, ferryboats, inclined plane vehicles, jitneys, Públicos, and trolleybuses. "Commuter Rail" includes commuter rail locomotives, commuter rail passenger coaches, and commuter rail self-propelled passenger cars. "Fixed-Route Buses" includes articulated buses, double-decker buses, school buses, and over-the-road buses.*

[2] *Source for "Special Service Vehicles" is the FTA, Fiscal Year Trends Report on the Use of Section 5310, Elderly Persons and Persons with Disabilities program funds, 2002.*

Source: National Transit Database.

Exhibit 2-19 shows the composition of the Nation's urban transit road vehicle fleet in 2012. More than one-third of these vehicles, or 41 percent, are full-sized motor buses. Additional information on trends in the number and condition of vehicles over time is included in Chapter 3. Vans, as presented here, are the familiar 10-seat passenger vans. Articulated buses are the long vehicles articulated for better maneuverability on city streets. Full-sized buses are the standard 40-foot, 40-seat city buses. Mid-sized buses are in the 30-foot, 30-seat range. Small buses, typically built on truck chassis ("cutaways"), are shorter and seat around 20 people.

Exhibit 2-19 Composition of Urban Transit Road Vehicle Fleet, 2012

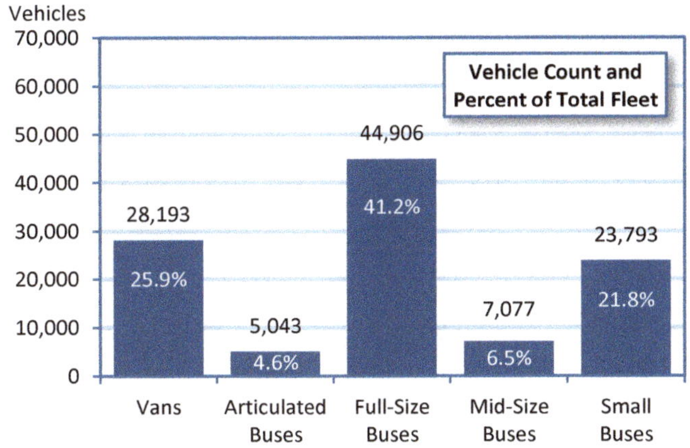

Source: Transit Economic Requirements Model and National Transit Database.

Track, Stations, and Maintenance Facilities

Maintenance facility counts are broken down by mode and by size of urbanized area for directly operated service in *Exhibit 2-20*. Modes such as hybrid rail, demand-response taxi, and Público are not included because all service is purchased. Chapter 3 includes data on the age and condition of these facilities.

A single facility can be used by more than one mode. In these cases, the count of facilities is prorated based on the number of peak vehicles for each mode.

As *Exhibit 2-21* shows, transit providers operated 12,617 miles of track and served 3,281 stations in 2012. The Nation's rail system mileage is dominated by the longer distances generally covered by commuter rail. Light and heavy rail typically operate in more densely developed areas and have more stations per track mile.

Exhibit 2-20 Maintenance Facilities for Directly Operated Services, 2012

Maintenance Facility Type[1]	Population Category: Over 1 Million	Population Category: Under 1 Million	Total
Heavy Rail	59	0	59
Commuter Rail	46	0	46
Light Rail	35	1	36
Streetcar Rail	9	2	11
Other Rail[2]	4	5	9
Fixed-Route Bus	305	281	586
Commuter Bus	24	6	30
Bus Rapid Transit	2	0	3
Demand Response	52	116	168
Vanpool	5	5	9
Ferryboat	8	1	9
Trolleybus	4	1	5
Total Urban Maintenance Facilities	**553**	**418**	**971**
Rural Transit[3]		727	727
Total Maintenance Facilities	**553**	**1,145**	**1,698**

[1] Includes owned and leased facilities.
[2] Alaska railroad, automated guideway, cable car, inclined plane, and monorail.
[3] Vehicles owned by operators receiving funding from FTA as directed by 49 USC Section 5311. These funds are for transit services in areas with populations of less than 50,000 (Section 5311, Status of Rural Public Transportation, 2000; Community Transportation Association of America, April 2001).

Source: National Transit Database.

Exhibit 2-21 Transit Rail Mileage and Stations, 2012

Urbanized Area Track Mileage	
Heavy Rail	2,274
Commuter Rail	7,738
Light Rail	1,419
Hybrid Rail	173
Streetcar Rail	286
Other Rail and Tramway[1]	728
Total Urbanized Area Track Mileage	**12,617**
Urbanized Area Transit Rail Stations Count	
Heavy Rail	1,044
Commuter Rail	1,234
Light Rail	794
Hybrid Rail	49
Streetcar Rail	85
Other Rail and Tramway[1]	75
Total Urbanized Area Transit Rail Stations	**3,281**

[1] Includes Alaska railroad, automated guideway, cable car, inclined plane, monorail, and aerial tramway.

Source: National Transit Database.

Rural Transit Systems (Section 5311 Providers)

The FTA first instituted rural data reporting to NTD in 2006. In 2012, 1,703 transit operators reported providing rural service; additionally, 235 urban agencies reported providing rural service. Together, these agencies reported 518 million unlinked passenger trips and 625 million vehicle revenue miles. These data include the more than 2 million unlinked passenger trips that 124 Indian Tribes provided. Rural systems provide both traditional fixed-route bus and demand-response services, with 1,108 demand-response services, 56 demand taxi services, 60 commuter bus services, 6 ferryboat services, 515 fixed-route bus services, and 21 vanpool services. They reported 22,225 vehicles in 2012. *Exhibit 2-22* shows the number of rural transit vehicles in service in 2010 and 2012.

Exhibit 2-22 Rural Transit Vehicles, 2010 and 2012[1]

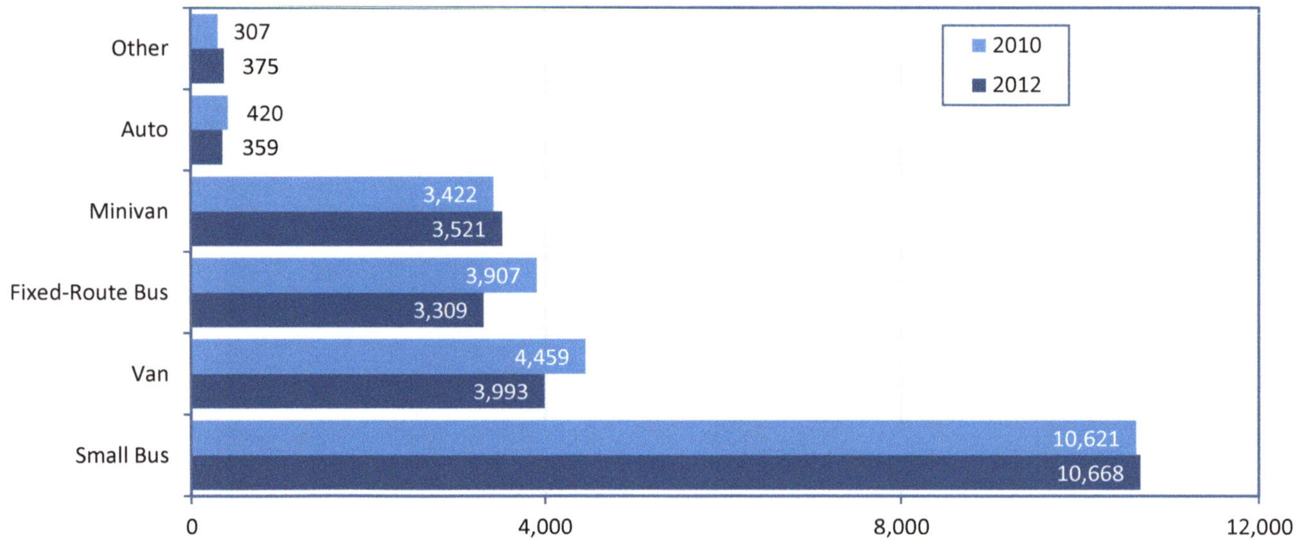

[1] *Other includes ferryboat, over-the-road bus, school bus, sport utility vehicle, and other similar vehicles.*

Source: National Transit Database.

Transit System Characteristics for Americans with Disabilities and the Elderly

The ADA is intended to ensure that persons with disabilities have access to the same facilities and services as other Americans, including transit vehicles and facilities. This equality of access is brought about by upgrading transit vehicles and facilities on regular routes, providing demand-response transit service for those individuals who still cannot use regular transit service, and operating special service vehicles by private entities and some public organizations, often with the assistance of FTA funding.

The overall percentage of transit vehicles that are ADA compliant has not significantly changed in recent years. In 2012, 77.6 percent of all transit vehicles reported in NTD were ADA compliant. Although this percentage has decreased slightly from 79.3 percent in 2010, it has increased substantially from the 73.3 percent reported for 2000. The percentage of vehicles compliant with the ADA for each mode is shown in *Exhibit 2-23.*

Exhibit 2-23 Urban Transit Operators' ADA Vehicle Fleets by Mode, 2012

Transit Mode	Active Vehicles	ADA-Compliant Vehicles	Percentage of Active Vehicles that are ADA Compliant
Rail			
Heavy Rail	11,422	10,988	96.2%
Commuter Rail	7,263	3,960	54.5%
Light Rail	1,981	1,826	92.2%
Alaska Railroad	63	23	36.5%
Automated Guideway/Monorail	156	156	100.0%
Cable Car	38	0	0.0%
Inclined Plane	8	6	75.0%
Hybrid Rail	44	24	54.5%
Streetcar	316	100	31.6%
Total Rail	**21,291**	**17,083**	**80.2%**
Nonrail			
Fixed-Route Bus	62,204	61,524	98.9%
Demand Response	30,846	26,013	84.3%
Vanpool	13,537	144	1.1%
Ferryboat	145	118	81.4%
Trolleybus	572	572	100.0%
Público	2,873	0	0.0%
Bus Rapid Transit	90	90	100.0%
Commuter Bus	1,994	1,928	96.7%
Demand Response Taxi	6,142	895	14.6%
Total Nonrail	**118,403**	**91,284**	**77.1%**
Total All Modes	**139,694**	**108,367**	**77.6%**

Source: National Transit Database.

In addition to the services urban and rural transit operators provide, the most recent American Public Transportation Association fact book indicates that approximately 4,800 nonprofit providers operate in rural and urban areas. These providers are eligible to receive funding from

FTA for Transportation for Persons with Disabilities and the Elderly. This funding supports "special" transit services (i.e., demand-response). Nonprofit providers include religious organizations, senior citizen centers, rehabilitation centers, nursing homes, community action centers, sheltered workshops, and coordinated human services transportation providers.

The ADA requires that new transit facilities and alterations to existing facilities be accessible to the disabled. In 2012, 78.7 percent of total transit stations were ADA compliant, an increase from the 76.0 percent compliant in 2010. Earlier data for this parameter might not be comparable to data provided in this current report due to improvements in reporting quality. *Exhibit 2-24* presents the number of urban transit ADA stations and percentage of total stations by mode.

Exhibit 2-24 Urban Transit Operators' ADA-Compliant Stations by Mode, 2012

Transit Mode	Total Stations	ADA-Compliant Stations	Percentage of Stations that are ADA Compliant
Rail			
Heavy Rail	1,044	542	51.9%
Commuter Rail	1,234	822	66.6%
Light Rail	794	725	91.3%
Alaska Railroad	10	10	100.0%
Automated Guideway/ Monorail	57	56	98.2%
Inclined Plane	8	7	87.5%
Hybrid Rail	49	49	100.0%
Street Car	85	41	48.2%
Total Rail	**3,281**	**2,252**	**68.6%**
Nonrail			
Fixed-Route Bus	1,355	1,337	98.7%
Ferryboat	94	89	94.7%
Trolleybus	5	5	100.0%
Bus Rapid Transit	7	7	100.0%
Commuter Bus	195	195	100.0%
Total Nonrail	**1,656**	**1,633**	**98.6%**
Total All Modes	**4,937**	**3,885**	**78.7%**

Source: National Transit Database.

Under the ADA, FTA was given responsibility for identifying key rail stations and facilitating the accessibility of these stations to disabled persons by July 26, 1993. Rail stations identified as "key" have the following characteristics:

- The number of passengers boarding exceeds the average number of passengers boarding on the rail system as a whole by at least 15 percent.
- The station is a major point where passengers shift to other transit modes.
- The station is at the end of a rail line, unless it is close to another accessible station.
- The station serves a "major" center of activities, including employment or government centers, institutions of higher education, and major health facilities.

Although ADA legislation required all key stations to be accessible by July 26, 1993, the DOT ADA regulation—Title 49 Code of Federal Regulations (CFR) Part 37.47(c)(2)—permitted the FTA

Administrator to grant extensions up to July 26, 2020, for stations that required extraordinarily expensive structural modifications to achieve compliance. Of the 680 key rail stations in 2010, 8 stations (1.2 percent) were under FTA-approved time extensions. The total number of key rail stations has changed slightly over the years as certain stations have closed. As of May 23, 2014, 680 stations were designated as key. Of these, 607 were accessible and fully compliant, 30 were accessible but not fully compliant, and 28 were self-certified as accessible. "Accessible but not fully compliant" means that these stations are functionally accessible (i.e., persons with disabilities, including wheelchair users, can make use of the station), but minor outstanding issues must be addressed for the station to be fully compliant; example issues include missing or misallocated signage and parking-lot striping errors. Fifteen key rail stations that are not yet compliant are in the planning, design, or construction stage. Of these, seven stations are under FTA-approved time extensions up to 2020 (as provided under 49 CFR §37.47[c][2]). FTA continues to focus its attention on the eight stations that are not accessible and are not under a time extension, and on the seven stations with time extensions that will be expiring in the coming years.

Transit System Characteristics: Alternative Fuel Vehicles

Exhibit 2-25 shows that the share of alternative fuel buses increased from 10.5 percent in 2002 to 22.8 percent in 2012. In 2012, 12.5 percent of buses used compressed natural gas, 8.7 percent used biodiesel, and 1.6 percent used liquefied natural gas or petroleum gas. Conventional fuel buses, which make up most of the U.S. bus fleet, used diesel fuel and gasoline. In 2012, hybrid buses made up 5.9 percent of urban bus fleets as shown in *Exhibit 2-26*. These hybrid vehicles are more efficient than conventional fuel buses, but they are not technically counted as alternative-fuel vehicles.

Exhibit 2-25 Percentage of Urban Bus Fleet Using Alternative Fuels, 2002–2012

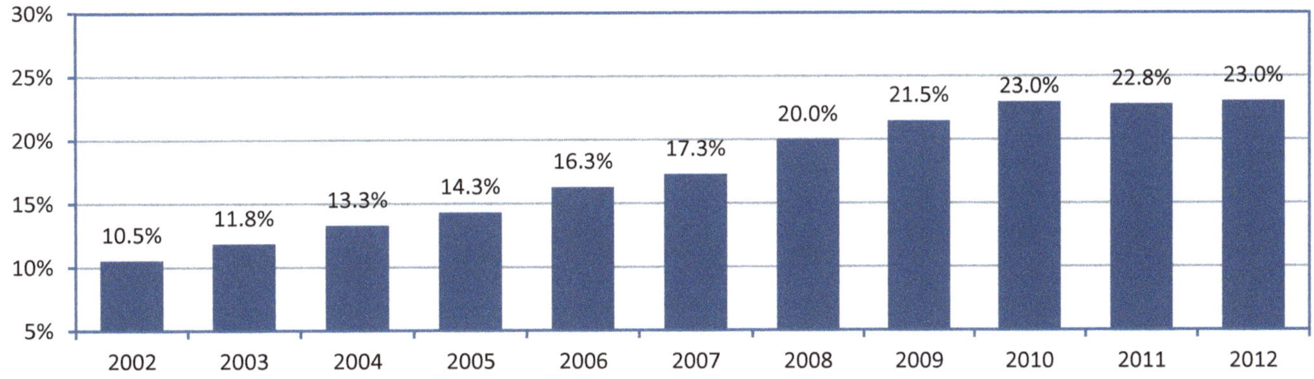

Source: National Transit Database.

Exhibit 2-26 Hybrid Buses as a Percentage of Urban Bus Fleet, 2007–2012

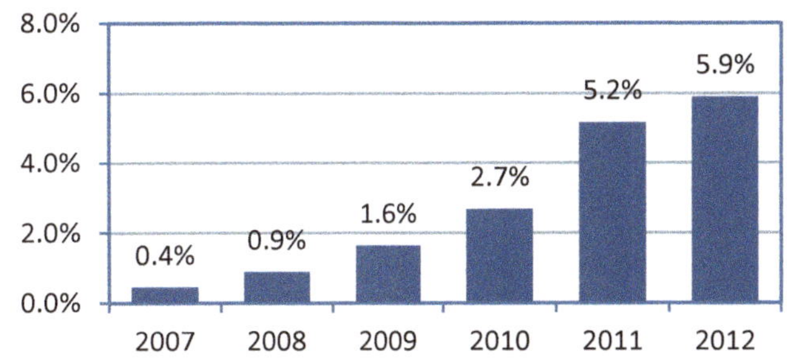

Source: National Transit Database.

chapter 3

System Conditions

Highway System Conditions

As referenced in the Introduction to Part I, a key feature of the Moving Ahead for Progress in the 21st Century Act (MAP-21) was the establishment of a performance- and outcome-based program, with the objective of having States invest resources in projects that collectively will make progress toward achieving national goals. For infrastructure condition, MAP-21 established a goal to maintain highway assets in a state of good repair.

Although there is broad consensus that the Nation's transportation infrastructure falls short of a state of good repair, no definition of the term has been uniformly accepted for all transportation assets. The condition of some asset types traditionally has been measured using multiple quantitative indicators, which owners of different transportation assets often weight differently during the assessment process. The condition of other assets has been measured using a single qualitative rating, which introduces subjectivity into the assessment process.

As part of its ongoing efforts to encourage the integration of Transportation Performance Management principles into project selection decisions and to implement related provisions in MAP-21, the Federal Highway Administration (FHWA) issued a Notice of Proposed Rulemaking that included a pavement and bridge performance measures rule (PM-2) on January 5, 2015. Some of the information presented in this section is influenced by the proposed performance measures for pavement and bridge condition presented in the Notice of Proposed Rulemaking; future editions of the C&P report will more fully integrate the final measures that emerge from this rulemaking process.

Data Sources

Pavement condition data are reported to FHWA through the Highway Performance Monitoring System (HPMS). Currently, HPMS requires reporting for Federal-aid highways only, which represent about 25 percent of the Nation's road mileage but carry more than 80 percent of the Nation's travel. States are not required to report on roads functionally classified as rural minor collectors, rural local, or urban local, which comprise the remaining 75 percent of the Nation's road mileage.

HPMS contains data on multiple types of pavement distresses. Data on pavement roughness are used to assess the pavement ride quality experienced by highway users. For some functional systems, States can report a general PSR (Pavement Serviceability Rating) value in place of an actual measurement of pavement roughness through the IRI (International Roughness Index). Other measures of pavement distress include pavement cracking, pavement rutting (surface depressions in the vehicle wheel path, generally relevant only to asphalt pavements), and pavement faulting (the vertical displacement between adjacent jointed sections on concrete pavements).

Condition data for all bridges on the Nation's roadways are reported to FHWA through the National Bridge Inventory (NBI). NBI reflects information the States, Federal agencies, and Tribal governments gather during periodic safety inspections of bridges. Most inspections occur once every 24 months. If a structure shows advanced deterioration, the frequency of inspections might increase so that the safety of the structure can be monitored more closely. Based on certain criteria, some bridges that are in satisfactory or better condition might be inspected between 24 and 48 months with prior FHWA approval. Approximately 83 percent of bridges are inspected every 24 months, 12 percent every 12 months, and 5 percent on a maximum 48-month cycle.

Bridge inspectors are trained to inspect bridges based on, as a minimum, the criteria in the National Bridge Inspection Standards. Routine inspections are required for all structures in the NBI database, 473,709 bridges and 133,589 culverts, with a span greater than 20 feet (6.1 meters) located on public roads.

The NBI database contains condition ratings on the three primary components of a bridge: deck, superstructure, and substructure. The bridge deck, supported by the superstructure, is the surface on which vehicles travel. The superstructure transfers the load of the deck and bridge traffic to the substructure, which provides support for the entire bridge. Such ratings are not reported for the culverts represented in the NBI, as culverts are self-contained units typically located under roadway fill, and thus do not have a deck, superstructure, or substructure. For culverts, a general condition rating is applied instead.

Summary of Current Highway and Bridge Conditions

The PM-2 Notice of Proposed Rulemaking proposed classifications of "Good," "Fair," and "Poor" to assess the conditions of pavements and bridges based on combinations of ratings for individual metrics. This chapter does not include statistics for those combinations, but some data are presented for the individual metrics that would factor into computing the statistics. *Exhibit 3-1* identifies criteria for "Good," "Fair," and "Poor" classifications for several individual metrics, based in part on the information laid out in the PM-2 Notice of Proposed Rulemaking. This chapter also references an additional term pertaining to pavement ride quality: "Acceptable" ride quality combines the "Good" and "Fair" categories referenced in *Exhibit 3-1*.

Condition of Pavements on Federal-aid Highways

As shown in *Exhibit 3-2*, approximately 36.4 percent of pavement miles on Federal-aid highways were rated as having good ride quality in 2012, 43.9 percent had fair ride quality, and 19.7 percent had poor ride quality.

When weighted by vehicle miles traveled (VMT) rather than miles of pavement, ride quality appears significantly better. In 2012, approximately 44.9 percent of VMT on Federal-aid highways was on pavements with good ride quality, while only 16.7 percent of VMT on Federal-aid highways was on pavements with poor ride quality. The differences between the mileage-based and VMT-weighted measures imply that, on average, the Nation's roadways with higher traffic volumes have better ride quality than those with lower traffic volumes. This result is positive from a system user perspective, as the VMT-weighted measures better reflect the experience of the individual driver.

Exhibit 3-1 Condition Rating Classifications Used in the 2015 C&P Report

Condition Metric	Rating Criteria	Good	Fair	Poor
Pavement Ride Quality [1]	The International Roughness Index (IRI) measures the cumulative deviation from a smooth surface in inches per mile.	IRI < 95	IRI 95 to 170	IRI > 170
Pavement Ride Quality (alternative)	For roads functionally classified as urban minor arterials, rural or urban major collectors, or urban minor collectors, States can instead report a Present Serviceability Rating (PSR) on a scale of 0 to 5.	PSR ≥ 3.5	PSR ≥ 2.5 and < 3.5	PSR < 2.5
Pavement Cracking	For asphalt pavements, cracking is measured as the percentage of the pavement surface in the wheel path in which interconnected cracks are present. For concrete pavements cracking is measured as the percent of cracked concrete panels in the evaluated section.	<5%	5% to 10%	>10%
Pavement Rutting (Asphalt Pavements only)	Rutting is measured as the average depth in inches of any surface depression present in the vehicle wheel path.	<0.20	0.20 to 0.40	>0.40
Pavement Faulting (Concrete Pavements only)	Faulting is measured as the average vertical displacement in inches between adjacent jointed concrete panels.	<0.05	0.05 to 0.15	>0.15
Bridge Deck Condition	Ratings are on a scale from 0 "Failed" to 9 "Excellent."	≥7	5 to 6	≤4
Bridge Superstructure Condition	Ratings are on a scale from 0 "Failed" to 9 "Excellent."	≥7	5 to 6	≤4
Bridge Substructure Condition	Ratings are on a scale from 0 "Failed" to 9 "Excellent."	≥7	5 to 6	≤4
Culvert Condition	Ratings are on a scale from 0 "Failed" to 9 "Excellent."	≥7	5 to 6	≤4

[1] The PM-2 NPRM sets a different standard for Fair versus Poor ride quality in areas with population over 1 million, setting the break point at 220 rather than 170. This report did not follow this approach, in order to better align with the definition of Acceptable ride quality traditionally used in this report, which includes pavements with IRI values <= 170 inches per mile.

Exhibit 3-2 Federal-Aid Highway Pavement Conditions, 2012

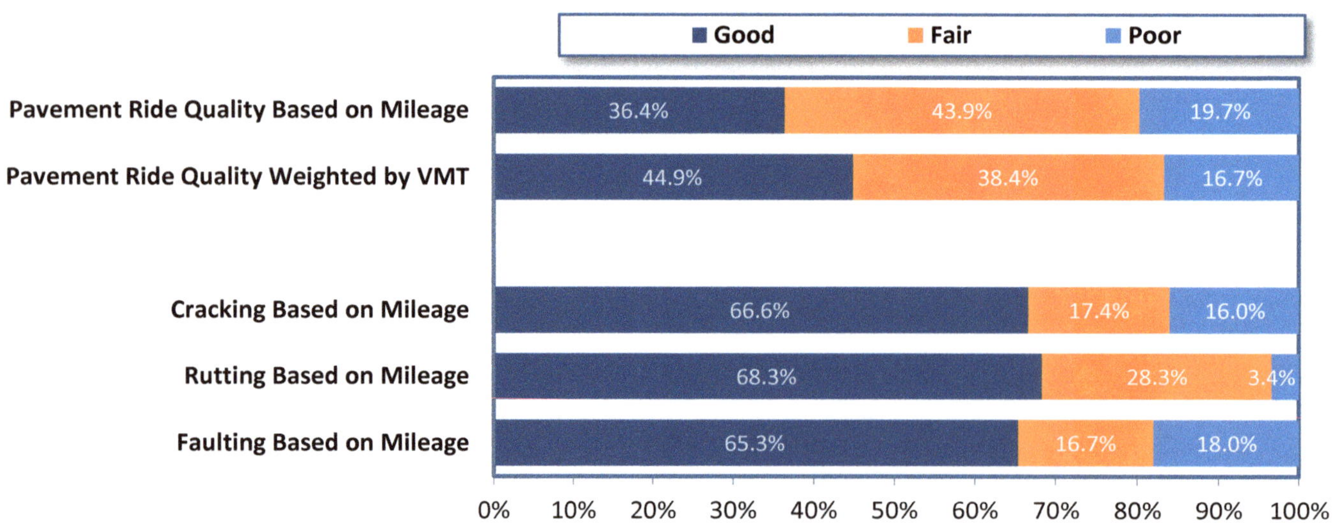

Source: Highway Performance Monitoring System.

In 2012, approximately 66.6 percent of pavements on Federal-aid highways had good cracking ratings, 68.3 percent had good rutting ratings (where applicable), and 65.3 percent had good faulting ratings (where applicable). Approximately 16.0 percent of pavements on Federal-aid highways had poor cracking ratings, 3.4 percent had poor rutting ratings, and 18.0 percent had poor faulting ratings.

Condition of Bridges – Systemwide

As shown in *Exhibit 3-3,* the decks of approximately 59.1 percent of bridges were rated as good condition in 2012; 4.9 percent were rated as poor condition. A higher percentage of bridge superstructures had a good rating (61.2 percent) and a higher percentage was rated as poor (5.2 percent). Bridge substructures were in the worst condition among the three primary bridge components, with only 58.0 percent rated as good and 6.7 percent rated as poor.

In 2012, approximately 64.2 percent of culverts were rated as good condition, while only 2.2 percent were rated as poor condition. Note that the analyses of future bridge investment presented in Part II of this report exclude culverts; costs associated with culverts are instead indirectly factored into the highway investment analyses.

Exhibit 3-3 Bridge and Culvert Conditions, 2012

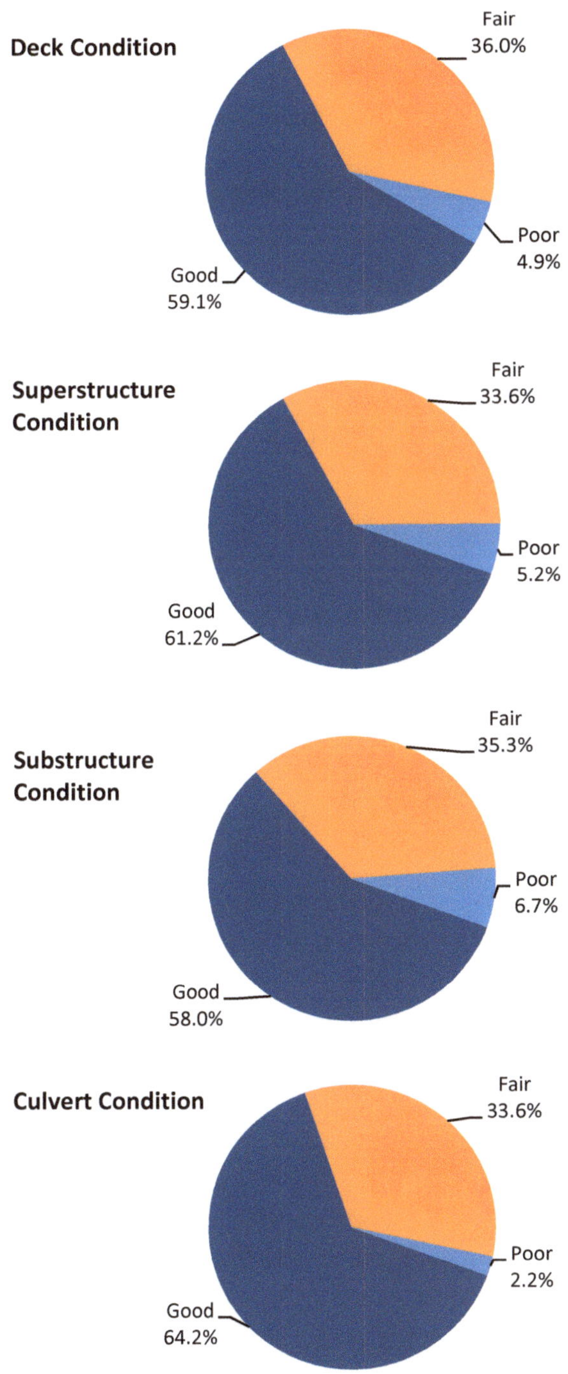

Source: National Bridge Inventory.

Trends in Pavement Ride Quality

Exhibit 3-4 details pavement ride quality on Federal-aid highways. The share of pavement mileage with "acceptable" ride quality decreased from 87.4 percent in 2002 to 80.3 percent in 2012. During the same period, the share of miles with pavement ride quality classified as good decreased from 46.6 percent to 36.4 percent.

Between 2008 and 2010, the percentage of pavement mileage with good quality declined from 40.7 percent to 35.1 percent, while the share of mileage with poor ride quality rose from 15.8 percent to 20.0 percent. These results should be interpreted with the understanding that HPMS guidance for reporting IRI changed, beginning with the 2009 data submittal. The revised instructions directed States to include measurements of roughness captured on bridges and railroad crossings; the previous instructions called for such measurements to be excluded from the reported values. This change would tend to increase the measured IRI on average, as the data should now reflect the bump experienced when driving over railroad tracks and the bumpiness associated with open-grated bridges and expansion joints on the bridge decks.

Exhibit 3-4 Pavement Ride Quality on Federal-Aid Highways, 2002–2012[1]

	2002	2004	2006	2008	2010	2012
By Mileage						
Good	46.6%	43.1%	41.5%	40.7%	35.1%	36.4%
Fair	40.8%	43.6%	42.7%	43.5%	44.9%	43.9%
Acceptable (Good + Fair)	87.4%	86.6%	84.2%	84.2%	80.0%	80.3%
Poor	12.6%	13.4%	15.8%	15.8%	20.0%	19.7%
Weighted By VMT						
Good	43.8%	44.2%	47.0%	46.4%	50.6%	44.9%
Fair	41.6%	40.7%	39.0%	39.0%	31.4%	38.4%
Acceptable (Good + Fair)	85.3%	84.9%	86.0%	85.4%	82.0%	83.3%
Poor	14.7%	15.1%	14.0%	14.6%	18.0%	16.7%

[1] Due to changes in data reporting instructions, data for 2010 and beyond are not fully comparable to data for 2008 and prior years.

Source: Highway Performance Monitoring System.

Weighting the ride quality data by VMT produces significantly different results. From 2002 to 2012, the share of VMT on Federal-aid highways with acceptable ride quality decreased from 85.3 percent to 83.3 percent, a much smaller decline than that observed above based on mileage. The percentage of Federal-aid highway VMT on pavements with good ride quality rose from 43.8 percent to 44.9 percent.

Although VMT-weighted figures more accurately reflect the typical conditions that highway users would experience over the full length of their trips, focusing on these statistics alone presents an incomplete picture of the current state of Federal-aid highways. The differences between the VMT-weighted and mileage-based data clearly suggest that ride quality on those Federal-aid highways that are relatively less traveled has been declining significantly over the past decade. These trends are visible in the data from 2002 to 2008, which predate the 2009 changes to the HPMS guidance, making clear that this finding is not simply a data anomaly but, instead, reflects changes in actual conditions.

Another source of recent data variability is that States have begun reporting ride quality data for shorter section lengths, which would tend to increase the variability of reported ratings. For example, a short segment of pavement in significantly better or worse condition than an adjacent segment is now more likely to be rated as good or poor, whereas before 2009 it might have been averaged with neighboring segments, yielding a rating of fair.

Pavement Ride Quality on the National Highway System

In 1998, the U.S. Department of Transportation (DOT) began establishing annual targets for pavement ride quality. Since 2006, the metric reflected in DOT performance-planning documents has been the share of VMT on pavements within the National Highway System (NHS) having good ride quality. Consequently, the discussion in this section focuses on VMT-weighted measures.

MAP-21 expanded the NHS to include most of the principal arterial mileage that was not previously included on the system. Although 2012 was the first year for which HPMS data were collected based on this expanded NHS, *Exhibit 3-5* includes estimates for 2010 that also were presented in the 2013 C&P Report. As a comparison of the actual 2010 values and these estimates reflects, expanding the NHS reduced the percentage of NHS VMT on pavements with good and acceptable ride quality. On average, the additional routes added to the NHS had rougher pavements than the routes that were already part of the NHS.

Exhibit 3-5 Percentages of National Highway System Vehicle Miles Traveled on Pavements With Good and Acceptable Ride Quality, 2002–2012

	2002	2004	2006	2008	2010[1]	2012
Based on NHS before MAP-21[2]						
Good (IRI < 95)	50%	52%	57%	57%	60%	
Acceptable (IRI ≤ 170)	91%	91%	93%	92%	93%	
Based on Current NHS						
Good (IRI < 95)					*54.7%*	57.1%
Acceptable (IRI ≤ 170)					*88.8%*	89.0%

[1] Italicized 2010 values shown for the Current NHS are estimates as presented in the 2013 C&P report. Exact values cannot be determined as the 2010 HPMS data were collected based on the pre-MAP-21 NHS.

[2] Values are shown as whole percentages to be consistent with how they were reported at the time in DOT performance planning documents.

Source: Highway Performance Monitoring System.

From 2010 to 2012, the share of VMT on NHS pavements with acceptable ride quality rose slightly from an estimated 88.8 percent to 89.0 percent. Over the same period, the share of NHS travel on pavements with good ride quality rose from an estimated 54.7 percent to 57.1 percent.

The estimated improvement between 2010 and 2012 represents a continuation of a longer-term trend. Based on data for the NHS as it existed at the time, from 2002 to 2010 the percentage of VMT on NHS pavements with acceptable ride quality rose from 91 percent to 93 percent; VMT on NHS pavements with good ride quality rose sharply from 50 percent to 60 percent over this same period.

NHS Pavement Ride Quality Trends

Exhibit 3-4 showed that for pavement ride quality on Federal-aid highways, the share of VMT on pavements with good ride quality improved from 2002 to 2012, while the share of mileage with good ride quality declined.

In contrast, the share of pavements with good ride quality for the NHS improved over this period regardless of the weighting method used. *Exhibit 3-5* shows that the share of NHS VMT on pavements with good ride quality increased from 50.0 percent in 2002 to 57.1 percent in 2012. This is the metric currently used in DOT performance planning documents.

The share of NHS mileage with good pavement ride quality, however, increased more slowly, from 57.4 percent in 2002 to 59.0 percent in 2012. The share of NHS lane miles with good pavement ride quality increased over this period from 56.7 percent to 59.4 percent. Under the PM-2 Notice of Proposed Rulemaking, pavement-related targets will be set based on lane mileage, rather than mileage or VMT.

Pavement Ride Quality by Functional Classification

Although changes in HPMS reporting procedures in 2009 make identifying trends over the full 10-year period shown in *Exhibit 3-6* more challenging, drawing some significant conclusions from the data is still possible. Rural Interstates have the best ride quality of all functional systems, with 78.6 percent of VMT on pavements having good ride quality, up from

72.2 percent in 2002. The share of urban Interstate System VMT on pavements with good ride quality from 2002 to 2012 rose sharply from 45.0 percent to 62.5 percent.

Exhibit 3-6 Percentages of Vehicle Miles Traveled on Pavements with Good and Acceptable Ride Quality by Functional System, 2002–2012

	2002	2004	2006	2008	2010[1]	2012
Functional System			**Percent Good**			
Rural Interstate	72.2%	73.7%	78.6%	79.0%	79.1%	78.6%
Rural Other Freeway and Expressway[2]					74.3%	72.8%
Rural Other Principal Arterial[2]					72.9%	67.4%
Rural Other Principal Arterial[2]	60.2%	61.0%	66.8%	68.4%		
Rural Minor Arterial	51.0%	51.5%	56.3%	56.2%	60.9%	57.7%
Rural Major Collector	42.4%	40.3%	39.8%	39.0%	41.4%	39.7%
Subtotal Rural	**58.0%**	**58.3%**	**62.2%**	**62.5%**	**64.6%**	**59.8%**
Urban Interstate	45.0%	49.4%	54.0%	55.7%	64.6%	62.5%
Urban Other Freeway and Expressway	33.6%	38.8%	45.3%	44.4%	53.3%	53.0%
Urban Other Principal Arterial	25.7%	26.5%	28.8%	26.9%	39.7%	30.3%
Urban Minor Arterial	34.1%	32.3%	33.6%	32.5%	28.8%	22.0%
Urban Collector[2]	35.5%	35.7%	34.1%	31.5%		
Urban Major Collector[2]					25.7%	19.0%
Urban Minor Collector[2]					8.6%	29.8%
Subtotal Urban	**34.9%**	**36.6%**	**39.5%**	**38.9%**	**44.0%**	**36.8%**
Total Good[3]	**43.8%**	**44.2%**	**47.0%**	**46.4%**	**50.6%**	**44.9%**
Functional System			**Percent Acceptable**			
Rural Interstate	97.3%	97.8%	98.2%	97.3%	91.1%	97.6%
Rural Other Freeway and Expressway[2]					93.7%	97.9%
Rural Other Principal Arterial[2]					93.0%	95.9%
Rural Other Principal Arterial[2]	96.2%	96.1%	97.0%	97.6%		
Rural Minor Arterial	93.8%	94.3%	95.1%	94.5%	87.3%	93.7%
Rural Major Collector	87.6%	88.5%	87.8%	88.3%	81.2%	85.5%
Subtotal Rural	**94.1%**	**94.5%**	**94.9%**	**94.8%**	**87.8%**	**92.8%**
Urban Interstate	89.6%	90.3%	92.7%	91.9%	89.8%	93.4%
Urban Other Freeway and Expressway	87.8%	87.7%	92.1%	91.4%	89.2%	91.9%
Urban Other Principal Arterial	71.0%	72.6%	73.8%	72.4%	76.4%	73.5%
Urban Minor Arterial	76.3%	73.8%	75.6%	75.5%	70.6%	69.8%
Urban Collector[2]	74.6%	72.6%	72.6%	72.0%		
Urban Major Collector[2]					67.0%	63.8%
Urban Minor Collector[2]					26.2%	59.7%
Subtotal Urban	**79.8%**	**79.7%**	**81.7%**	**81.0%**	**79.4%**	**78.1%**
Total Acceptable[3]	**85.3%**	**84.9%**	**86.0%**	**85.4%**	**82.0%**	**83.3%**

[1] HPMS pavement reporting requirements were modified in 2009 to include bridges; features such as open grated bridge decks or expansion joints can greatly increase the IRI for a given section.

[2] Beginning in 2010, the data reflect revised HPMS functional classifications. Rural Other Freeways and Expressways were split out of the Rural Other Principal Arterial category, and Urban Collect was split into Urban Major Collector and Urban Minor Collector.

[3] Totals shown reflect Federal-aid highways only and exclude roads classified as rural minor collector, rural local, or urban local for which pavement data are not reported in HPMS.

Source: Highway Performance Monitoring System.

The concept of classification of roadways was presented in Chapter 2. In general, roads with higher functional classifications, which carry higher volumes of traffic at higher speeds such as Interstates and principal arterials, have better ride quality than lower-ordered systems that carry low amounts of traffic, typically at lower speeds, such as collectors. Among the rural functional classifications, the percentage of VMT on pavements with good ride quality in 2012 ranged from 78.6 percent for rural Interstates to 39.7 percent for rural major collectors. A similar pattern is evident among most urban functional classifications, as the percentage of VMT on pavements with good ride quality in 2012 ranged from 62.5 percent for urban Interstates to 19.0 percent for urban major collectors. Urban minor collectors actually showed a higher percentage of VMT on pavements with good ride quality than did urban major collectors in 2012. This observation, however, could derive from the fact that some States have not yet fully adapted to the new functional classifications added to HPMS in 2009, so that the data on urban minor collectors might not be fully representative of the Nation as a whole.

As noted in Chapter 2, rural areas contain about 75 percent of national road miles, but support only about 33 percent of annual national VMT. Pavement conditions in urban areas thus have a greater impact on the VMT-weighted measure shown in *Exhibit 3-6* than do pavement conditions in rural areas. Pavement conditions are generally better in rural areas. The share of rural VMT on pavements with good ride quality rose slightly from 58.0 percent in 2002 to 59.8 percent in 2012, while the portion of urban VMT on pavements with good ride quality increased from 34.9 percent in 2002 to 36.8 percent in 2010. The share of VMT on pavements with acceptable ride quality decreased slightly from 2002 to 2012 in rural and urban areas.

Trends in Bridge Structural Deficiencies

Q&A What makes a bridge structurally deficient, and are structurally deficient bridges unsafe?

Structurally deficient bridges are not unsafe.

Bridges are considered structurally deficient if significant load-carrying elements are in poor condition due to deterioration or damage. They are also considered structurally deficient if the waterway opening of the bridge causes intolerable roadway traffic interruptions.

The classification of a bridge as structurally deficient does not mean that it is likely to collapse or that it is unsafe. Properly scheduled inspections can identify unsafe conditions; if the bridge is determined to be unsafe, the structure is closed. A structurally deficient bridge, when left open to traffic, typically requires significant maintenance and repair and eventual rehabilitation or replacement to address deficiencies. To remain in service, structurally deficient bridges often have lane closures or weight limits that restrict the gross weight of vehicles using the bridges to less than the maximum weight typically allowed by statute.

Bridges are considered structurally deficient if significant load-carrying elements are in poor condition due to deterioration, damage, or both. Structural deficiencies are determined by ratings for a bridge's deck or superstructure, or ratings for culverts. If the load-carrying capacity of a bridge does not meet current design standards and the situation cannot be mitigated through corrective actions short of replacing it, the bridge will be rated as structurally deficient. Bridges over rivers, streams, or channels convey the flow of water so that the roadway is not impacted by flooding. The size of the area or opening under the bridge through which the water is conveyed is a major factor in determining the amount of water that can be passed under the structure. If the size of the structure's hydraulic opening with respect to the passage of water under a bridge does not meet current criteria for potential submersion during a flood event, the bridge will be classified as structurally deficient if bridge replacement is the only option for addressing the situation.

The classification of a bridge as structurally deficient does not imply that the bridge is unsafe. Instead, the classification indicates the extent to which a bridge has deteriorated from its original condition when first built. Once a bridge is classified as structurally deficient, the bridge might experience reduced performance in the form of lane closures or load limits. If a bridge inspection determines a bridge to be unsafe, it is closed.

Exhibit 3-7 identifies the percentages of all bridges classified as structurally deficient based on the number of bridges, bridges weighted by deck area, and bridges weighted by average daily traffic. Chapter 2 provides an overview of growth in the number of bridges over time.

Exhibit 3-7 Structurally Deficient Bridges—Systemwide, 2002–2012

	2002	2004	2006	2008	2010	2012
Count						
Total Bridges	591,243	594,100	597,561	601,506	604,493	607,380
Structurally Deficient	84,031	79,971	75,422	72,883	70,431	66,749
Percent Structurally Deficient						
By Bridge Count	14.2%	13.5%	12.6%	12.1%	11.7%	11.0%
Weighted by Deck Area	10.4%	10.1%	9.6%	9.3%	9.1%	8.2%
Weighted by ADT	8.0%	7.6%	7.4%	7.2%	6.7%	5.9%

Source: National Bridge Inventory.

Based on raw bridge counts, approximately 11.0 percent of bridges were classified as structurally deficient in 2012—a 3.2-percentage point improvement from the 14.2 percent based on 2002 data. Weighted by deck area, the comparable share was 8.2 percent in 2012, a 2.2-percentage point improvement from 10.4 percent based on 2002 data. Although 11.0 percent of the Nation's bridges are structurally deficient, only 5.9 percent of ADT (average daily traffic) crossed a structurally deficient bridge. ADT measures the total volume of vehicular traffic on a bridge divided by the 365 days in a year.

Structurally Deficient Bridges by Owner

As discussed in Chapter 2, the owner of a road or bridge is responsible for its operation and maintenance. Many local governments have established an interagency agreement with their respective State governments to assume operation and maintenance. Such agreements do not transfer ownership nor do they negate the responsibilities of the bridge owners. Owners must ensure that the operation and maintenance of their bridges comply with Federal and State requirements. Additionally, the National Bridge Inspection Standards specify that each State is responsible for inspecting all bridges in that State except for tribally or federally owned bridges. Similarly, Federal agencies and Tribal governments are responsible for inspecting or causing to be inspected all bridges in their jurisdiction, respectively.

Bridge deficiencies by ownership are examined in *Exhibit 3-8*. State and local governments own 98.3 percent of the Nation's bridges. Of the relatively few privately owned bridges for which data are reported in NBI—0.2 percent of the total number of bridges—31.6 percent were classified as structurally deficient in 2012. Of the 1.5 percent of bridges Federal agencies own, 7.6 percent were classified as structurally deficient. In terms of structural deficiency, State-owned and locally owned bridges differ significantly, as 7.0 percent of State-owned bridges were structurally deficient in 2012, compared with 14.8 percent of locally owned bridges.

Exhibit 3-8 Structurally Deficient Bridges by Owner, 2012[1]

	Federal	State	Local	Private/Other[2]	Total
Counts					
Total Bridges	8,930	292,830	304,235	1,385	607,380
Structurally Deficient Bridges	679	20,531	45,101	438	66,749
Percentages					
Total Inventory Owned	1.5%	48.2%	50.1%	0.2%	100.0%
Structurally Deficient Bridges	7.6%	7.0%	14.8%	31.6%	11.0%

[1] These data only reflect bridges for which inspection data were submitted to the NBI.

[2] An unknown number of privately owned bridges are omitted.

Source: National Bridge Inventory.

Structurally Deficient Bridges on the National Highway System

Exhibit 3-9 identifies the percentage of bridges on the NHS classified as structurally deficient based on the number of bridges, bridges weighted by deck area, and bridges weighted by ADT. The 2012 data shown in the exhibit reflect the NHS before it was expanded under MAP-21. Bridge data for the expanded NHS will be reflected in the next C&P report because MAP-21 was passed in the middle of 2012.

Exhibit 3-9 Structurally Deficient Bridges on the National Highway System, 2002–2012

	2002	2004	2006	2008	2010	2012
Count						
Total Bridges	114,544	115,103	115,202	116,523	116,669	117,485
Structurally Deficient Bridges	6,712	6,617	6,339	6,272	5,902	5,237
Percentage Structurally Deficient						
By Bridge Count	5.9%	5.7%	5.5%	5.4%	5.1%	4.5%
Weighted by Deck Area	8.6%	8.9%	8.4%	8.2%	8.3%	7.1%
Weighted by ADT	7.1%	6.8%	6.6%	6.4%	6.0%	5.1%

Source: National Bridge Inventory.

In 2012, approximately 4.5 percent of NHS bridges were classified as structurally deficient. The comparable values weighted by deck area and by ADT were 7.1 percent and 5.1 percent, respectively. These results suggest an above-average concentration of deficiencies on heavily traveled and larger bridges.

FHWA has adopted deck-area weighting for use in agency performance planning in recognition of the significant logistical and financial challenges that might be involved in addressing deficiencies on larger bridges. Between 2002 and 2012, the share of structurally deficient bridges on the NHS weighted by deck area declined from 8.6 percent to 7.1 percent. The 1.2-percentage point improvement between 2010 and 2012 was the largest decline during this period.

Structurally Deficient Bridges on the STRAHNET

The STRAHNET (Strategic Highway Network) system is a key subset of NHS. The physical composition of this system was described in Chapter 2, and the condition of the pavement portion was presented earlier in this chapter. The share of structurally deficient bridges decreased from 5.4 percent in 2002 to 4.2 percent in 2012. These data are shown in *Exhibit 3-10*.

Exhibit 3-10 Structurally Deficient Bridges on the Strategic Highway Network, 2002–2012

	2002	2004	2006	2008	2010	2012
Total Bridges	79,852	72,046	73,003	73,771	68,529	68,118
Structurally Deficient Bridges	4,320	3,640	3,645	3,659	3,355	2,890
Percentage of Bridges Structurally Deficient	5.4%	5.1%	5.0%	5.0%	4.9%	4.2%

Source: National Bridge Inventory.

Structurally Deficient Bridges by Functional Classification

As shown in *Exhibit 3-11*, the percentage of structurally deficient bridges on the Nation's rural roadways decreased from 15.6 percent in 2002 to 12.3 percent in 2012. Over this same period, the share of structurally deficient bridges on the Nation's urban roadways decreased from 9.5 percent to 7.5 percent.

Exhibit 3-11 Structurally Deficient Bridges by Functional Class, 2002–2012

	Percentages of Structurally Deficient Bridges by Year					
Functional System	2002	2004	2006	2008	2010	2012
Rural						
Interstate	4.1%	4.3%	4.3%	4.5%	4.5%	4.1%
Other Principal Arterial	5.5%	5.4%	5.1%	4.9%	4.5%	3.9%
Minor Arterial	8.7%	8.4%	8.3%	8.1%	7.3%	6.6%
Major Collector	12.3%	11.7%	11.2%	10.5%	10.2%	9.7%
Minor Collector	14.0%	13.5%	12.7%	12.4%	12.1%	11.4%
Local	22.0%	20.7%	19.1%	18.3%	17.9%	17.2%
Subtotal Rural	**15.6%**	**14.8%**	**13.9%**	**13.3%**	**12.9%**	**12.3%**
Urban						
Interstate	6.5%	6.3%	6.0%	5.9%	5.4%	4.7%
Other Freeway and Expressway	6.4%	6.1%	5.8%	5.5%	5.0%	4.3%
Other Principal Arterial	9.6%	9.2%	8.7%	8.6%	8.2%	7.6%
Minor Arterial	10.9%	10.3%	10.0%	9.8%	9.1%	8.5%
Collector	11.6%	11.1%	11.0%	10.8%	9.9%	9.2%
Local	12.1%	11.5%	11.1%	10.8%	10.3%	9.8%
Subtotal Urban	**9.5%**	**9.1%**	**8.8%**	**8.6%**	**8.1%**	**7.5%**
Total	**14.2%**	**13.5%**	**12.6%**	**12.1%**	**11.7%**	**11.0%**

Source: National Bridge Inventory.

Among the individual functional classes in 2012, rural local bridges continue to have the highest percentage of structural deficiencies, 17.2 percent. Rural Interstate bridges, however, had the lowest percentage of structural deficiencies, 4.1 percent.

Structurally Deficient Bridges by Age

Exhibit 3-12 identifies the age composition of all highway bridges in the Nation. As of 2012, approximately 36.1 percent of the Nation's bridges were between 26 and 50 years old. For NHS bridges, 49.4 percent were in this age range, while 63.0 percent of the Interstate bridges fell into this age range.

Approximately 69.3 percent of all bridges are 26 years old or older. The percentages of NHS and Interstate bridges in this group are 74.0 percent and 83.4 percent, respectively. Most bridges are 26 to 50 years old. The large number of bridges in this age range has implications in terms of long-term bridge rehabilitation and replacement strategies. The need for such actions could be concentrated within certain periods rather than being spread out evenly. Several other variables such as maintenance practices and environmental conditions, however, also influence when future capital investments might be needed.

Exhibit 3-12 Bridges by Age, 2012

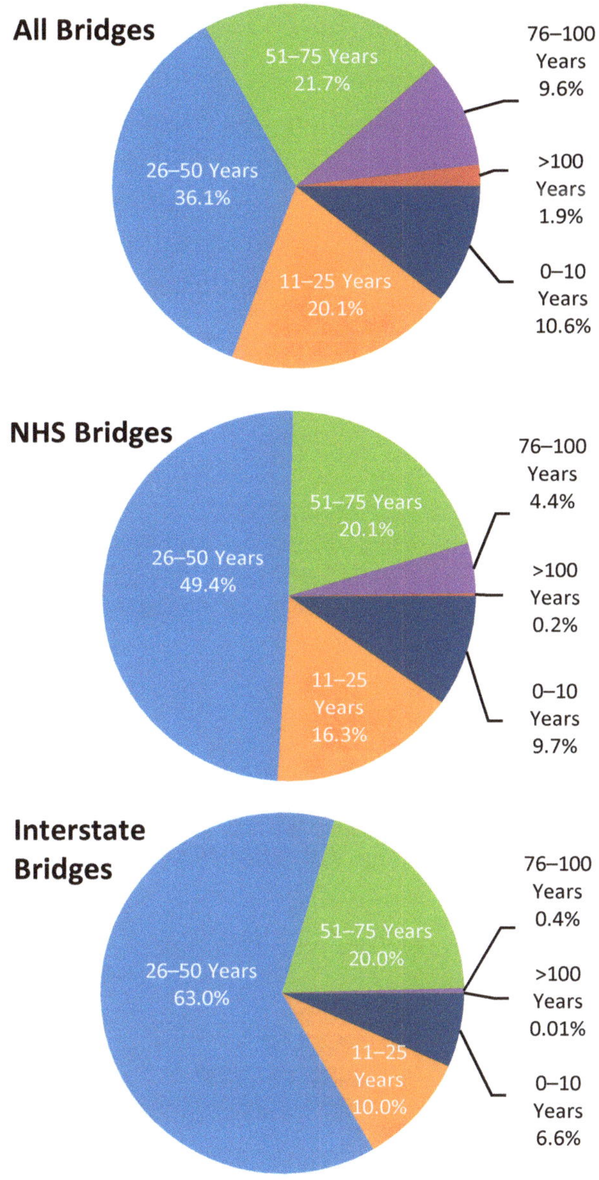

Source: National Bridge Inventory.

Exhibit 3-13 identifies the distribution of structurally deficient bridges within the age ranges presented in *Exhibit 3-12*. The percentage of bridges classified as structurally deficient generally tends to rise as bridges age. Although only 8.2 percent of bridges in the 26-to-50 year group are structurally deficient, the percentage is 18.0 percent for bridges 51 to 75 years of age and 29.2 percent for bridges 76 to 100 years of age. Similar patterns are evident in the data for NHS and Interstate System bridges, although the overall percentage of structurally deficient bridges for these systems is lower than for the national bridge population.

The age of a bridge structure is one indicator of its serviceability, or condition under which a bridge is still considered useful. A combination of several factors, however, influences the serviceability of a structure, including the original design; the frequency, timeliness, effectiveness, and appropriateness of the maintenance activities implemented over the life of the structure; the loading to which the structure has been subjected during its life; the climate of the area where the structure is located; and any additional stresses from events such as flooding to which the structure has been subjected. As an example, two structures built at the same time using the same design standards and in the same climate can have very different serviceability levels. The first structure might have had increased heavy truck traffic, lack of preventive maintenance of the deck or the substructure, or lack of rehabilitation work. The second structure could have had the same increases in heavy truck traffic but received timely preventive maintenance activities on all parts of the structure and proper rehabilitation activities. In this example, the first structure would have a low serviceability level, while the second structure would have a high serviceability level.

Exhibit 3-13 Percentages of Structurally Deficient Bridges by Age, 2012

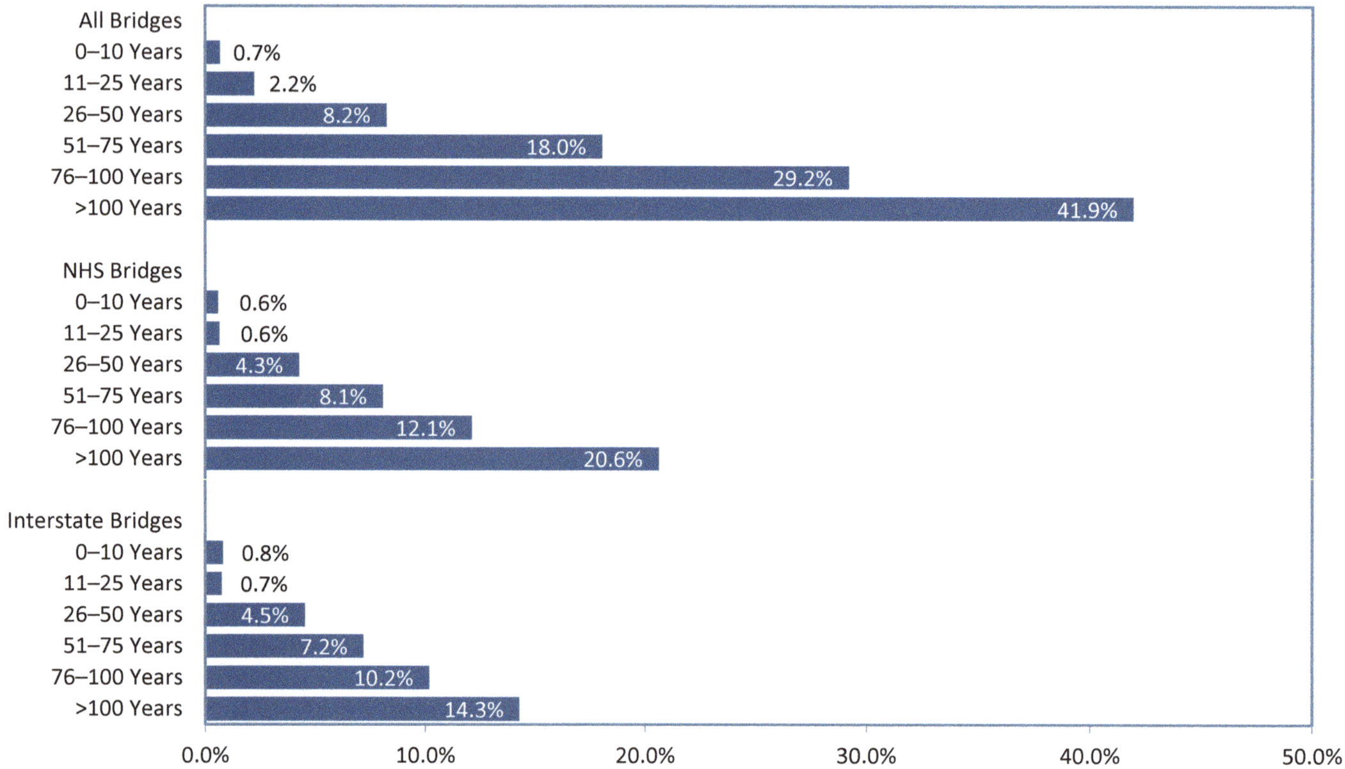

Source: National Bridge Inventory.

Geometric Design Standards

Design standards and best practices for the Nation's roadways have improved over the years. Design standards are intended to improve travel throughout the network by facilitating the movement of passengers and goods through the network. Traveling at higher and safer speeds mitigates congestion and the loss of productivity that occurs from spending more time in a vehicle.

Design standards for both roads and bridges have evolved. Even though standards have improved, however, some facilities have not been updated to meet existing standards. That facilities have been built to lower standards or to outdated standards does not imply that they are poorly maintained.

Roadway Alignment

The term "roadway alignment" refers to the curvature and grade of a roadway, that is, the extent to which it swings from side to side and points up or down. The term "horizontal alignment" relates to curvature (how sharp the curves are), while the term "vertical alignment" relates to gradient (how steep a slope is). Alignment adequacy affects the level of service and safety of the highway system. Inadequate alignment can result in speed reductions and impaired sight distance. Trucks are particularly affected by inadequate vertical alignment with regard to speed. Alignment adequacy is evaluated on a scale from Code 1 (best) to Code 4 (worst).

Alignment adequacy is more important on roads with higher travel speeds or higher volumes (e.g., the Interstate System). Because alignment generally is not a major issue in urban areas, only rural alignment statistics are presented in this section. The amount of change in roadway alignment over time is gradual and occurs only during major reconstruction of existing roadways. New roadways are constructed to meet current vertical and horizontal alignment criteria and, therefore, generally have no alignment problems except under extreme conditions.

As shown in *Exhibit 3-14,* in 2012, approximately 85.2 percent of rural Interstate System miles are classified as Code 1 for horizontal alignment and 86.6 percent as Code 1 for vertical alignment. In contrast, the percentages of rural minor arterial miles classified as Code 1 for horizontal and vertical alignment, respectively, are only 69.8 percent and 67.7 percent.

Exhibit 3-14 Rural Alignment by Functional Class, 2012[1]

	Code 1	Code 2	Code 3	Code 4
Horizontal				
Interstate	85.2%	0.1%	1.2%	13.4%
Other Freeway and Expressway	63.9%	1.3%	1.4%	33.3%
Other Principal Arterial	73.2%	7.9%	2.7%	16.3%
Minor Arterial	69.8%	4.7%	2.0%	23.5%
Major Collector	68.1%	1.5%	0.7%	29.7%
Vertical				
Interstate	86.6%	11.2%	1.9%	0.3%
Other Freeway and Expressway	79.7%	17.5%	1.9%	1.0%
Other Principal Arterial	74.2%	19.1%	4.4%	2.3%
Minor Arterial	67.7%	19.8%	8.1%	4.3%
Major Collector	90.3%	6.7%	0.9%	2.0%

Code 1 All curves and grades meet appropriate design standards.

Code 2 Some curves or grades are below design standards for new construction, but curves can be negotiated safely at prevailing speed limits. Truck speed is not substantially affected.

Code 3 Infrequent curves or grades occur that impair sight distance or severely affect truck speeds. May have reduced speed limits.

Code 4 Frequent grades occur that impair sight distance or severely affect truck speeds. Generally, curves are unsafe or uncomfortable at prevailing speed limit, or the speed limit is severely restricted due to the design speed limits of the curves.

[1] Values are based on State-reported information and have not been fully validated. The percentage of Horizontal Alignment with Code 4 is significantly higher than that reported in 2008 and prior years. The percentage of Vertical Alignment with Code 1 for Major Collector is also significantly higher than that reported in prior years.

Source: Highway Performance Monitoring System.

Lane Width

Lane width affects capacity and safety. Narrow lanes have less capacity and can affect the frequency of crashes. As with roadway alignment, lane width is more crucial on functional classifications that have higher travel volumes.

Currently, higher functional systems such as the Interstate System are expected to have 12-foot lanes. As shown in *Exhibit 3-15*, approximately 98.7 percent of rural Interstate System miles and 98.6 percent of urban Interstate System miles had minimum 12-foot lane widths in 2010.

In 2012, approximately 53.8 percent of urban collectors have lane widths of 12 feet or greater, but approximately 18.7 percent have 11-foot lanes and 20.0 percent have 10-foot lanes; the remaining 5.2 percent have lane widths of 9 feet or less. Among rural major collectors, 43.1 percent have lane widths of 12 feet or greater, but approximately 26.1 percent have 11-foot lanes and 22.8 percent have 10-foot lanes. Roughly 6.0 percent of rural major collector mileage has lane widths of 9 feet or less.

Exhibit 3-15 Lane Width by Functional Class, 2012

	≥12 foot	11 foot	10 foot	9 foot	<9 foot
Rural					
Interstate	98.7%	1.2%	0.1%	0.0%	0.0%
Other Freeway and Expressway	97.7%	2.3%	0.0%	0.0%	0.0%
Other Principal Arterial	91.2%	6.9%	1.4%	0.3%	0.1%
Minor Arterial	71.6%	18.9%	8.5%	0.8%	0.2%
Major Collector	43.1%	26.0%	22.7%	6.0%	2.1%
Urban					
Interstate	98.6%	1.0%	0.3%	0.1%	0.0%
Other Freeway and Expressway	95.9%	3.3%	0.8%	0.0%	0.0%
Other Principal Arterial	82.6%	12.1%	4.7%	0.3%	0.3%
Minor Arterial	67.0%	18.5%	11.7%	1.8%	1.0%
Collector	53.8%	18.7%	20.0%	5.2%	2.4%

Source: Highway Performance Monitoring System.

Functionally Obsolete Bridges

A functionally obsolete bridge is not an unsafe bridge. Functional obsolescence is generally determined by the geometrics of a bridge in relation to the geometrics that current design standards require. In contrast to structural deficiencies, which typically result from deterioration of the bridge components, functional obsolescence generally results from changing traffic demands on the structure. The classification of functionally obsolete is determined by the NBI appraisal ratings for structural evaluation, waterway adequacy, deck geometry, alignment of the approach roadway, and underclearances. Appraisal ratings are used to compare existing characteristics of a bridge to the current standards used for highway and bridge design. Existing bridges constructed before the establishment of more stringent design standards are more likely to be classified functionally obsolete when compared to newer bridges.

Facilities, including bridges, are designed to conform to the design standards in place at the time they are designed. Over time, design requirements improve. For example, a bridge designed in the 1930s would have shoulder widths that conform with 1930s design standards. Current design standards, however, are based on different criteria, and current safety standards require wider bridge shoulders. The difference between the required, current-day shoulder width and the shoulder width designed in the 1930s represents a deficiency. The magnitudes of such deficiencies determine whether a bridge is classified as functionally obsolete.

Of note is whether a bridge has issues that would warrant its classification as both structurally deficient and functionally obsolete. A bridge cannot be classified as both functionally obsolete and structurally deficient. If a functionally obsolete bridge has a structurally deficient component, it is classified as a structurally deficient bridge. To avoid double counting, the standard NBI data reporting convention is to identify it as structurally deficient only. Such bridges are excluded from the statistics on functionally obsolete bridges presented in this section.

Across the system on a national basis, the share of functionally obsolete bridges by bridge count decreased from 15.4 percent in 2002 to 14.0 percent in 2012, as shown in *Exhibit 3-16*. When considering ADT, the share of functionally obsolete bridges decreased from 22.0 percent in 2002 to 21.3 percent in 2012.

Exhibit 3-16 Functionally Obsolete Bridges—Systemwide, 2002–2012

	2002	2004	2006	2008	2010	2012
Count						
Total Bridges	591,243	594,100	597,561	601,506	604,493	607,380
Functionally Obsolete	90,823	90,076	89,591	89,189	85,858	84,748
Percent Functionally Obsolete						
By Bridge Count	15.4%	15.2%	15.0%	14.8%	14.2%	14.0%
Weighted by Deck Area	20.4%	20.5%	20.3%	20.5%	19.8%	20.1%
Weighted by ADT	22.0%	21.9%	21.9%	22.2%	21.5%	21.3%

Source: National Bridge Inventory.

Exhibit 3-17 provides the share of functionally obsolete bridges on the NHS. The share of functionally obsolete bridges in NHS based on bridge count decreased from 17.2 percent in 2002 to 16.2 percent in 2012. The share of functionally obsolete bridges based on ADT decreased from 20.0 percent in 2002 to 19.5 percent in 2012.

Exhibit 3-17 Functionally Obsolete Bridges on the National Highway System, 2002–2012

	2002	2004	2006	2008	2010	2012
Count						
Total Bridges	114,544	115,103	115,202	116,523	116,669	117,485
Functionally Obsolete	19,667	19,408	19,368	19,707	19,061	19,075
Percent Functionally Obsolete						
By Bridge Count	17.2%	16.9%	16.8%	16.9%	16.3%	16.2%
Weighted by Deck Area	21.1%	20.9%	20.8%	21.4%	20.3%	21.0%
Weighted by ADT	20.0%	19.8%	20.1%	20.5%	19.7%	19.5%

Source: National Bridge Inventory.

Most functionally obsolete bridges are located in urban environments. As shown in *Exhibit 3-18*, urban minor arterials had the highest share of functionally obsolete bridges at 28.2 percent. In the rural setting, Interstate bridges had the highest share of functionally obsolete bridges at 11.6 percent. The disparities between the urban and rural settings could be because urban environments are generally densely populated and have higher traffic volumes.

Exhibit 3-18 Functionally Obsolete Bridges by Functional Class, 2002–2012

Functional System	Percentages of Functionally Obsolete Bridges by Year					
	2002	2004	2006	2008	2010	2012
Rural						
Interstate	12.9%	12.8%	12.0%	11.8%	11.6%	11.6%
Other Principal Arterial	10.3%	9.9%	9.4%	9.3%	8.5%	8.3%
Minor Arterial	12.0%	11.6%	11.0%	10.6%	10.2%	9.7%
Major Collector	11.3%	11.0%	10.5%	10.1%	9.3%	8.9%
Minor Collector	12.3%	12.1%	11.9%	11.4%	10.6%	10.4%
Local	13.5%	13.2%	12.8%	12.4%	11.7%	11.3%
Subtotal Rural	**12.5%**	**12.2%**	**11.7%**	**11.4%**	**10.7%**	**10.4%**
Urban						
Interstate	23.0%	23.3%	23.6%	23.9%	23.0%	22.9%
Other Freeway and Expressway	23.5%	23.2%	23.1%	22.9%	22.0%	22.1%
Other Principal Arterial	25.4%	25.4%	24.5%	24.5%	23.8%	23.4%
Minor Arterial	29.3%	29.3%	29.4%	29.3%	28.6%	28.2%
Collector	28.1%	28.6%	28.7%	28.5%	28.1%	27.4%
Local	21.4%	22.0%	21.9%	21.4%	20.5%	20.7%
Subtotal Urban	**24.9%**	**25.1%**	**25.0%**	**24.9%**	**24.2%**	**24.0%**
Total	**15.4%**	**15.2%**	**15.0%**	**14.8%**	**14.2%**	**14.0%**

Source: National Bridge Inventory.

Although bridge design standards have evolved over the past several decades, the standards are not necessarily followed when bridge owners are constructing new bridges. As shown in *Exhibit 3-19*, 20.7 percent of the functionally obsolete bridges on the Interstate System are between the ages of 0 and 10 years. That portion is the second highest share compared to 25.1 percent of the functionally obsolete bridges on the Interstate System aged 51 to 75 years. Although bridge owners ideally would follow current bridge standards, certain situations might prevent them from completely adhering to the standards.

Exhibit 3-19 Percentages of Functionally Obsolete Bridges by Age, 2012

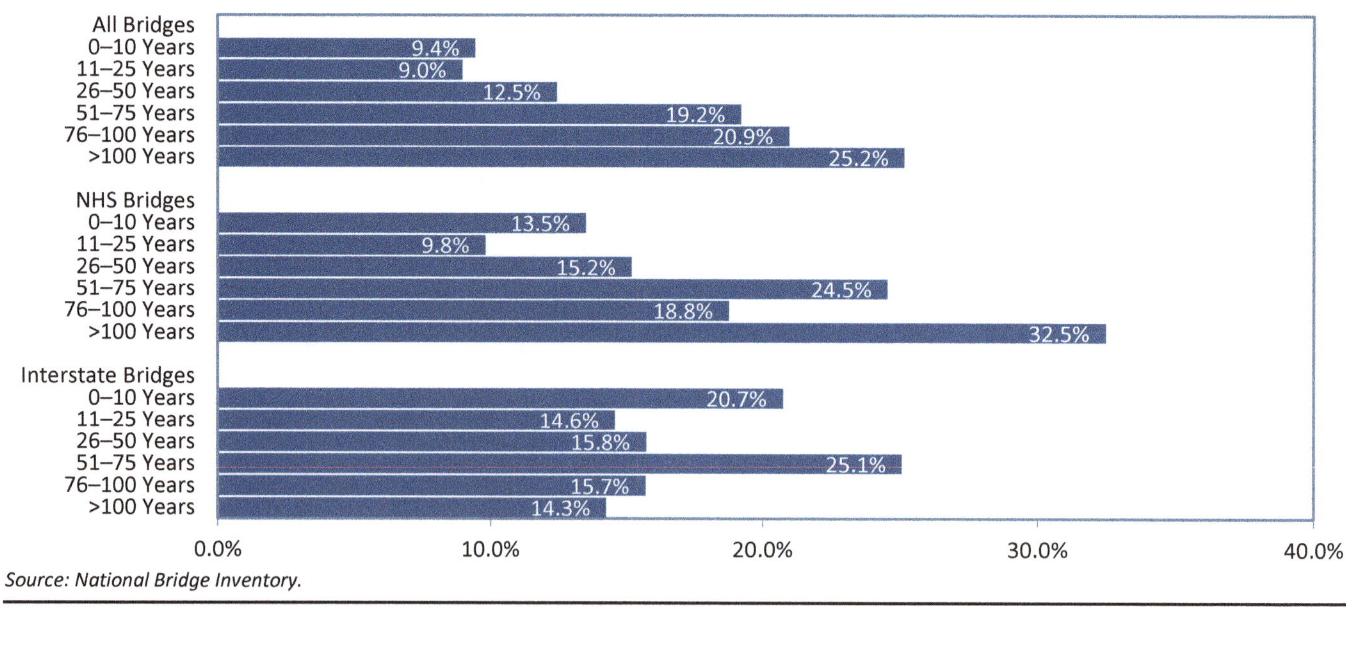

Source: National Bridge Inventory.

How do total bridge deficiencies differ between the National Highway System and the Nation as a whole?

Exhibit 3-20 Bridge Deficiencies: Systemwide vs. National Highway System, 2012

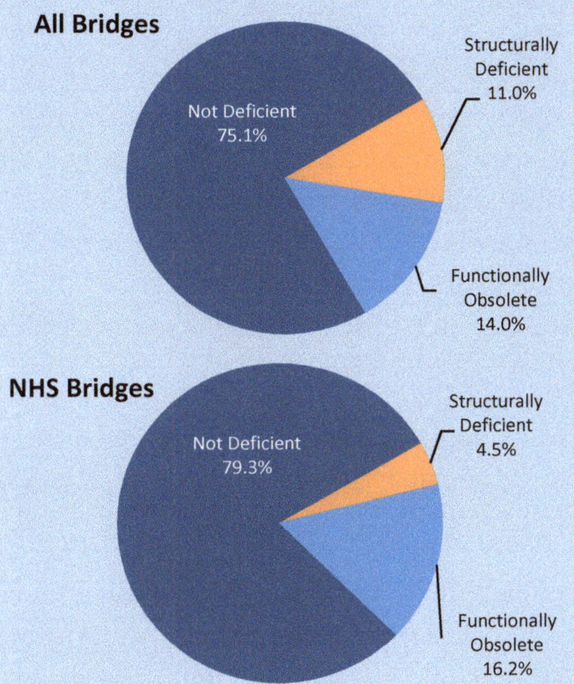

Source: National Bridge Inventory.

Previous editions of the C&P report focused on total bridge deficiencies, combining the structurally deficient bridges with functionally obsolete bridges. Although the number of functionally obsolete bridges remains a concern, FHWA has shifted its focus toward structurally deficient bridges in light of programmatic changes under MAP-21. Consequently, this chapter places greater emphasis on structural deficiencies.

Exhibit 3-20 compares the total share of deficient bridges for NHS with all bridges. In 2012, 75.1 percent of the Nation's bridges were not classified as deficient. Approximately 11.0 percent of the Nation's bridges were classified as structurally deficient, and 14.0 percent were classified as functional obsolete, for a total of approximately 24.9 percent deficient.

Among NHS bridges, 79.3 percent were not classified as deficient. Approximately 4.5 percent of NHS bridges were classified as structurally deficient, and 16.2 percent were classified as functionally obsolete, summing to 20.7 percent deficient. Thus, NHS bridges are much less likely to be classified as structurally deficient than non-NHS bridges, but are more likely to be classified as functionally obsolete.

Factors Affecting Pavement and Bridge Performance

Environmental conditions can significantly influence the deterioration of pavements and bridges due to continuous exposure. Pavement and bridge deterioration accelerates on facilities with high traffic volumes. Also, the use of a facility by large numbers of heavy trucks impacts its useful life. Deterioration could be mitigated through reconstruction, rehabilitation, or preventive maintenance. Deterioration can happen rapidly because the impacts of traffic and the environment are cumulative. If no action is taken, deterioration of the pavement and bridges could continue until they can no longer safely support traffic loads.

Constructing new facilities or major rehabilitation is a relatively expensive undertaking. Such actions might not be economically justified until a pavement section or bridge has deteriorated to a poor condition. Such considerations are reflected in the investment scenarios presented in Part II of this report. Those scenarios show that, even if all cost-beneficial investments were made, at any given time a certain percentage of pavements would not meet the criteria for acceptable.

Preventive maintenance actions are less expensive than rehabilitation and can be used to maintain and improve the quality of a pavement section or a bridge. Preventive maintenance actions, however, are less enduring than reconstruction or rehabilitation actions. Preventive maintenance actions are important in extending the useful life of a pavement section or bridge but cannot completely address deterioration over the long term. More aggressive actions would eventually need to be taken to preserve pavement and bridge quality.

Implications of Pavement and Bridge Conditions for Highway Users

Pavement and bridge conditions directly affect vehicle operating costs because deteriorating pavement and bridge decks increase wear and tear on vehicles and repair costs. Poor pavement can also affect travel time costs if road conditions force drivers to reduce speed. Additionally, poor pavement can increase the frequency of crash rates. Highway user costs are discussed in more detail in Chapter 7. Poor bridge conditions could create scenarios in which weight limits force freight trucks to seek alternative routes because they cannot cross a bridge on the most direct route. In worst-case scenarios, a bridge could be closed, forcing all traffic to use alternative routes.

Poor pavement conditions on higher functional classification roadways, such as the Interstate System, tend to result in higher user costs because of vehicle speed. For example, a vehicle hitting a pothole at 55 miles per hour on an Interstate highway could accelerate wear and tear faster than hitting the same pothole at 25 miles per hour.

Although poor pavement and bridge conditions can influence individual users, poor conditions could affect an entire network. Roads with a higher functional classification are meant to facilitate traffic's moving at higher speeds to reduce travel times. Drivers slowing to avoid poor pavement and bridge conditions could create congestion at peak travel times. Congestion increases travel times and slows the movement of freight traffic. The reduction in travel speed would add to the cost of the delivery of goods.

Strategies to Achieve State of Good Repair

Although the Nation's infrastructure system could be rehabilitated to a state of good repair with more investment, FHWA recognizes that stakeholders have limited resources when constructing or repairing roads and bridges. Limited resources—both staff and budgets—at transportation agencies across the country create the need to work more efficiently and focus on technologies and processes that produce the best results.

Improving project delivery continues to be a priority for FHWA. Projects that are delivered faster and more efficiently can minimize the disruption to stakeholders that construction causes. Through the agency's Every Day Counts[1] initiative, FHWA is partnering with State DOTs and stakeholders to identify and rapidly deploy proven but underutilized innovations to shorten the project delivery process, enhance roadway safety, reduce congestion, and improve environmental sustainability.

Bridge replacement projects create considerable traffic disruptions over long periods. Stakeholders might be reluctant to repair or replace a bridge due to its potential impact on traffic. New methodologies enable stakeholders to construct a new bridge off site and perform replacement activities in a consolidated timeframe. Several accelerated bridge construction initiatives are identified below.

- **Geosynthetic reinforced soil integrated bridge system (GRS-IBS).** Although utilizing traditional equipment and materials, a GRS-IBS makes use of alternating layers of compacted granular fill material and fabric sheets of geotextile reinforcement to provide support. The technology is particularly advantageous in the construction of small bridges (less than 140 feet long), reducing construction time, and generating cost savings of 25 to 60 percent compared to conventional construction methods. It facilitates design flexibility conducive to construction under variable site conditions, including soil type, weather, utilities and other obstructions, and proximity to existing structures.

Geosynthetic Reinforced Soil Integrated Bridge System

Defiance County, Ohio, used GRS-IBS to build a bridge in just 6 weeks, compared to the months required for traditional construction methods.[1] The county saved nearly 25 percent on the project, not only because of the reduced labor costs resulting from shorter construction time and simpler construction, but also because fewer materials were required for the GRS bridge abutments. GRS-IBS technology also helped Clearfield County, Pennsylvania, build a bridge on a school bus route in just 35 days, saving months of time and 50 percent on costs.[2] A project to build a bridge built using GRS-IBS technology in St. Lawrence County, New York realized a 60-percent cost savings.[3]

[1] Federal Highway Administration, Every Day Counts, GRS-IBS Case Studies, www.fhwa.dot.gov/everydaycounts/technology/grs_ibs/casestudies.cfm.
[2] Randy Albert, Pennsylvania Department of Transportation, "Every Day Counts," EDC Forum, www.fhwa.dot.gov/everydaycounts/forum/post.cfm?id=27.
[3] Federal Highway Administration, Every Day Counts, GRS-IBS Case Studies, www.fhwa.dot.gov/everydaycounts/technology/grs_ibs/casestudies.cfm.

- **Prefabricated bridge elements and systems (PBES).** With PBES, prefabricated components are constructed off site and moved to the work zone for rapid installation, reducing the level of traffic disruption typically associated with bridge replacement. In some cases, PBES makes removing the old bridge overnight possible, while putting the new bridge in place the next day. Because PBES components are usually fabricated under controlled conditions, weather has less impact on the quality and duration of the project.

Prefabricated Bridge Elements and Systems

The Massachusetts DOT used prefabricated bridge elements on a project to replace 14 bridge superstructures on I-93 in Medford, shrinking a 4-year bridge replacement project to just one summer. The agency built the bridge superstructures in sections off site and installed them on weekends during 55-hour windows to minimize impact on travelers.

In addition to delivering bridge projects faster, FHWA is also delivering pavement innovations to prolong a road's lifespan while providing stakeholders cost savings. These efforts include:

- **Intelligent compaction.** When pavement cracks prematurely, a potential cause is improper compaction during construction. Intelligent compaction—using global positioning system-based mapping and real-time monitoring to control the compaction process—improves the quality, uniformity, and lifespan of pavements.

- **Warm-mix asphalt (WMA).** Composed in various fashions, WMA enables construction crews to produce and place asphalt on a road at lower temperatures than is possible using conventional hot-mix methods. In most cases, the lower temperatures result in significant cost savings because fuel consumption during WMA production is typically 20 percent lower. WMA production also generates fewer emissions, making conditions for workers healthier, and can extend the construction season, enabling agencies to deliver projects faster.

By cost effectively repairing and replacing roads and bridges with those having longer lifespans, localities can repair or replace a facility to a state of good repair. Stakeholders also will be able to maintain facilities at a high level for a longer period. Localities, in turn, can focus the cost savings from a previous project to other areas of need on the road network

---

[1] FHWA launched Every Day Counts (EDC) in cooperation with the American Association of State Highway and Transportation Officials (AASHTO) to speed up the delivery of highway projects and to address the challenges presented by limited budgets. EDC is a State-based model to identify and rapidly deploy proven but underutilized innovations to shorten the project delivery process, enhance roadway safety, reduce congestion, and improve environmental sustainability. EDC-1 occurred in 2011–2012, followed by EDC-2 in 2013–2014, and EDC-3 in 2015–2016.

Transit System Conditions

Ideally, the condition and performance of the U.S. transit infrastructure should be evaluated by how well it supports the objectives of the transit agencies that operate it. These objectives include providing safe, fast, cost-effective, reliable, and comfortable service that takes people where they want to go. The degree to which transit service meets these objectives, however, is difficult to quantify and involves trade-offs that are outside the scope of Federal responsibility. This section reports on the quantity, age, and physical condition of transit assets—factors that determine how well the infrastructure can support an agency's objectives and set a foundation for consistent measurement. Transit assets include vehicles, stations, guideway, rail yards, administrative facilities, maintenance facilities, maintenance equipment, power systems, signaling systems, communication systems, and structures that carry elevated or subterranean guideway. Chapter 5 addresses issues relating to the operational performance of transit systems.

FTA uses a numerical rating scale ranging from 1 to 5, detailed in *Exhibit 3-21*, to describe the relative condition of transit assets. A rating of 4.8 to 5.0, or "excellent," indicates that the asset is in nearly new condition or lacks visible defects. The midpoint of the "marginal" rating (2.5) is the threshold below which the assets are considered not in a state of good repair. At the other end of the scale, a rating of 1.0 to 1.9, or "poor," indicates that the asset needs immediate repair and does not support satisfactory transit service.

Exhibit 3-21 Definitions of Transit Asset Conditions

Rating	Condition	Description
Excellent	4.8–5.0	No visible defects, near-new condition.
Good	4.0–4.7	Some slightly defective or deteriorated components.
Adequate	3.0–3.9	Moderately defective or deteriorated components.
Marginal	2.0–2.9	Defective or deteriorated components in need of replacement.
Poor	1.0–1.9	Seriously damaged components in need of immediate repair.

Source: Transit Economic Requirements Model.

FTA uses the Transit Economic Requirements Model (TERM) to estimate the condition of transit assets for this report. This model consists of a database of transit assets and deterioration schedules that express asset conditions principally as a function of an asset's age. Vehicle condition is based on the vehicle's maintenance history and an estimate of the major rehabilitation expenditures in addition to vehicle age; the conditions of wayside control systems and track are based on an estimate of use (revenue miles per mile of track) in addition to age. For the purposes of this report, the state of good repair is defined using TERM's numerical condition rating scale. Specifically, this report considers an asset to be in a state of good repair when the physical condition of that asset is at or above a condition rating value of 2.5 (the midpoint of the marginal range). An entire transit system would be in a state of good repair if all of its assets have an estimated condition value of 2.5 or higher. The State of Good Repair benchmark presented in Chapter 8 represents the level of investment required to attain and maintain this definition of a state of good repair by rehabilitating or replacing all assets having estimated condition ratings that are less than this minimum condition value. FTA is currently developing a broader definition of state of good repair to use as a basis for administering MAP-21 grant programs and requirements that are intended to foster better infrastructure reinvestment practices across the industry. This definition might not be the same as the one used in this report.

FTA has estimated typical deterioration schedules for vehicles, maintenance facilities, stations, train control systems, electric power systems, and communication systems through special on-site engineering surveys. Transit vehicle conditions also reflect the most recent information on vehicle age, use, and level of maintenance from the National Transit Database (NTD); the information used in this edition of the C&P report is from 2012. Age information is available on a vehicle-by-vehicle basis from NTD and for all other assets is collected through special surveys. Average maintenance expenditures and major rehabilitation expenditures by vehicle are also available on agency and modal bases. When calculating conditions, FTA assumes agency maintenance and rehabilitation expenditures for a particular mode are the same average value for all vehicles the agency operates in that mode. Because agency maintenance expenditures can fluctuate from year to year, TERM uses a 5-year average.

The deterioration schedules applied for track and guideway structures are based on special studies. Appendix C presents a discussion on the methods used to calculate deterioration schedules and the sources of data on which deterioration schedules are based.

Condition estimates in each edition of the C&P report are based on up-to-date asset inventory information that reflects updates in TERM's asset inventory data. Annual data from NTD were used to update asset records for the Nation's transit vehicle fleets. In addition, updated asset inventory data were collected from 30 of the Nation's largest rail and fixed-route bus transit agencies to support analysis of nonvehicle needs. Because these data are not collected annually, providing accurate time series analysis of nonvehicle assets is not possible. FTA is working to develop improved data in this area. Appendix C provides a more detailed discussion of TERM's data sources. *Exhibit 3-22* shows the distribution of asset conditions, by replacement value, across major asset categories for the entire U.S. transit industry.

Exhibit 3-22 Distribution of Asset Physical Conditions by Asset Type for All Modes

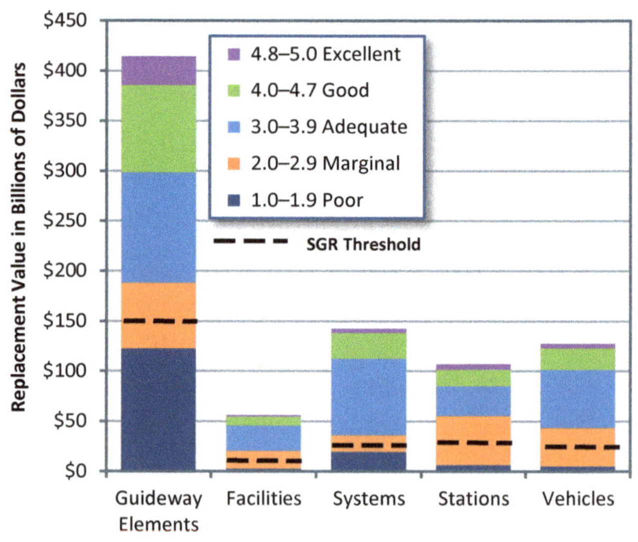

Source: Transit Economic Requirements Model.

Condition estimates for assets are weighted by the replacement value of each asset. This weighting accounts for the fact that assets vary substantially in replacement value. For example, a $1-million railcar in poor condition is a much bigger problem than a $1-thousand turnstile in similar condition. To illustrate the calculation involved, consider: The cost-weighted average of a $100 asset in condition 2.0 and a $50 asset in condition 4.0 would be (100 × 2.0 + 50 × 4.0)/(100 + 50) = 2.67. The unweighted average would be (2 + 4)/2 = 3.

The Replacement Value of U.S. Transit Assets

The total replacement value of the transit infrastructure in the United States for 2012 was estimated at $847.5 billion (in 2102 dollars). These estimates, presented in *Exhibit 3-23*, are based on asset inventory information in TERM. They exclude the value of assets that belong to special service operators that do not report to NTD. Rail assets totaled $748.1 billion, or roughly 88 percent of all transit assets. Nonrail assets were estimated at $85.1 billion. Joint assets totaled $14.2 billion; joint assets are those that serve more than one mode within a single agency and can include administrative facilities, intermodal transfer centers, agency communications systems (e.g., telephone, radios, and computer networks), and vehicles that agency management uses (e.g., vans and automobiles).

Exhibit 3-23 Estimated Replacement Value of the Nation's Transit Assets, 2012

	Replacement Value (Billions of 2012 Dollars)			
Transit Asset	Nonrail	Rail	Joint Assets	Total
Maintenance Facilities	$22.2	$26.3	$7.6	$56.2
Guideway Elements	$7.0	$406.4	$1.1	$414.4
Stations	$3.8	$102.3	$0.4	$106.6
Systems	$4.8	$133.5	$4.3	$142.6
Vehicles	$47.3	$79.6	$0.8	$127.7
Total	**$85.1**	**$748.1**	**$14.2**	**$847.5**

Source: Transit Economic Requirements Model.

Bus Vehicles (Urban Areas)

Bus vehicle age and condition are reported according to vehicle type for 2002 to 2012 in *Exhibit 3-24*. When measured across all vehicle types, the average age of the Nation's bus fleet has remained essentially unchanged since 2002. Similarly, the average condition rating for all bus types (calculated as the weighted average of bus asset conditions, weighted by asset replacement value) is also relatively unchanged, remaining near the bottom of the adequate range for the past 10 years. The percentage of vehicles below the state of good repair replacement threshold (condition 2.5) has remained at 10–12 percent for this same period. Note that, although this observation holds across all vehicle types, the proportion of full-size buses (the vehicle type that supports most fixed-route bus services) declined from 15.2 percent in 2008 to 12.3 percent in 2012. This reduction likely reflects impacts of transit-related spending under the American Recovery and Reinvestment Act. The Nation's bus fleet has grown at an average annual rate of roughly 2 percent over the past 10 years, with most of this growth concentrated in three vehicle types: large, 60-foot articulated buses; small buses less than 25 feet long (frequently dedicated to flexible-route bus services); and vans. The large increase in the number of vans reflects both the needs of an aging population (paratransit services) and an increase in the popularity of vanpool services. In contrast, the number of full- and medium-sized buses has remained relatively flat since 2002.

Exhibit 3-24 Urban Transit Bus Fleet Count, Age, and Condition, 2002–2012

	2002	2004	2006	2008	2010	2012
Articulated Buses						
Fleet Count	2,799	3,074	3,445	4,302	4,896	5,043
Average Age (Years)	7.2	5.0	5.3	6.3	6.5	7.0
Average Condition Rating	3.3	3.5	3.5	3.3	3.2	3.1
Below Condition 2.50 (Percent)	16.6%	5.0%	2.1%	2.6%	3.7%	5.3%
Full-Size Buses						
Fleet Count	46,573	46,139	46,714	45,985	45,441	44,906
Average Age (Years)	7.5	7.2	7.4	7.9	7.8	8.0
Average Condition Rating	3.2	3.2	3.2	3.1	3.1	2.9
Below Condition 2.50 (Percent)	13.1%	12.3%	11.3%	15.2%	12.5%	12.3%
Mid-Size Buses						
Fleet Count	7,269	7,114	6,844	7,009	7,218	7,077
Average Age (Years)	8.4	8.1	8.2	8.3	8.1	7.4
Average Condition Rating	3.1	3.1	3.1	3.1	3.1	3.0
Below Condition 2.50 (Percent)	14.1%	13.2%	14.2%	12.4%	12.5%	8.2%
Small Buses						
Fleet Count	14,857	15,972	16,156	19,366	19,493	23,793
Average Age (Years)	4.5	4.6	5.1	5.1	5.2	5.2
Average Condition Rating	3.4	3.5	3.4	3.4	3.4	3.3
Below Condition 2.50 (Percent)	8.8%	10.1%	10.3%	11.6%	10.2%	13.1%
Vans						
Fleet Count	17,147	18,713	19,515	26,823	28,531	28,193
Average Age (Years)	3.2	3.3	3.0	3.2	3.4	3.8
Average Condition Rating	3.7	3.8	3.8	3.8	3.7	3.6
Below Condition 2.50 (Percent)	7.2%	6.7%	8.4%	8.0%	8.2%	4.1%
Total Fixed-Route Bus						
Total Fleet Count	88,645	91,012	92,674	103,485	105,579	109,012
Weighted Average Age (Years)	6.2	6.0	6.0	6.1	6.1	6.2
Weighted Average Condition Rating	3.2	3.3	3.3	3.1	3.0	3.2
Below Condition 2.50 (Percent)	11.8%	10.6%	10.4%	12.0%	10.5%	9.8%

Sources: Transit Economic Requirements Model and National Transit Database.

Exhibits 3-25 and *3-26* present the age distribution of the Nation's transit buses and vans, minivans, and autos, respectively. Note here that full-size buses and vans account for the highest proportion (roughly 67 percent) of the Nation's rubber-tire transit vehicles. Moreover, although most vans are retired by age 7 and most buses by age 15, roughly 5 to 20 percent of these fleets remain in service well after their typical retirement ages.

A distinction should be made between "small buses" and cutaways. By definition, small buses are 30-foot long vehicles operating mostly as fixed route. Cutaways are buses less than 30 feet in length, operating mostly as demand response.

Exhibit 3-25 Age Distribution of Fixed-Route Buses (Urban Areas), 2012

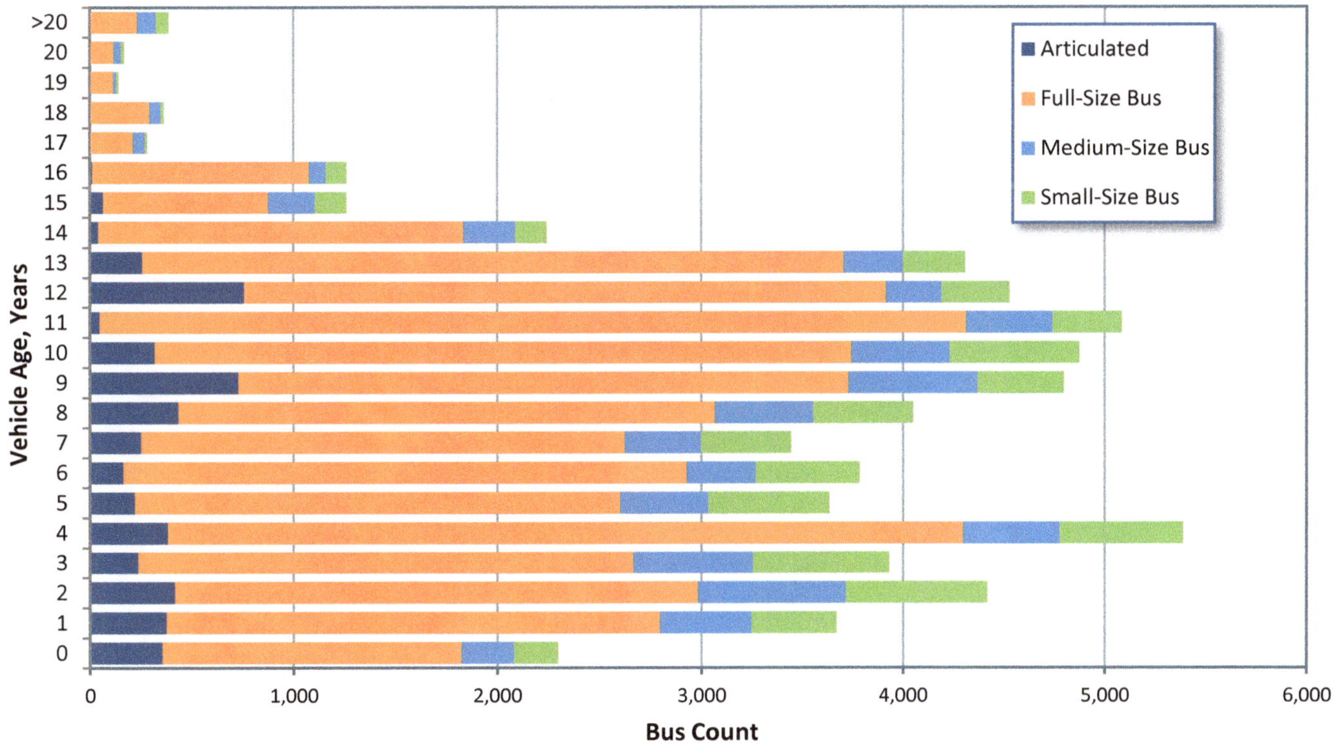

Source: Transit Economic Requirements Model and National Transit Database.

Exhibit 3-26 Age Distribution of Vans, Minivans, Autos, and Cutaways (Urban Areas), 2012

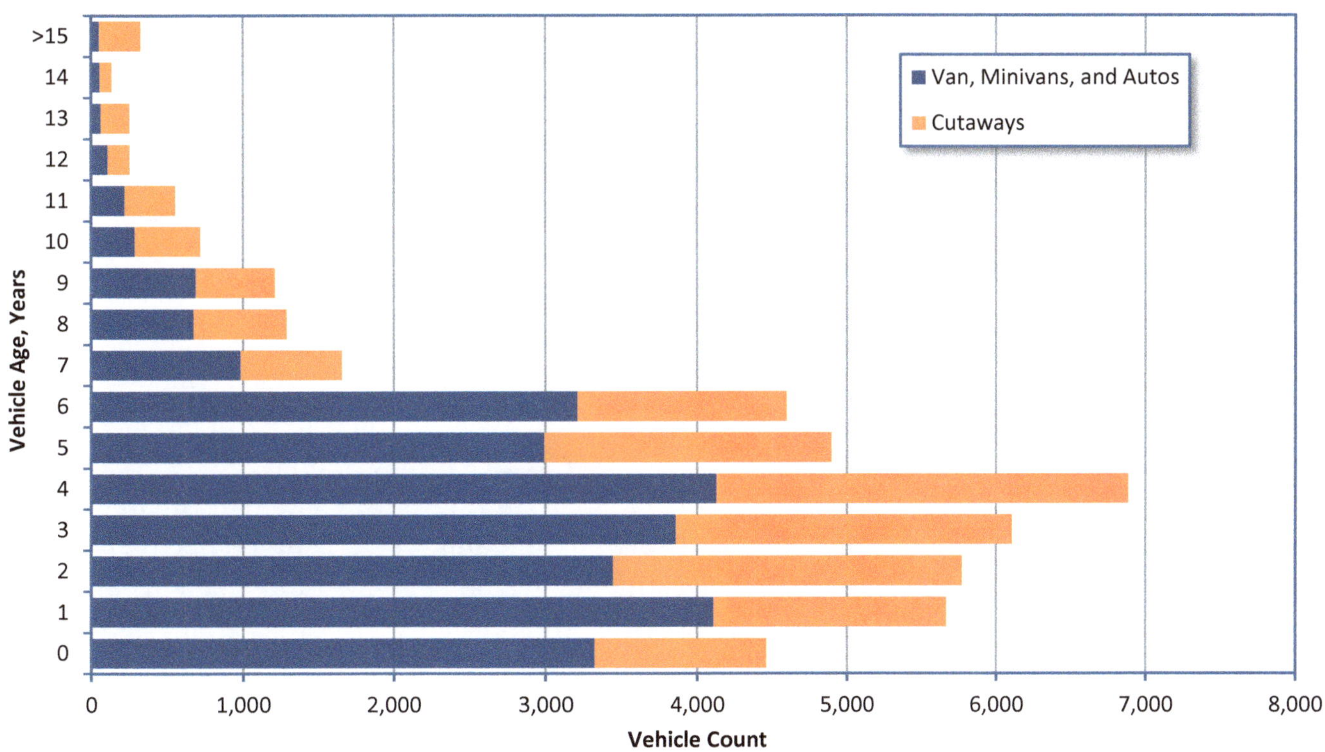

Source: Transit Economic Requirements Model and National Transit Database.

Other Bus Assets (Urban Areas)

The more comprehensive capital asset data described above enable reporting of a more complete picture of the overall condition of bus-related assets. *Exhibit 3-27* shows TERM estimates of current conditions for the major categories of fixed-route bus assets. Vehicles comprise roughly half of all fixed-route bus assets, and maintenance facilities make up another third. Roughly one-third of bus maintenance facilities are rated below condition 3.0, compared to roughly one-half for bus, paratransit, and vanpool vehicles.

Exhibit 3-27 Distribution of Estimated Asset Conditions by Asset Type for Fixed-Route Bus

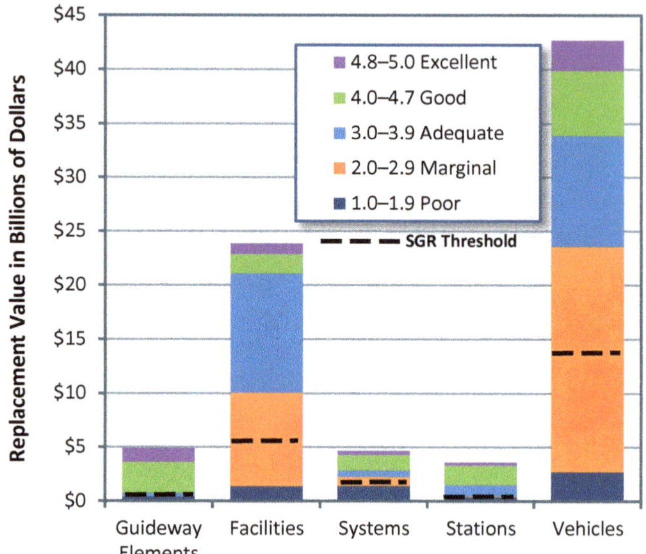

Source: Transit Economic Requirements Model.

Rail Vehicles

NTD compiles annual data on all rail vehicles; these data are shown in *Exhibit 3-28*, broken down by major category of rail vehicles. Measured across all rail vehicle types, the average age of the Nation's rail fleet has remained essentially unchanged, between 19 and 20 years old, since 2004. The average condition of all rail vehicle types (calculated as the weighted average of vehicle conditions, weighted by vehicle replacement cost) is also relatively unchanged, remaining near 3.5 since 2002. The percentage of vehicles below the state of good repair replacement threshold (condition 2.5) has remained between 3.6 and 4.6 percent since 2002. Note that, although this observation holds across all vehicle types, the analysis suggests that most vehicles in lesser condition occur in the light and heavy rail fleets. Most light rail vehicles with an estimated condition of less than 2.5, however, are historic streetcars and trolley cars with an average age of 75 years. Given their historic vehicle status, the estimated condition of these vehicles (determined primarily by age) should be viewed as a rough approximation.

From 2002 to 2012, the Nation's rail transit fleet grew at an average annual rate of roughly 1.3 percent. This rate of growth was largely due to the rate of increase in the heavy rail fleet (which represents slightly more than half the total fleet and grew at an average annual rate of 0.4 percent over this period). In contrast, the annual rate of increase in commuter rail and light rail fleets has been appreciably higher, averaging approximately 2.1 percent and 3.2 percent, respectively. These higher growth rates reflect recent rail transit investments in small and medium-sized urban areas where the size and population density do not justify the greater investment needed for heavy rail construction.

Exhibit 3-28 Urban Transit Rail Fleet Count, Age, and Condition, 2002–2012

	2002	2004	2006	2008	2010	2012
Commuter Rail Locomotives						
Fleet Count	709	710	740	790	822	877
Average Age (Years)	17.2	17.8	16.7	19.6	19.4	17.8
Average Condition Rating	3.7	3.7	4.0	3.6	3.6	3.7
Below Condition 2.50 (Percent)	0.0%	0.0%	0.0%	0.0%	0.0%	1.8%
Commuter Rail Passenger Coaches						
Fleet Count	2,985	3,513	3,671	3,539	3,711	3,758
Average Age (Years)	19.2	17.7	16.8	19.9	19.1	20.2
Average Condition Rating	3.7	3.8	4.1	3.6	3.7	3.6
Below Condition 2.50 (Percent)	0.0%	0.0%	0.0%	0.0%	0.0%	0.4%
Commuter Rail Self-Propelled Passenger Coaches						
Fleet Count	2,389	2,470	2,933	2,665	2,659	2,930
Average Age (Years)	27.1	23.6	14.7	18.9	19.7	19.7
Average Condition Rating	3.5	3.7	3.8	3.7	3.7	3.6
Below Condition 2.50 (Percent)	0.0%	0.0%	0.0%	0.0%	0.0%	0.0%
Heavy Rail						
Fleet Count	11,093	11,046	11,075	11,570	11,648	11,587
Average Age (Years)	19.8	19.8	22.3	21.0	18.8	19.9
Average Condition Rating	3.4	3.4	3.3	3.3	3.4	3.4
Below Condition 2.50 (Percent)	6.1%	5.6%	5.5%	6.1%	5.2%	3.7%
Light Rail[1]						
Fleet Count	1,637	1,884	1,832	2,151	2,222	2,241
Average Age (Years)	17.9	16.5	14.6	17.1	18.1	14.6
Average Condition Rating	3.5	3.6	3.7	3.6	3.5	3.6
Below Condition 2.50 (Percent)	11.8%	9.3%	6.4%	7.1%	6.9%	6.3%
Total Rail						
Total Fleet Count	18,813	19,623	20,251	20,715	21,062	21,393
Weighted Average Age (Years)	20.4	19.5	19.3	20.1	18.9	19.3
Weighted Average Condition Rating	3.5	3.5	3.6	3.5	3.5	3.5
Below Condition 2.50 (Percent)	4.6%	4.1%	3.6%	4.2%	3.6%	2.8%

[1] Excludes vintage streetcars.

Source: Transit Economic Requirements Model and National Transit Database.

Exhibit 3-29 presents the age distribution of the Nation's heavy rail, light rail, and commuter rail transit vehicles. Heavy rail vehicles account for more than half the Nation's rail fleet, whereas light rail, a mode typically found in smaller rail markets, accounts for only 11 percent of rail vehicles. At the same time, roughly one-third of rail and commuter vehicles are more than 25 years old—with close to 3,000 heavy and commuter rail vehicles exceeding 35 years in age. Comparing the results in *Exhibit 3-29* with the age distribution of transit buses and vans in *Exhibit 3-25* and *Exhibit 3-26* is instructive; a comparatively clear pattern of preferred retirement age is evident in the bus and van vehicle type but no such pattern is shown in the rail vehicle results. *Exhibit 3-30* presents the age distribution of the Nation's hybrid rail, streetcar, and other rail transit vehicles.

Exhibit 3-29 Age Distribution of Rail Transit Vehicles, 2012

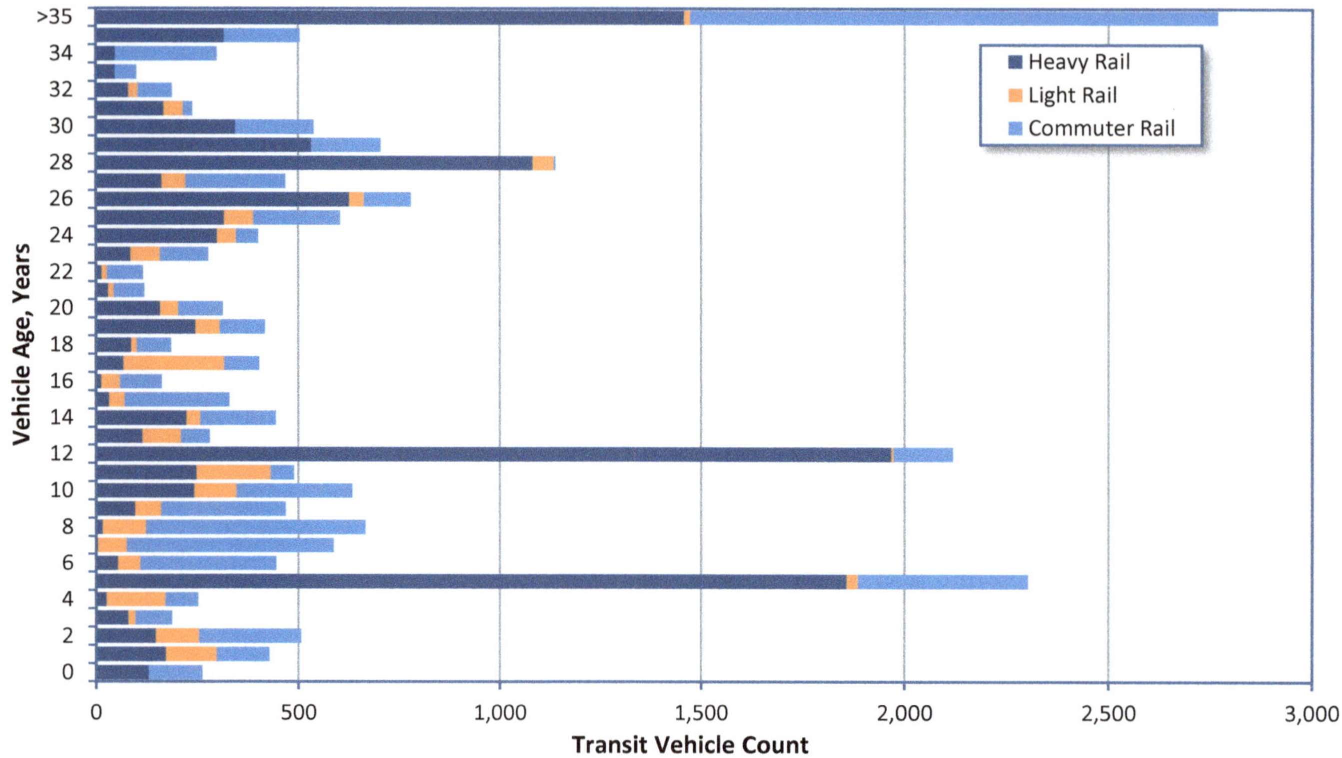

Source: Transit Economic Requirements Model and National Transit Database.

Exhibit 3-30 Age Distribution of Rail Transit Vehicles, 2012

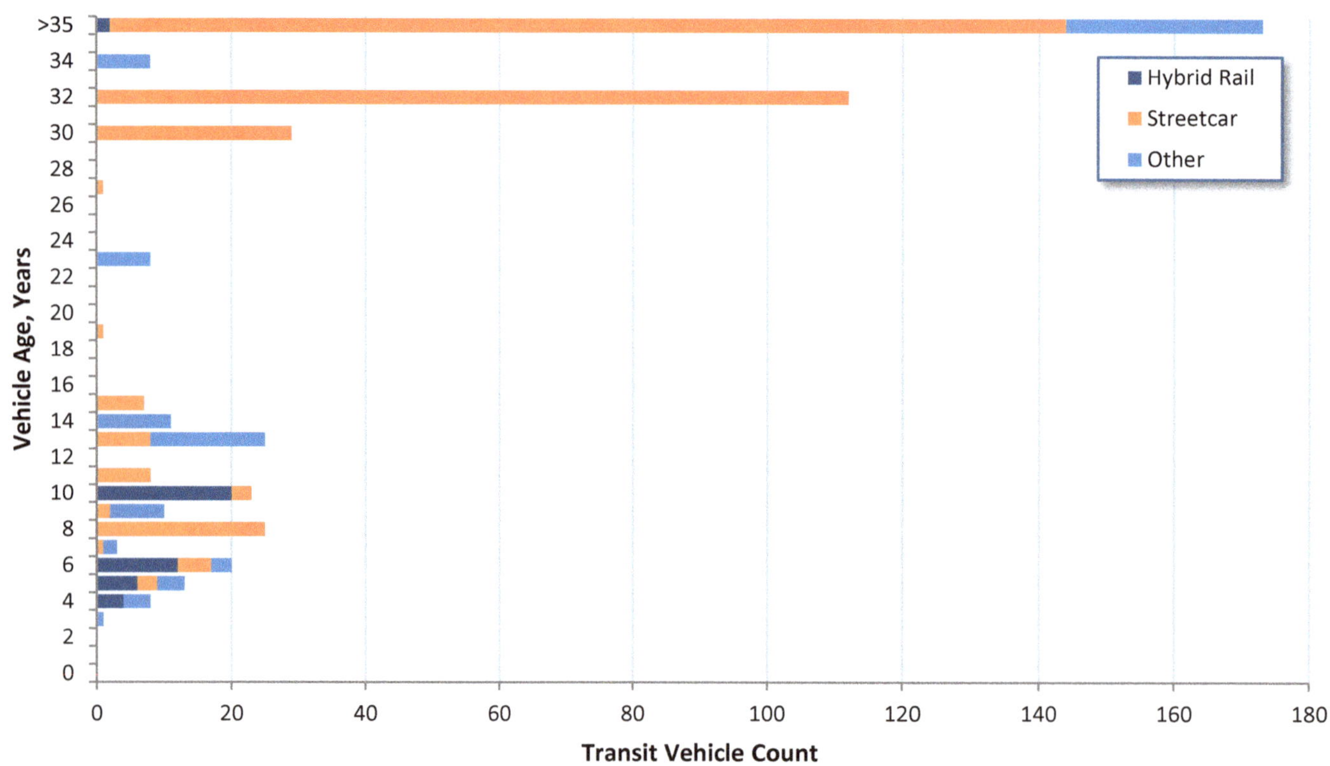

Source: Transit Economic Requirements Model and National Transit Database.

Other Rail Assets

Assets associated with nonvehicle transit rail can be divided into four general categories: guideway elements, systems, stations, and facilities. TERM estimates of the condition distribution for each category are shown in *Exhibit 3-31.*

Exhibit 3-31 Distribution of Asset Physical Conditions by Asset Type for All Rail

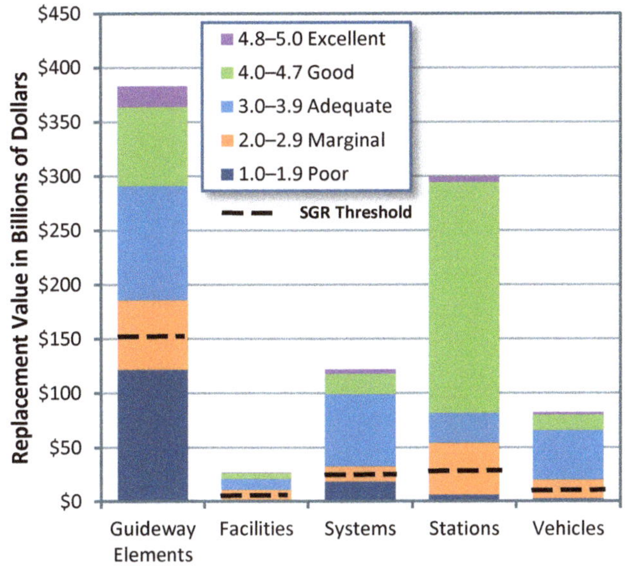

Source: Transit Economic Requirements Model.

The largest category by replacement value is guideway elements. These elements consist of tracks, ties, switches, ballasts, tunnels, and elevated structures. The replacement value of this category is $382.8 billion, of which $121.6 billion is rated below condition 2.0 (32 percent) and $64.1 billion is rated between conditions 2.0 and 3.0. The relatively large proportion of guideway and systems assets rated below condition 2.0 and the magnitude of the $140-billion investment required to replace or keep them in a state of good repair represent major challenges to the transit rail industry.

Although maintaining these assets is among the largest expenses associated with operating rail transit, FTA does not collect detailed data on these elements, in part because the elements are difficult to categorize into discrete sections having common life expectancies. Service life for track, for example, highly depends on the amount of use it receives and its location.

Systems, which consist of power, communication, and train control equipment, have a replacement value of $121.4 billion, of which $18.3 billion is rated below condition 2.0 (15 percent) and $14 billion is rated between conditions 2.0 and 3.0. This category is another for which many assets are difficult to characterize in terms of standard types and life expectancies. As a result, FTA has only limited data from which to make needs projections.

Stations have a replacement value of $299.3 billion. Only $5.9 billion is rated below condition 2.0 and $48.4 billion is rated between conditions 2.0 and 3.0.

Facilities, principally consisting of maintenance and administration buildings, have a replacement value of $26.4 billion. The value rated below condition 2.0 is $1.3 billion, and between conditions 2.0 and 3.0 is $9.3 billion.

Rail transit consists of heavy rail (urban dedicated guideway), light rail, hybrid rail, streetcar (in mixed traffic), and commuter rail (suburban passenger rail) modes. Almost half of rail transit vehicles are in heavy rail systems. Heavy rail represents $255.2 billion (84 percent) of the total transit rail replacement cost of $303.2 billion. Heavy rail serves some of the Nation's oldest and largest transit systems (Boston, New York, Washington, San Francisco, Philadelphia, and Chicago).

The condition distribution of heavy rail assets, which represent the largest share of U.S. rail transit assets, is shown in *Exhibit 3-32*.

Exhibit 3-33 shows the average age and condition of nonvehicle transit assets for fixed-route bus and rail modes reported for 2012.

While *Exhibit 3-31* depicts the replacement value of national transit assets by category for rail modes, *Exhibit 3-33* provides additional data such as average fleet age, average condition, and percent of assets below the state of good repair threshold (rating below 2.5).

Exhibit 3-32 Distribution of Asset Physical Conditions by Asset Type for Heavy Rail

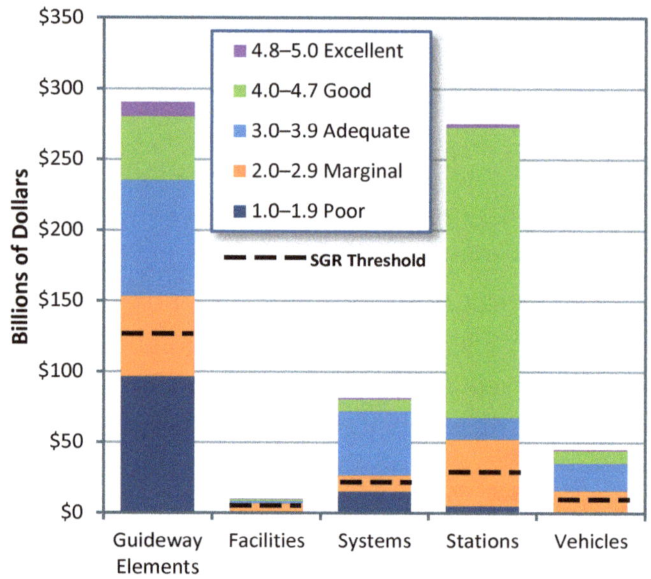

Source: Transit Economic Requirements Model.

The data reinforce the analysis of *Exhibit 3-23*, namely that assets for rail modes represent major challenges to the transit rail industry. For example, the average condition of all rail asset categories is near the upper bound of the marginal rating; 35 percent of guideway elements—the asset category having the highest replacement value—are below a state of good repair. Stations have the largest percentage of assets below a state of good repair, 38 percent.

Exhibit 3-33 Non-Vehicle Transit Assets: Age and Condition, 2012

Category	Mode Type	Average Age	Average Condition	Percent Below Condition 2.5
Facilities	Rail	35.4	3.2	24%
	Fixed-Route Bus	30.4	3.2	7%
	All	**32.8**	**3.2**	**15%**
Guideway Elements	Rail	64.3	3.1	35%
	Fixed-Route Bus	24.0	4.5	6%
	All	**63.5**	**3.1**	**35%**
Stations	Rail	57.8	3.0	38%
	Fixed-Route Bus	22.9	4.0	12%
	All	**57.3**	**3.0**	**37%**
Systems	Rail	31.6	3.3	17%
	Fixed-Route Bus	23.6	3.5	17%
	All	**31.1**	**3.3**	**17%**

Source: Transit Economic Requirements Model.

Rural Transit Vehicles and Facilities

Rural systems operate buses, vans, or other small passenger vehicles (see Chapter 2). Data on the numbers and ages of rural vehicles and the number of maintenance facilities are now compiled in NTD, enabling FTA to report more accurately on rural transit conditions and on the 727 rural maintenance facilities in 2012. The age distributions of rural transit vehicles for buses and for vans, minivans, autos, and cutaways are summarized in *Exhibit 3-34* and *Exhibit 3-35*, respectively. The relative small average fleet age of fixed-route buses is due to the large proportion of cutaway vehicles, which are usually built based on incomplete vans with added specific equipment required for transit service. These vehicles have 4 to 5 years average useful life and need frequent replacement.

Exhibit 3-34 Age Distribution of Rural Transit Vehicles for Fixed-Route Buses, 2012

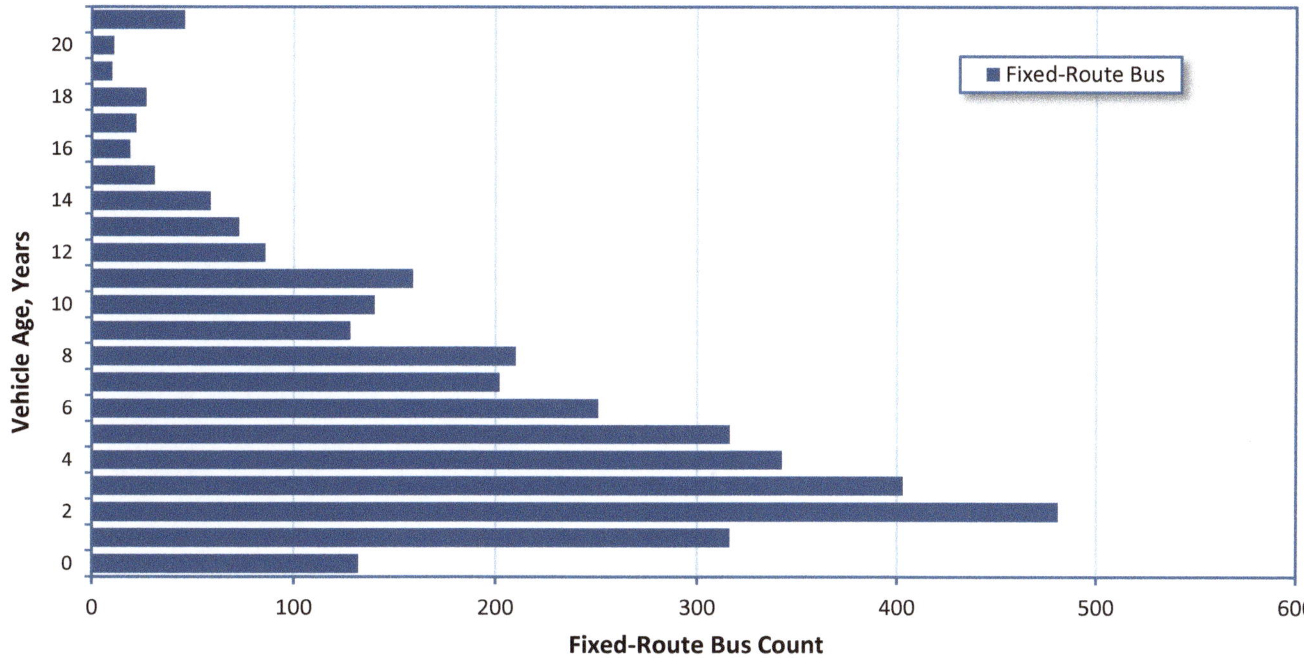

Source: Transit Economic Requirements Model and National Transit Database.

Exhibit 3-35 Age Distribution of Rural Transit Vehicles for Vans, Minivans, Autos, and Cutaways, 2012

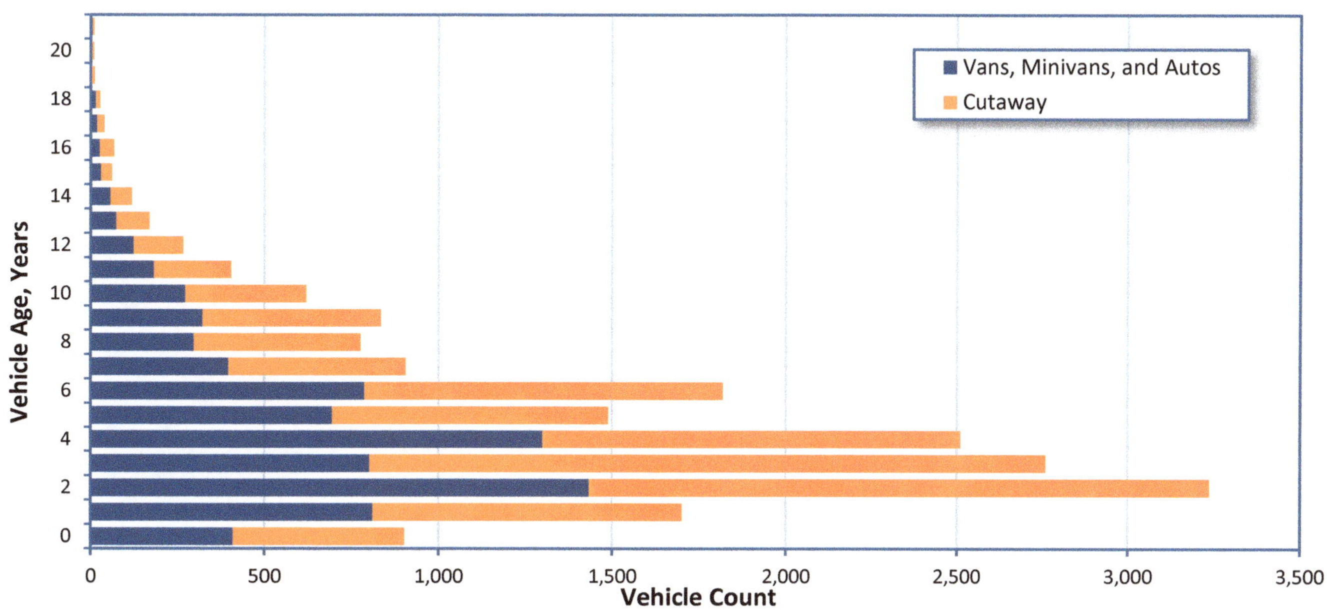

Source: Transit Economic Requirements Model and National Transit Database.

For 2012, data reported to NTD indicated that 17.0 percent of rural buses, 6.0 percent of cutaways, and 41.4 percent of rural vans were past their FTA minimum life expectancy (12 years for buses, 7 to 10 for cutaways, and 4 for vans). The rural transit fleet had an average age of 5.2 years in 2012; buses, with an average age of 8.1 years, were older than vans and cutaways, which each had an average age of 4.6 years.

chapter 4

Safety

Highway Safety

Safety is the U.S. Department of Transportation's (DOT) top priority. Three operating administrations within DOT have specific responsibilities for addressing highway safety. The Federal Highway Administration (FHWA) focuses on infrastructure safety design and operations. The National Highway Traffic Safety Administration (NHTSA) oversees vehicle safety standards and administers driver behavior programs. The Federal Motor Carrier Safety Administration (FMCSA) works to reduce crashes, injuries, and fatalities involving large trucks and buses.

This balance of coordinated efforts enables these DOT Administrations to concentrate on their areas of expertise while working toward a single goal and encourages a more unified effort. Coupled with a comprehensive focus on shared, reliable safety data, collectively these organizations ensure that the Federal effort is implemented to its greatest potential.

This chapter provides data on fatalities and injuries and details on FHWA safety programs. FHWA provides technical assistance and expertise to Tribal, State, and local governments for researching, designing, and implementing safety improvements in roadway infrastructure. FHWA also supports improvements in safety elements as part of road and bridge construction and system preservation projects. The Highway Safety Improvement Program (HSIP) is FHWA's main infrastructure safety funding program. It includes a performance-driven, strategic approach to achieve significant reductions in fatalities and serious injuries on all public roads for all road users, including pedestrians and bicyclists. HSIP also helps States improve their roadway safety data. Additionally, HSIP supports railway-highway safety through set-aside funding. Use of HSIP funds is driven by a statewide coordinated plan developed in cooperation with a broad range of multidisciplinary stakeholders that provides a comprehensive framework for safety. This data-driven State Strategic Highway Safety Plan (SHSP) defines State safety goals and integrates the four "E's"—engineering, education, enforcement, and emergency services. SHSP guides States and their collection of data in the use of HSIP and other funds to resolve safety problems and save lives.

On March 15, 2016, FHWA published Final Rules for HSIP and for Safety Performance Management Measures (PM-1). The HSIP Final Rule updates the existing HSIP requirements to be consistent with the Moving Ahead for Progress in the 21st Century Act (MAP-21) and the Fixing America's Surface Transportation (FAST) Act. The PM-1 Final Rule adds specific safety performance measure requirements for the purpose of carrying out the HSIP to assess serious injuries and fatalities on all public roads. The Safety PM Final Rule establishes five performance measures as the 5-year rolling averages for: (1) Number of Fatalities, (2) Rate of Fatalities per 100 million Vehicle Miles Traveled (VMT), (3) Number of Serious Injuries, (4) Rate of Serious Injuries per 100 million VMT, and (5) Number of Non-motorized Fatalities and Non-motorized Serious Injuries.

Overall Fatalities and Injuries

Statistics discussed in this section are drawn primarily from the Fatality Analysis Reporting System (FARS). NHTSA, which has a cooperative agreement with States to provide information on fatal crashes, maintains FARS. FARS is a nationwide census providing DOT, Congress, and the American public data regarding fatal motor vehicle traffic crashes. FARS data are combined with exposure data from other sources to produce fatal crash rates. The exposure data most frequently used are estimates of vehicle miles traveled (VMT) that FHWA collects through the Highway Performance Monitoring System. This system provides a standard, recognized database that covers all regions of the United States.

In addition to FARS, NHTSA estimates serious injuries nationally through the National Automotive Sampling System General Estimates System. Datasets in this system provide a statistically produced annual estimate of total nonfatal injury crashes. Safety statistics in this section, compiled in 2014, represent a "snapshot in time" during the preparation of this report, which is why they might not precisely correspond to other, more recently completed reports.

In 2012, 31,006 fatal crashes occurred in the United States. In this same year, approximately 1.63 million nonfatal injury crashes and 3.95 million property damage-only crashes occurred. The total estimated number of crashes in 2012, as *Exhibit 4-1* shows, was 5.62 million. All three crash types have significantly declined from 2002 to 2012. The number of fatal crashes in 2012 showed the first increase since 2005, up to a nearly identical fatal crash count in 2009. Similarly, the number of injury crashes rose in 2012, up to a number last observed in 2008.

Exhibit 4-1 Crashes by Severity, 2002–2012

	Crash Severity							
	Fatal		Injury		Property Damage Only		Total Crashes	
Year	Number	Percent	Number	Percent	Number	Percent	Number	Percent
2002	38,491	0.6	1,928,984	30.5	4,348,233	68.8	6,315,708	100.0
2003	38,477	0.6	1,924,912	30.4	4,364,566	69.0	6,327,955	100.0
2004	38,444	0.6	1,861,617	30.1	4,280,966	69.3	6,181,027	100.0
2005	39,252	0.6	1,816,105	29.5	4,303,993	69.9	6,159,350	100.0
2006	38,648	0.6	1,745,924	29.2	4,188,641	70.1	5,973,213	100.0
2007	37,435	0.6	1,711,304	28.4	4,275,269	71.0	6,024,008	100.0
2008	34,172	0.6	1,630,420	28.1	4,146,254	71.4	5,810,846	100.0
2009	30,862	0.6	1,517,075	27.6	3,957,243	71.9	5,505,180	100.0
2010	30,296	0.6	1,542,104	28.5	3,847,045	71.0	5,419,445	100.0
2011	29,867	0.6	1,529,968	28.7	3,777,994	70.8	5,337,829	100.0
2012	31,006	0.6	1,634,180	29.1	3,949,858	70.3	5,615,044	100.0

Source: Fatality Analysis Reporting System/National Center for Statistics and Analysis, NHTSA.

Exhibit 4-2 displays trends in motor vehicle fatality counts, fatality rates, injury counts, and injury rates. The motor vehicle fatality count rose to above 51,000 in 1980, and then dropped to less than 43,000 in 1983, coinciding with the recession occurring in the early 1980s. The fatality count declined to less than 40,000 in 1992 for the first time in decades, but remained above 40,000 every year from 1993 through 2007. *Exhibit 4-2* shows significant declines in fatality counts in recent years. In 2005, 43,510 traffic deaths occurred. Between 2005 and 2011, the number of fatalities declined 25 percent, to 32,479 in 2011. In 2012, the number of fatalities rose for the first year since 2005, up 4 percent to 33,782 in 2012. Of note is that the large decline in fatalities from 2005 through 2011 included the timing of the implementation of FHWA's HSIP and the occurrence of the largest recession (2007–2009) since World War II.

Exhibit 4-2 Summary of Fatality and Injury Rates, 1966–2012

Year	Number of Motor Vehicle Fatalities	Resident Population (Thousands)	Fatality Rate per 100,000 Population	Vehicle Miles Travelled (Millions)	Fatality Rate per 100 Million VMT	Number of Motor Vehicle Injuries	Injury Rate per 100,000 Population	Injury Rate per 100 Million VMT
1966	50,894	196,560	25.89	925,345	5.50			
1968	52,725	200,706	26.27	1,013,942	5.20			
1970	52,627	205,052	25.67	1,110,274	4.74			
1972	54,589	209,896	26.01	1,269,512	4.30			
1974	45,196	213,854	21.13	1,291,314	3.50			
1976	45,523	218,035	20.88	1,400,708	3.25			
1978	50,331	222,585	22.61	1,543,896	3.26			
1980	51,091	227,225	22.48	1,525,104	3.35			
1982	43,945	231,664	18.97	2,496,875	1.76			
1984	44,257	235,825	18.77	1,722,062	2.57			
1986	46,087	240,133	19.19	1,836,135	2.51			
1988	47,087	244,499	19.26	2,029,612	2.32	3,416,000	1,397	168
1990	44,599	249,439	17.88	2,144,183	2.08	3,231,000	1,295	151
1992	39,250	254,995	15.39	2,242,857	1.75	3,070,000	1,204	137
1994	40,716	260,327	15.64	2,353,526	1.73	3,266,000	1,255	139
1996	42,065	265,229	15.86	2,482,202	1.69	3,483,000	1313	140
1998	41,501	270,248	15.36	2,628,148	1.58	3,192,000	1181	121
2000	41,945	281,422	14.90	2,749,803	1.53	3,189,000	1133	116
2002	43,005	288,369	14.91	2,855,756	1.51	2,926,000	1015	102
2003	42,884	290,810	14.75	2,890,893	1.48	2,889,000	993	100
2004	42,836	293,655	14.59	2,962,513	1.45	2,788,000	949	94
2005	43,510	296,410	14.68	2,989,807	1.46	2,699,000	911	90
2006	42,708	299,398	14.26	3,014,116	1.42	2,575,000	860	85
2007	41,259	301,621	13.68	3,029,822	1.36	2,491,000	826	82
2008	37,423	304,060	12.31	2,973,509	1.26	2,346,000	772	79
2009	33,883	307,007	11.04	2,953,501	1.15	2,217,000	722	75
2010	32,999	308,746	10.69	2,967,266	1.11	2,239,000	725	75
2011	32,479	311,592	10.42	2,950,402	1.10	2,217,000	712	75
2012	33,782	314,112	10.75	2,968,815	1.14	2,362,000	752	80

Source: Fatality Analysis Reporting System/National Center for Statistics and Analysis, NHTSA.

In addition to the fatality counts shown in the left column of *Exhibit 4-2*, fatality rates are shown for two different measures of exposure: rates expressed in terms of population and in terms of VMT. To account for amount of travel on the road, the fatality rate is most often expressed in terms of VMT. Fatality rate per VMT provides a metric that enables transportation professionals to consider fatalities in terms of the additional exposure associated with driving more miles. The fatality rates per population shown in *Exhibit 4-2* are often stratified to examine in more depth how demographic variables, such as male drivers aged 16–20 versus male drivers aged 21–44, influence fatality rates.

The fatality rate per population was 22.48 per 100,000 population in 1980. This rate dropped to 17.88 in 1990 and to 14.90 in 2000. In 2012, the rate further declined to 10.75. The rate of 10.75 in 2012 was less than half the fatality rate in 1980.

Q&A What do 2015 traffic fatality data show?

Although this report focuses primarily on prior years of data, NHTSA has issued 2015 FARS data. During 2015, 35,092 people died in crashes on U.S. roadways, an increase from 32,744 in 2014. The 7.2-percent increase is the largest percentage increase in nearly 50 years. The largest percentage increase previously was an 8.1-percent increase from 1965 to 1966. The estimated number of traffic-related injuries also increased from 2014 to 2015, rising from 2.34 to 2.44 million injured people. The fatality rate per 100 million VMT increased to 1.12 from 1.08 in 2014. The 2014 rate was the lowest since NHTSA began collecting fatality data through FARS in 1975. VMT increased by 3.5 percent from 2014 to 2015, the largest increase since 1992, nearly 25 years ago.

The number of passenger car and light-truck occupant fatalities increased by 1,391 (a 6.6-percent increase) from 2014 to 2015, and is at its highest since 2009. This increase accounted for 59 percent of the overall increase in fatalities. Pedestrian fatalities increased by 466 (a 9.5-percent increase), and are at their highest number since 1996. Bicyclist fatalities increased by 89 (a 12.2-percent increase), and are at their highest level since 1995.

More information is available at https://crashstats.nhtsa.dot.gov/Api/Public/ViewPublication/812318.

The fatality rate, expressed in terms of VMT, in 1966 was 5.50 deaths per 100 million (M) VMT. That rate has remained less than 5.00 since 1970, and less than 4.00 since 1974. Due to significant progress in traffic safety in the United States, the motor vehicle fatality rate has continued to decline. The rate was less than 3.00 in 1982; it has remained less than 2.00 since 1992. In 2003, the rate dropped below 1.50 and continued to drop from 1.46 in 2005 to 1.36 in 2007, to 1.26 in 2008, and to 1.15 in 2009. A historic low of 1.10 was reached in 2011, before the rate climbed slightly to 1.14 in 2012 (see *Exhibit 4-2*).

Also shown in *Exhibit 4-2* are the national estimates for people nonfatally injured in motor vehicle crashes. A historic low of 2,217,000 injured was reached in 2009 and again in 2011, with an injury rate of 75 per 100 M VMT in both years. In 2012, the injury count rose slightly to 2,362,000 and the rate rose to 80 per 100 M VMT. Fatalities and injuries declined in almost all segments of the population, including passenger vehicle drivers, passenger vehicle occupants, large-truck occupants, pedestrians, young drivers, and drivers involved in alcohol-impaired driving fatalities.

DOT attributes the overall decline in roadway fatalities over the past several years to multiple factors, including an increase in the HSIP spending rate and roadway infrastructure improvements such as Safety Edge, Innovative Intersection and Interchange Geometrics, and High Friction Surface Treatments. The improvements in infrastructure are some of the innovative technologies being deployed as part of FHWA's Every Day Counts initiative discussed in Chapter 3.

The trends since 1980 of the fatality counts and fatality rates, as discussed above and shown in *Exhibit 4-2*, are displayed graphically in *Exhibits 4-3* and *4-4*. *Exhibit 4-3* shows the number of motor vehicle fatalities from 1980 to 2012. *Exhibit 4-4* shows the motor vehicle fatality rates per 100 M VMT from 1980 to 2012.

Exhibit 4-3 Fatalities, 1980–2012

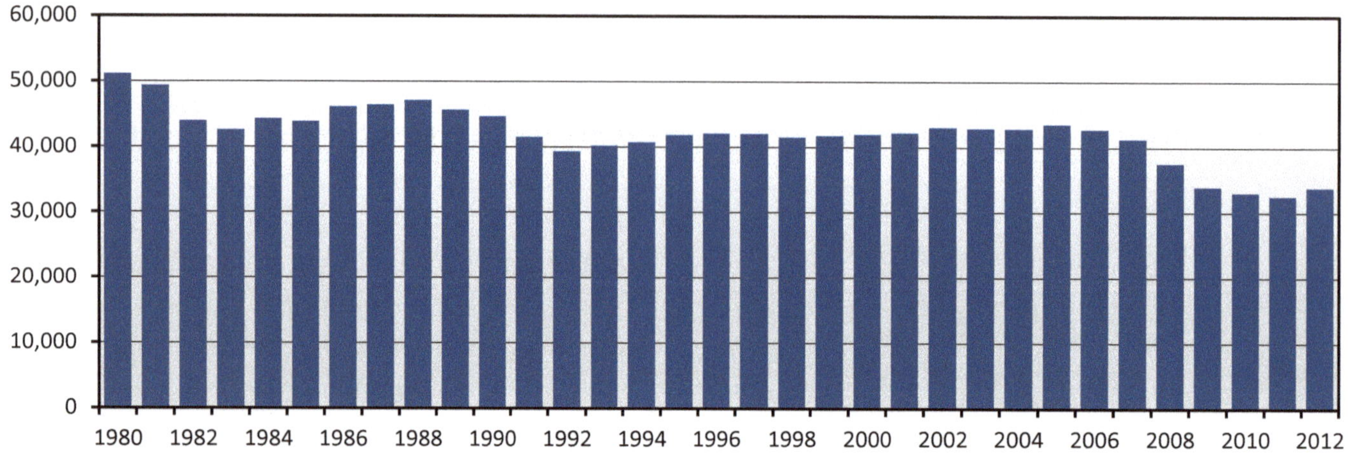

Source: Fatality Analysis Reporting System/National Center for Statistics and Analysis, NHTSA.

Exhibit 4-4 Fatality Rates, 1980–2012

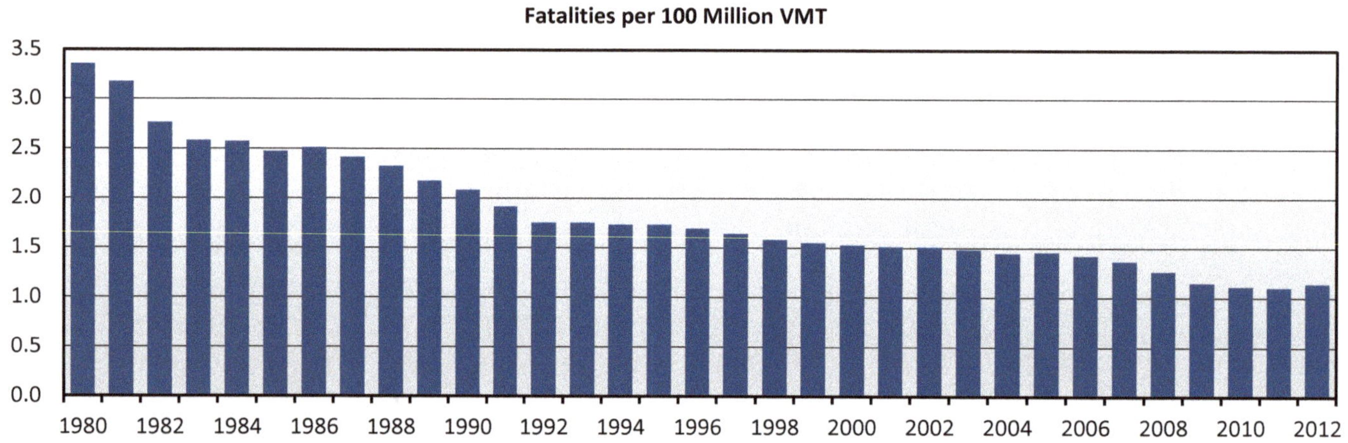

Source: Fatality Analysis Reporting System/National Center for Statistics and Analysis, NHTSA.

Fatalities by Roadway Functional Class

The previous section presents overall counts and rates of both fatalities and injuries. This section focuses on how fatality counts and fatality rates differ between rural and urban roadway functional class. *Exhibit 4-5* shows fatality counts and *Exhibit 4-6* displays fatality rates for 2002 through 2012.

As shown in *Exhibit 4-5*, the number of fatalities peaked in 2005, and then declined to 32,479 in 2011. In 2012, fatalities from urban crashes accounted for 45.9 percent of all fatalities, while those resulting from rural crashes accounted for 54.1 percent. From 2002 to 2012, the number of fatalities on urban roads decreased from 17,013 to 15,296, a reduction of 10.1 percent. The peak in urban fatalities occurred in 2006, at 18,791. Over the same period, the number of fatalities on rural roads decreased from 25,896 (in 2002) to 18,170 (in 2012), a reduction of 29.8 percent. Rural fatalities peaked in 2002.

Exhibit 4-5 Fatalities by Functional System, 2002–2012

Functional System	2002	2003	2004	2005	2006	2007	2008	2009	2010	2011	2012	% Change 2002–2012
Rural Areas (under 5,000 in population)												
Interstate	3,298	3,144	3,227	3,248	2,887	2,677	2,422	2,045	2,113	1,969	1,814	-45.0%
Other Principal Arterial	4,894	5,042	5,167	4,821	4,554	4,786	4,395	4,652	3,986	4,050	4,082	-16.6%
Minor Arterial	4,467	4,678	5,043	4,483	4,346	4,186	3,507	2,957	3,015	2,989	3,465	-22.4%
Major Collector	6,014	5,793	5,568	5,757	5,675	5,637	5,084	4,568	4,171	4,182	4,203	-30.1%
Minor Collector	2,003	1,837	1,787	1,635	1,650	1,487	1,421	1,342	1,143	989	955	-52.3%
Local	5,059	4,366	4,162	4,443	4,294	4,327	4,060	3,626	3,540	3,454	3,456	-31.7%
Unknown Rural	161	97	225	200	240	154	98	133	121	136	195	21.1%
Subtotal Rural	**25,896**	**24,957**	**25,179**	**24,587**	**23,646**	**23,254**	**20,987**	**19,323**	**18,089**	**17,769**	**18,170**	**-29.8%**
Urban Areas (5,000 or more in population)												
Interstate	2,482	2,482	2,482	2,734	2,663	2,685	2,300	2,049	2,124	2,159	2,160	-13.0%
Other Freeway and Expressway	1,506	1,591	1,673	1,735	1,690	1,497	1,538	1,321	1,232	1,277	1,137	-24.5%
Other Principal Arterial	5,124	5,067	4,847	5,364	5,447	5,021	4,504	4,005	4,294	4,142	4,500	-12.2%
Minor Arterial	3,218	3,684	3,573	3,836	3,807	3,596	3,128	2,829	2,945	2,858	3,023	-6.1%
Collector	1,151	1,323	1,385	1,426	1,513	1,467	1,256	1,158	1,069	1,137	1,267	10.1%
Local	3,497	3,528	3,290	3,458	3,622	3,612	3,461	3,098	2,978	2,969	3,170	-9.4%
Unknown Urban	35	90	211	74	49	30	31	41	17	33	39	11.4%
Subtotal Urban	**17,013**	**17,765**	**17,461**	**18,627**	**18,791**	**17,908**	**16,218**	**14,501**	**14,659**	**14,575**	**15,296**	**-10.1%**
Unknown Rural or Urban	96	144	76	296	271	97	218	59	251	135	95	-1.0%
Total Highway Fatalities	**43,005**	**42,866**	**42,716**	**43,510**	**42,708**	**41,259**	**37,423**	**33,883**	**32,999**	**32,479**	**33,561**	**-22.0%**

Sources: Fatality Analysis Reporting System/National Center for Statistics and Analysis, NHTSA, 2002-2011 Final and 2012 Annual Report File (ARF).

These declines varied greatly by roadway functional class. For example, urban interstate fatalities dropped by 13 percent from 2002 to 2012, while urban collector road fatalities rose by 10.1 percent. Rural interstate fatalities dropped by 45.0 percent from 2002 to 2012, compared to a drop in rural minor arterial roadway fatalities of only 22.4 percent.

Exhibit 4-6 shows the fatality rates per 100M VMT for urban and rural functional systems between 2002 and 2012. Urban Interstate highways were the safest functional system, with a fatality rate of 0.45 in 2012. Among urban roads, Interstate highways (26.9 percent decline) and Other Freeways and Expressways (35.8 percent decline) recorded the sharpest declines in fatality rates during this 11-year period. Among rural roads, the sharpest declines occurred in minor collectors (44.8 percent decline) and Interstates (37.5 percent decline). The decreases in overall fatality rate observed in urban areas (21.7 percent decline) and rural areas (19.1 percent decline) from 2002 to 2012 are due in part to a combination of safety countermeasures and programs DOT and State partners introduced.

Exhibit 4-6 Fatality Rates by Functional System, 2002–2012

Functional System	Fatalities (per 100 Million VMT) 2002	2004	2006	2008	2010	2012	% Change 2002–2012
Rural Areas (under 5,000 in population)							
Interstate	1.18	1.21	1.12	1.00	0.86	0.74	-37.5%
Other Principal Arterial	1.90	2.14	1.96	1.98	1.77	1.83	-3.8%
Minor Arterial	2.53	2.99	2.67	2.31	2.00	2.33	-7.9%
Major Collector	2.82	2.77	2.94	2.73	2.37	2.39	-15.2%
Minor Collector	3.26	2.97	2.84	2.58	2.14	1.80	-44.8%
Local	3.63	3.14	3.22	3.08	2.67	2.66	-26.8%
Subtotal Rural	**2.30**	**2.35**	**2.28**	**2.12**	**1.84**	**1.86**	**-19.1%**
Urban Areas (5,000 or more in population)							
Interstate	0.61	0.57	0.56	0.48	0.44	0.45	-26.9%
Other Freeway and Expressway	0.79	0.80	0.78	0.69	0.56	0.51	-35.8%
Other Principal Arterial	1.25	1.08	1.17	0.97	0.94	0.99	-21.2%
Minor Arterial	0.95	0.99	1.01	0.83	0.79	0.81	-14.3%
Collector	0.81	0.85	0.87	0.72	0.59	0.71	-12.9%
Local	1.46	1.29	1.36	1.28	1.09	1.15	-21.3%
Subtotal Urban	**0.98**	**0.93**	**0.95**	**0.82**	**0.74**	**0.77**	**-21.7%**
Total Highway Fatality Rate	**1.51**	**1.45**	**1.42**	**1.26**	**1.11**	**1.13**	**-25.1%**

Sources: Fatality Analysis Reporting System/National Center for Statistics and Analysis, NHTSA, 2002-2011 Final and 2012 Annual Report File (ARF).

Despite the overall decreases in fatality rates on both urban and rural functional systems, rural roads remain far more dangerous than urban roads, evidenced by a fatality rate on rural roads (1.86 per 100 M VMT) that is 2.42 times higher than the fatality rate on urban roads (0.77 per 100 M VMT). Several factors collectively comprise this rural road safety challenge, including roadway, behavioral, and emergency services issues.

The fatality rate for rural local roads (2.66) in 2012 was more than 3.5 times higher than that for rural Interstates (0.74). Similarly, the fatality rate for urban local roads (1.15) was more than 2.5 times higher than the fatality rate for urban Interstates (0.45). Addressing the challenges associated with non-Interstate roads can be made more difficult by the diversity of ownership; States maintain Interstate roads, while the State or a variety of local organizations, including cities and counties, maintain other roads.

Locally Owned Road Safety

More than 30,000 local agencies own and operate 75 percent of the Nation's roadways. Agency practitioners have varying levels of transportation safety expertise and often perform several duties in addition to transportation safety. FHWA developed *Road Safety 365: A Workshop for Local Governments*, to help local practitioners routinely identify safety issues along their roadways and provide ideas on how to address them.

Vision: Toward Zero Deaths and Serious Injuries on the Nation's Roadways

The DOT strategic goal on safety is "Improve public health and safety by reducing transportation-related fatalities and injuries for all users, working toward no fatalities across all modes of travel."

To help accomplish this goal, FHWA oversees HSIP, a core Federal-aid program, which has as its goal to achieve a significant reduction in traffic fatalities and serious injuries on all public roads, including non-State-owned public roads and roads on Tribal lands. HSIP requires a data-driven, strategic approach to improving highway safety on all public roads that focuses on performance. By improving data and promoting analysis and evaluation, implementing programs based on current highway safety knowledge, and conducting research to expand that knowledge base, FHWA continues to move toward zero deaths on the Nation's roadways.

FHWA coordinates with States as they develop SHSPs. As a major component and requirement of HSIP, an SHSP is a statewide-coordinated safety plan, developed by a State Department of Transportation in cooperation with a broad range of safety stakeholders. An SHSP analyzes highway safety problems, identifies a State's key safety needs, and guides decisions toward strategies and investments with the most potential to save lives and prevent injuries. The SHSP enables highway safety programs and partners in the State to work together to align goals, leverage resources, and collectively address the State's safety challenges.

To support their SHSPs, States must have a safety data system to identify problems and analyze countermeasures on all public roads; adopt strategic and performance-based goals; advance data collection, data analysis, and data integration capabilities; determine priorities for correcting the identified safety problems; and establish evaluation procedures.

Improved Data

FHWA promotes improved data, analysis methods, and evaluation capabilities, which collectively make a major contribution toward advancements in highway safety. Better data and enhanced ways to analyze the data produce valuable information for local, State, national, and private transportation safety stakeholders. These improvements also help members of the highway safety community reduce traffic fatalities, injuries, and property damage-only crashes.

The FHWA Roadway Safety Data Program works to develop, evaluate, and deploy life-saving countermeasures; advances the use of scientific methods and data-driven decisions; and promotes an integrated, multidisciplinary approach to safety. The program helps improve safety data and expand capabilities for analysis and evaluation. The effectiveness of safety programs is directly linked to the availability and analysis of reliable crash and roadway data.

Improved Safety Analysis Tools

FHWA also provides and supports a wide range of data and safety analysis tools for State and local practitioners. These tools are designed to help practitioners understand safety problems on their roadways, link crashes to their roadway environments, and select and apply appropriate countermeasures. The tools' capabilities range from simple to complex. Some provide general information; others enable complex analysis of crashes under specific conditions or with specific roadway features.

One valuable safety analysis tool is the Highway Safety Manual (HSM), published by the American Association of State Highway Transportation Officials and developed by cooperative research initiated by FHWA. The document's primary focus is the introduction and development of analytical tools for predicting the impact of transportation project and program decisions on road safety. The HSM provides improved information and tools that facilitate roadway planning, design, operations, and maintenance decisions based on precise consideration of their safety consequences. The second edition, developed in 2015, will continue to communicate valuable highway safety research and knowledge gained over many years.

Along with the HSM and various implementation tools, cooperative research initiated by FHWA has developed other safety analysis tools, such as the Safety Analyst, Interactive Highway Safety Design Model, and the Crash Modification Factors Clearinghouse. These tools greatly advance the abilities of State and local highway agencies to incorporate explicit, quantitative consideration of safety into their planning and project development decision-making.

Legislative Mandates

The MAP-21 reauthorizing legislation identifies the need for improved and more robust safety data for better safety analysis to support the development of States' HSIPs and SHSPs. MAP-21 builds on and refines many of the highway, transit, bicycle, and pedestrian programs and policies FHWA administers.

MAP-21 supports DOT's determined safety agenda. It continues the successful HSIP, doubling funding for infrastructure safety and strengthening the linkage among safety programs at FHWA, NHTSA, and FMCSA. It also continues to build on other aggressive safety efforts, including the Department's fight against distracted driving and its push to improve transit and motor carrier safety.

The FAST Act maintains a strong focus on safety, keeping intact the established structure of the various highway-related safety programs, while providing a predictable level of authorized funding over a 5-year period. The primary features of the current HSIP are retained, including the requirement for a comprehensive, data-driven SHSP that defines State safety goals and describes a program of strategies to improve safety.

FHWA published the HSIP and Safety Performance Management Measures (PM-1) Final Rules in the *Federal Register* on March 15, 2016, with an effective date of April 14, 2016.

The HSIP Final Rule updates the existing HSIP requirements under 23 CFR 924 to be consistent with MAP-21 and the FAST Act, and clarifies existing program requirements. Specifically, the HSIP Final Rule contains three major policy changes: SHSP Updates, HSIP Report Content and Schedule, and the Subset of the Model Inventory of Roadway Elements (MIRE).

The Safety PM Final Rule adds Part 490 to Title 23 of the Code of Federal Regulations to implement the performance management requirements under 23 U.S.C. 150, including the specific safety performance measure requirements for the purpose of carrying out the HSIP to assess serious injuries and fatalities on all public roads. The Safety PM Final Rule establishes five performance measures as the 5-year rolling averages for: (1) Number of Fatalities, (2) Rate of Fatalities per 100 million Vehicle Miles Traveled (VMT), (3) Number of Serious Injuries, (4) Rate of Serious Injuries per 100 million VMT, and (5) Number of Non-motorized Fatalities and Non-motorized Serious Injuries. The Safety PM Final Rule also establishes the process for State Departments of Transportation (DOTs) and Metropolitan Planning Organization (MPOs) to establish and report their safety targets, and the process that FHWA will use to assess whether State DOTs have met or made significant progress toward meeting their safety targets. The Safety PM Final Rule also establishes a common national definition for serious injuries.

Together, these regulations will improve data, foster transparency and accountability, and allow safety progress to be tracked at the national level. They will inform State DOT and MPO planning, programming, and decision-making to support reductions in fatalities and serious injuries.

Focused Approach to Safety Program

When it occurs, a crash is generally the result of numerous contributing factors. Roadway, vehicle, driver, passenger, and non-occupant factors all have an impact on the safety of the Nation's highway system. FHWA collaborates with other agencies to understand more clearly the relationship among all contributing factors and to address crosscutting ones, but focuses on infrastructure design and operation to address roadway factors.

FHWA examined crash data to identify the most common crash types relating to roadway characteristics. FHWA established three focus areas to address these factors: roadway departures, intersections, and pedestrian crashes. These three areas were selected because they account for more than three-quarters of overall fatalities and represent an opportunity to significantly reduce the number of fatalities and serious injuries. FHWA manages the Focused Approach to Safety Program to address the most critical safety challenges surrounding these crashes. Through this program, FHWA focuses its technical assistance and resources on States and cities with high fatality counts and fatality rates in one or more of these three categories.

In 2012, roadway departure, intersection, and pedestrian fatalities accounted for 52.2 percent, 21.7 percent, and 14.1 percent, respectively, of total highway fatalities. Note that these three categories overlap. For example, when a roadway departure crash includes a pedestrian's being fatally struck, that crash would be accounted for in both the roadway departure and the pedestrian-related crash categories below. *Exhibit 4-7* shows how the number of crashes for these crash types has changed between 2002 and 2012.

Exhibit 4-7 Highway Fatalities by Crash Type, 2002–2012

Crash Type	2002	2003	2004	2005	2006	2007	2008	2009	2010	2011	2012	% Change 2002–2012
Roadway Departure-Related[1,2]	25,415	25,576	22,340	22,863	22,665	22,180	19,878	18,052	17,423	16,973	17,532	-31.0%
Intersection-Related[1,3]	9,273	9,362	9,176	9,238	8,850	8,703	7,809	7,278	7,313	6,995	7,279	-21.5%
Pedestrian-Related[1]	4,851	4,774	4,675	4,892	4,795	4,699	4,414	4,109	4,302	4,457	4,743	-2.2%

[1] Some fatalities may overlap; for example, some intersection-related fatalities may involve pedestrians.

[2] Definition for roadway departure crashes was modified beginning in 2004.

[3] Definition for Intersection crashes was modified beginning in 2010.

Sources: Fatality Analysis Reporting System/National Center for Statistics and Analysis, NHTSA, 2002-2011 Final and 2012 Annual Report File (ARF).

Because the fatalities shown in *Exhibit 4-7* can involve a combination of factors, FHWA has developed targeted programs that include collaborative and comprehensive efforts to address all three of these areas. The Focused Approach to Safety Program works to address the most critical safety challenges by devoting additional effort to high-priority States and targeting technical assistance and resources. More information is available at http://safety.fhwa.dot.gov/fas/.

In 2012, FHWA issued the *Guidance Memorandum on Promoting the Implementation of Proven Safety Countermeasures*. This guidance considers the latest safety research to advance a group of countermeasures that have shown great effectiveness in improving safety. The nine countermeasures address the three focus areas of the Focused Approach to Safety Program. This combined approach is designed to provide consistency in safety programming, target limited resources to problem areas, and implement safety countermeasures that are likely to yield the greatest results in reducing the number of crash-related fatalities and injuries. More information on this approach can be found at http://safety.fhwa.dot.gov/provencountermeasures/.

Roadway Departures

In 2012, the number of roadway departure fatalities was 17,532, which accounted for 52.2 percent of all traffic fatalities. A roadway departure crash is defined as a nonintersection crash that occurs after a vehicle crosses an edge line or a center line, or otherwise leaves the traveled way. In some cases, a vehicle crossed the center line and struck another vehicle, hitting it head on or sideswiping it. In other cases, the vehicle left the roadway and struck one or more constructed or natural objects, such as utility poles, embankments, guardrails, trees, or parked vehicles.

Roadway Departure Focus States and Countermeasures

Roadway Departure Focus States are eligible for additional resources and assistance based on their average number of roadway departure fatalities over 3 years and the roadway departure fatality rate per vehicle miles traveled (VMT). In addition, FHWA considers the urban and rural roadway percentages within these States and the ratio of their actual roadway departure fatality rate versus the expected roadway departure fatality rate per VMT based on national urban and rural rates.

FHWA currently offers roadway departure technical assistance in the form of crash data analysis and implementation plan development to State highway agencies that have a particularly high number of roadway departure fatalities. Roadway Departure Implementation Plans have been developed in many States. Each plan is designed to address State-specific safety issues related to roadway departure on both State and local roadways to the extent that relevant data can be obtained and are appropriate based on consultation with State and local agencies and the FHWA Division Office.

FHWA works with participating Roadway Departure Focus States to develop individual data analysis packages focused on crash history and roadway attributes and to identify strategies for use in reducing roadway departure crashes. The plans identify cost-effective countermeasures, deployment levels, and funding needs to reduce the number and severity of roadway departure crashes in the State by a targeted amount consistent with Strategic Highway Safety Plan goals. The final plan quantifies the costs and benefits of a roadway departure-focused initiative and provides a systematic process for implementation.

Three proven safety countermeasures for reducing roadway departure crashes are:

- Longitudinal rumble strips and stripes on two-lane rural roads – Milled or raised elements on the pavement intended to alert inattentive drivers through vibration and sound that their vehicles have left the travel lane;
- Enhanced delineation and friction for horizontal curves – Signs and pavement deployed to warn the driver in advance of the curve, with pavement friction to enable the tires to change the vehicle's direction without skidding; and
- Safety Edge – Technology that shapes the edge of a paved roadway in a way that eliminates tire scrubbing, a phenomenon that contributes to losing control of a vehicle.

Intersections

Of the 33,561 fatalities that occurred in 2012, about 21.7 percent occurred at intersections. Of these, 37.5 percent were rural and 62.5 percent were urban, as shown in *Exhibit 4-8*.

The United States has more than 3 million intersections, both signalized (controlled by traffic signals) and non-signalized (controlled by stop or yield signs). Many factors can contribute to unsafe conditions at intersections. Road designs or traffic signals might need to be upgraded to account for current traffic levels. Approximately one-third of signalized intersection fatalities involve red-light running.

Exhibit 4-8 Intersection-Related Fatalities by Functional System, 2012

Functional System	Fatalities	
	Count	% of Total
Rural Areas (under 5,000 in population)		
Principal Arterials	781	10.8%
Minor Arterials	624	8.6%
Collectors (Major and Minor)	749	10.4%
Locals	557	7.7%
Subtotal Rural Areas	**2,711**	**37.5%**
Urban Areas (5,000 or more in population)		
Principal Arterials	1,928	26.7%
Minor Arterials	1,115	15.4%
Collectors (Major and Minor)	396	5.5%
Locals	1,082	15.0%
Subtotal Urban Areas	**4,521**	**62.5%**
Total Highway Fatalities[1]	**7,232**	**100.0%**

[1] Total excludes 47 intersection-related fatalities not identified by functional class.

Sources: Fatality Analysis Reporting System/National Center for Statistics and Analysis, NHTSA, 2012 Annual Report File (ARF).

Intersection Focus States and Countermeasures

Intersection Focus States are eligible for additional resources and assistance based on their average number of intersection fatalities over 3 years. In addition, FHWA considers the urban and rural roadway percentages within these States and the ratio of their actual intersection fatality rate versus the expected intersection fatality rate per vehicle miles traveled based on national urban and rural rates.

As part of the Focused Approach to Safety, FHWA works with States to develop Intersection Safety Implementation Plans. These plans include the specific activities, countermeasures, strategies, deployment levels, implementation steps, and estimates of funds necessary to achieve intersection safety improvement—a component of a State's Strategic Highway Safety Plan goals. FHWA also assists those States through webinars, technical support, and training courses.

FHWA promotes three proven countermeasures associated specifically with intersection safety:

- Roundabouts – A modern circular intersection defined by a set of specific operational principles designed to create a low-speed environment, high operational performance, and a reduction of conflict points;
- Corridor access management –A set of techniques useful for controlling access to highways, major arterials, and other roadways and that result in improved movement of traffic, reduced crashes, and fewer vehicle conflicts; and
- Backplates with retroreflective border – A device added to traffic signals to improve the visibility of the illuminated face of the signal.

In addition, two countermeasures promoted for pedestrian safety can also improve intersection safety: pedestrian hybrid beacons (pedestrian activated warning devices) and road diets (lane reductions or road rechannelizations).

Pedestrians, Bicyclists, and Other Nonmotorists

While the third of the FHWA focus areas referenced above was pedestrian crashes, the PM-1 Final rule establishes performance measures for the broader category of nonmotorists, which includes pedestrians, bicyclists, skateboarders, roller skaters, and others using nonmotorized forms of transportation. The requirement for reporting on these measures will lead to more data being available in the future, which will be reflected in future editions of this report.

In 2012, 33,561 persons were fatally injured in motor vehicle crashes, of which 17 percent were nonmotorists. *Exhibit 4-9* shows that in 2012, 4,750 pedestrians, 726 bicyclists, and 216 other/unknown nonmotorists were killed, totaling 5,692 nonmotorist fatalities.

Nonmotorist fatalities declined 16.6 percent from 5,864 in 2005 to 4,888 in 2009, yet increased each year since 2009, to 5,110 in 2010, to 5,339 in 2011, and to 5,692 in 2012. Pedestrian fatalities rose from 4,108 in 2009 to 4,750 in 2012. Bicyclist fatalities rose from 628 in 2009 to 726 in 2012.

Roadway designs that accommodate all users, referred to as "complete streets," help reduce fatalities and injuries. Such roadway designs feature sidewalks, raised medians, turning access controls, better bus stop placement, better lighting, and traffic calming measures. Instituting policies that accommodate all roadway users ensures that every transportation project becomes a comprehensive safety project. These policies have the added benefit of making walking and biking more attractive options and of enhancing the aesthetic quality and commercial activity on local streets.

Exhibit 4-9 Pedestrian and Other Nonmotorist Traffic Fatalities, 2002–2012

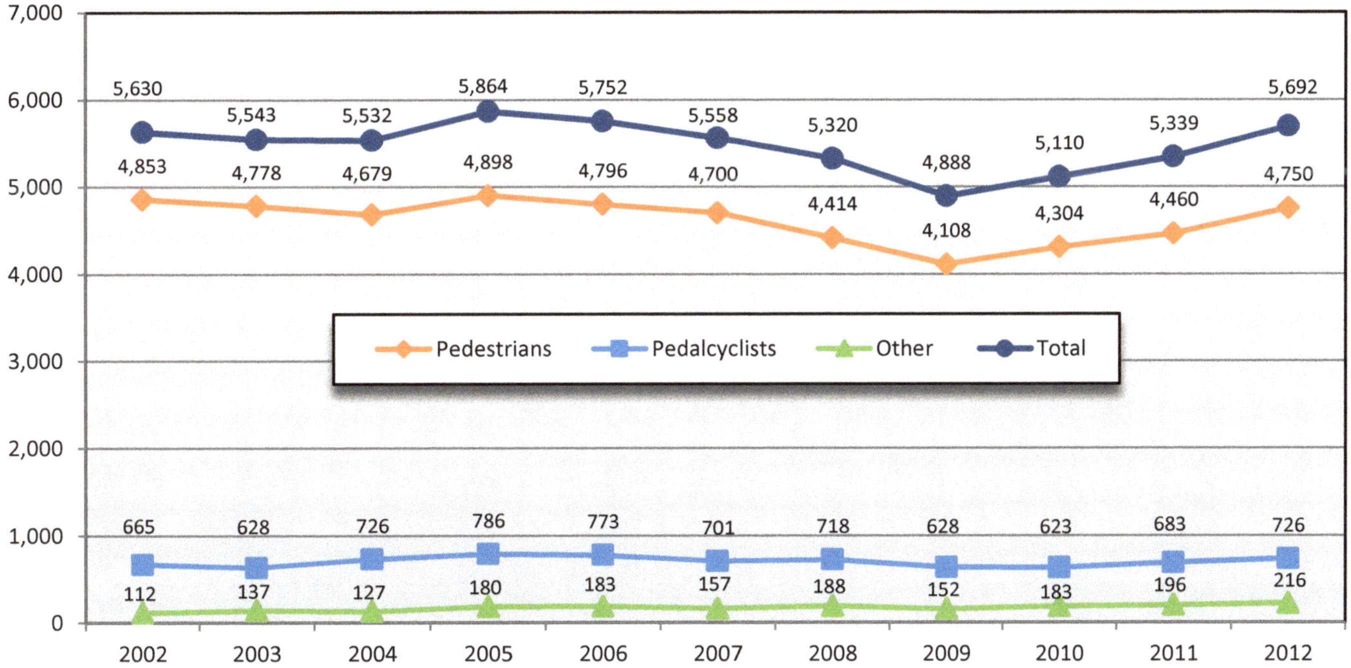

Sources: Fatality Analysis Reporting System/National Center for Statistics and Analysis, NHTSA, 2002-2011 Final and 2012 Annual Report File (ARF).

Pedestrian and Bicyclist Safety Focus States and Cities and Countermeasures

In July 2014, FHWA expanded the pedestrian focus area to include bicyclist and other nonmotorist fatalities. This change was incorporated into the Focused Approach to Safety Program in 2015.

FHWA designates focus States and focus cities for the pedestrian and bicycle focus area. States and cities are eligible to participate as pedestrian and bicycle focus States and cities based on the number of pedestrian and bicyclist fatalities or the pedestrian and bicyclist fatality rate per population over a 3-year period.

FHWA's Office of Safety is aggressively working to reduce pedestrian and bicyclist fatalities by providing resources to focus States and cities. Focused Approach has helped raise awareness of pedestrian and bicyclist safety problems and generate momentum for addressing pedestrian and bicyclist issues. Focused Approach has provided course offerings, conference calls, Web conferences, data analysis, and technical assistance for development of Pedestrian and Bicyclist Safety Action Plans. These plans help State and local officials determine where to begin addressing pedestrian and bicyclist safety issues.

Focused Approach offers free technical support and training courses to focus States and cities and free bimonthly webinars on a comprehensive, systemic approach to preventing pedestrian and bicyclist crashes. Training is available at a cost to nonfocus States and cities through the Pedestrian and Bicycle Information Center, made possible by the National Highway Institute.

FHWA is also promoting three proven countermeasures associated specifically with pedestrian safety:

- Median and pedestrian crossing islands in urban and suburban areas – A refuge area in the middle of the roadway, enhancing pedestrian crossing visibility and reducing the speed of vehicles approaching the crossing.
- Pedestrian hybrid beacons – Pedestrian-activated warning device located on the roadside or on mast arms over midblock pedestrian crossings.
- Road diets – A classic roadway reconfiguration that involves converting an undivided four-lane roadway into three lanes comprising two through-lanes and a center two-way left turn lane.

Transit Safety

This section summarizes national trends in safety and security incidents such as injuries, fatalities, and related performance ratios reported in the National Transit Database (NTD). Safety data are reported by mode and type of service. In December 2011, the NTD safety data included 16 modes. In January 2012, new modes were added to NTD, including

- streetcar – previously reported as light rail,
- hybrid rail – previously reported as light rail,
- commuter bus – previously reported as motor bus,
- bus rapid transit – previously reported as motor bus, and
- demand-response-taxi – previously reported as demand response.

NTD does not compile safety data for commuter rail systems, which is managed and collected by the Federal Railroad Administration (FRA). This section presents statistics and counts of basic aggregate data such as injuries and fatalities for those systems. For 2012, data were received from 49 rail transit systems, more than 750 urban fixed-route bus providers, and 1,357 rural agencies. Reported events occurred on transit property or vehicles, involved transit vehicles, or affected persons using public transportation systems.

Q&A **What sort of events result in a recorded transit incident?**

A transit agency records an incident for any event occurring on transit property, on board or involving transit vehicles, or to persons using the transit system that results in one of the following:

- One or more confirmed fatalities within 30 days of the incident;
- One or more injuries requiring immediate transportation away from the scene for medical attention;
- Total property damage to transit property or private property exceeding $25,000;
- Evacuation for life safety reasons;
- Mainline derailment (i.e., occurring on a revenue service line, regardless of whether the vehicle was in service or out of service); or
- Fire.

Additionally, a transit agency records an incident whenever certain security situations occur on transit property, such as:

- Robbery, burglary, or theft;
- Rape;
- Arrest or citation, such as for trespassing, vandalism, fare evasion, or assault;
- Cyber security incident;
- Hijacking; or
- Nonviolent civil disturbance that disrupts transit service.

Incidents, Fatalities, and Injuries

A transit agency records an incident for a variety of events occurring on transit property or vehicles, involving transit vehicles, or affecting persons using the transit system. Included among these events is any that results in significant property damage, one or more reported injuries, one or more reported fatalities, or some combination thereof. From 2002 to 2007, the definition of significant property damage was total property damage exceeding $7,500 (in current-year dollars, not indexed to inflation); this threshold increased to $25,000 in 2008.

Injuries and fatalities data in NTD are reported by type of person involved in incidents. Passengers are defined as persons traveling, boarding, or alighting a transit vehicle. Patrons are individuals who are in a rail station or at a bus stop but are not necessarily boarding a transit vehicle. Employees are individuals who work for the transit agency, including both staff and contractors. Public includes pedestrians, occupants of other vehicles, and other persons.

Q&A **What types of injuries and fatalities are reported?**

Person types are defined as

- Passengers: Individuals on board a transit vehicle or boarding or alighting a transit vehicle.
- Patrons: Individuals waiting for or leaving transit at stations; in mezzanines; on stairs, escalators, or elevators; in parking lots; or on other transit-controlled property.
- Public: All others who come into contact with the transit system, including pedestrians, automobile drivers, and trespassers.
- Workers: Transit agency employees or contractors engaged in operations or maintenance but not construction of new transit infrastructure.
- Suicides: Individuals who come into contact with the transit system intending to harm themselves.

Any event for which an injury or fatality is reported is considered an incident. An injury is reported when a person has been transported immediately from the scene for medical care. A transit-related fatality is reported for any death occurring within 30 days of a transit incident that is confirmed to be a result of that incident. These statistics, however, do not include fatalities resulting from medical emergencies on transit vehicles.

Incidents also are recorded when property damage exceeds $25,000 regardless of whether the incident resulted in injuries or fatalities.

Since 2008, nationwide, collisions involving transit vehicles and pedestrians, bicyclists, motorists, and individuals waiting in stations, at stops, and rail grade crossings have resulted in approximately 140 fatalities per year. *Exhibit 4-10* shows data on fatalities, excluding suicides, both in total fatalities and fatalities per 100 million passenger miles traveled (PMT) for heavy rail, light rail, demand response, and fixed-route bus. From 2002 to 2011, the number of fatalities per 100 million PMT has remained relatively static, but it increased significantly in 2012.

Public transit interaction with pedestrians, cyclists, and motorists at rail grade crossings, pedestrian crosswalks, and intersections largely influences overall transit safety performance. Most fatalities and injuries result from interaction with the public on busy city streets, trespassing on transit right-of-way and facilities, and suicide. Pedestrian fatalities accounted for 23 percent of all transit fatalities in 2012.

Exhibit 4-11 shows the transit fatality rate by person type between 2002 and 2012. Transit workers and passengers typically account for the lowest fatality rate by person type. In 2012, worker fatalities accounted for 2 percent of all fatalities. In response to recent events in 2013 involving roadway workers, the National Transportation Safety Board issued a series of safety recommendations to support needed improvements in this area. The FTA responded in December 2013 by issuing *Safety Advisory 14-1: Right-of-Way Worker Protection*.

Exhibit 4-10 Annual Transit Fatalities Excluding Suicides, 2002–2012[1]

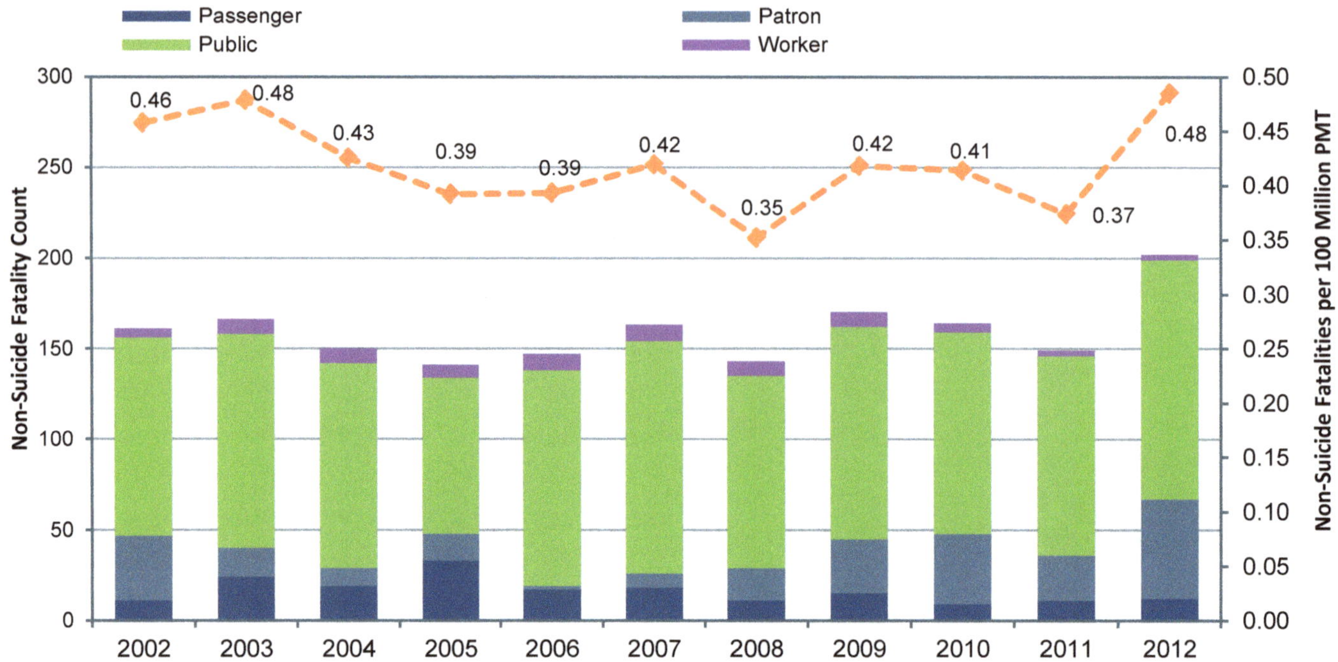

[1] *Exhibit includes data for DR, HR, LR, and MB. Fatality totals include both directly operated (DO) and purchased-transportation (PT) service types.*

Source: National Transit Database—Transit Safety and Security Statistics and Analysis Reporting.

Exhibit 4-11 also highlights the relatively few fatalities in transit per passenger mile. Suicides steadily increased to a peak of 79 in 2011, then decreased to 63 in 2012. On average, suicides and persons who are not transit passengers or patrons (usually pedestrians and drivers) account for approximately 75 percent of all public transportation fatalities. This situation creates distinct challenges for public transportation agencies and FTA, because the causes of these fatalities are largely beyond the control of transit operators. In 2013, FTA, in partnership with Operation Lifesaver, made grant funds available to transit and local government agencies to develop safety education and public awareness initiatives for rail transit to ensure that people are safe near trains, tracks, and at crossings. Such awareness is increasingly important for drivers and pedestrians as rail transit expands into new communities across the country. To receive a grant, projects must provide a 25-percent match and focus on safety education or public awareness initiatives in communities with rail transit systems (commuter rail, light rail, and streetcar) using Operation Lifesaver-approved materials. (2014 Annual Report: *The U.S. Department of*

Exhibit 4-11 Transit Fatality Rates by Person Type per 100 Million PMT, 2002–2012[1]

Year	Passenger	Patron	Public	Worker	Suicide
2002	0.03	0.10	0.31	0.01	0.04
2003	0.10	0.04	0.34	0.02	0.04
2004	0.06	0.03	0.33	0.02	0.04
2005	0.09	0.04	0.23	0.02	0.02
2006	0.05	0.01	0.31	0.02	0.03
2007	0.04	0.02	0.32	0.02	0.06
2008	0.03	0.04	0.25	0.02	0.06
2009	0.04	0.07	0.28	0.03	0.12
2010	0.02	0.09	0.27	0.01	0.13
2011	0.03	0.06	0.26	0.01	0.19
2012	0.02	0.11	0.31	0.01	0.14

[1] Exhibit includes data for all transit modes excluding commuter rail.

Source: National Transit Database.

Transportation's (DOT) Status of Actions Addressing the Safety Issue Areas on the National Transportation Safety Board's Most Wanted List.)

Exhibit 4-12 shows fatalities for the transit industry that include suicide data. The number and rate of suicides increased each year through 2011 and decreased in 2012.

Exhibit 4-12 Annual Transit Fatalities Including Suicides, 2002–2012[1]

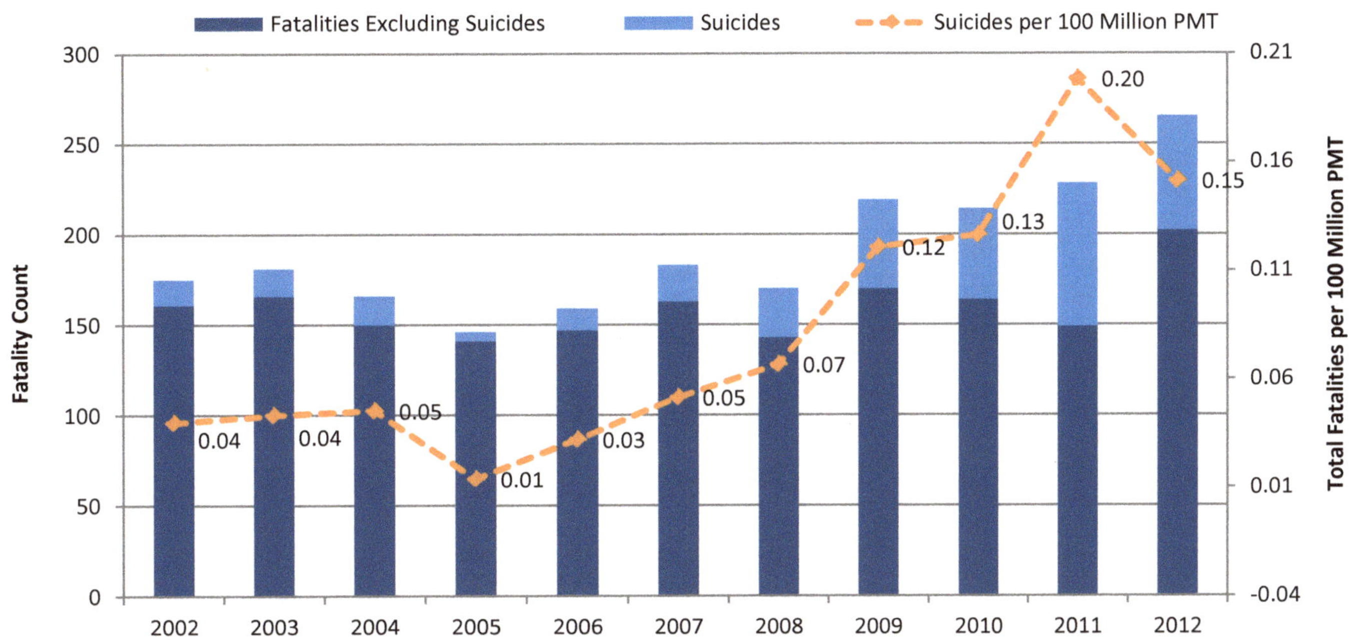

[1] *Exhibit includes data for DR, HR, LR, and MB. Fatality totals include both directly-operated (DO) and purchased-transportation (PT) service types.*

Source: National Transit Database—Transit Safety and Security Statistics and Analysis Reporting.

Exhibit 4-13 shows transit injury rates by person type. Although transit incident occurrences and impacts fluctuate from year to year, the occurrence of injuries to transit persons appears to be decreasing since 2009. Transit agencies, however, are becoming increasingly concerned with the recent increase in patron fatalities: In 2011, patron fatalities accounted for 21 percent of total fatalities, up from a low of 4 percent in 2007.

Exhibit 4-13 Transit Injury Rates by Person Type per 100 Million PMT, 2002–2012[1]

Year[2]	Passenger	Patron	Public	Worker	Suicide
2002	34.95	7.03	7.72	3.19	0.05
2003	30.03	8.85	9.91	3.30	0.03
2004	29.93	10.44	10.22	2.99	0.00
2005	28.55	9.07	8.35	2.62	0.00
2006	31.25	9.20	8.01	3.12	0.07
2007	33.58	7.35	8.79	4.76	0.04
2008	20.33	19.57	14.61	3.63	0.04
2009	22.13	20.79	15.49	3.43	0.05
2010	26.45	19.51	11.42	3.30	0.09
2011	20.42	16.90	13.09	3.23	0.09
2012	21.02	15.82	12.87	3.24	0.11

[1] Exhibit includes data for all transit modes excluding commuter rail.

[2] Beginning for calendar year 2008, the reporting threshold for a reportable injury changed from two people to one person.

Source: National Transit Database.

Exhibit 4-14 shows fatalities per 100 million PMT for fixed-route bus and demand response (including suicides). The fatality rate for demand response is more volatile than for fixed-route bus. This observation is not unexpected, as fewer people use demand response and even one or two more fatalities in a year can make the rate jump significantly. Fatality rates have not changed significantly for fixed-route bus. Note that the absolute number of fatalities is not comparable across modes because of the wide range of passenger miles traveled on each mode.

Exhibit 4-14 Annual Transit Fatalities Excluding Suicides by Highway Mode per 100 Million PMT, 2002–2012[1]

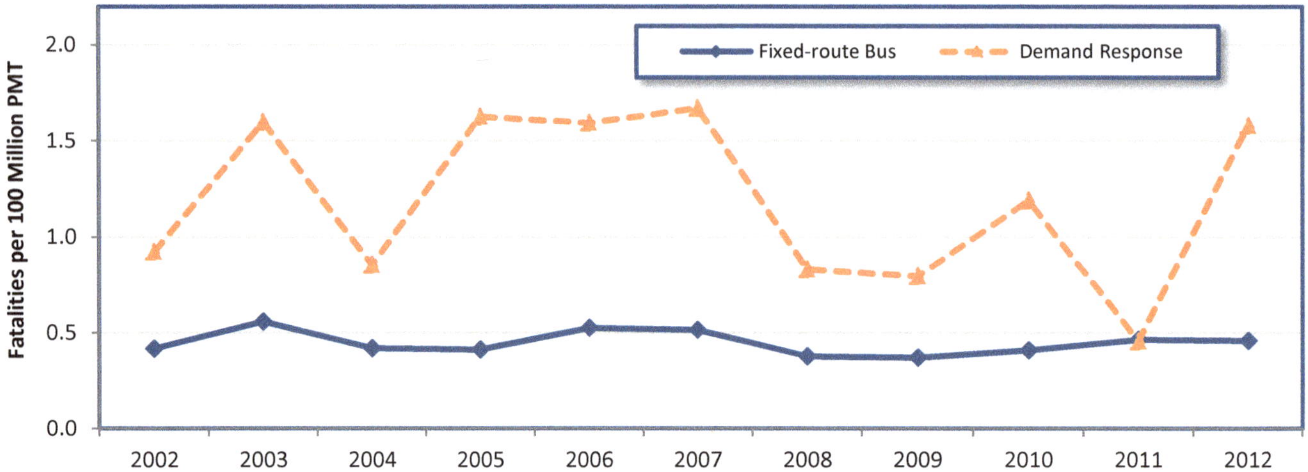

[1] Fatality totals include both DO and PT service types.

Source: National Transit Database.

Exhibit 4-15 shows fatalities per 100 million PMT for heavy rail and light rail (including suicides). Heavy-rail fatality rates remained relatively stable from 2009 through 2012. Suicides represent a large share of fatalities for heavy rail—45 percent in 2012. Light rail experienced more incidents than heavy rail as many systems are streetcars operating in non-dedicated guideways and generally pick up passengers from roadside stops rather than from station platforms.

Exhibit 4-15 Annual Transit Fatalities Excluding Suicides by Rail Mode per 100 Million PMT, 2002–2012[1]

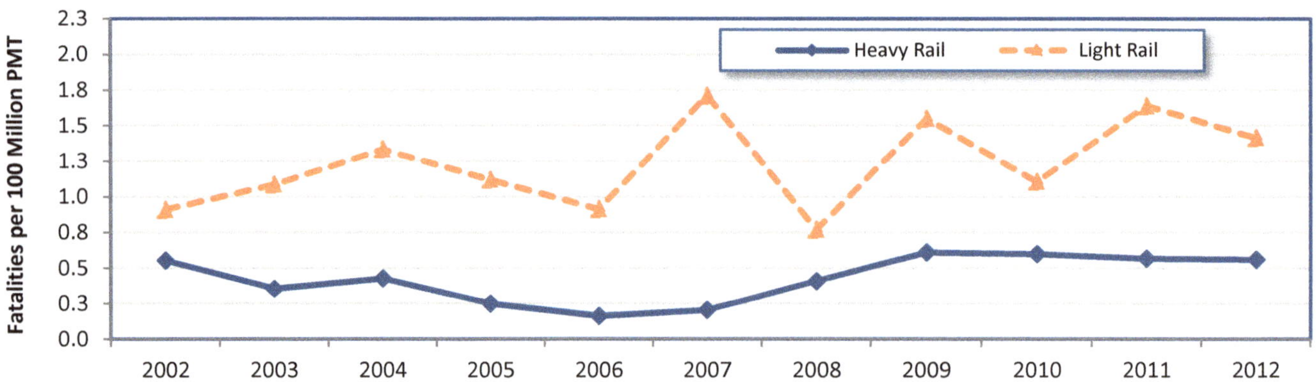

[1] Fatality totals include both DO and PT service types. Rail modes include heavy rail and light rail.

Source: National Transit Database.

The analysis that follows is by mode, which includes all major modes reported in NTD with the exception of commuter rail. Safety data for commuter rail are included in FRA's Rail Accident/Incident Reporting System (RAIRS). The RAIRS database records fatalities that occurred because of a commuter rail collision, derailment, or fire. The database also includes a category called "not otherwise classified," which includes fatalities that occurred because of a slip, trip, or fall. Before 2011, RAIRS did not include a separate category for suicides, which *are* reported in NTD for all modes. Therefore, for comparative purposes, suicides are excluded from this analysis.

Exhibit 4-16 shows incidents and injuries per 100 million PMT reported in NTD for the two main highway modes in transit, fixed-route bus and demand response, and two main rail modes, heavy rail and light rail. Commuter rail is presented separately as those data were collected according to different definitions in RAIRS. The data in *Exhibit 4-16* suggest that the incidents in highway modes (fixed-route bus and demand response) decreased between 2004 and 2012. Injuries for demand response remained flat. Data for rail modes show decreasing trend in incidents per 100 million PMT for light rail but no trend in injuries (either increasing or decreasing) per 100 million PMT. Both incidents and injuries per 100 million PMT for heavy rail showed increasing trends.

Exhibit 4-16 Transit Incidents and Injuries by Mode per 100 Million PMT, 2004–2012

Analysis Parameter	2004	2005	2006	2007	2008	2009	2010	2011	2012
Incidents per 100 Million PMT									
Fixed-Route Bus	66.21	65.63	69.62	66.86	54.12	58.22	55.27	47.82	45.03
Heavy Rail	43.75	39.44	42.86	43.49	53.25	52.82	54.32	49.29	48.64
Light Rail	59.51	66.13	60.67	61.29	48.44	44.67	37.14	39.50	35.25
Demand Response	292.25	326.79	375.15	404.13	204.16	194.77	171.66	151.82	141.47
Injuries per 100 Million PMT									
Fixed-Route Bus	68.06	63.80	62.63	68.88	66.89	72.27	71.96	65.03	61.98
Heavy Rail	33.53	26.68	32.86	31.55	43.95	45.77	46.83	41.88	42.07
Light Rail	41.49	36.36	35.38	43.67	48.34	47.99	42.11	42.86	36.05
Demand Response	148.61	160.14	213.73	236.46	234.50	215.20	196.03	175.72	168.52

Source: National Transit Database.

Exhibit 4-17 shows the number of fatalities, and the fatality rate, for commuter rail. These data were obtained from FRA's RAIRS (suicides not included). In 2012, 201 fatalities (excluding suicides) were recorded in NTD for all modes except commuter rail. Fatalities per 100 million PMT (excluding suicides and commuter rail) was 0.46. For commuter rail, however, the total number of fatalities in 2012 was 86, with a fatality rate of 0.77—significantly higher than the national aggregate rate (0.46). The national rate with suicides included is 0.6, which is less than the rate for commuter rail.

Exhibits 4-18 and *4-19* show the number of commuter rail incidents and the injuries per 100 million PMT, respectively. Although commuter rail has a very low number of incidents per PMT, commuter rail incidents are far more likely to result in fatalities than incidents occurring on any other mode. One contributing factor might be that the average speed of commuter rail vehicles is considerably higher than the average speeds of other modes (except vanpools). The number of both incidents and injuries declined from 2007 to 2008, steadily increased to a peak in 2010, and then declined again between 2011 and 2012.

Exhibit 4-17 Commuter Rail Fatalities, 2002–2012

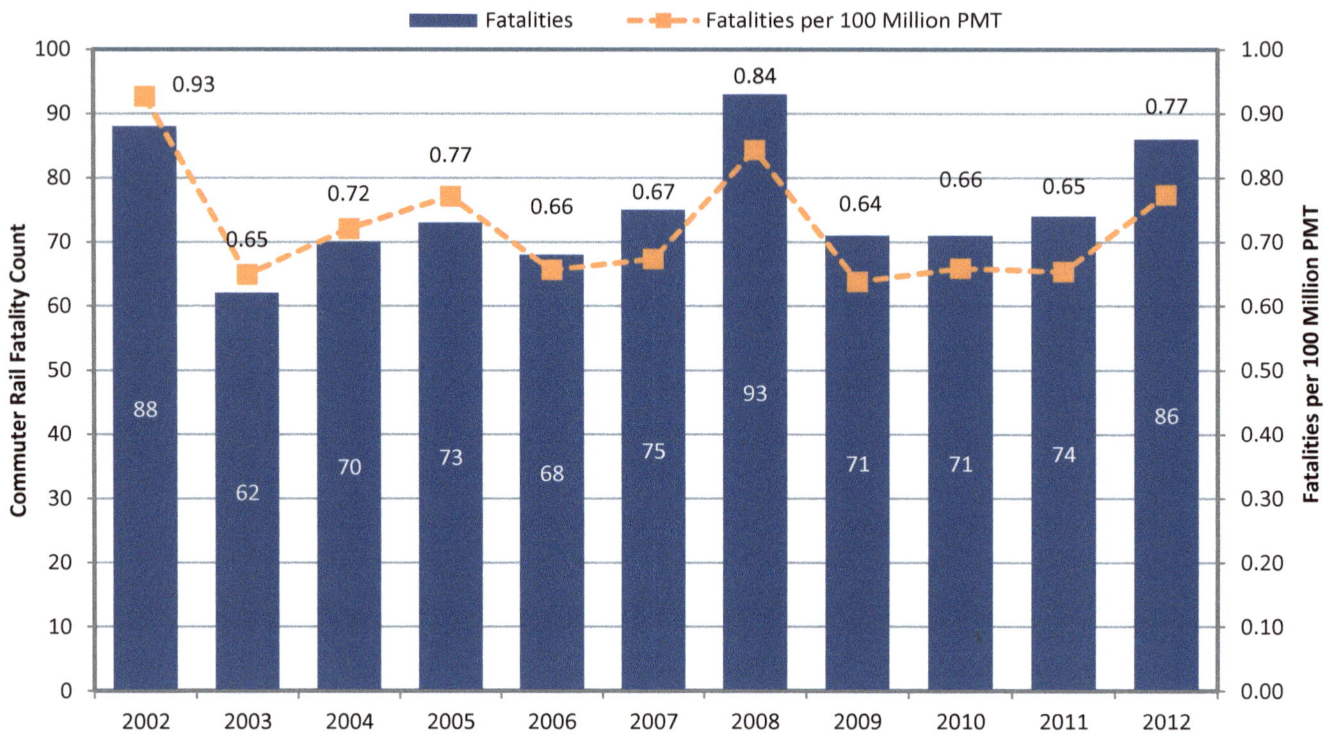

Source: Federal Railroad Administration Rail Accident/Incident Reporting System.

Exhibit 4-18 Commuter Rail Incidents, 2002–2012

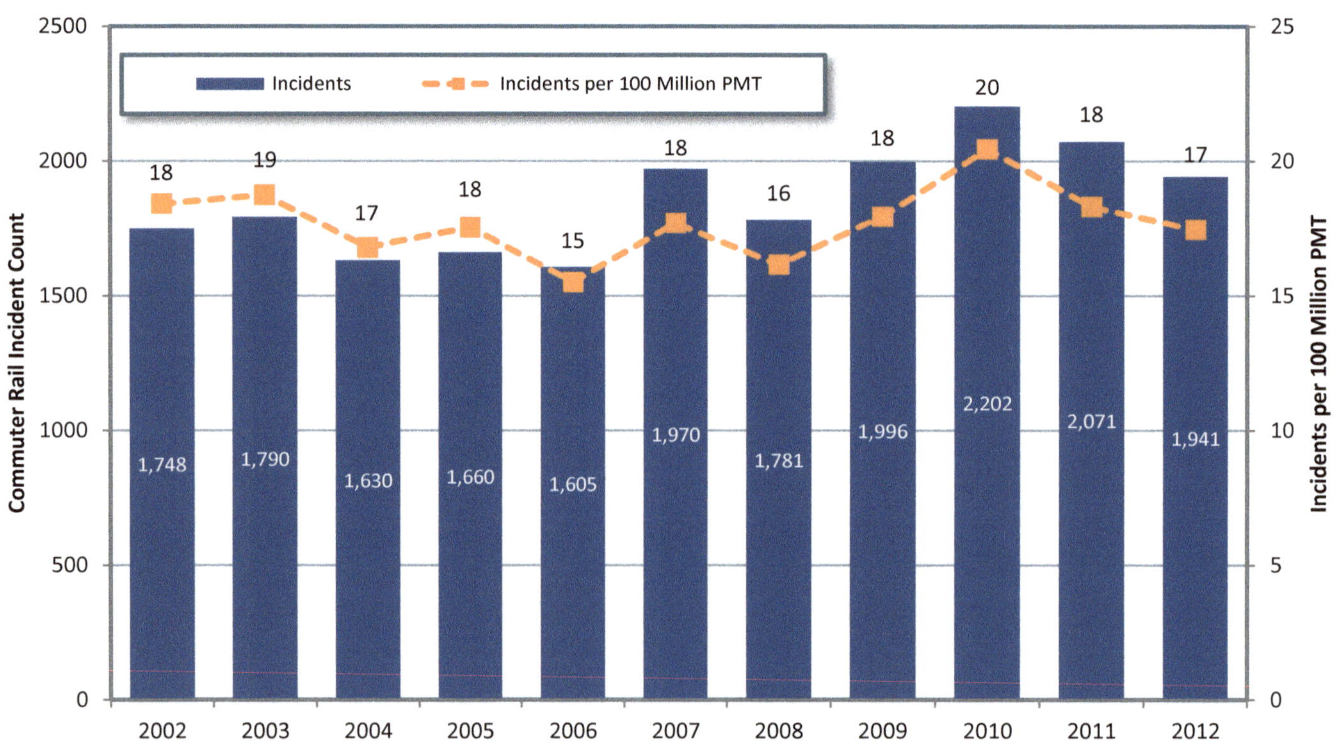

Source: Federal Railroad Administration Rail Accident/Incident Reporting System.

Exhibit 4-19 Commuter Rail Injuries, 2002–2012

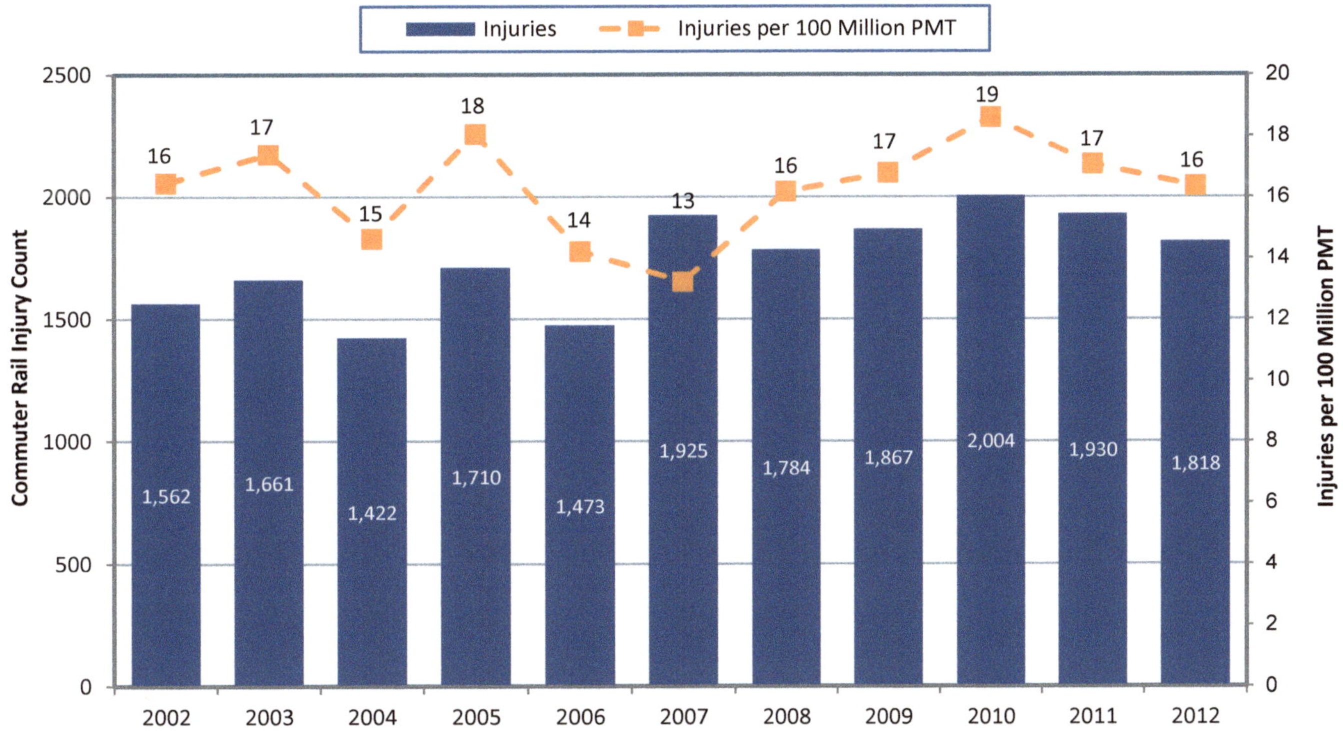

Source: Federal Railroad Administration Rail Accident/Incident Reporting System.

chapter 5

System Performance

Highway System Performance

Transportation is the backbone of the U.S. economy. Not only does the Nation's transportation system move people and goods, it also enables Americans to access unique economic, social, and cultural opportunities. In *Transportation for a New Generation, a Strategic Plan for Fiscal Years 2014–18*, DOT outlines the strategic goals and objectives for the Nation's transportation system. Among the strategic goals are achieving a state of good repair and ensuring safety, which are addressed in Chapters 3 and 4, respectively. Additional goals for economic competitiveness, quality of life, and environmental sustainability are addressed in this chapter.

- Economic Competitiveness – Promote transportation policies and investments that bring lasting and equitable economic benefits to the Nation and its citizens.
- Quality of Life in Communities – Foster improved quality of life in communities by integrating transportation policies, plans, and investments with coordinated housing and economic development policies to increase transportation choices and access to transportation services for all.
- Environmental Sustainability – Advance environmentally sustainable policies and investments that reduce carbon and other harmful emissions from transportation sources.

Economic Competitiveness

Transportation enables economic activity, quality of life, connected communities, and access to education, opportunities, and services. Both rural and urban centers require reliable multimodal transportation systems to create thriving, healthy, and environmentally sustainable communities; promote centers of economic activity; support efficient goods movement and strong financial benefits; and attract a strong workforce. The economic vitality of communities, especially in rural States, increasingly depends on the ability of businesses to access markets, not only throughout the United States, but also globally.

An efficient freight transportation system that connects population centers, economic activity, production, and consumption is critical to maintaining the competitiveness of our economy. Freight movements in the United States range from the shipment of farm products across town to the shipment of electronic components across the world. Nearly 52 million tons of freight worth more than $46 billion currently moves through the U.S. transportation system each day. Freight tonnage is forecast to increase by 1.7 percent annually to 28.25 billion tons by 2040. The value of freight moved is expected to increase faster than the weight (tonnage) is expected to grow, by 3 percent annually, from 18.0 trillion in 2013 to $39.3 trillion dollars in 2040.

> **Q&A** **Where can I find more recent information regarding freight trends?**
>
> Updates to some of the freight performance maps and tables presented in this chapter can be found at: http://ops.fhwa.dot.gov/Freight/freight_analysis/perform_meas/fpmdata/index.htm

By 2050, the U.S. population is projected to increase to 439 million from 310 million in 2010. The U.S. gross domestic product (GDP) is expected to almost triple from $14 trillion in 2010 to $41 trillion by 2050. Growth in exports of goods and services, which represented 19 percent of GDP in 2012, is expected to continue. More goods will be transported by land from within the country to airports and seaports and across national borders. Clearly, based on these forecasts, the movement of people and goods both within, and to and from, the United States will continue to increase. As a result, the transportation sector needs to continue to enable economic growth and job creation. The Nation must make strategic investments that enable people and goods to move more efficiently—with full use of the existing capacity across all transportation modes—to retain our economic competitiveness. In the past, a highly developed U.S. transportation system was instrumental in allowing GDP per capita to grow faster domestically than abroad. Other countries have increased their investments in transportation infrastructure, however, and closed the gap with the United States.

The strategic objectives for the Economic Competitiveness goal include:

- Improve the contribution of the transportation system to the Nation's productivity and economic growth by supporting strategic, multimodal investment decisions and policies that reduce costs, increase reliability and competition, satisfy consumer preferences more efficiently, and advance U.S. transportation interests worldwide.
- Increase access to foreign markets by eliminating transportation-related barriers to international trade through Federal investments in transportation infrastructure, international trade and investment negotiations, and global transportation initiatives and cooperative research, thereby providing additional opportunities for American business and creating export-related jobs.
- Improve the efficiency of the Nation's transportation system through transportation-related research, knowledge sharing, and technology transfer.
- Foster the development of a dynamic and diverse transportation workforce through partnerships with the public sector, private industry, and educational institutions.

Congestion Definition

Congestion, which can be recurring or nonrecurring, occurs when traffic demand approaches or exceeds the available capacity of the system. "Recurring" congestion (also known as "bottlenecks") refers to congestion taking place at roughly the same place and time every day, usually during peak traffic periods due to insufficient infrastructure or physical capacity, such as roadways too narrow to accommodate the demand.

"Nonrecurring" congestion is caused by temporary disruptions that render part of the roadway unusable. Factors that trigger nonrecurring congestion include traffic incidents, bad weather construction work, poor traffic signal timing, and special events. About half the total congestion on roadways is recurring, and half is nonrecurring.

No definition or measurement of exactly what constitutes congestion has been universally accepted. Generally, transportation professionals examine congestion from several perspectives, such as delays and variability. Increased traffic volumes and additional delays caused by crashes, poor weather, special events, or other nonrecurring incidents lead to increased travel times. This report examines congestion through indicators of duration (travel time, congestion hours, planning time, delay time) and severity (cost).

Congestion Measures

FHWA generates the Freight Performance Measures and quarterly Urban Congestion Reports. (Freight performance measures are addressed in detail later in this chapter.) The Urban Congestion Reports characterize emerging traffic congestion and reliability trends at the national and city levels using probe-based travel time data for 52 urban areas in the United States with populations above 1,000,000 in 2010. The reports address mobility, congestion, and reliability using three traffic system performance indicators: Travel Time Index, Congested Hours, and Planning Time Index. These indicators are estimated from FHWA's National Performance Management Research Data Set (NPMRDS).

The NPMRDS is a compilation of observed average travel times, date/time, direction, and location for freight, passenger, and other traffic. It covers data for the National Highway System (NHS) and 5-mile radii of arterials at border crossings. Passenger data are collected from mobile phones, portable navigation devices, and vehicle transponders. The American Transportation Research Institute accumulates fleet system data, with travel times reported in 5-minute bins by traffic segment. Monthly historical data sets then become available by the middle of the following month. FHWA provides this data set to States and metropolitan planning organizations (MPOs) for use in their performance measurement activities. (Note: The NPMRDS data are available only for 2012 onward; data from the first year—2012—are limited to the Interstate Highway System.)

Travel Time Index

The Travel Time Index is a performance indicator used to examine congestion. This index is calculated as the ratio of travel time required to make a trip during the congested peak period to travel time for the same trip during the off-peak period in noncongested conditions. The value of Travel Time Index is always greater than or equal to 1, and a greater value indicates a higher degree of congestion. For example, a value of 1.30 indicates that a 60-minute trip on a road that is not congested would take 78 minutes (30 percent longer) during the period of peak congestion.

Exhibit 5-1 indicates that the average driver spent 29 percent more time during the congested peak time compared with traveling the same distance during the noncongested period (i.e., the Travel Time Index was 1.29).

Exhibit 5-1 Travel Time Index for 52 Urban Areas, 2012

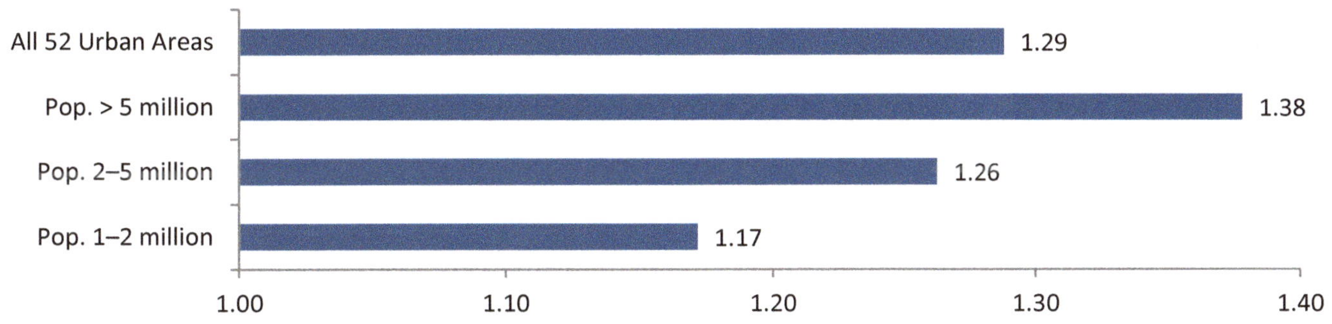

Sources: Travel Time index weighted by VMT over 52 urban areas based on the Urban Congestion Reports. Population from United States Census Bureau 2014 Metropolitan Statistical Areas Population Estimates for 2010.

Congestion occurs in urban areas of all sizes. Residents in large metropolitan areas tend to experience more severe congestion, and smaller urban areas usually experience better mobility. For example, a trip that normally takes 60 minutes on the Interstate Highway System during off-peak time would have taken 70.3 minutes (17 percent longer, or Travel Time Index 1.17) on average during the peak period for an urban area with population between 1 and 2 million. The same trip would take an average of 75.7 minutes (26 percent longer, or Travel Time Index 1.26) in a medium-sized urban area with 2–5 million population and an average of 82.7 minutes (Travel Time Index 1.38) in a metropolis with more than 5 million residents.

Road congestion also varies slightly over the course of a year. The Travel Time index increased from the first to the second quarter of 2012, and then declined slightly in the third quarter for urban areas with populations above 5 million (see *Exhibit 5-2*).

Exhibit 5-2 Quarterly Travel Time Index for 52 Urban Areas, 2012

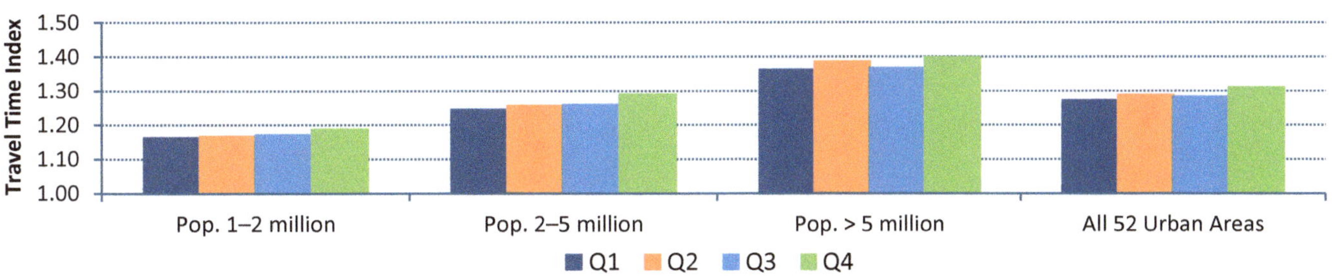

Source: Weighted average from NPMRDS; travel time weighted by VMT. Travel Time Index weighted by VMT over 52 urban areas was based on the Urban Congestion Reports. Population was obtained from United States Census Bureau 2014 Metropolitan Statistical Areas Population Estimates for 2010.

The Travel Time Index grew steadily across all four quarters for urban regions with populations less than 5 million. The quarterly trend for other urban regions was less consistent, but regardless of population size, the Travel Time Index increased in the fourth quarter relative to the first quarter.

Congested Hours

Congested Hours is another performance indicator that is used in the Urban Congestion Report. NPMRDS is used to calculate congested hours per day for the 52 major urban areas in the United

States. Similar to results for the Travel Time Index, more hours of congestion were observed in larger urban areas (see *Exhibit 5-3*).

Exhibit 5-3 Congested Hours per Weekday for 52 Urban Areas, 2012

Population Group	1st Quarter	2nd Quarter	3rd Quarter	4th Quarter	2012
Pop. 1–2 million	3.45	3.43	3.55	3.80	3.55
Pop. 2–5 million	4.48	4.38	4.50	4.95	4.58
Pop. > 5 million	5.98	5.95	5.97	6.28	6.05
All 52 Urban Areas	4.83	4.78	4.87	5.23	4.93

Source: Weighted average from NPMRDS; travel time weighted by VMT.

Congested Hours in Minneapolis/St. Paul

The Minnesota Department of Transportation derived its congestion data using 3,000 surveillance detectors in roadways and field observations on Twin Cities Freeways. Based on the traffic conditions in October (a "normal" traffic month), 758 miles of urban freeways were evaluated to measure the miles congested during the morning and afternoon commutes, Monday through Friday. The Department defined congested sections as those operating at speeds below 45 miles per hour at any time during the morning and afternoon peak periods.

The results show that most congestion lasted less than 2 hours, and less than 30 miles of freeway experienced severe congestion (duration greater than 3 hours) (see *Exhibit 5-4*). More miles, however, were reported to have moderate (duration of 2–3 hours) to severe (duration greater than 3 hours) congestion in recent years. Additionally, more freeways were congested in the morning peak period than in the afternoon.

Exhibit 5-4 Miles by Duration of Congestion: Minneapolis/St. Paul, 2002–2012

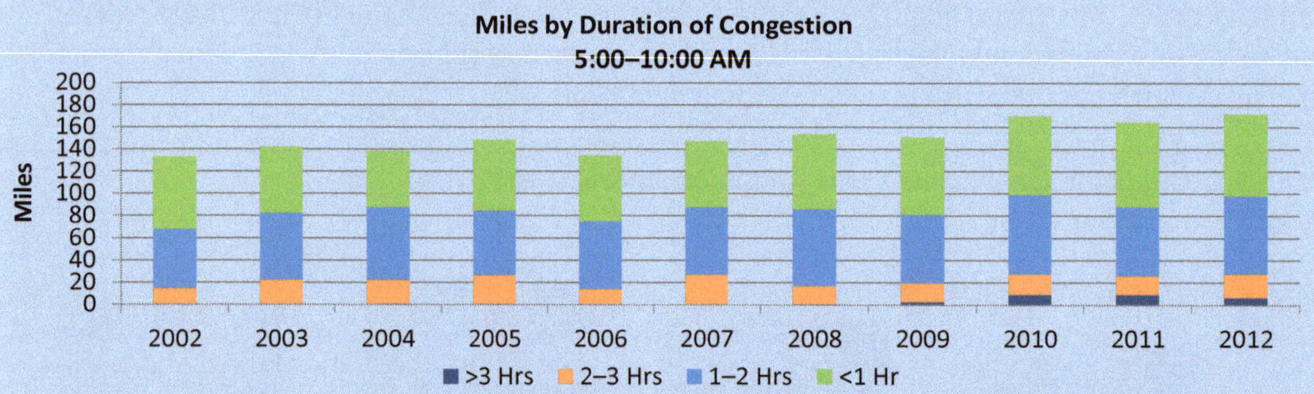

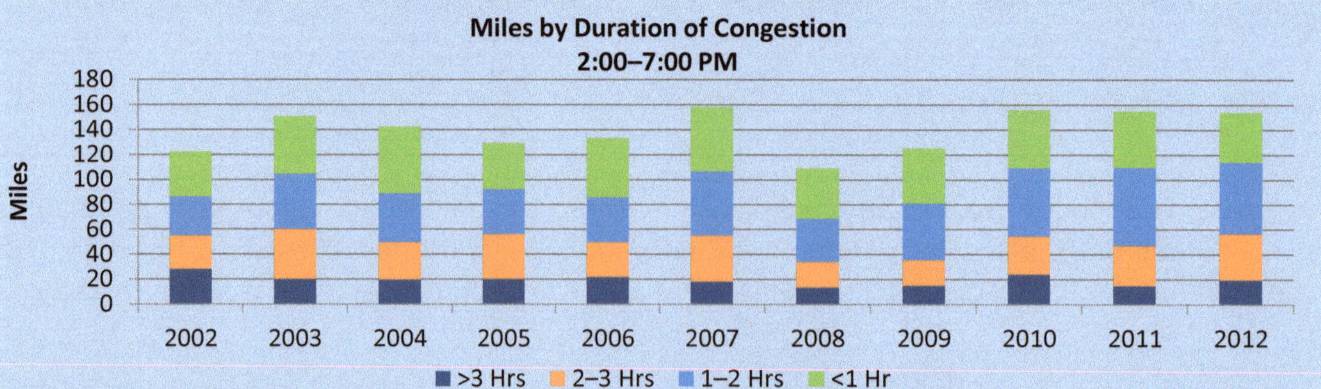

Source: Metropolitan Freeway System 2012 Congestion Report (Minnesota Department of Transportation, 2012).

In 2012, roads in very large urban areas experienced 6.05 hours of congestion on an average day, which is 70 percent higher than the 3.55 hours in a typical medium-sized urban area with population between 1 and 2 million. Congested Hours exhibited a similar pattern across different sizes of urban centers, usually dropping slightly in the second quarter and rising strongly afterwards.

Planning Time (Reliability)

Most travelers are less tolerant of unexpected delays than everyday congestion. Although drivers dislike everyday congestion, they may have an option to alter their schedules to accommodate it, or are otherwise able to factor it into their travel choices. Unexpected delays, however, often have larger consequences. Travelers also tend to remember the situations when they spent more time in traffic because of unanticipated disruptions, rather than the average time for a trip throughout the year.

Compared with simple average measures of congestion, like the Travel Time Index or Congested Hours, measures of travel time reliability provide a different perspective of improved travel. Users familiar with a route (such as commuters) can anticipate how bad traffic is during those few poor days and plan their trips accordingly. Such travelers reach their destinations on time more often or with fewer significant delays. Hence, measures of travel time reliability more accurately represent a commuter's experience than a simple average travel time.

Transportation reliability measures primarily compare high-delay days with average-delay days. The simplest methods usually identify days that exceed the 95th percentile in terms of travel times and estimate the severity of delay on specific routes during the heaviest traffic days of each month. The Planning Time Index is defined for the purpose of this report as "the ratio of travel time on the worst day of the month compared to the time required to make the same trip at 'normal travel time.'" More precisely, it is the ratio of the 95th percentile of travel time and the 50th percentile of travel time (i.e., the median). For example, a Planning Time Index of 1.60 means that, for a trip that takes 60 minutes in light traffic, a traveler should budget a total of 96 (60 × 1.60) minutes to ensure on-time arrival for 19 times of 20 trips (95 percent of the trips).

The Planning Time Index is particularly useful because it can be compared directly to the Travel Time Index (a measure of average congestion) on similar numeric scales. The Planning Time Index is usually higher than the Travel Time Index. This difference is because, in most cases, travel time follows a normal distribution (bell curve). Statistically, the mean of travel time (Travel Time Index) is close to the median (50th percentile), and the median is always less than the 95 percentile value used to determine the Planning Time Index.

Exhibit 5-5 indicates that ensuring on-time arrival 95 percent of the time in 2012 required planning for 2.51 times the travel time that would be necessary under median traffic conditions (i.e., the Planning Time Index was 2.51). Similar to average travel time during congested periods (Travel Time Index), travel time reliability is worse, on average, in larger urban areas than in smaller urban areas. The average Planning Time Index was 2.89 in major cities with more than 5 million residents, which is 39 percent higher than the index for small urban areas with populations between 1 and 2 million (Planning Time Index 2.09).

Exhibit 5-5 Planning Time Index for 52 Urban Areas (95th Percentile)

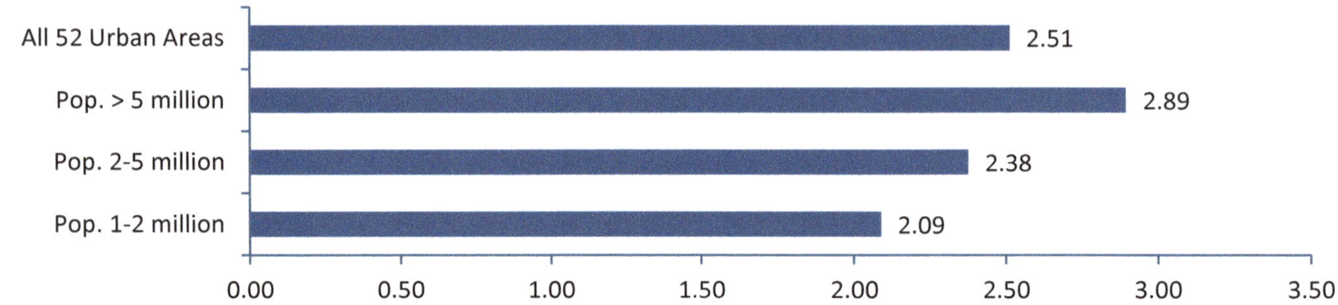

Sources: Weighted average from NPMRDS; travel time weighted by VMT. Planning Time Index weighted by VMT over 52 urban areas was based on the Urban Congestion Reports. Population was obtained from United States Census Bureau 2014 Metropolitan Statistical Areas Population Estimates for 2010.

Congestion in Atlanta

The Georgia Regional Transportation Authority calculated several mobility measures to track highway system performance.

The freeway travel index is calculated as the weighted average of the travel time indices for each freeway segment with vehicle miles traveled used as the weight. As with the simple Travel Time Index, the higher the weighted Travel Time Index, the worse the congestion. The average morning peak-period Travel Time Index barely increased from 1.24 in 2009 to 1.25 in 2010, and during the afternoon peak period the Travel Time Index worsened from 1.32 to 1.35 (see *Exhibit 5-6*).

Exhibit 5-6 Congestion in Atlanta, 2009–2010

	Morning Peak (7:45–8:45 a.m.)		Afternoon Peak (5:00–6:00 p.m.)	
Time Index	2009	2010	2009	2010
Freeway Travel Time Index	1.24	1.25	1.32	1.35
Freeway Planning Time Index	1.67	1.68	1.91	1.98
Freeway Buffer Time Index	36.0	34.4	43.2	46.1

Source: 2011 Transportation MAP Report: A Snapshot of Atlanta's Transportation System Performance (Georgia Regional Transportation Authority, 2012).

The freeway Planning Time Index at the 95th percentile provides a benchmark for the travel time reliability of the road network. Compared with the 2009 base year, planning time index in 2010 increased marginally during the morning peak period, but the drop in road reliability was more noticeable during the afternoon peak period.

The buffer time index is another measure of travel reliability. It represents the extra time (or buffer) that a traveler would need to add to the time for a congested trip to arrive on time consistently 19 of 20 times (95 percent of the trips). The Buffer Time Index is expressed as a percentage of the average congested trip time. So, for the same trip that takes an average of about 8.6 minutes, a traveler should allow for a buffer of 87 percent (16 minutes = 8.6 × 1.87) if he or she wants to be on time 19 of 20 times. A deeper decline in buffer time index is observed for the afternoon peak period in the Atlanta area.

Congestion Trends

Although the NPMRDS is currently FHWA's official data source for measuring congestion and the Urban Congestion Report is the official program for measuring congestion, the data used in the current edition started in 2012. Hence, examining other data sources is necessary to observe trends over a longer period. The *2015 Urban Mobility Scorecard*, developed by the Texas Transportation Institute, provides time series data for selected congestion measures starting in

1982. The report includes data for all 471 U.S. urbanized areas, including small urbanized areas with populations less than 500,000. The report's estimated congestion trends are based on the speed data provided by INRIX®, which contains historical traffic information from more than 1.5 million global positioning system (GPS)-enabled vehicles and mobile devices for every 15-minute period every day for all major U.S. metropolitan areas.

Although the Texas Transportation Institute produces measures of congestion similar to those generated from the NPMRDS, the measures differ in geographic coverage and are calculated using a different method. Consequently, the Texas Transportation Institute's values for measures such as the Travel Time Index deviate somewhat from those presented above for 2012 based on NPMRDS data.

Exhibit 5-7 shows changes in the national average of the Travel Time Index since 2002 for all urbanized area categories. The Travel Time Index rose steadily until 2008 and started to increase again after a brief drop during the Nation's recent economic recession. By 2012, the Travel Time Index had risen close to its prerecession level across different sizes of urban area, indicating that congestion had worsened since 2009. Urbanized areas with higher populations have longer travel times. For example, in 2012, the Travel Time Index was 1.13 in small urbanized areas from 2002 to 2012, 1.18 in medium-sized urbanized areas, 1.23 in large urbanized areas, and 1.32 in very large metropolitan areas.

Exhibit 5-7 Travel Time Index for All Urbanized Areas, 2002–2012

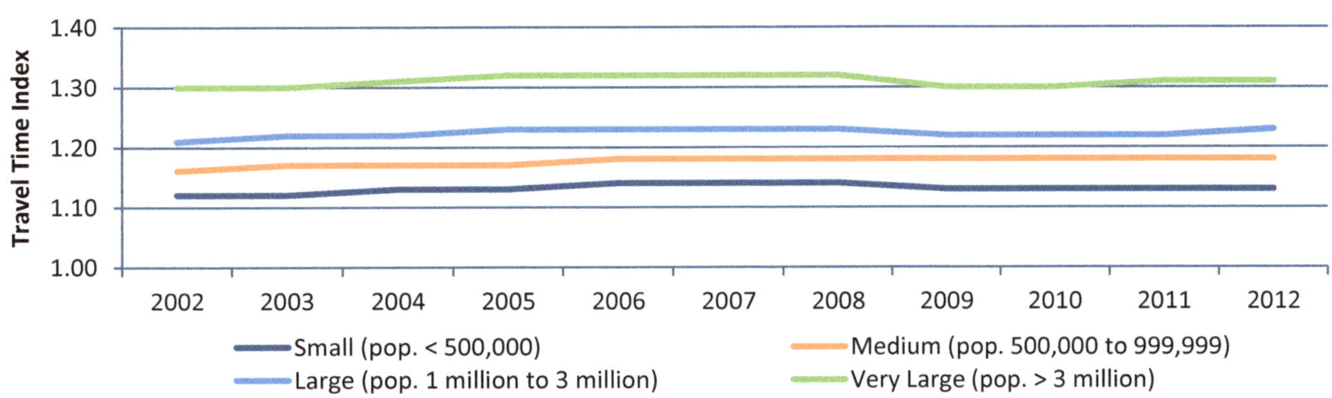

Source: Texas Transportation Institute (2015), population based on the U.S. Census Bureau estimates.

Cost of Delay

Congestion adversely affects the American economy and results in a massive waste of time, fuel, and money. When travel time increases or reliability decreases, businesses need to increase average inventory levels to compensate, leading to higher overall costs. Congestion imposes an economic drain on businesses, and the resulting increased costs negatively affect producer and consumer prices.

Although automobile and truck congestion currently imposes a relatively small cost on the GDP (about 0.8 percent of GDP), the cost of congestion is growing faster than GDP. If current trends continue, congestion is expected to impose a larger proportional cost in the future. The cost of congestion has risen almost 5 percent per year over the past 25 years, almost double the growth rate of GDP.

As shown in *Exhibit 5-8*, the Texas Transportation Institute estimates that each auto commuter averaged an extra 41 hours traveling during the peak traveling period in 2012. Together, congestion wastes 6.7 billion hours of travel time for the society collectively. Combining wasted time with approximately 3 billion gallons of wasted fuel, the total cost of congestion was estimated to reach $154 billion in 2012. (The Texas Transportation Institute assumed an average cost of time of $17.67 per hour, which differs from the value used in the analyses reflected in Part II of this report.)

Exhibit 5-8 National Congestion Measures, 2002–2012

Year	Delay per Commuter (Hours)	Total Delay (Billions of Hours)	Total Cost (Billions of 2014 Dollars)
2002	39	5.6	$124
2003	40	5.9	$128
2004	41	6.1	$136
2005	41	6.3	$143
2006	42	6.4	$149
2007	42	6.6	$154
2008	42	6.6	$152
2009	40	6.3	$147
2010	40	6.4	$149
2011	41	6.6	$152
2012	41	6.7	$154

Source: Texas Transportation Institute, 2015.

Total delay time increased from 5.6 billion hours in 2002 to 6.7 billion hours in 2012. Total costs rose at an average annual rate of 1.9 percent per year from 2002 to 2012. The estimated total cost of delay declined during the most recent recession but by 2012 had risen to the 2007 pre-recession level.

Travel Delays in Puget Sound of Washington State

Washington State Department of Transportation used maximum throughput speeds to measure delays relative to the highway's most efficient operating condition. Maximum throughput is achieved when vehicles travel at speeds between 42 and 51 miles per hour (below the posted speed of 60 miles per hour). At maximum throughput speeds, highways are operating at peak efficiency because more vehicles are passing through the segment than when they are traveling at posted speeds. This situation occurs because drivers operating at maximum throughput speeds can travel more safely with a shorter distance between vehicles than at posted speeds.

Maximum throughput speeds vary from one highway segment to another, depending on prevailing roadway design (roadway alignment, lane width, slope, shoulder width, pavement conditions, presence or absence of median barriers) and traffic conditions (traffic composition, conflicting traffic movements, heavy truck traffic, etc.). The maximum throughput speed is not static and depends on traffic conditions.

On an average weekday, each Washingtonian spent an estimated extra 4 hours and 30 minutes delayed due to traffic in 2012, which is below the prerecession levels in 2007 (see *Exhibit 5-9*). Despite a decline in statewide travel delay, congestion still caused drivers to waste 30.9 million hours in 2012 due to increased travel time. Combined with increased vehicle operating expense, total travel costs of delay reached $780 million in 2012.

Exhibit 5-9 Annual Delay: Washington State, 2007–2012[1]

Annual Delay Statewide	2007	2008	2009	2010	2011	2012
Per Person Travel Delay (Hours)	5.4	5.3	4.2	4.7	4.8	4.5
Total Travel Delay (Millions of Hours)	35.1	34.8	28.1	31.6	32.5	30.9
Cost of Delay (Millions of Dollars)	$931	4890	$721	$800	$821	$780

[1] The annual delay is defined as total hours of annual travel delay divided by total population in the State.

Source: The 2012 Corridor Capacity Report (Washington Department of Transportation 2013).

Freight Performance

When travel time increases or reliability decreases, businesses need to adjust average inventory levels to compensate for delays in receipt and shipment of goods. This situation leads to higher overall operating costs, which imposes an economic drain on business and a rise in producer and consumer prices. Although congestion might minimally affect the overall economy relative to other factors, the *2012 Urban Mobility Report* estimates costs of overall truck congestion to be $27 billion per year. Such inefficiency increases production costs and consumer prices, and contributes to businesses' moving their operations and jobs to locations where they can achieve more efficient supply chains, resulting in regional and national job losses.

Freight Performance Measurement (FPM)

FHWA has been collecting and analyzing data for freight-significant Interstate corridors since 2002. FHWA continues to collect travel time information on key Interstates and domestic freight corridors, at border crossings, in metropolitan areas, and at intermodal connectors. The objectives of the current FPM research program are to expand on the existing data sources, further develop and refine methods for analyzing data, derive national measures of congestion and reliability, analyze freight bottlenecks and intermodal connectors, and develop data products and tools that will help DOT, FHWA, and State and local transportation agencies address surface transportation congestion. FHWA sponsors research to develop performance measure approaches and tools and provides a national travel time data set (which includes freight and passenger traffic data) to States and metropolitan planning organizations to support performance measurement and management programs. Additionally, FHWA partners with other operating administrations, Federal agencies, and international agencies to evaluate and advance multimodal freight performance for North American corridors and critical supply chains.

Effect of Congestion on Freight Travel

FHWA monitors performance indicators for the freight system as part of its Freight Performance Measure (FPM) program to analyze impacts of congestion and determine the operational capacity and efficiency of key freight routes in the United States.

FHWA measures freight highway congestion using truck probe data from more than 600,000 trucks equipped with GPS. These trucks provide billions of position signals that FHWA analyzes to determine truck freight performance, both for routine monitoring and for ad hoc analysis to understand truck movements and impacts, such as when an incident compromises highway network reliability. Having used these data since 2002, FHWA actively seeks to increase the number of probes to improve data availability. FHWA estimates that the current number of probes represents approximately 30 percent of the truck population for Classes 6, 7, and 8 (i.e., trucks with gross vehicle weight exceeding 19,500 pounds). In addition to the FPM truck probe data, FHWA uses information from the Freight Analysis Framework tool for tonnage and volume flows.

FPM's routine monitoring of truck freight performance is principally for monitoring congestion, using measures of travel time reliability and speed for corridors, border crossings, urban areas, freight intermodal connections, and freight bottlenecks. FHWA produces quarterly performance monitoring reports that provide insight into these areas. More information is available on FHWA's website at http://ops.fhwa.dot.gov/freight/freight_analysis/perform_meas/. Specifically, FHWA produces a Freight Movement Efficiency Index (FMEI) that combines measures of speeds and travel times for intermodal locations, urban areas, bottlenecks, and border crossings. FHWA monitors travel times for the top 25 freight corridors in the United States.

FHWA has found that much of the current congestion negatively influencing truck carrier operations happens on a recurring basis during peak periods, particularly in and near major metropolitan areas. The map in *Exhibit 5-10* shows the location of this peak-period congestion on high-volume truck portions of the NHS in 2011. Overall, peak-period congestion created stop-and-go conditions on 5,800 miles of the NHS and caused traffic to travel below posted speed limits on an additional 4,500 miles of the high-volume truck portions of the NHS.

Exhibit 5-10 Peak-Period Congestion on the High-Volume Truck Portions[1] of the National Highway System, 2011[2,3]

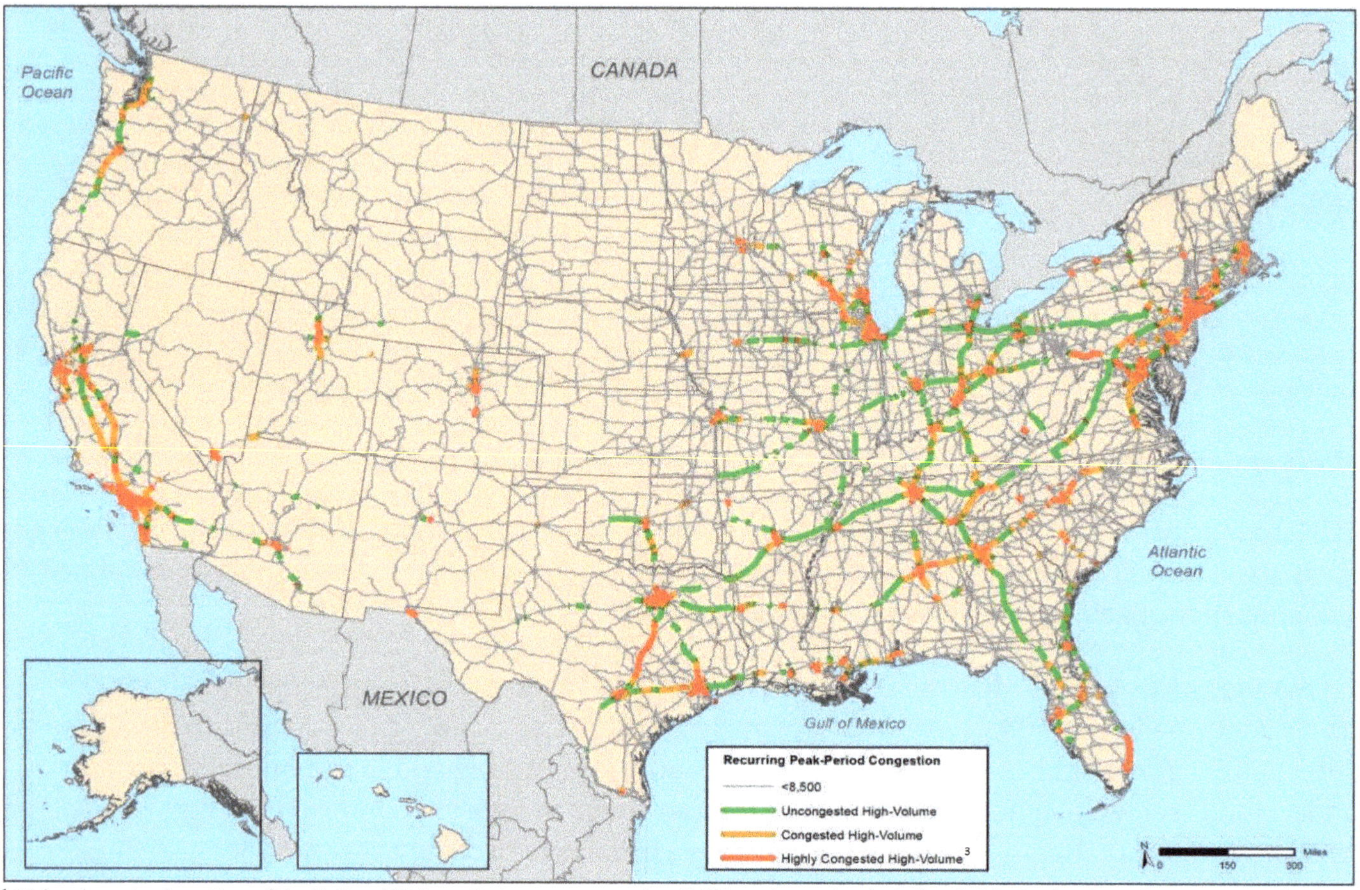

[1] High-volume truck portions of the National Highway System carry more than 8,500 trucks per day, including freight-hauling long-distance trucks, freight-hauling local trucks, and other trucks with six or more tires.

[2] The volume/service flow ratio is estimated using the procedures outlined in the HPMS Field Manual, Appendix N. NHS mileage as of 2011, prior to MAP-21 system expansion.

[3] Highly congested segments are stop-and-go conditions with volume/service flow ratios greater than 0.95. Congested segments have reduced traffic speeds with volume/service flow ratios between 0.75 and 0.95.

Sources: U.S. Department of Transportation, Federal Highway Administration, Office of Freight Management and Operations, Freight Analysis Framework, version 3.4, 2013.

Exhibits 5-11 and *5-12* show some of the results of FHWA's analyses using truck probe data indicating the most congested, freight-significant locations in the United States and average truck travel speeds on Interstate highways, respectively. Reduced travel speeds for trucks most commonly occur in large metropolitan areas. They can also occur at international border crossings and gateways, in mountainous areas that require trucks to climb steep inclines, and in areas frequently prone to poor visibility driving conditions.

Exhibit 5-11 Top 25 Congested Freight-Significant Locations, 2013[1]

Ranking[2]	Location[3]	Average Speed[4]	Peak-Hour Speed	Non-Peak-Hour Speed	Peak/Off-Peak Ratio
1	Fort Lee, NJ: I-95 at NJ 4	36	30	38	1.25
2	Chicago, IL: I-290 at I-90/I-94	30	23	33	1.42
3	Atlanta, GA: I-285 at I-85 (North)	42	30	49	1.61
4	Cincinnati, OH: I-71 at I-75	47	39	50	1.27
5	Houston, TX: I-45 at US 59	39	29	44	1.52
6	Houston, TX: I-610 at US 290	42	34	46	1.34
7	St. Louis, MO: I-70 at I-64 (West)	43	39	45	1.14
8	Diamond Bar, CA: CA 60 at CA 57	47	39	50	1.27
9	Louisville, KY: I-65 at I-64/I-71	47	41	49	1.21
10	Austin, TX: I-35	36	22	43	1.93
11	Chicago, IL: I-90 at I-94 (North)	35	21	41	1.94
12	Dallas, TX: I-45 at I-30	42	33	46	1.39
13	Houston, TX: I-10 at I-45	46	36	50	1.38
14	Atlanta, GA: I-75 at I-285 (North)	48	37	52	1.39
15	Denver, CO: I-70 at I-25	43	37	46	1.26
16	Houston, TX: I-10 at US 59	47	36	52	1.46
17	Lynwood, CA: I-710 at I-105	45	36	49	1.37
18	Baton Rouge, LA: I-10 at I-110	44	36	48	1.33
19	Bloomington, MN: I-35W at I-494	46	36	50	1.40
20	Seattle, WA: I-5 at I-90	38	29	42	1.47
21	Hartford, CT: I-84 at I-91	47	37	51	1.36
22	Houston, TX: I-45 at I-610 (North)	48	38	52	1.36
23	Decatur, GA: I-20 at I-285 (East)	49	44	51	1.18
24	Auburn, WA: WA 18 at WA 167	48	42	51	1.23
25	Atlanta, GA: I-20 at I-285 (West)	50	45	52	1.15

[1] Using data associated with the FHWA-sponsored Freight Performance Measures (FPM) initiative, the American Transportation Research Institute (ATRI) provides a yearly analysis to quantify the impact of traffic congestion on truck-borne freight at 250 specific locations throughout the United States.

[2] The ranking analysis factors in the number of trucks using a particular highway facility and the impact that congestion has on average commercial vehicle speed in each of the 250 study areas. These data represent truck travel during weekdays at all hours of the day in 2014.

[3] These locations were identified over several years through reviews of past research, available highway speed and volume data sets, and surveys of private and public sector stakeholders.

[4] Average speeds below a free flow of 55 miles per hour indicate congestion.

Source: American Transportation Research Institute (ATRI), Congestion Impact Analysis of Freight Significant Highway Locations, 2013.

Exhibit 5-12 Average Truck Speeds on Selected Interstate Highways, 2012

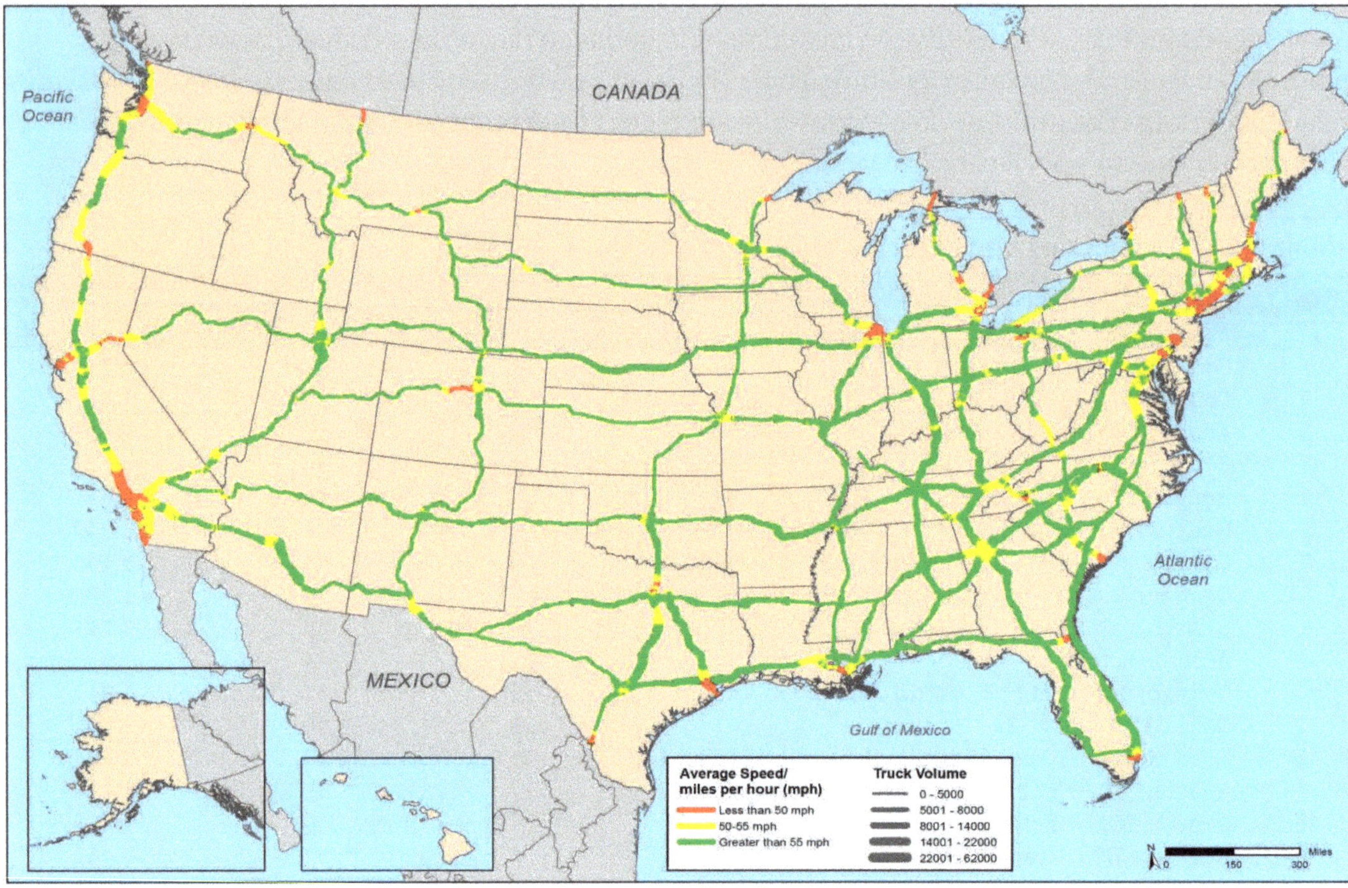

Sources: U.S. Department of Transportation, Federal Highway Administration, Office of Freight Management and Operations, Freight Performance Measurement Program, 2013.

To understand freight performance on critical freight routes, FHWA monitors performance using the truck probe data on the top 25 domestic freight corridors. As noted earlier in this section, FHWA uses a derivative of the truck probe data, the NPMRDS, to monitor these corridors using the Planning Time Index to evaluate average speeds.

Determination of Top 25 Domestic Freight Corridors

To determine the top 25 domestic freight corridors, FHWA used its Freight Analysis Framework (FAF 3.4) data to identify the top 10 percent of the FAF highway segments by tonnage. *Exhibit 5-13* identifies the corridors with the most freight tonnage, that is, the top 10 percent. The corridors that handle the top 10 percent of U.S. freight tonnage are shown in thick, dark blue lines on the map at the top of the exhibit, while all other corridors are shown in thin, lighter blue lines.

Exhibit 5-13 FAF Network Commodity Tonnage

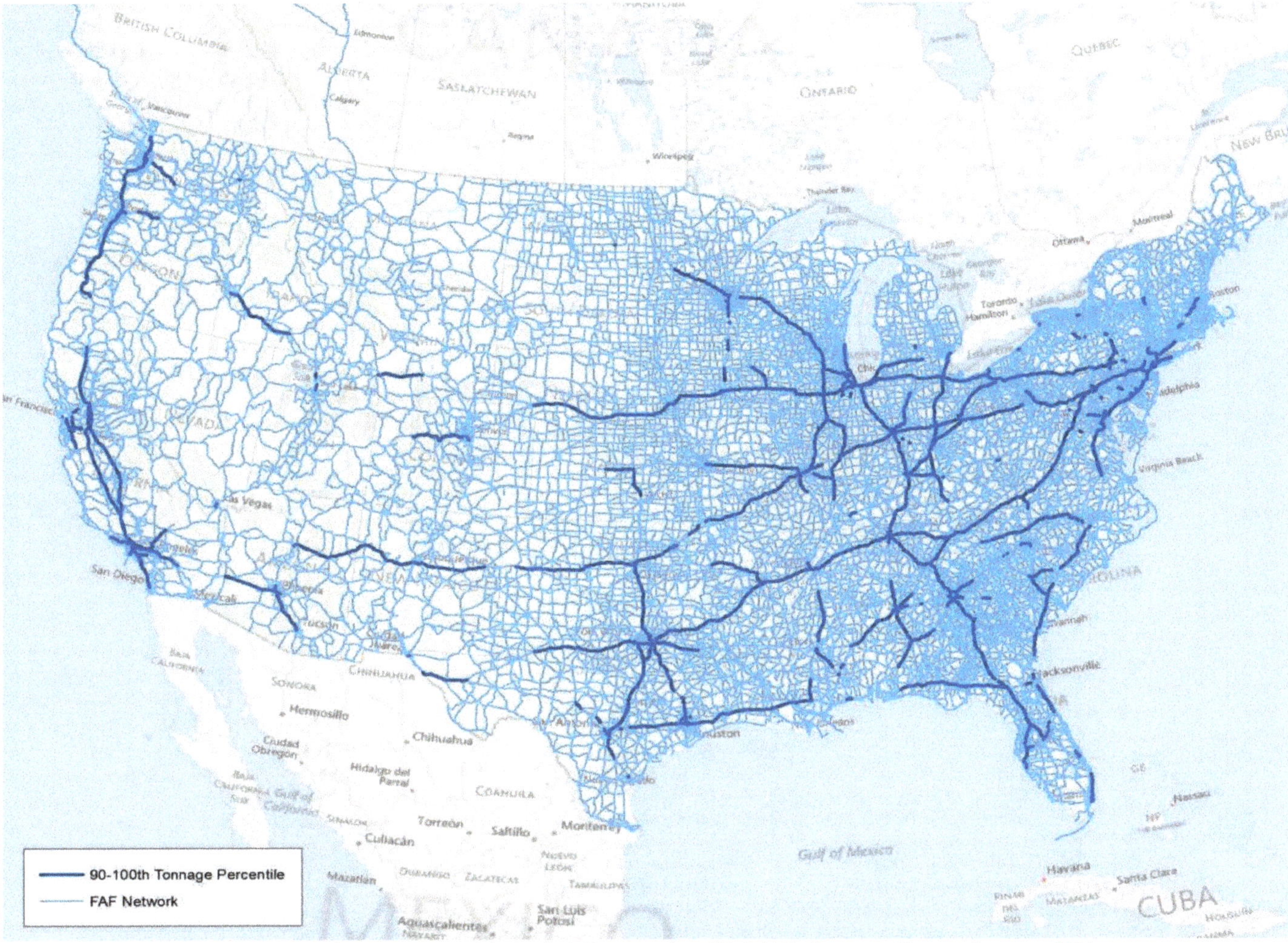

Source: FHWA Freight Management and Operations, Freight Analysis Framework and Freight Performance Measure Program, 2014.

From the network shown in *Exhibit 5-13*, FHWA connected segments with the highest tonnage and with known freight generators (land uses or groups of land uses that generate high freight transportation volumes, such as truck terminals, intermodal rail yards, water ports, airports, warehouses and distribution centers, or large manufacturing facilities) or population centers (origins and destinations) to identify 25 corridors that have the greatest freight movement. These corridors are illustrated in *Exhibit 5-14*.

Exhibit 5-14 Top 25 Intercity Truck Corridors

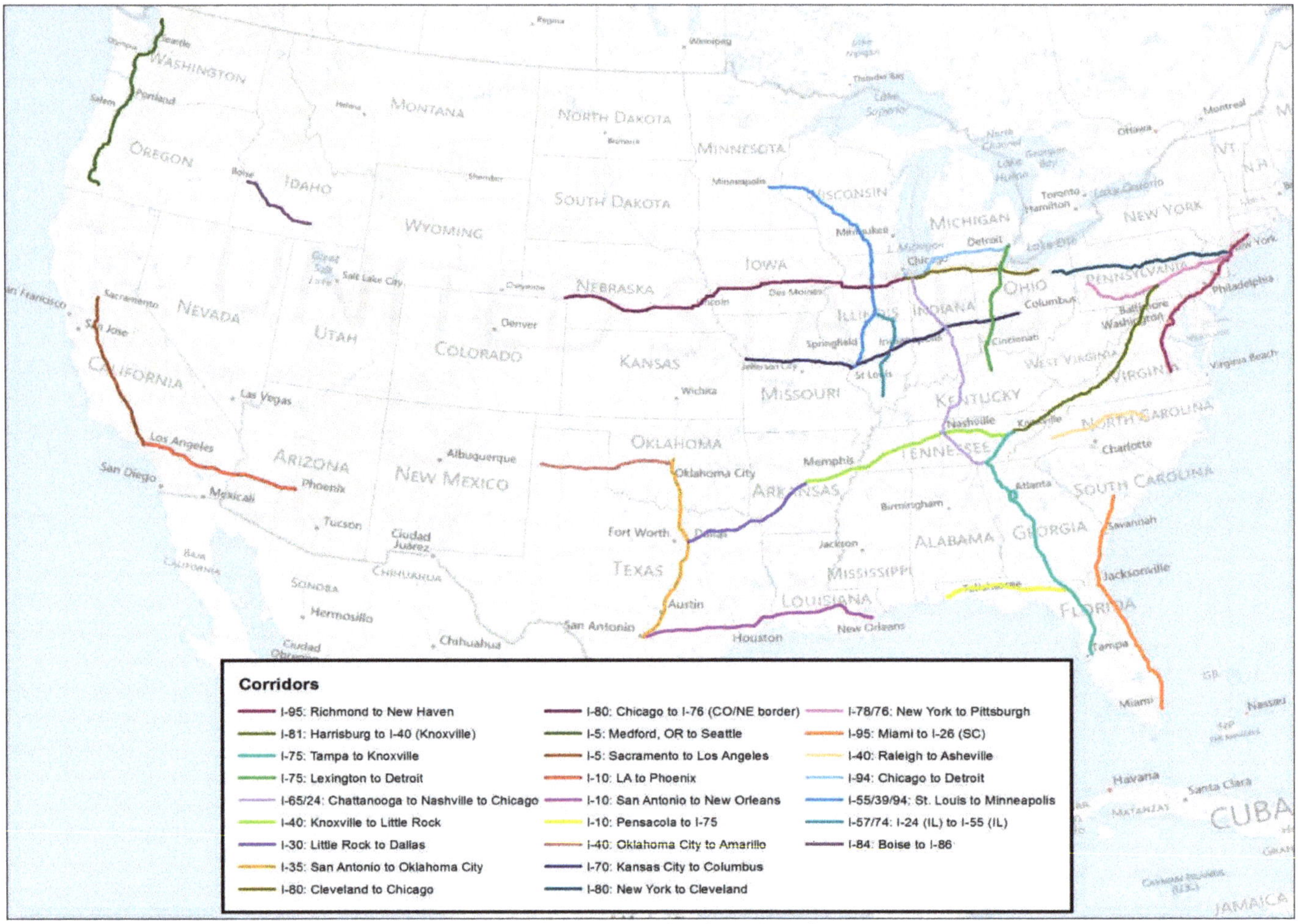

Source: FHWA Freight Management and Operations, Freight Analysis Framework and Freight Performance Measure Program, 2014.

The NPMRDS truck probe data also measure corridor-level travel time reliability. Travel time reliability is derived from measured average speeds of commercial vehicles for the top 25 domestic freight corridors annually. *Exhibit 5-15* shows the Planning Time Index for the 25 most significant intercity truck corridors in the United States.

Exhibit 5-15 Travel Time Reliability Planning Time Index for the Top 25 Intercity Truck Corridors in the United States, 2011–2014

	Planning Time Index (95th PCTL/50th PCTL)			
Freight Corridor	2011	2012	2013	2014
1. I-5: Medford, OR to Seattle	1.31	1.34	1.37	1.41
2. I-5/CA 99: Sacramento to Los Angeles	1.28	1.33	1.34	1.33
3. I-10: Los Angeles to Tucson	1.24	1.21	1.26	1.27
4. I-10: San Antonio to New Orleans	1.23	1.28	1.30	1.31
5. I-10: Pensacola to I-75	1.06	1.06	1.06	1.07
6. I-30: Little Rock to Dallas	1.21	1.15	1.14	1.17
7. I-35: Laredo to Oklahoma City	1.24	1.24	1.28	1.30
8. I-40: Oklahoma City to Flagstaff	1.10	1.12	1.11	1.11
9. I-40: Knoxville to Little Rock	1.17	1.18	1.20	1.24
10. I-40: Raleigh to Asheville	1.11	1.12	1.14	1.15
11. I-55/I-39/I-94: St. Louis to Minneapolis	1.15	1.13	1.14	1.14
12. I-57/I-74: I-24 (IL) to I-55 (IL)	1.09	1.12	1.15	1.14
13. I-70: Kansas City to Columbus	1.21	1.18	1.20	1.20
14. I-65/I-24: Chattanooga to Nashville to Chicago	1.26	1.26	1.29	1.34
15. I-75: Tampa to Knoxville	1.16	1.16	1.20	1.21
16. I-75: Lexington to Detroit	1.26	1.24	1.26	1.30
17. I-78/I-76: New York to Pittsburgh	1.18	1.20	1.20	1.21
18. I-80: New York to Cleveland	1.23	1.19	1.19	1.20
19. I-80: Cleveland to Chicago	1.18	1.14	1.17	1.21
20. I-80: Chicago to I-76 (CO/NE border)	1.13	1.12	1.12	1.12
21. I-81: Harrisburg to I-40 (Knoxville)	1.11	1.12	1.11	1.11
22. I-84: Boise to I-86	1.14	1.08	1.09	1.14
23. I-94: Chicago to Detroit	1.09	1.08	1.10	1.15
24. I-95: Miami to I-26 (SC)	1.17	1.18	1.21	1.23
25. I-95: Richmond to New Haven	1.62	1.59	1.69	1.85

Source: NPMRDS truck probe data.

In *Exhibit 5-15*, values greater than 1.00 illustrate travel time variability in the given corridors. Higher numbers indicate greater variability, and the portions of the numbers after the decimal points can be treated as percentages. As an example, for number 25, the I-95 corridor between Richmond and New Haven, the Travel Time Reliability Planning Time Index in 2011 was 1.62, meaning travel times were 62 percent longer on heavy travel days, compared to normal days, for drivers traveling the entire length of the corridor. More unpredictable travel times are problematic for truck drivers and freight receivers because they have a harder time optimizing the transportation portion of their supply chains.

Finally, the NPMRDS truck probe data are used to determine the average speed for the top 25 domestic highway freight corridors. The average speeds shown in *Exhibit 5-16* serve as an indicator of congestion for each corridor and should not be interpreted as the average speed expected at any location on any given corridor.

Exhibit 5-16 Average Travel Speeds for the Top 25 Intercity Truck Corridors in the United States, 2011–2014

Freight Corridor	Average Speed (24/7)			
	2011	2012	2013	2014
1. I-5: Medford, OR to Seattle	56.64	56.33	56.12	54.94
2. I-5/CA 99: Sacramento to Los Angeles	56.19	56.05	56.11	55.99
3. I-10: Los Angeles to Tucson	59.53	59.42	59.42	58.60
4. I-10: San Antonio to New Orleans	61.79	61.45	61.77	60.82
5. I-10: Pensacola to I-75	64.69	63.90	64.03	63.99
6. I-30: Little Rock to Dallas	61.78	62.64	62.82	62.13
7. I-35: Laredo to Oklahoma City	61.06	61.45	61.05	59.76
8. I-40: Oklahoma City to Flagstaff	63.99	63.86	64.15	64.31
9. I-40: Knoxville to Little Rock	62.34	62.24	62.14	61.53
10. I-40: Raleigh to Asheville	62.42	62.36	62.32	61.62
11. I-55/I-39/I-94: St. Louis to Minneapolis	62.00	62.37	62.16	62.10
12. I-57/I-74: I-24 (IL) to I-55 (IL)	62.86	62.71	62.56	62.76
13. I-70: Kansas City to Columbus	61.51	61.94	61.81	61.50
14. I-65/I-24: Chattanooga to Nashville to Chicago	60.97	61.04	60.85	59.57
15. I-75: Tampa to Knoxville	62.74	62.47	62.39	61.67
16. I-75: Lexington to Detroit	60.18	60.76	60.66	59.30
17. I-78/I-76: New York to Pittsburgh	59.59	59.94	59.88	59.34
18. I-80: New York to Cleveland	60.78	61.12	61.13	60.68
19. I-80: Cleveland to Chicago	61.86	62.26	61.99	61.57
20. I-80: Chicago to I-76 (CO/NE border)	62.96	63.16	63.36	63.39
21. I-81: Harrisburg to I-40 (Knoxville)	62.38	62.42	62.60	62.60
22. I-84: Boise to I-86	61.81	62.53	62.53	62.43
23. I-94: Chicago to Detroit	59.89	60.54	59.95	58.74
24. I-95: Miami to I-26 (SC)	63.07	62.63	62.48	61.77
25. I-95: Richmond to New Haven	55.36	55.52	54.70	51.72

Source: NPMRDS truck probe data.

Quality of Life

Fostering quality of life is a continued goal of DOT. DOT's Strategic Plan for Fiscal Years 2014-2018 addresses the strategic goal to "Foster improved quality of life in communities by integrating transportation, policies, plans, and investments with coordinated housing and economic development policies to increase transportation choices and access to transportation services for all."

To achieve this goal, DOT will strive to:

- Expand convenient, safe, and affordable transportation choices for all users by directing Federal investments in infrastructure toward projects that more efficiently meet transportation, land use, goods movement, and economic development goals developed through integrated planning approaches.
- Ensure Federal transportation investments benefit all users by emphasizing greater public engagement, fairness, equity, and accessibility in transportation investment plans, policy guidance, and programs.

Building quality of life in communities involves a multiagency approach, so DOT is collaborating across lines of authority to leverage related Federal investments. The Interagency Partnership for Sustainable Communities includes DOT (https://www.sustainablecommunities.gov/), the U.S. Department of Housing and Urban Development, and the U.S. Environmental Protection Agency. Through this Partnership, DOT has provided grants and technical assistance to ensure that its policies and investments promote quality of life; developed and provided tools for communities to assess, plan, and design sustainable communities; increased flexibility to use Federal funds; promoted safe and accessible transportation choices for all users; supported disaster recovery and resiliency planning in impacted communities; and convened leaders at all levels to share lessons learned by communities and to engage stakeholders to help shape partnership efforts.

Strategies to Increase Access to Convenient and Affordable Transportation Choices

DOT's FY 2014–2018 Strategic Plan identifies the following strategies to increase access to convenient and affordable transportation choices:

- Continue to encourage States and metropolitan planning organizations to consider the impact of transportation investments on local land use, affordable housing, scenic and historic resources, access to recreation, people, and goods movement;
- Continue to invest in high-speed and intercity passenger rail to complement highway, transit, and aviation networks and encourage projects that improve transit connectivity to intercity and high-speed rail, airports, roadways, and walkways;
- Increase the capacity and reach of public transportation, improve the quality of service, and increase travel time reliability through deployment of advanced technologies and significant gains in the state of good repair of transit infrastructure; and
- Advocate for transportation investments that strategically improve community design and function by providing an array of safe transportation options, such as vanpools, smart paratransit, car sharing, bike sharing, and pricing strategies that, in conjunction with transit services, reduce single-occupancy driving.

Measuring Quality of Life

Progress is being made on measuring the impact of transportation investments on livability. Several tools, such as the Sustainable Communities Indicator Catalog, Infrastructure Voluntary Evaluation Sustainability Tool (INVEST), and the Community Vision Metrics Web Tool have been developed to measure the impact of transportation investments on quality of life in communities.

Livability Defined

The terms "Quality of life" and "livability" are used interchangeably in this report. Livability in transportation concerns tying the quality and location of transportation facilities to broader opportunities, such as access to good jobs, affordable housing, quality schools, and safer streets and roads.

Communities can measure progress toward quality of life goals using the Sustainable Communities Indicator Catalog. Indicators in the catalog focus on the relationships among land use, housing, transportation, human health, and the environment. The user can choose an indicator type related to housing, transportation, or land use and identify the geographic scale; level of urbanization and issues of concern such as access to equity, affordability, community, and sense of place; economic competitiveness; environmental quality; and public health. The tool provides a summary of how the indicators chosen relate to quality of life, an approach to measuring the indicator, and a case study of a community that uses the chosen indicator (see *Exhibit 5-17*).

Exhibit 5-17 Examples of Sustainable Community Indicators

Indicator Name	Indicator Topic	Issue of Concern	Level of Urbanization	Geographic Scale
Intersection density	Land use, transportation	Access and equity, community and sense of place, environmental quality, public health	Rural, suburban, urban	Neighborhood/ corridor, project
Access to transit: percentage of jobs within walking distance of transit service	Land use, transportation	Access and equity, Affordability, economic competitiveness, environmental quality	Rural, suburban, urban	County, municipality, region
City fleet: gas mileage	Transportation	Economic competitiveness, environmental quality	Rural, suburban, urban	County, municipality, region
Walkability	Land use, transportation	Access and equity, community and sense of place, environmental quality, public health	Rural, suburban, urban	County, municipality, neighborhood/ corridor
Fuel consumption/ purchase	Transportation	Economic competitiveness, environmental quality	Rural, suburban, urban	County, municipality, region
Access to safe parks and recreation areas: percentage of residents within walking distance of recreation land	Housing, land use, transportation	Access and equity, community and sense of place, public health	Suburban, urban	County, municipality, neighborhood/ corridor, project, region
Access to healthy food options	Housing, land use, transportation	Access and equity, public health	Rural, suburban, urban	County, municipality, neighborhood/ corridor, region
Bike parking per capita	Land use, transportation	Access and equity, community and sense of place, environmental quality, public health	Rural, suburban, urban	County, municipality, neighborhood/ corridor, project, region
Access to transit: Percentage of population within walking distance of frequent transit service	Housing, land use, transportation	Access and equity, affordability, environmental quality	Rural, suburban, urban	County, municipality, region
Percentage of population served by transit	Housing, land use, transportation			

Source: Partnership for Sustainable Communities, https://cms.sustainablecommunities.gov/indicators/discover.

FHWA has developed the Web-based INVEST tool that allows decision makers to evaluate and improve sustainable practices in their transportation projects and programs. The tool has a collection of voluntary best practices, called criteria, designed to help transportation agencies integrate sustainability into their programs (policies, processes, procedures, and practices) and projects. INVEST considers the full life cycle of projects and has three modules to self-evaluate the entire life cycle of transportation services, including System Planning (SP), Project Development, and Operations and Maintenance. Each module, based on a separate collection of criteria, can be evaluated separately. More information on INVEST is available at www.sustainablehighways.org.

Sustainable Communities Indicator Catalog – Pedestrian Infrastructure Indicator

The City of Indianapolis has used the pedestrian infrastructure indicator. The City's Office of Sustainability along with the Indianapolis Bicycle Advocacy/INDYCOG, and Health by Design conducted a bicycle and pedestrian documentation count. The purpose of the count was to provide the City with data on the total number of people walking and biking in their city. Volunteers were located in various areas around Indianapolis, including the downtown area, where they counted bicyclists in bike lanes and pedestrians on sidewalks for 2 hours. The results were used as benchmarks for the City of Indianapolis and the Office of Sustainability. The City will continue the counting exercise biannually in the spring and fall. By investing in infrastructure and affording citizens options, the City has confirmed residents are using the bicycle and pedestrian facilities. The City will continue to encourage residents to take advantage of the bicycle and pedestrian infrastructure improvements.

The SP module in INVEST has several quality-of-life-related items that are used in scoring. Examples of quality-of-life-related criteria in the SP module include:

- **SP-01 Integrated Planning: Economic Development and Land Use** – Integrate statewide and metropolitan Long Range Transportation Plans (LRTP) with statewide, regional, and local land use plans and economic development forecasts and goals. Proactively encourage and facilitate sustainability through the coordination of transportation, land use, and economic development planning.
- **SP-03 Integrated Planning: Social** – The agency's LRTP is consistent with and supportive of the community's vision and goals. When considered from an integrated perspective, these plans, goals, and visions provide support for sustainability principles. The agency applies context-sensitive principles to the planning process to achieve solutions that balance multiple objectives to meet stakeholder needs.
- **SP-04 Integrated Planning: Bonus** – The agency has a continuing, cooperative, and comprehensive (3-C) transportation planning process. Planners and professionals from multiple disciplines and agencies (e.g., land use, transportation, economic development, energy, natural resources, community development, equity, housing, and public health) work together to incorporate and apply all three sustainability principles when preparing and evaluating plans.
- **SP-05 Access and Affordability** – Enhance accessibility and affordability of the transportation system for all users by multiple modes.
- **SP-07 Multimodal Transportation and Public Health** – Expand travel choices and modal options by enhancing the extent and connectivity of multimodal infrastructure. Support and enhance public health by investing in active transportation modes.

Quality of Life Performance Indicators in Transportation Planning

The Community Vision Metrics Web Tool enables practitioners to search for quality-of-life indicators relevant to their specific circumstances, community, and quality-of-life goals to track the success of plans and projects in their communities. The indicators can be used to compare the status of different places or track change over time for an issue of importance. This information helps people understand the results of policies, identify where progress has been made, and highlight changes or disparities that are inconsistent with community goals. The tool includes specific quality-of-life areas of interest such as community amenities, community engagement, economics, housing, land use, housing, public health, and safety.

INVEST Use by KACTS

Kittery Area Comprehensive Transportation System (KACTS) is the metropolitan planning organization (MPO) for the Maine portion of the urbanized areas of Kittery-Portsmouth and Dover-Rochester, New Hampshire. KACTS used the INVEST System Planning (SP) module to score their approved 2010 Long Range Transportation Plan (LRTP) and used the results to identify opportunities to highlight and more fully integrate sustainability principles in their 2014 LRTP. After drafting the 2014 LRTP, KACTS used the SP module to evaluate the draft plan and compare the results with the 2010 LRTP. KACTS recognized that the new plan should be more informative and useful for the public to illustrate their sustainability-related practices, partnerships, policies, and programs more clearly.

Key outcomes noted in using INVEST were as follows:

- The criteria in the SP module helped enrich and improve the draft KACTS LRTP.
- The collaborative approach to scoring resulted in productive conversations about the LRTP and elucidated ways to increase the public visibility of KACTS.
- The exercise helped KACTS engage their partners more directly in the planning process and the connections of specific activities to broader outcomes.
- The SP module's emphasis on performance measures was very useful in helping KACTS prepare for performance management requirements stemming from the Moving Ahead for Progress in the 21st Century Act.
- KACTS has recommended improvements to INVEST so that it can consider the work of a small MPO more appropriately.

Location Affordability Portal

The Location Affordability Portal provides individuals with reliable, user-friendly data and resources on combined housing and transportation costs. This portal helps consumers, policy makers, and developers make more informed decisions about where to live, work, and invest. Vignettes are included to show how families and organizations can use the portal to make such decisions. The Location Affordability Portal features two tools: the Location Affordability Index (LAI) and My Transportation Calculator.

The LAI was developed to help individuals, planners, developers, and researchers gain a complete understanding of the costs of living in a given location by accounting for variations among households, neighborhoods, and region. All of these factors influence affordability. The LAI provides estimates of the percentage of a family's income dedicated to the combined cost of housing and transportation in a given location. Users can choose from among eight different family profiles—defined by household income, size, and number of commuters—and observe the affordability landscape for each one in a neighborhood, city, or region.

The My Transportation Cost Calculator enables a user to customize information from the LAI by entering basic information about their family's income, housing, cars, and travel patterns. The customized estimates offer a more thorough understanding of an individual's or household's transportation costs, how much they vary in different locations, and how much they are influenced by individual choices. This enables users to make more informed decisions about where to live and work.

The University of Florida's Southeastern Transportation Research, Innovation, Development and Education (STRIDE) Center used the Community Vision Metrics Web Tool during five workshops in the southeastern United States to help localities develop performance measures for use in transportation and comprehensive planning. The tool was used to identify context specific to quality-of-life indicators. Criteria to help participants critically evaluate the performance indicators were selected through the Community Vision Metrics Web Tool. Participants at all five workshops commented on the importance of identifying measures relevant to both the planning process and quality-of-life outcomes. The STRIDE report concluded that the Community Vision Metrics Web Tool provides an important starting point for practitioners to begin investigating quality-of-life indicators that can be used in the planning process. The report noted that the tool is essential for taking the first step toward evaluating performance measures.

Environmental Sustainability

The FY 2014-2018 DOT Strategic Plan includes the strategic goal to advance environmentally sustainable policies and investments that reduce carbon and other harmful emissions from transportation sources and increase resilience to climate change.

To achieve this goal, the DOT will undertake efforts to:

- Reduce oil dependence and carbon emissions through research and deployment of new technologies, including alternative fuels, and by promotion of more energy-efficient modes of transportation.
- Avoid and mitigate transportation-related impacts to climate, ecosystems, and communities by helping partners make informed project planning decisions through an analysis of acceptable alternatives, balancing the need to obtain sound environmental outcomes with demands to accelerate project delivery.
- Promote infrastructure resilience and adaptation to extreme weather events and climate change through research, guidance, technical assistance, and direct federal investment.

Climate Change Resilience, Adaptation, and Mitigation

Climate change and extreme weather events present significant and growing risks to the safety, reliability, and sustainability of the Nation's transportation infrastructure and operations. The impacts of a changing climate, such as higher temperatures, sea level rise, and changes in seasonal precipitation and intensity of rain events, are affecting the life cycle of transportation systems and are expected to intensify. Sea level rise coupled with storm surges can inundate coastal roads, necessitate more emergency evacuations, and require costly (and sometimes recurring) repairs to damaged infrastructure. Inland flooding from unusually heavy downpours can disrupt traffic, damage culverts, and reduce service life. High heat can degrade materials, resulting in shorter replacement cycles and higher maintenance costs. Although transportation infrastructure is designed to handle a broad range of impacts based on historic climate, preparing for climate change and extreme weather events is critical to protecting the integrity of the transportation system.

Given the long life span of transportation assets, planning for system preservation and safe operation under current and future conditions constitutes responsible risk management. In December 2014, FHWA issued Order 5520-Transportation System Preparedness and Resilience to Climate Change and Extreme Weather Events. The Order states that FHWA's policy is to strive to identify the risks of climate change and extreme weather events to current and planned transportation systems and that the agency will work to integrate consideration of these risks into its planning, operations, and policies.

With over a fourth of the climate change-causing greenhouse gas (GHG) emissions in the United States coming from the transportation sector, FHWA is committed to reducing GHG pollution from vehicles traveling on our Nation's highways. FHWA is establishing resources to help State DOTs and local agencies better analyze GHGs and energy use, weigh GHG reduction strategies, and integrate climate change considerations into the transportation planning process.

Greenhouse Gas Emissions

Transportation is the leading consumer of U.S. petroleum and a major source of GHG emissions. In 2013, tailpipe emissions from the U.S. transportation sector directly accounted for over 31 percent of total U.S. carbon pollution and 27 percent of total U.S. GHG emissions. On-road vehicles (including cars, light-duty trucks, and freight trucks) are the primary source of transportation GHGs, accounting for more than 80 percent of the sector total and almost one-quarter of the total across all sectors. Other sources of transportation GHGs include aircraft, rail, ships and boats, pipelines, and lubricants (see *Exhibit 5-18*).

Exhibit 5-18 Transportation-related Greenhouse Gas Emissions By Mode, 2013

Transportation Type	1990	2005	2010	2011	2012	2013
On-Road Transportation						
Light-Duty Vehicles	992.3	1264.5	1132.6	1106.4	1094.2	1086.7
Medium- and Heavy-Duty Trucks	231.1	409.8	403	401.3	401.4	407.7
Buses	8.4	12.1	15.9	16.9	18	18.3
Motorcycles	1.8	1.7	3.7	3.6	4.2	4
Total On-Road	**1233.6**	**1688.1**	**1555.2**	**1528.2**	**1517.8**	**1516.7**
Non-Road Transportation						
Commercial Aircraft	110.9	133.9	114.3	115.6	114.3	115.4
Other Aircraft	78.3	59.6	40.4	34.2	32.1	34.7
Ships and Boats	44.9	45.2	45	46.7	40.4	39.6
Rail	39	53.3	46.5	48.1	46.8	47.5
Pipelines	36	32.2	37.1	37.8	40.3	47.7
Lubricants	11.8	10.2	9.5	9	8.3	8.8
Total Transportation	**1554.4**	**2022.5**	**1848.1**	**1819.7**	**1799.8**	**1810.3**
Total, All Sectors	**6301.1**	**7350.2**	**6989.8**	**6776.6**	**6545.1**	**6673.0**

Source: Inventory of U.S. Greenhouse Gas Emissions and Sinks 1990–2013, Table 2-13 (transportation sources) and Table 2-1 (U.S. total).

On-road vehicles also have been a major contributor to the net change in U.S. GHG emissions, especially between 1990 and 2005 when on-road GHGs increased by 37 percent, compared with 11 percent for all other sources across the U.S. economy. Both on-road and economy-wide emissions were driven significantly lower by the recession of 2007–2009, and by 2012, on-road GHGs were roughly 9 percent below 2005 levels. This decrease reflected declining per capita passenger VMT, increased consumer preference for smaller passenger vehicles (resulting from higher fuel prices), and improvements in new vehicle fuel economy resulting from Phase I light-duty CAFE (Corporate Average Fuel Economy) standards. On-road GHGs in 2013 were virtually unchanged from 2012 levels. Light-duty GHGs decreased by 0.7 percent, reflecting further improvements in new vehicle fuel economy that were offset in part by an increase in light-duty VMT. Truck GHG emissions increased by 1.6 percent, reflecting a 2.2-percent increase in truck VMT and a slight improvement in overall truck fuel efficiency.

Climate Mitigation Tools and Resources

FHWA has developed several tools and resources to help State DOTs and local agencies better analyze GHG emissions and energy use, calculate GHG reduction strategies, and integrate climate change considerations into the transportation planning process.

- **Carbon Estimator (ICE) Tool**—FHWA created a spreadsheet tool to help practitioners gauge life-cycle energy and GHG emissions from transportation infrastructure, including roads, bridges, transit facilities, and bike/pedestrian infrastructure. The tool also is intended to help weigh the emissions benefits of alternative construction and maintenance practices. The tool can be found at: http://www.fhwa.dot.gov/environment/climate_change/mitigation/publications_and_tools/carbon_estimator/.
- **Handbook for Estimating GHG Emissions in the Transportation Planning Process**—This handbook is a reference for State DOTs and MPOs to document available tools, methods, and data sources that can be used to generate GHG emission inventories, forecasts, and analyses of GHG plans and mitigation strategies. The handbook can be found at: http://www.fhwa.dot.gov/environment/climate_change/mitigation/publications/ghg_handbook/index.cfm.
- **Energy and Emissions Reduction Policy Analysis Tool (EERPAT)**—EERPAT was developed for State DOTs to model many inputs and policy scenarios to support strategic transportation and visioning, including GHG emissions reduction alternatives. State DOTs can use the tool to analyze GHG reduction scenarios and alternatives for use in the transportation planning process, climate action plan development, and scenario planning exercises for meeting State GHG reduction targets and goals. FHWA piloted the tool at four State DOTs (Colorado, Washington, Vermont, and Maryland). The pilot studies helped assess the sensitivity of EERPAT to various mitigation strategies and identified future enhancements to the model that might be needed. The tool can be found at: http://www.planning.dot.gov/FHWA_tool/.
- **A Performance-Based Approach to Addressing Greenhouse Gas Emissions in Transportation Planning**—This handbook is a resource for State DOTs and MPOs interested in addressing GHG emissions through performance-based planning and programming. It

discusses techniques for integrating GHG emissions in such planning, considerations for selecting relevant GHG performance measures, and ways of using GHG performance measures to support investment choices and enhance decision-making. The handbook can be found at http://www.fhwa.dot.gov/environment/climate_change/mitigation/publications_and_tools/ghg_planning/index.cfm.

Greenhouse Gas/Energy Analysis Demonstration Projects

In fall 2014, FHWA funded one State DOT and three metropolitan planning organizations (MPOs) to perform a planning-level GHG/energy analysis. The effort was undertaken to encourage State DOTs and MPOs to incorporate GHG and energy considerations in the transportation planning process and to use several new FHWA study tools and methods. The study approach and focus varied by organization based on their individual needs and interests, but each effort will improve the assessment and quantification of transportation-related GHG emissions for use in the transportation planning process.

Massachusetts DOT used the FHWA funding to analyze and quantify GHG emissions benefits from current activities and to estimate the impact of a set of potential future policies and strategies designed to help the State meet their GHG targets and goals. The project is using FHWA's Energy and Emissions Reduction Policy Analysis Tool.

The **Delaware Valley Regional Planning Council** is updating an evaluation of electric vehicle ownership. The Council is developing a spreadsheet tool to determine the changes in energy use and GHG emissions associated with different deployment scenarios of electric vehicles and compressed natural gas vehicles. Other transportation agencies around the country can use the scenarios to help reduce vehicle-related emissions and energy use.

The **East-West Gateway Council of Governments** is estimating GHG emissions from on-road vehicles at the regional and subregional scales and analyzing future emissions for multiple policy and land use scenarios. The project includes an analysis of the feasibility of corridor-level GHG analysis on the I-70 corridor. The review will increase the agency's capacity to integrate GHG considerations into decision-making processes and programs, advance the agency's transportation and sustainability goals, and serve as a case study for other regions.

The **Southern California Association of Governments** is undertaking an effort to advance methods of analyzing GHG emissions generated from multimodal transit trips, including first-last mile access and egress from transit stations. The findings will be used to prioritize the most effective transportation and land-use planning strategies for optimizing GHG reductions achieved from transit investments.

Building Partnerships to Improve Resilience

FHWA is partnering with State DOTs, MPOs, and Federal Land Management Agencies to pilot approaches for conducting vulnerability assessments of climate change and extreme weather for transportation infrastructure and to analyze options for adapting and improving resiliency.

Since 2010, FHWA has worked with 24 climate resilience pilots in two rounds. In the first round of pilot projects, FHWA funded five partnerships, including State DOTs, MPOs, and other agencies to test a draft framework for conducting vulnerability and risk assessments of transportation infrastructure given the projected impacts of climate change. FHWA used the experiences of these five pilots and other studies to update the draft framework. In 2012, FHWA formed 19 more partnerships with States and MPOs to use and build on the framework and to address previous gaps, such as evaluations of inland area impacts and actionable adaptation solutions.

FHWA has also worked with Federal, State, and local transportation agencies as part of four cooperative projects in the Gulf Coast, Northeast, New Mexico, and Southeast. Each area's approach differed and contributed significantly to the Agency's understanding of potential climate change impacts on its transportation assets and to the body of knowledge of the transportation community as a whole.

Central New Mexico Climate Change Scenario Planning Project

The transportation planning body for the Albuquerque, New Mexico region—the Mid Region Council of Governments (MRCOG)—embarked on a planning effort to test the impact of different transportation and land use scenarios on community goals. Federal grant funding and technical assistance enabled the region to integrate into the scenario planning an examination of strategies to reduce greenhouse gas emissions and improve resilience to climate change impacts, such as wildfires and flooding.

The goals of this Central New Mexico Climate Change Scenario Planning Project (CCSP) were to help the region improve sustainability through its metropolitan transportation plan and to demonstrate a process that could be replicated in other regions of the country (especially inland areas) for using scenario planning to respond to the challenges of climate change in conjunction with other community goals. The CCSP successfully integrated climate change consideration into the region's scenario planning process, and this analysis was then incorporated into the 2040 metropolitan transportation plan. The project enabled MRCOG to introduce the idea to stakeholders that some growth patterns are more sustainable and are more robust to climate change impacts than others are. In addition, the project helped make connections between local and Federal agencies with diverse missions and helped supply basic climate data for the Central New Mexico region that multiple sectors can now use. The CCSP also developed an integration plan that provides guidance to MRCOG in implementing several of the GHG reduction and climate resilience strategies discussed in the scenario-planning project.

Climate Resilience and Adaptation Tools and Resources

FHWA is working with Federal, State, and local partners by furnishing tools and resources to enable transportation agencies to increase the resilience of the transportation system to climate change. FHWA has designed an interactive online framework for use as a guide to assess the vulnerability of transportation assets to climate change and extreme weather events. The results of recent FHWA pilot and research projects informed this Virtual Framework for Vulnerability assessment. Each step of the framework includes case studies, videos, and other associated resources. The Virtual Framework, which includes several vulnerability assessment tools, can be found here: http://www.fhwa.dot.gov/environment/climate_change/adaptation/adaptation_framework/.

- **Climate Data Processing Tool (CMIP)**—CMIP processes data sets that are publicly available, large, and complicated into local temperature and precipitation projections tailored to transportation practitioners.
- **Sensitivity Matrix**—This spreadsheet tool documents the sensitivity of roads, bridges, airports, ports, pipelines, and rail to 11 climate impacts.
- **Vulnerability Assessment Scoring Tool (VAST)**—VAST is a spreadsheet tool that guides the user through conducting a quantitative, indicator-based vulnerability screen. The tool is intended for agencies assessing the vulnerability of their transportation system components to climate stressors.

Gulf Coast Study

The groundbreaking DOT Gulf Coast Study produced tools and lessons learned that transportation agencies across the country are using to assess vulnerabilities and build resilience to climate change. Phase 1 of the study, completed in 2008, examined the impacts of climate change on transportation infrastructure at a regional scale. Phase 2, completed in early 2015, focused on the Mobile, Alabama region with the goal of enhancing regional decision makers' ability to understand potential impacts on specific critical components of infrastructure and to gauge adaptation options. In Mobile, DOT assessed the vulnerability of the most critical transportation assets to climate change impacts and then cultivated risk management tools to help transportation system planners, owners, and operators determine which systems and assets to protect and how. The methods and tools developed under Phase 2 are intended to be replicable in other regions throughout the country. Reports include (1) synthesis of lessons learned and methods applied, (2) criticality assessment, (3) climate projections and sensitivity assessment, (4) vulnerability assessment, and (5) engineering assessment of adaptation options. All of the reports can be found here: http://www.fhwa.dot.gov/environment/climate_change/adaptation/ongoing_and_current_research/gulf_coast_study/index.cfm.

Transit System Performance

Basic goals all transit operators share include minimizing travel times, making efficient use of vehicle capacity, and providing reliable performance. The Federal Transit Administration (FTA) collects data on average speed, how full the vehicles are on average (utilization), and how often they break down (mean distance between failures) to characterize how well transit service meets these goals. These data are reported here; safety data are reported in Chapter 4.

Customer satisfaction issues that are more subjective, such as how easy accessing transit service is (accessibility) and how well that service meets a community's needs, are harder to measure. Data from the FHWA 2009 National Household Travel Survey, reported here, provide some insights, but are not available on an annual basis and so do not support time series analysis.

FTA Livable Communities Outcomes and Performance Measures	
Modal Network	***Demand Response***
1. Increased access to convenient and affordable transportation choices	▪ Increase the number of transit boardings reported by urbanized area transit providers from 10.0 billion in 2011 to 10.5 billion in 2016. ▪ Increase the number of transit boardings reported by rural area transit providers from 141 million in 2011 to 160 million in 2016. ▪ Increase transit's market share among commuters to work in at least 10 of the top 50 urbanized areas by population, as compared to 2010 market-share levels.
2. Improved access to transportation for people with disabilities and older adults	▪ Increase the number of key transit rail stations verified as accessible and fully compliant from 522 in 2010 to 560 in 2016.

The following analysis presents data on average operating speeds, average number of passengers per vehicle, average percentage of seats occupied per vehicle, average distance traveled per vehicle, and mean distance between failures for vehicles. Average speed, seats occupied, and distance between failures address efficiency and customer service issues; passengers per vehicle and miles per vehicle are primarily effectiveness and efficiency measures, respectively. Financial efficiency metrics, including operating expenditures per revenue mile or passenger mile, are discussed in Chapter 6.

The National Transit Database (NTD) includes urban data reported by mode and type of service. As of December 2010, NTD contained data for 16 modes. Beginning in January 2011, new modes were added to the NTD urban data, including

- streetcar rail – previously reported as light rail,
- hybrid rail – previously reported as light rail and commuter rail,
- commuter bus – previously reported as motor bus,

- bus rapid transit – previously reported as motor bus, and
- demand-response taxi – previously reported as demand response.

Data from NTD are presented for each new mode for analyses specific to 2012. For NTD time series analysis, however, streetcar rail and hybrid rail are included as light rail, commuter bus and bus rapid transit as fixed-route bus, and demand response-taxi as demand response.

Average Operating (Passenger-Carrying) Speeds

Average vehicle operating speed is an approximate measure of the speed transit riders experience; it is not a measure of the operating speed of transit vehicles between stops. More specifically, average operating speed is a measure of the speed passengers experience from the time they enter a transit vehicle to the time they exit it, including dwell times at stops. It does not include the time passengers spend waiting or transferring. Average vehicle operating speed is calculated for each mode by dividing annual vehicle revenue miles by annual vehicle revenue hours for each agency in each mode, as reported to NTD. When an agency contracts with a service provider or provides the service directly, the speeds for each service within a mode are calculated and weighted separately. *Exhibit 5-19* presents the results of these average speed calculations.

Exhibit 5-19 Average Speeds for Passenger-Carrying Transit Modes, 2012[1]

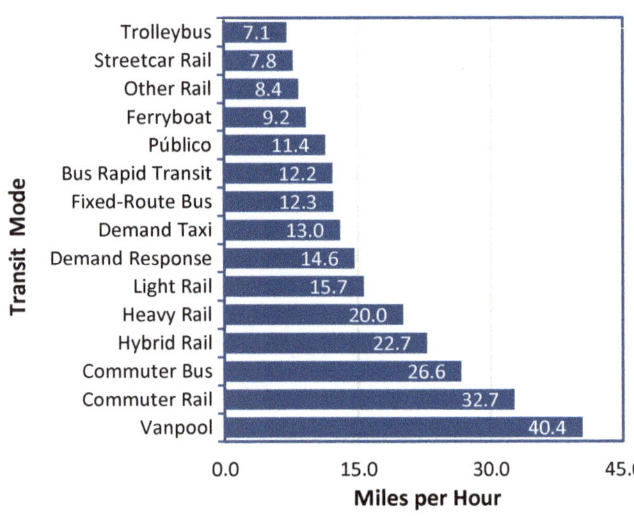

[1] The "other rail" transit mode includes Alaska railroad, monorail/automated guideway, cable car, and inclined plane.

Source: National Transit Database.

The number and distance between stops and the time required for boarding and alighting of passengers strongly influence the average speed of a transit mode. Fixed-route bus service, which typically makes frequent stops, has a relatively low average speed. In contrast, commuter rail has sustained high speeds between infrequent stops, and thus a relatively high average speed. Vanpools also travel at high speeds, usually with only a few stops at each end of the route. Modes using exclusive guideway can offer more rapid travel time than similar modes that do not. Heavy rail, which travels exclusively on dedicated guideway, has a higher average speed than streetcar, which often shares its guideway with mixed traffic. These average speeds have not changed significantly over the past decade.

Vehicle Use

Vehicle Occupancy

Exhibit 5-20 shows vehicle occupancy by mode for selected years from 2002 to 2012. Vehicle occupancy is calculated by dividing passenger miles traveled (PMT) by vehicle revenue miles (VRMs), resulting in the average passenger load in a transit vehicle. Vehicle occupancy has changed little between 2002 and 2012, indicating sustained ridership levels across all types of transit. In 2010–2012, average passenger load for all major transit modes increased, especially heavy rail (8.7 percent) and light rail (6.3 percent), which indicates increased demand in large urbanized areas.

Exhibit 5-20 Unadjusted Vehicle Occupancy: Passenger Miles per Vehicle Revenue Mile, 2002–2012

Mode	2002	2004	2006	2008	2010	2012
Rail						
Heavy Rail	22.6	23.0	23.2	25.7	25.3	27.5
Commuter Rail	36.7	36.1	36.1	35.7	34.2	35.0
Light Rail[1]	23.9	23.7	25.5	24.1	23.7	25.2
Other Rail[2]	8.4	10.4	8.4	9.3	10.7	8.1
Nonrail						
Fixed-Route Bus[3]	10.5	10.0	10.8	10.8	10.7	11.2
Demand Response/ Demand Taxi	1.2	1.3	1.3	1.2	1.2	1.2
Ferryboat	112.1	119.5	130.7	118.1	119.3	125.2
Trolleybus	14.1	13.3	13.9	14.3	13.6	14.3
Vanpool	6.4	5.9	6.3	6.3	6.0	6.1
Other Nonrail[4]	7.9	5.8	7.8	8.2	7.4	10.6

[1] Includes light rail, streetcar rail, and hybrid rail.

[2] Includes Alaska railroad, monorail/automated guideway, cable car, and inclined plane.

[3] Includes bus, commuter bus, and bus rapid transit.

[4] Includes Público.

Source: National Transit Database.

An important metric of vehicle occupancy is weighted average seating capacity utilization. This average is calculated by dividing passenger load by the average number of seats in the vehicle (or passenger car for rail modes). The weighting factor is the number of active vehicles in the fleet. The weighted average seating capacity for some modes are vanpool, 10; heavy rail, 53; light rail, 65; trolleybus, 48; ferryboat, 473; commuter rail, 113; fixed-route bus, 39; and demand response, 11.

As shown in *Exhibit 5-21*, the average seating capacity utilization ranges from 11.2 percent for demand response to 59.9 percent for vanpools. At first glance, the data seem to indicate excess seating capacity for all modes. Several factors, however, explain these apparent low utilization rates. For example, the low utilization rate for fixed-route bus, which operates in large and small urbanized areas, can be explained partially by low average passenger loads in urbanized areas with low ridership. For heavy rail and light rail, factors could include (1) high passenger demand in one direction, and small or very small demand in the opposite direction during peak periods;

and (2) sharp drops in loads beyond segments of high demand, with limited room for short turns, and other factors.

Exhibit 5-21 Average Seat Occupancy Calculations for Passenger-Carrying Transit Modes, 2012[1,2]

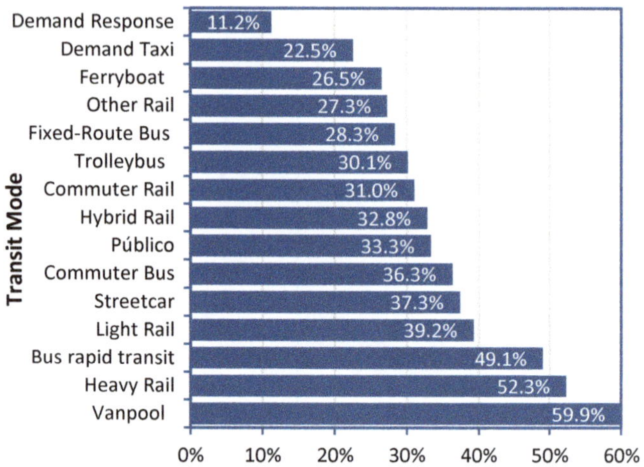

[1] *The "other rail" transit mode includes Alaska railroad, monorail/automated guideway, cable car, and inclined plane.*

[2] *Some modes also have substantial standing capacity that is not considered here, but which can allow the measure of the percentage of seats occupied to exceed 100 percent for a full vehicle.*

Source: National Transit Database; does not include agencies who qualified and opted to use the small systems waiver of the National Transit Database.

Vehicles also tend to be relatively empty at the beginning and ends of their routes. For many commuter routes, a vehicle that is crush-loaded (i.e., filled to maximum capacity) on part of the trip ultimately might achieve only an average occupancy of around 35 percent (as shown by analysis of the Washington Metropolitan Area Transit Authority peak-period data). Revenue miles per active vehicle (service use), defined as average distance traveled per vehicle in service, can be measured by the ratio of VRMs per active vehicles in the fleet. *Exhibit 5-22* provides vehicle service use by mode for selected years from 2002 to 2012. Heavy rail, generally offering long hours of frequent service, had the highest vehicle use during this period. Vehicle service use for vanpool and demand response shows an increasing trend. Vehicle service use for other nonrail modes appears to be relatively stable over the past few years with no apparent trends in either direction.

Exhibit 5-22 Vehicle Service Utilization: Vehicle Revenue Miles per Active Vehicle by Mode[1]

							Average Annual Rate of Change
Mode[2]	**2002**	**2004**	**2006**	**2008**	**2010**	**2012**	**2012/2002**
Rail							
Heavy Rail	55.1	57.0	57.2	57.7	56.6	55.8	0.13%
Commuter Rail	43.9	41.1	43.0	45.5	45.1	43.7	-0.04%
Light Rail[3]	41.1	39.9	39.9	44.1	42.5	42.2	0.25%
Nonrail							
Fixed-Route Bus[4]	29.9	30.2	30.2	30.3	29.7	29.4	-0.14%
Demand Response[5]	21.1	20.1	21.7	21.3	20.0	20.5	-0.26%
Ferryboat	24.4	24.9	24.8	21.9	24.9	22.1	-1.02%
Vanpool	13.6	14.1	13.7	14.3	15.5	15.3	1.17%
Trolleybus	20.3	21.1	19.1	18.7	20.4	19.8	-0.28%

[1] 2012 data does not include agencies who qualified and opted to use the small systems waiver of the National Transit Database.

[2] Rail category does not include Alaska railroad, cable car, inclined plane, and automated guideway/monorail; nonrail category does not include Público and jitney.

[3] Includes light rail, hybrid rail, and streetcar rail.

[4] Includes bus, bus rapid transit, and commuter bus.

[5] Includes demand response and demand response taxi.

Source: National Transit Database.

Frequency and Reliability of Service

The frequency of transit service varies considerably according to location and time of day. Transit service is more frequent in urban areas and during rush hours—namely, where and when the demand for transit is highest. Studies have found that transit passengers consider the time spent waiting for a transit vehicle to be less well spent than the time spent traveling in a transit vehicle. The higher the degree of uncertainty in waiting times, the less attractive transit becomes as a means of transportation—and it will attract fewer users. To minimize this problem, many transit systems have implemented in recent years technologies to track vehicle location (automatic vehicle location systems) that, combined with accessed operating speeds, enable agencies to estimate the amount of time required for arrival of vehicles at stations and stops. This information is displayed in platforms and bus stops in real time. By knowing the waiting time, passengers are less frustrated and might be more willing to use transit.

Transit System Resiliency

Transit systems practice resiliency by operating through all but the worst weather on a daily basis. Most are instrumental in community emergency-response plans. Dispatchers and vehicle operators receive special training for these circumstances. All bus systems maintain a small fleet of spare buses that enables them to schedule maintenance activities while maintaining regular service levels. These "spare buses" also can be used to replace damaged vehicles on short notice. Rail systems have contingency plans for loss of key assets and most can muster local resources to operate bus bridges in emergencies. Operationally speaking, transit providers are some of the most resilient community institutions. Much transit infrastructure, however, has not yet been upgraded to address changing climactic patterns. FTA does not collect systematic data on these upgrades, but significant grant money has been made available for transit systems to upgrade their structures and guideways to be more resistant to extreme precipitation events, sea level rise, storm surge, heat waves, and other environmental stressors. This is particularly evident in the aftermath of Superstorm Sandy. Addressing such issues is a common use of FTA grant funds.

Exhibit 5-23 shows findings on wait times from the 2009 FHWA National Household Travel Survey, the most recent nationwide survey of this information. The survey found that 44.5 percent of passengers who ride transit wait 5 minutes or less and 73.2 percent wait 10 minutes or less. The survey also found that 8.0 percent of passengers wait 21 minutes or more. Several factors influence passenger wait times, including the frequency and reliability of service and passengers' awareness of timetables. These factors are also interrelated. For example, passengers might intentionally arrive earlier for service that is infrequent, compared with equally reliable services that are more frequent. Overall, waiting times of 5 minutes or less are clearly associated with good service that is either frequent, reliably provided according to a schedule, or both. Wait times of 5 to 10 minutes are most likely consistent with adequate levels of

Exhibit 5-23 Distribution of Passengers by Wait Time

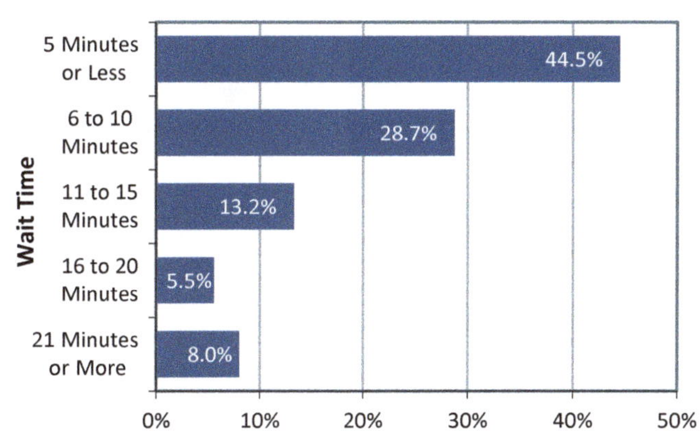

Source: National Household Travel Survey, FHWA, 2009.

service that are both reasonably frequent and generally reliable. Wait times of 21 minutes or more indicate that service is likely less frequent or less reliable.

Access to transit service varies by location. *Exhibit 5-24* shows the share of working-age residents that have access to transit in 100 selected metropolitan areas. The study evaluated census block groups and counted block groups with at least one transit stop within three-fourths of a mile of their population-weighted centroid as having access. Cities in the western United States tend to enjoy higher rates of coverage, while those in the Southeast tend to have a lower percentage of residents with access to transit.

Exhibit 5-24 Share of Working-Age Residents with Access to Transit, 100 Metropolitan Areas

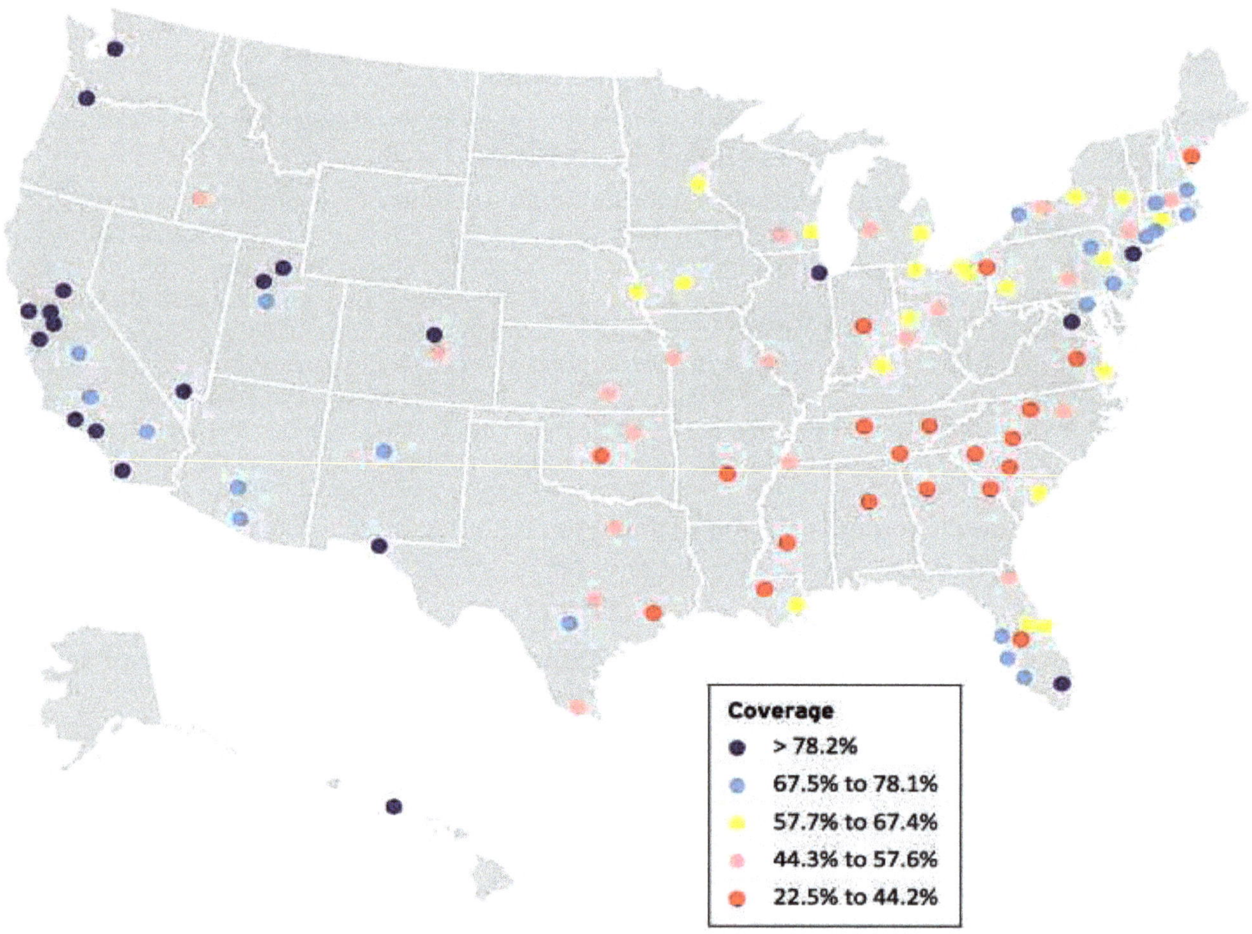

Source: Brookings Institution, Missed Opportunity: Transit and Jobs in Metropolitan America, May 2011 report citing Brookings Institution analysis of transit agency data and Nielson Pop-Facts 2010 data.

Of note is that accessibility to transit depends to some extent on geographical constraints such as mountains, deserts, and other natural obstacles. These constraints affect western cities more than they do eastern cities, yet western cities enjoy higher rates of accessibility.

Mean distance between failures is shown in *Exhibit 5-25*. The mean distance between failures is calculated by the ratio of VRMs per mechanical (major) and other (minor) failures. FTA does not collect data on delays due to guideway conditions, which would include congestion for roads and slow zones (due to system or rail problems) for track. Miles between failures for all modes combined decreased between 2004 and 2006 by 13 percent. Between 2006 and 2012, the ratio increased steadily to reach a level similar to 2004. The trend for fixed-route bus is nearly identical to all modes combined.

Exhibit 5-25 Mean Distance between Failures, Directly Operated Service, 2004–2012[1]

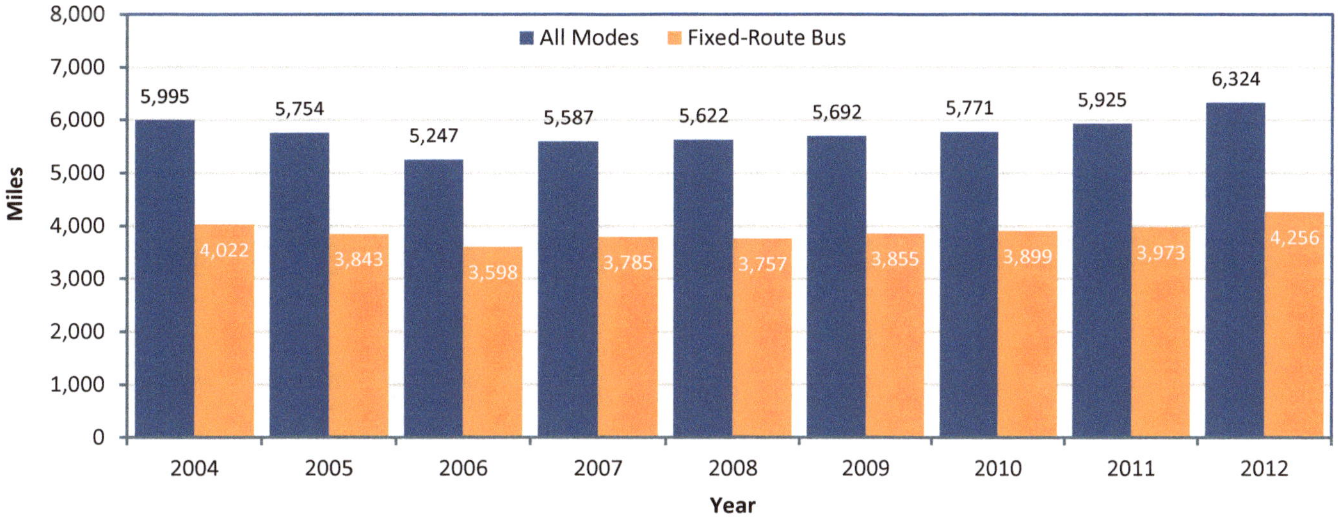

[1] Includes both major and minor failures. Does not include agencies who qualified and opted to use the small systems waiver of the National Transit Database. Years 2002 and 2003 not included due to questionable data.

Source: National Transit Database.

System Coverage: Urban Directional Route Miles

The extent of the Nation's transit system is measured in directional route miles, or simply "route miles." Route miles measure the distance covered by a transit route. Even though transit routes might use the same road or track, but in the opposite direction, they are counted separately. Data associated with route miles are not collected for demand-response and vanpool modes because these transit modes do not travel along specific predetermined routes. Route mile data are also not collected for jitney services because these transit modes often have highly variable route structures.

Exhibit 5-26 shows directional route miles by mode over the past 10 years. Growth in both rail (27.3 percent) and nonrail (6.2 percent) route miles is evident over this period. The average 6.0-percent rate of annual growth for light rail clearly outpaces the rate of growth for all other modes due to the large increase in new systems in the past 10 years.

System Capacity

Exhibit 5-27 provides reported VRMs for both rail and nonrail modes. These numbers are interesting because they show the actual number of miles each mode travels in revenue service. VRMs that fixed-route bus services and rail services provide both show consistent growth, with light-rail and vanpool miles growing somewhat faster than the other modes. Overall, the number of VRMs has increased by 15.6 percent since 2002, with an average annual rate of change of 1.4 percent.

Exhibit 5-26 Transit Urban Directional Route Miles, 2002–2012[1]

							Average Annual Rate of Change
Transit Mode	**2002**	**2004**	**2006**	**2008**	**2010**	**2012**	**2012/2002**
Rail	**9,484**	**9,782**	**10,865**	**11,270**	**11,735**	**12,072**	**+2.4%**
Commuter Rail[2]	6,923	6,968	7,930	8,219	8,590	8,682	+2.3%
Heavy Rail	1,572	1,597	1,623	1,623	1,617	1,622	+0.3%
Light Rail[3]	960	1,187	1,280	1,397	1,497	1,724	+6.0%
Other Rail[4]	30	30	31	30	30	44	+4.1%
Nonrail[5]	**225,820**	**216,619**	**223,489**	**212,801**	**237,580**	**239,957**	**+0.6%**
Fixed-Route Bus[6]	224,838	215,571	222,445	211,664	236,434	238,806	+0.6%
Ferryboat	513	623	620	682	690	695	+3.1%
Trolleybus	468	425	424	456	456	456	-0.3%
Total	**235,304**	**226,401**	**234,354**	**224,071**	**249,314**	**252,029**	**+1.5%**
Percent Nonrail	96.0%	95.7%	95.4%	95.0%	95.3%	95.2%	

[1] 2012 data does not include agencies who qualified and opted to use the small systems waiver of the National Transit Database.

[2] Includes Alaska railroad.

[3] Includes light rail, streetcar rail, and hybrid rail.

[4] Includes monorail/automated guideway, inclined plane, and cable car.

[5] Excludes jitney, Público, and vanpool

[6] Includes bus, bus rapid transit, and commuter bus.

Source: National Transit Database.

Exhibit 5-27 Rail and Nonrail Vehicle Revenue Miles, 2002–2012

	Miles (Millions)						Average Annual Rate of Change
Transit Mode	**2002**	**2004**	**2006**	**2008**	**2010**	**2012**	**2012/2002**
Rail	**925**	**963**	**997**	**1,054**	**1,056**	**1,061**	**1.4%**
Heavy Rail	603	625	634	655	647	638	0.6%
Commuter Rail	259	269	287	309	315	318	2.1%
Light Rail[1]	60	67	73	86	92	99	5.1%
Other Rail[2]	3	2	3	3	2	6	7.4%
Nonrail	**2,502**	**2,586**	**2,674**	**2,841**	**2,863**	**2,900**	**1.5%**
Fixed-Route Bus[3]	1,864	1,885	1,910	1,956	1,917	1,892	0.1%
Demand Response/Demand Taxi	525	561	607	688	718	759	3.8%
Vanpool	71	78	110	157	181	207	11.3%
Ferryboat	3	3	3	3	3	3	0.6%
Trolleybus	13	13	12	11	12	11	-1.4%
Other Nonrail[4]	26	46	32	25	32	27	0.5%
Total	**3,427**	**3,549**	**3,671**	**3,895**	**3,920**	**3,960**	**1.4%**

[1] Includes light rail, streetcar rail, and hybrid rail.

[2] Includes Alaska railroad, monorail/automated guideway, cable car, and inclined plane.

[3] Includes bus, commuter bus, and bus rapid transit.

[4] Includes Público.

Source: National Transit Database.

Transit system capacity, particularly in cross-modal comparisons, is typically measured by capacity-equivalent VRMs. This parameter measures the distances transit vehicles travel in revenue service and adjusts them by the passenger-carrying capacity of each transit vehicle type, with the average carrying capacity of fixed-route bus vehicles representing the baseline. To calculate capacity-equivalent VRMs, the number of revenue miles for a vehicle is multiplied by the bus-equivalent capacity of that vehicle. Thus, a heavy rail car that seats 2.5 times more people than a full-size bus provides 2.5 capacity-equivalent miles for each revenue mile it travels.

Exhibit 5-28 shows the 2012 capacity-equivalent factors for each mode. Unadjusted VRMs for each mode are multiplied by a capacity-equivalent factor to calculate capacity-equivalent VRMs. These factors are equal to the average full-seating and full-standing capacities of vehicles in active service for each transit mode divided by the average full-seating and full-standing capacities of all motor bus vehicles in active service. The average capacity of the national motor bus fleet changes slightly from year to year as the proportion of large, articulated, and small buses varies. The average capacity of the bus fleet in 2012 was 39 seated and 23 standing, or 62 riders.

Exhibit 5-28 Capacity-Equivalent Factors by Mode[1]

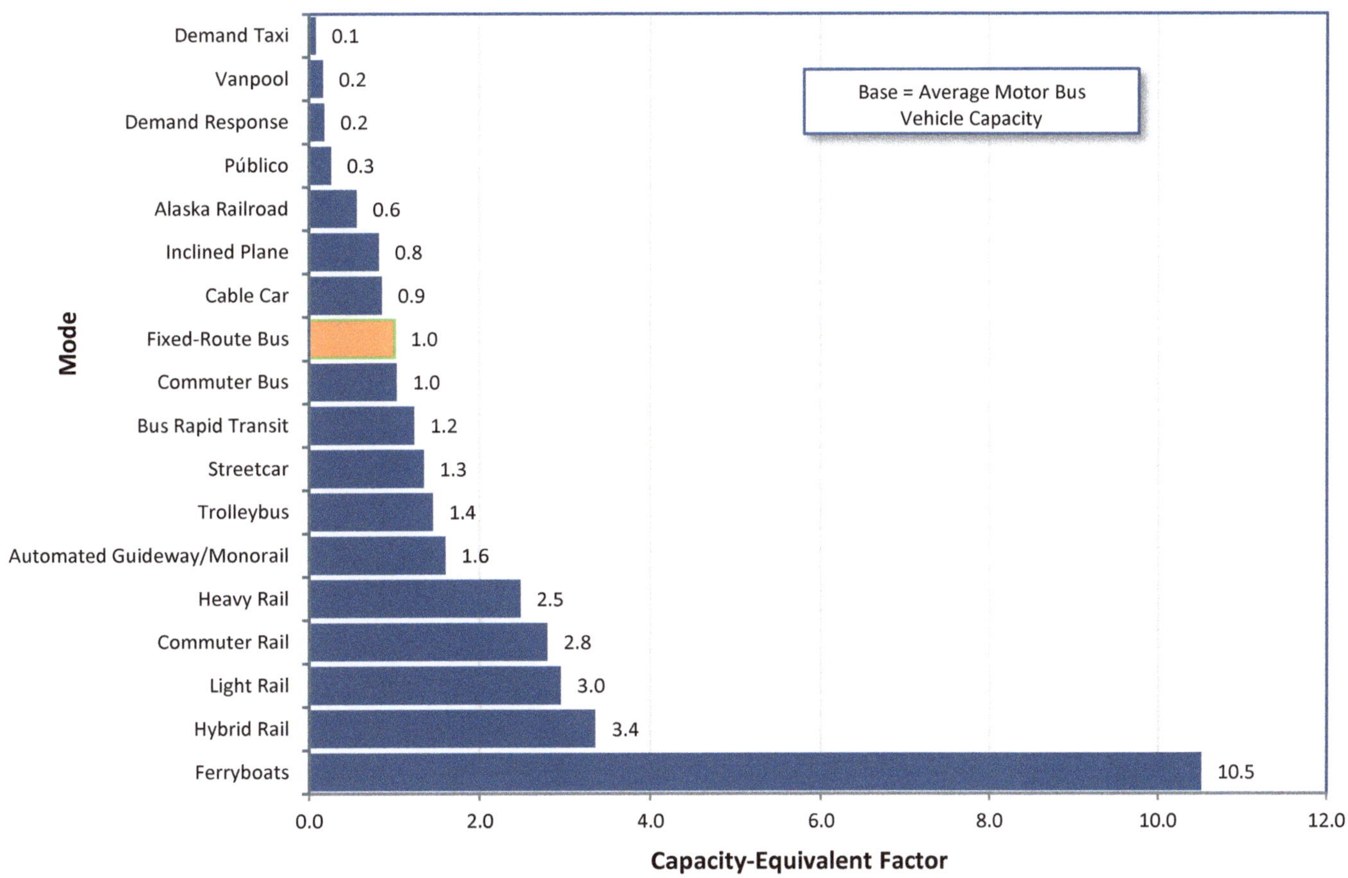

[1] Data do not include agencies who qualified and opted to use the small systems waiver of the National Transit Database.

Source: National Transit Database.

Exhibit 5-29 shows total capacity-equivalent VRMs. Vanpools show the most rapid expansion in capacity-equivalent VRMs from 2002 to 2012, followed by light rail, demand response, and commuter rail. Annual VRMs for monorail/automated guideway more than doubled, resulting in an increase in capacity-equivalent VRMs for the other rail category. Total capacity-equivalent revenue miles increased from 4,311 million in 2002 to 5,003 million in 2012, an increase of 16 percent.

Exhibit 5-29 Capacity-Equivalent Vehicle Revenue Miles, 2002–2012[1]

							Average Annual Rate of Change
Transit Mode	**2002**	**2004**	**2006**	**2008**	**2010**	**2012**	**2012/2002**
Rail	**2,274**	**2,413**	**2,681**	**2,799**	**2,714**	**2,728**	**1.8%**
Heavy Rail	1,469	1,546	1,648	1,621	1,599	1,580	0.7%
Commuter Rail	652	685	832	940	860	888	3.1%
Light Rail[2]	150	179	197	235	252	252	5.3%
Other Rail[3]	3	3	4	3	3	8	9.4%
Nonrail	**2,037**	**2,064**	**2,118**	**2,152**	**2,131**	**2,275**	**1.1%**
Fixed-Route Bus[4]	1,864	1,885	1,910	1,956	1,917	2,052	1.0%
Demand Response/ Demand Taxi	100	101	121	115	124	132	2.8%
Vanpool	15	15	22	27	30	34	8.7%
Ferryboat	32	32	37	32	35	34	0.5%
Trolleybus	20	20	19	16	17	16	-1.7%
Other Nonrail[5]	7	12	10	6	8	7	0.3%
Total	**4,311**	**4,478**	**4,800**	**4,951**	**4,845**	**5,003**	**1.5%**

[1] 2012 data does not include agencies who qualified and opted to use the small systems waiver of the National Transit Database.
[2] Includes light rail, streetcar rail, and hybrid rail.
[3] Includes Alaska railroad, monorail/automated guideway, cable car, and inclined plane.
[4] Includes bus, commuter bus, and bus rapid transit.
[5] Includes Público.

Source: National Transit Database.

Ridership

The two primary measures of transit ridership are unlinked passenger trips and passenger miles traveled (PMT). An unlinked passenger trip, sometimes called a boarding, is defined as a journey on one transit vehicle. PMT is calculated based on unlinked passenger trips and estimates of average trip length. Either measure provides an appropriate time series because average trip lengths, by mode, have not changed substantially over time. Comparisons across modes, however, might differ substantially, depending on which measure is used due to large differences in the average trip length for the various modes.

Exhibits 5-30 and *5-31* show the distribution of unlinked passenger trips and PMT by mode. In 2012, urban transit systems provided 10.4 billion unlinked trips and 55.2 billion PMT across all modes. Heavy rail and fixed-route bus modes continue to be the largest segments of both measures. Commuter rail supports relatively more PMT due to its greater average trip length (23.7 miles compared to 4.0 for fixed-route bus, 4.7 for heavy rail, and 5.0 for light rail).

Exhibit 5-30 Unlinked Passenger Trips (Total in Millions and Percent of Total) by Mode, 2012[1]

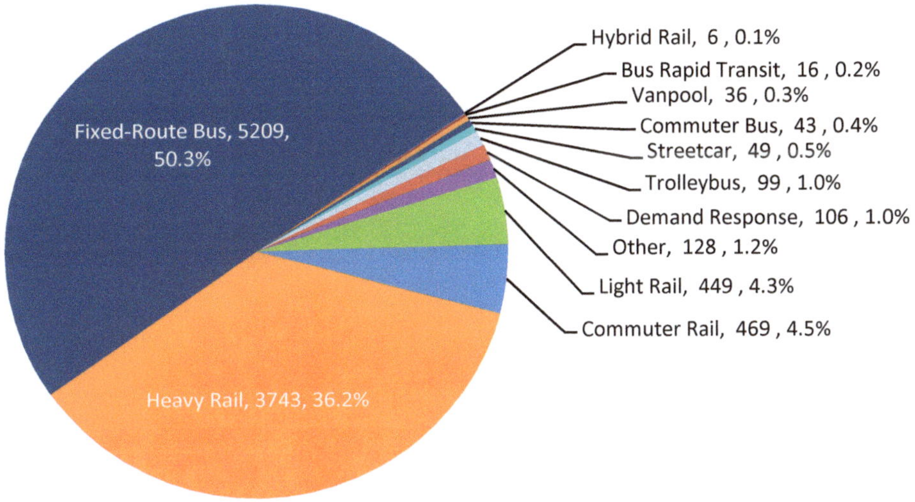

[1] "Other" includes Alaska railroad, cable car, ferryboat, inclined plane, monorail/automated guideway, and Público; "demand response" includes demand response and demand response taxi.

Source: National Transit Database.

Exhibit 5-31 Passenger Miles Traveled (Total in Billions and Percent of Total) by Mode, 2012[1,2]

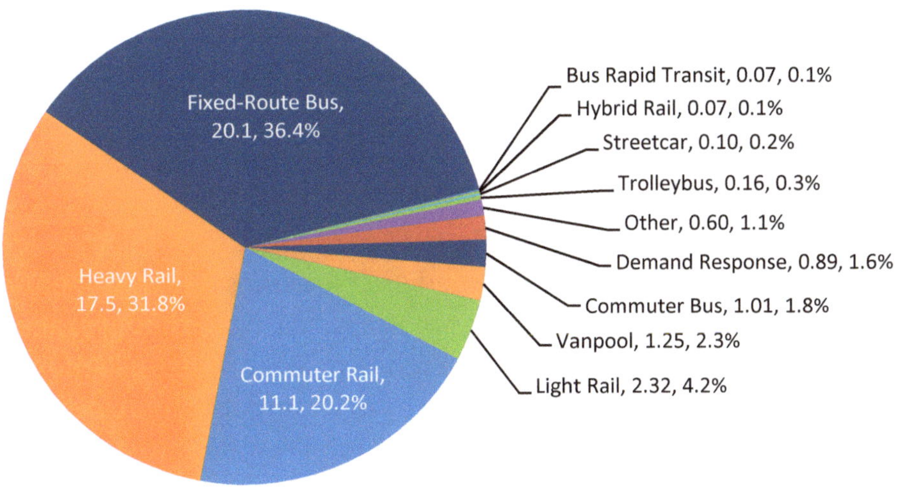

[1] 2012 data does not include agencies who qualified and opted to use the small systems waiver of the National Transit Database.

[2] "Other" includes Alaska railroad, cable car, ferryboat, inclined plane, monorail/automated guideway, and Público; "demand response" includes demand response and demand response taxi.

Source: National Transit Database.

Exhibit 5-32 provides total PMT for selected years between 2002 and 2012, showing steady growth in all major modes. Demand response, light-rail, and vanpool modes grew at the highest rates. Growth in demand response (up 3.1 percent per year) might be a response to demand from the growing number of elderly citizens. Light rail (up 5.7 percent per year) enjoyed increased capacity during this period due to expansions and addition of new systems. The rapidly increasing popularity of vanpools (up 10.7 percent per year), particularly the surge between 2006 and 2008 (up 20 percent per year), can be partially attributed to rising gas prices—regular gasoline sold for more than $4 per gallon in July of 2008. FTA has also encouraged vanpool reporting during this period, successfully enrolling numerous new vanpool systems to report to NTD.

Exhibit 5-32 Transit Urban Passenger Miles, 2002–2012[1]

Transit Mode	2002	2004	2006	2008	2010	2012	Average Annual Rate of Change 2012/2002
Rail	**24,617**	**25,667**	**26,972**	**29,989**	**29,380**	**31,176**	**2.4%**
Heavy Rail	13,663	14,354	14,721	16,850	16,407	17,516	2.5%
Commuter Rail	9,500	9,715	10,359	11,032	10,774	11,121	1.6%
Light Rail[2]	1,432	1,576	1,866	2,081	2,173	2,489	5.7%
Other Rail[3]	22	22	25	26	26	50	8.6%
Nonrail	**21,328**	**20,879**	**22,533**	**23,723**	**23,247**	**23,993**	**1.2%**
Fixed-Route Bus[4]	19,527	18,921	20,390	21,198	20,570	21,142	0.8%
Demand Response[5]	651	704	753	844	874	887	3.1%
Vanpool	455	459	689	992	1,087	1,255	10.7%
Ferryboat	301	357	360	390	389	402	2.9%
Trolleybus	188	173	164	161	159	162	-1.5%
Other Nonrail[6]	206	265	176	138	169	145	-3.4%
Total	**45,945**	**46,546**	**49,504**	**53,712**	**52,627**	**55,169**	**1.7%**
Percent Rail	53.6%	55.1%	54.5%	55.8%	55.8%	56.5%	

[1] 2012 data does not include agencies who qualified and opted to use the small systems waiver of the National Transit Database.

[2] Includes light rail, streetcar rail, and hybrid rail.

[3] Includes Alaska railroad, monorail/automated guideway, cable car, and inclined plane.

[4] Includes bus, commuter bus, and bus rapid transit.

[5] Includes demand response and demand response taxi.

[6] Público.

Source: National Transit Database.

Exhibit 5-33 depicts average passenger trip length (defined as passenger miles traveled per unlinked passenger trips) versus revenue speed, defined as train miles per train hours for rail, and vehicle revenue miles per vehicle revenue hours for nonrail modes. Note that average passenger trip length is the average distance traveled of one unlinked trip. Most riders use more than one mode to commute from origin to destination (linked trip), which might include other transit modes, car, or other modes such as bicycle, walking, etc. Therefore, the average trip length of an individual mode as depicted in *Exhibit 5-33* is the lower bound of the total average distance traveled. The total trip distance is a function of a linked trip factor that varies from mode to mode and is not available in NTD.

Demand-response and vanpool systems are modes with linked factors close to 1, that is, the average trip length of one unlinked trip should be close to the total length of the linked trip. This is because vanpools and demand response are "by-demand" modes, and the routes can be set up to optimize the proximity from the origin and destination.

Exhibit 5-33 Transit Urban Average Unlinked Passenger Trip Length vs. Average Revenue Speed for Selected Modes

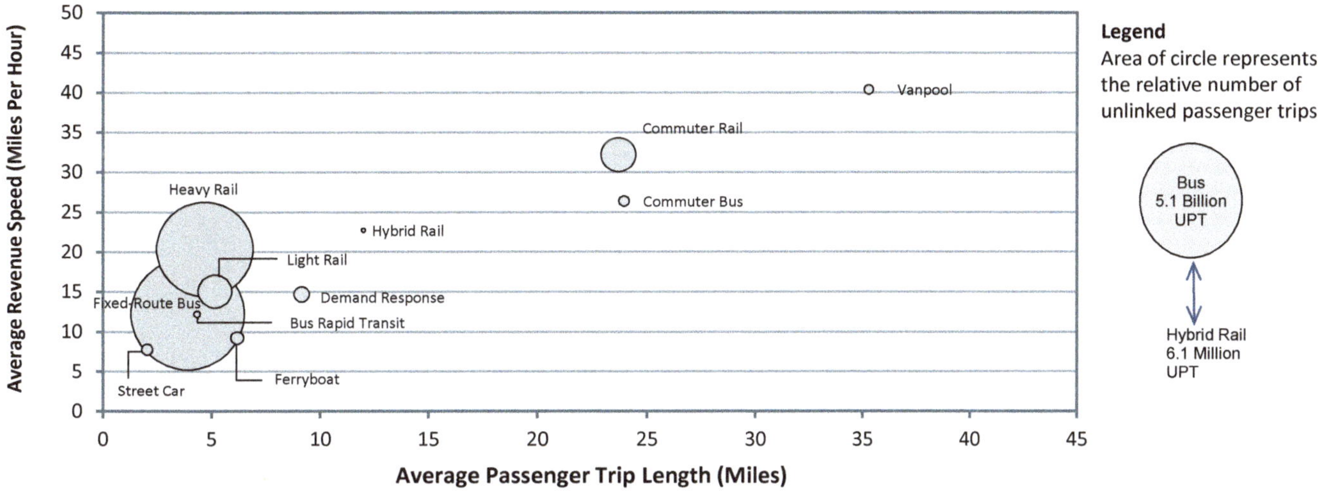

Source: National Transit Database.

Commuter bus and commuter rail, on the other hand, are fixed-route modes, and a high percentage of commuters require other modes to reach their final destinations. Additionally, commuter bus and commuter rail are not as fast as vanpools due to more frequent stops near areas of attraction and generation of trips, among other factors.

Several modes (heavy rail, light rail, fixed-route bus, bus rapid transit, streetcar, and ferryboat) cluster within a narrow range for average passenger trip length (less than 5 miles) and a wider range for average revenue speed (10 to 20 miles per hour). Heavy rail and light rail have higher average speed than nonrail modes for operating in exclusive right-of-ways. The modes in this cluster serve areas with high population density and significant average number of boarding and alighting per station or stop, which results in shorter average trip lengths than modes with a commuter orientation. These modes should have similar link factors but smaller than commuter rail and commuter bus.

Operating Characteristics of Hybrid Rail

Hybrid rail, introduced in 2011, was reported prior to 2011 as commuter rail and light rail. Hybrid rail has quite different operating characteristics than commuter rail and light rail. It has higher average station density (stations per track mileage) than commuter rail and a lower average station density than light rail. This results in revenue speeds that are lower than commuter rail and higher than light rail. Hybrid rail has smaller average peak-to-base ratio (number of trains during peak service per number of trains during midday service) than commuter rail, which indicates higher demand at off-peak hours.

Exhibit 5-34 shows the complex relationship among an index of rolling 12-month transit ridership, gasoline prices, and employment rates.

On the most basic level, the effectiveness of transit operations can be gauged by the demand for transit services. People choose to use transit if they perceive that it meets their needs as well as, or better than, the alternatives. These choices occur in an economic context in which the need for transportation and the cost of that transportation are constantly changing due to factors that have very little to do with the characteristics of transit.

Exhibit 5-34 Transit Ridership vs. Employment, 2006–2012

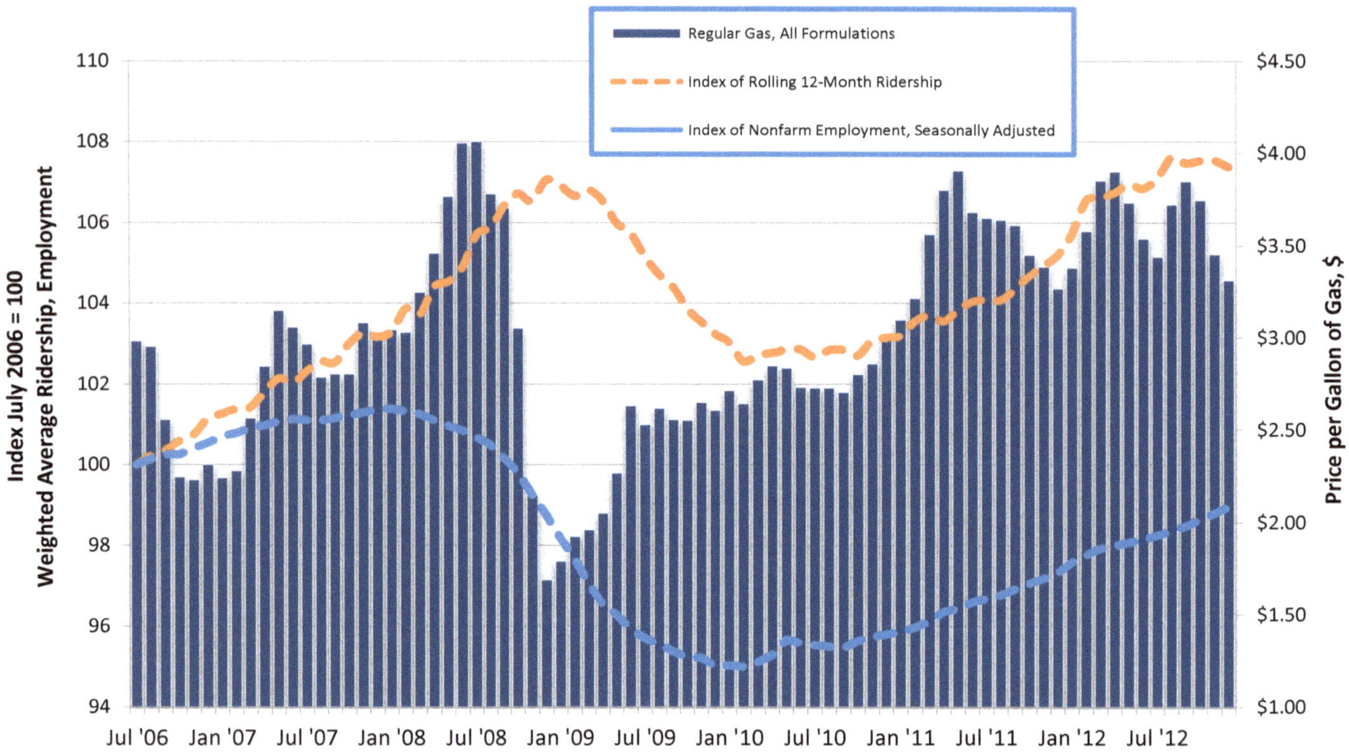

Source: National Transit Database, U.S. Energy Information Administration's Gas Pump Data History, and Bureau of Labor Statistics' Employment Data.

The relationship between employment and transit is well established. According to the May 2007 American Public Transportation Association report, *A Profile of Public Transportation Passenger Demographics and Travel Characteristics Reported in On-Board Surveys*: "Commuting to work is the most common reason a person rides public transportation, accounting for 59.2 percent of all transit trips reported in on-board surveys." The corollary of this statement is that transit ridership should decrease during periods of high unemployment. In fact, until 2008, the correlation between transit ridership and employment levels was so strong that FTA corrected ridership to account for employment levels. From early 2007 through summer of 2008, however, transit ridership increased in the absence of employment growth. This anomaly could be due to dramatic increases in the price of gas during this period; gas prices increased in average from around $2.35 per gallon to more than $4.00 per gallon. Since the start of 2009, gas prices have eased and then grown again gradually, but without influencing transit ridership in the same way (perhaps due to a concurrent decline in employment). Since 2010, ridership has once again been tracking employment levels but has retained some of its 2007–2008 gains. In December of 2012, transit ridership was up 7 percent over its July 2006 level while employment was still down 1 percent from its July 2006 level.

chapter 6

Finance

Highway Finance

This chapter presents data and analyses for finance trends for highways and transit across all levels of government and sources of funding. The revenue sources for investments in highways and bridges are discussed first, followed by the details of total highway expenditures and, more specifically, capital outlays. A separate section presents data on transit system funding, highlighting trends in revenues, capital, and operating expenditures.

The classification of the revenue and expenditure items in this section is based on definitions contained in *A Guide to Reporting Highway Statistics* (http://www.fhwa.dot.gov/policyinformation/hss/guide/guide.pdf), which is the instructional manual for States providing financial data for the *Highway Statistics* (http://www.fhwa.dot.gov/policyinformation/statistics.cfm) publication.

Revenue Sources for Highways

The revenue collected in 2012 from all levels of government for highways and bridges was $216.6 billion, as illustrated in *Exhibit 6-1*. Of the total revenues generated, the Federal government contributed $42.8 billion; State governments, $106.3 billion; and local governments, $67.5 billion.

These revenues were raised from user charges (motor-fuel tax, motor-vehicle taxes and fees, and tolls) and several other sources (General Fund appropriations, other taxes, investment income, and debt financing). In 2012, the overall split between user charges and other sources was about even, at 48.6 percent versus 51.4 percent. The reliance on different sources, however, differs significantly by level of government.

User charges, in particular motor-fuel taxes, account for most of the Federal revenues raised for highways—80 percent in 2012. User charges also account for most of the revenues that State governments raise. In 2012, State governments raised $106.3 billion of highway funding, of which $66.7 billion (about two-thirds), derived from State-imposed fees on highway users. Funding from other sources ($39.5 billion) included $12.4 billion from bond sale proceeds. In contrast, the revenues that local governments raise for highways derive mainly from sources other than user charges. This difference is partly because many States prohibit local governments from imposing taxes on motor fuel or motor vehicles and, where allowed, these taxes are often capped at low rates. The source on which local governments rely most heavily is general fund appropriations, which in 2012 accounted for nearly half, or $31.4 billion, of the total $67.5 billion in revenue raised. The next largest sources were property taxes and bond sale proceeds, at $10.3 billion and $8.9 billion. User charges generated only $4.6 billion of revenue.

Exhibit 6-1 Government Revenue Sources for Highways, 2012

Source	Federal	State	Local	Total	Percent
	Highway Revenue, Billions of Dollars				
User Charges[1]					
Motor-Fuel Taxes	$28.1	$32.8	$1.0	$61.9	28.6%
Motor-Vehicle Taxes and Fees	$5.8	$22.1	$1.8	$29.7	13.7%
Tolls	$0.0	$11.8	$1.8	$13.5	6.2%
Subtotal	**$33.8**	**$66.7**	**$4.6**	**$105.2**	**48.6%**
Other					
Property Taxes and Assessments	$0.0	$0.0	$10.3	$10.3	4.8%
General Fund Appropriations[2]	$6.1	$6.7	$31.4	$44.1	20.4%
Other Taxes and Fees	$0.4	$8.9	$5.5	$14.8	6.9%
Investment Income and Other Receipts[3]	$2.5	$11.6	$6.8	$20.9	9.6%
Bond Issue Proceeds	$0.0	$12.4	$8.9	$21.3	9.8%
Subtotal	**$9.0**	**$39.5**	**$62.9**	**$111.4**	**51.4%**
Total Revenues	**$42.8**	**$106.3**	**$67.5**	**$216.6**	**100.0%**
Funds Drawn From (or Placed in) Reserves[3]	$4.6	($0.4)	$0.6	$4.8	2.2%
Total Expenditures Funded During 2012	**$47.4**	**$105.8**	**$68.1**	**$221.3**	**102.2%**

[1] Amounts shown represent only the portion of user charges that are used to fund highway spending; a portion of the revenue generated by motor-fuel taxes, motor-vehicle taxes and fees, and tolls is used for mass transit and other nonhighway purposes. Gross receipts generated by user charges totaled $133.6 billion in 2012.

[2] The $6.1 billion shown for Federal reflects $3.0 billion of the funding authorized for use on highways by the Recovery Act. The remainder supported expenditures by the FHWA and other Federal agencies that were not paid for from the Highway Trust Fund.

[3] The $2.5 billion figure shown for Federal includes $2.4 billion transferred from the balance of the Leaking Underground Storage Tank Fund to the Highway Account of the Highway Trust Fund.

Sources: Highway Statistics 2012, Table HF-10, and unpublished FHWA data.

Q&A **Do the user charges reflected in *Exhibit 6-1* include all revenues generated by motor-fuel taxes, motor-vehicle taxes and fees, and tolls in 2012?**

No. The $105.2 billion identified as highway-user charges in *Exhibit 6-1* represents only 78.7 percent of total highway-user revenue, defined as all revenue generated by motor-fuel taxes, motor-vehicle taxes, and tolls. *Exhibit 6-2* shows that combined highway-user revenue collected in 2012 by all levels of government totaled $133.6 billion.

In 2012, $17.9 billion of highway-user revenue was used for transit, and $10.6 billion was used for other purposes, such as ports, schools, collection costs, and general government activities. The $0.3 billion shown as Federal highway-user revenue used for other purposes reflects the difference between total collections in 2012 and the amounts deposited into the Highway Trust Fund during Fiscal Year 2012. Much of this difference is attributable to the proceeds from the deposits of the 0.1-cent portion of the Federal motor-fuel tax into the Leaking Underground Storage Tank trust fund.

Exhibit 6-2 Disposition of Highway-User Revenue by Level of Government, 2012

	Revenue, Billions of Dollars			
	Federal	State	Local	Total
Highways	$33.8	$66.7	$4.6	$105.2
Transit	$6.1	$10.6	$1.1	$17.9
Other	$0.3	$10.2	$0.1	$10.6
Total Collected	**$40.3**	**$87.6**	**$5.7**	**$133.6**

Source: Highway Statistics 2012, Table HF-10, (revised).

The $6.1 billion shown as Federal highway-user revenue used for transit includes deposits into the Transit Account of the Highway Trust Fund and deposits into the Highway Account of the Highway Trust Fund that States elected to use for transit purposes.

As shown in *Exhibit 6-1*, all levels of government combined spent $221.3 billion for highways in 2012. The net difference of $4.8 billion between the total revenues generated during the year and the expenditures during the year reduced reserves available for use in future years. For example, the $4.6-billion difference between total Federal revenues and expenditures represents the decrease in the cash balance of the Highway Account of the Highway Trust Fund (HTF) in 2012. Although individual State and local governments might have increased or decreased their cash balances, the net national balance was an increase in reserves for State governments of $0.4 billion and a decrease for local governments of $0.6 billion.

The total proceeds to the Highway Account of the HTF from dedicated excise taxes and other receipts have been less than expenditures for many years. Funds were transferred from the Federal General Fund to the Highway Account in 2008, 2009, and 2010 to keep the account solvent. In 2012, $2.4 billion was transferred from the balance of the Leaking Underground Storage Tank Fund to the Highway Account; these are identified as "Investment Income and Other Receipts" in *Exhibit 6-1*, although the original source of these funds was revenues generated in prior years from a 0.1-cent tax on motor fuels.

Q&A **How long has it been since excise tax revenues deposited into the Highway Account exceeded expenditures?**

The last time that annual net receipts credited to the Highway Account of the Highway Trust Fund exceeded annual expenditures from the Highway Account was in 2000. As shown in *Exhibit 6-3* , for each year since 2000, total annual receipts to the Highway Account from excise taxes and other income (such as interest income and motor-carrier safety fines and penalties) have been lower than the annual expenditures from the Highway Account (including amounts transferred to the Transit Account).

Exhibit 6-3 Highway Trust Fund Highway Account Receipts and Outlays, Fiscal Years 2000–2014

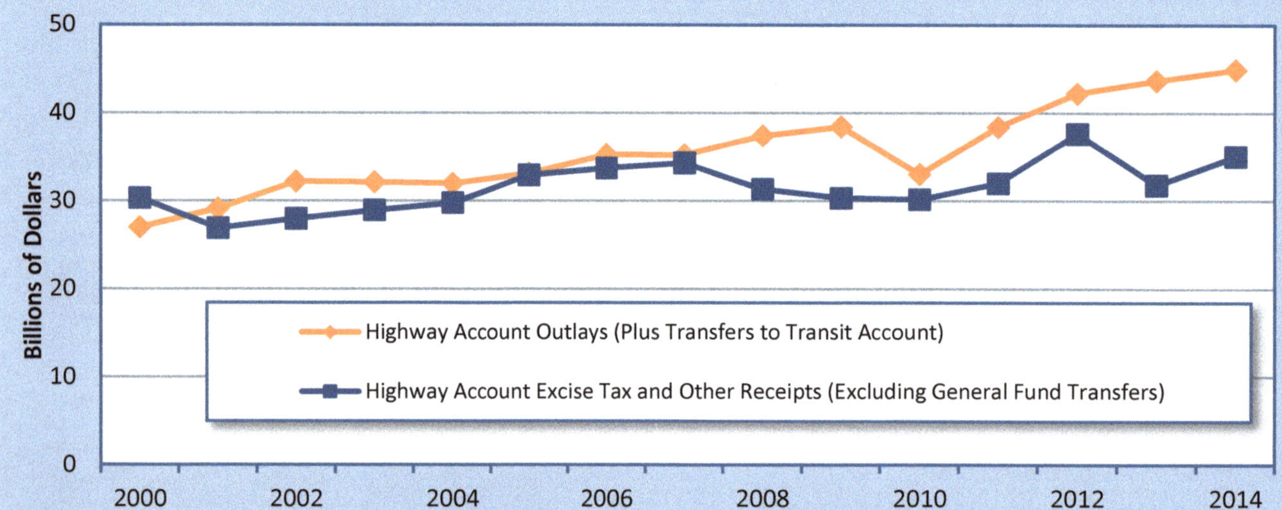

Source: Highway Statistics, various years, Tables FE-210 and FE-10.

To help maintain a positive cash balance in the Highway Trust Fund, transfers from the General Fund to the Highway Trust Fund were legislatively mandated in Fiscal Years 2008, 2009, 2010, 2013, and 2014. In Fiscal Years 2012 and 2014, funds were transferred from the balance of the Leaking Underground Storage Tank Fund to the Highway Trust Fund; the original source of these funds was revenues generated in previous years from a 0.1-cent portion of the Federal tax on motor fuels.

The Investment Income and Other Receipts category in *Exhibit 6-1* includes development fees and special district assessments and private-sector investment in highways to the extent that such investment is captured in State and local accounting systems.

Financing for highways comes from both public and private sectors. The private sector has increasingly been instrumental in the delivery of highway infrastructure, but the public sector still provides the vast majority of funding. The financial statistics presented in this chapter are drawn predominantly from State reports based on State and local accounting systems. Figures in these systems can include some private-sector investment; where so, these amounts are generally classified as "Other Receipts." For additional information on private-sector investment in highways, see http://www.fhwa.dot.gov/ipd/p3.

Revenue Trends

Since passage of the Federal-Aid Highway Act of 1956 and establishment of the HTF, user charges such as motor-fuel and motor-vehicle tax receipts have consistently provided most of the combined revenues raised for highway and bridge programs by all levels of government. After 2006, when user revenues flattened and transfers were made to keep the HTF solvent, the share of user revenues, excluding tolls, fell below 50 percent.

Exhibit 6-4 shows the trends in revenues used for highways by source for all levels of government from 2002 to 2012. From 2010 to 2012, total revenues generated for highways declined from $228.3 billion to $216.6 billion. This decrease was driven by reductions in General Fund appropriations and bond issue proceeds; all other sources of revenue increased between these 2 years. The 2010 funding levels for both General Fund appropriations and bond issue proceeds were atypically high and could reflect the actions of governments to fund transportation to support jobs and to take advantage of low construction prices and interest rates during a period of recession. Some decrease in General Fund appropriations between 2010 and 2012 reflects the phasedown of the American Recovery and Reinvestment Act, which was enacted in February 2009 and provided additional funds for transportation and other programs. Of the funds authorized under the Recovery Act, $11.9 billion was expended for highway purposes in 2010, which dropped to $3.0 billion in 2012.

The revenues generated by user charges increased from 2010 to 2012; combined motor-fuel and motor-vehicle tax revenues rose by $7.5 billion, while toll revenues rose by $3.8 billion after many years of little growth.

From 2002 to 2012, total revenues for highways have increased at an annual rate of 4.9 percent. The increase in motor-fuel and motor-vehicle revenues was 2.3 percent, the lowest among the funding sources, even though these revenues increased by 9.0 percent from 2010 to 2012. The increase in General Fund appropriations averaged 8.1 percent per year, despite the decline between 2010 and 2012 of 28.2 percent. Investment income and other receipts increased at an average annual rate of 10.0 percent over the 10-year period, with a 2010-to-2012 increase of 32.5 percent.

Exhibit 6-4 Government Revenue Sources for Highways, 2002–2012

Source	Highway Revenue, Billions of Dollars 2002	2004	2006	2008	2010	2012	Annual Rate of Change 2012/2002
Motor-Fuel and Motor-Vehicle Taxes	$73.1	$76.4	$85.4	$84.7	$84.1	$91.6	2.3%
Tolls	$6.6	$6.6	$8.3	$9.1	$9.7	$13.5	7.5%
Property Taxes and Assessments	$6.5	$7.5	$9.0	$9.0	$10.1	$10.3	4.7%
General Fund Appropriations	$20.3	$23.6	$28.3	$40.0	$61.5	$44.1	8.1%
Other Taxes and Fees	$7.5	$7.9	$10.1	$12.2	$13.5	$14.8	7.0%
Investment Income & Other Receipts	$8.1	$7.6	$9.7	$16.6	$15.8	$20.9	10.0%
Bond Issue Proceeds	$12.7	$15.8	$18.3	$20.9	$33.7	$21.3	5.2%
Total Revenues	**$134.8**	**$145.3**	**$169.0**	**$192.6**	**$228.3**	**$216.6**	**4.9%**

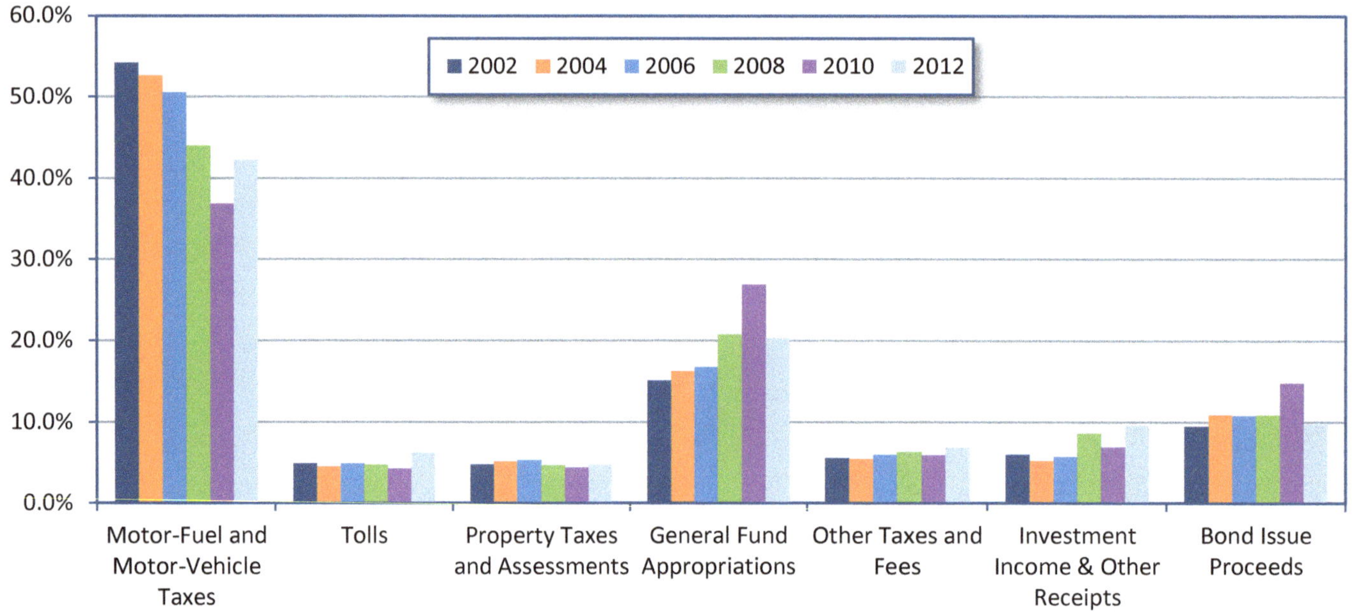

Source: Highway Statistics, various years, Tables HF-10A and HF-10.

The graph at the bottom of *Exhibit 6-4* shows the percentage share of each funding source by year for 2002–2012. Until 2012, the share of revenues from user charges, excluding tolls, had declined from more than 50 percent to less than 40 percent. This trend changed in 2012, when the share of these user charges increased to more than 40 percent.

Exhibit 6-5 shows the change in the share of highway revenue derived from user charges by level of government. The share declined for all levels of government combined from 2002 to 2010, and then rebounded slightly from 2010 to 2012. At the Federal level, the decline from 2007 to 2010 can be attributed in part to General Fund transfers to the HTF and to General Funds provided through the Recovery Act. Since 2010, the percentage of Federal highway revenue derived from user charges has increased from 49.1 percent to 79.1 percent. The State user revenue share also increased from 55.3 percent to 62.8 percent from 2010 to 2012. User charges have continued to decline as a share of local government revenue between these 2 years.

Exhibit 6-5 Percentages of Highway Revenue Derived from User Charges, Each Level of Government, 2002–2012

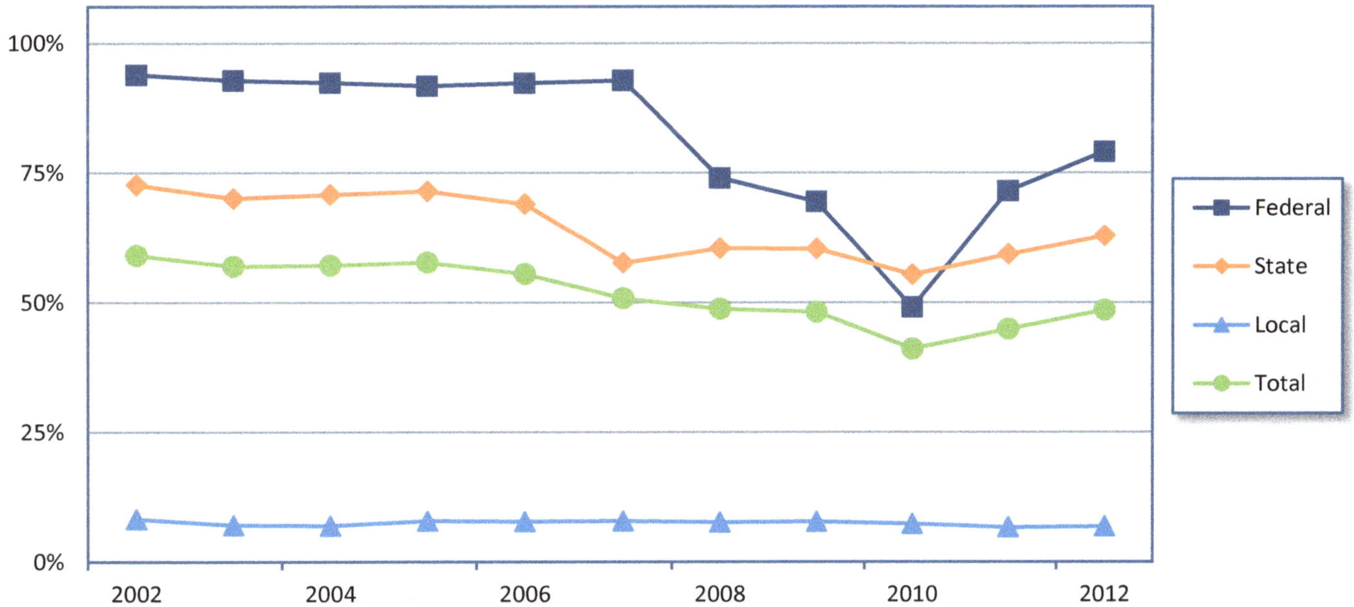

Source: Highway Statistics, various years, Tables HF-10A and HF-10.

Revenue Trend Details

Federal Motor-Fuel and Vehicle Taxes: The $7.5-billion increase in motor-fuel and vehicle-tax revenue for highways from 2010 to 2012 was largely driven by a $5.4-billion increase at the Federal level. Between these 2 years, gross revenues from all categories of receipts for the Highway Account of the Highway Trust Fund increased: gasoline tax revenue by **2.8 percent,** diesel and special motor-fuel tax revenue by **7.3 percent,** tire tax revenue by **32.0 percent,** truck and trailers tax revenue by **146.8 percent**, and use tax revenue by **93.9 percent**.

State Motor-Fuel Tax: State motor-fuel tax revenue from 2002 to 2012 increased about 18 percent, from $27.8 billion to $32.8 billion. This increase occurred every year except from 2007 to 2008, when it decreased 14 percent from $34.8 billion to $29.9 billion and from 2009 to 2010, when it decreased 1 percent. The first, larger reduction in revenue appears to have come in the midst of the decade's second recession. This reduction is the case for many of the revenue sources discussed.

Toll Revenue: Although imposing tolls is not a new idea, States are revisiting their use due in part to technological advances that make toll deployment and operation easier. About 29 States have some type of toll road covering about 5,100 miles of road (FHWA, *Toll Facilities in the United States*, 2013).

State General Funds: Another source of revenue is money taken from State General Funds, which increased 43 percent from $4.7 billion in 2002 to $6.7 billion in 2012. States received a respite from using General Funds with the passage of the American Recovery and Reinvestment Act starting in Fiscal Year 2010. Since its passage, States have received about $39 billion in total transportation assistance, with about $26 billion going toward highway infrastructure (www.recovery.gov).

Local General Funds: After State motor-fuel and vehicle taxes, local General Funds are the next largest source of non-Federal highway transportation dollars for the States. Local governments—including counties, townships and municipalities—provide about 30 percent of total surface transportation funding (National Conference of State Legislatures, *Transportation Governance and Finance,* 2011).

Highway Expenditures

Highway expenditures by all levels of government combined totaled $221.3 billion in 2012, as shown in *Exhibit 6-1*. *Exhibit 6-6* breaks down the Federal, State, and local expenditures by type. The rows "Funding Sources for Capital Outlay" and "Funding Sources for Total Expenditures" indicate the level of government that provided the funding for those expenditures. These expenditures represent cash outlays, not authorizations or obligations of funds. (The terms "expenditures," "spending," and "outlays" are used interchangeably in this report).

Exhibit 6-6 Direct Expenditures for Highways by Expending Agency and Type, 2012

	Highway Expenditures (Billions of Dollars)				
	Federal	State	Local	Total	Percent
Expenditures by Type					
Capital Outlay	**$1.1**	**$76.0**	**$28.0**	**$105.2**	**47.5%**
Noncapital Expenditures					
Maintenance	$0.3	$15.3	$19.5	$35.1	15.9%
Highway and Traffic Services	$0.0	$6.5	$6.4	$12.9	5.8%
Administration	$1.8	$8.9	$5.4	$16.0	7.2%
Highway Patrol and Safety	$0.0	$9.3	$8.6	$17.8	8.1%
Interest on Debt	$0.0	$7.9	$3.7	$11.6	5.2%
Subtotal	**$2.1**	**$47.9**	**$43.5**	**$93.5**	**42.2%**
Total, Current Expenditures	**$3.2**	**$123.9**	**$71.5**	**$198.7**	**89.8%**
Bond Retirement	**$0.0**	**$8.7**	**$14.0**	**$22.6**	**10.2%**
Total, All Expenditures	**$3.2**	**$132.6**	**$85.5**	**$221.3**	**100.0%**
Funding Sources for Capital Outlay[1]					
Funded by Federal Government	*$1.1*	*$43.4*	*$0.8*	*$45.3*	*43.1%*
Funded by State or Local Governments	*$0.0*	*$32.6*	*$27.2*	*$59.9*	*56.9%*
Total	***$1.1***	***$76.0***	***$28.0***	***$105.2***	***100.0%***
Funding Sources for Total Expenditures[1]					
Funded by Federal Government	*$3.2*	*$43.4*	*$0.8*	*$47.4*	*21.4%*
Funded by State Governments	*$0.0*	*$86.2*	*$19.6*	*$105.8*	*47.8%*
Funded by Local Governments	*$0.0*	*$3.0*	*$65.0*	*$68.1*	*30.8%*
Total	***$3.2***	***$132.6***	***$85.4***	***$221.3***	***100.0%***

[1] Amounts shown in italics are provided to link this table back to revenue sources shown in Exhibit 6-1. These are nonadditive to the rest of the table, which classifies spending by expending agency.

Sources: Highway Statistics 2012, Table HF-10, and unpublished FHWA data.

Even though the Federal government funded $47.4 billion of highway expenditures in 2012, direct Federal spending on capital outlay, maintenance, administration, and research was only $3.2 billion (1.5 percent of all highway expenditures). The remaining $44.2 billion was in the form of transfers to State and local governments.

State governments combined $43.4 billion of Federal funds, $86.2 billion of State funds, and $3.0 billion of local funds to support direct expenditures of $132.6 billion (59.9 percent of all highway expenditures). Local governments directly spent $0.8 billion of Federal funds, $19.6 billion of State funds, and $65.0 billion of local funds on highways, totaling $85.4 billion (38.6 percent of all highway expenditures).

Types of Highway Expenditures

Definitions for selected expenditure category types referenced in this section are as follows:

- **Capital outlay:** highway improvements such as land acquisition and other right-of-way costs; preliminary and construction engineering; new construction, reconstruction, resurfacing, rehabilitation, and restoration; and installation of guardrails, fencing, signs, and signals.
- **Maintenance:** routine and regular expenditures required to keep the highway surface, shoulders, roadsides, structures, and traffic control devices in usable condition. These efforts include completing spot patching and crack sealing of roadways and bridge decks and maintaining and repairing highway utilities and safety devices, such as route markers, signs, guardrails, fence, signals, and highway lighting.
- **Highway and traffic services:** activities designed to improve the operation and appearance of the roadway, including items such as the operation of traffic control systems, snow and ice removal, highway beautification, litter pickup, mowing, toll collection, and air quality monitoring.
- **Current expenditures:** all highway expenditures except for bond retirement.
- **Noncapital expenditures:** all current expenditures except for capital outlay. (Includes interest payments on bonds).

As shown in *Exhibit 6-6*, $105.2 billion, or 47.5 percent, of spending by all levels of government on highways in 2012, was for capital outlays. Additional information on types of capital outlay and the distribution of capital outlay by type of highway facility is presented later in this chapter. Combined spending on maintenance and traffic services of $48.0 billion represented 21.7 percent on total highway expenditures.

Most Federal funding for highways is for capital outlay rather than noncapital expenditures, which State and local governments primarily fund. The Federal government funded 43.1 percent of capital outlay in 2012, but only 21.4 percent of total highway expenditures.

In terms of direct highway expenditures by expending agency, State expenditures represent a majority of total spending for most expenditure types. The exception is in the maintenance category; local governments spent $19.5 billion on maintenance in 2012, which is 53.9 percent of total maintenance spending by all levels of government combined. Local governments also spent $8.6 billion on highway patrol and safety expenditures, representing 48.3 percent of combined spending on these activities by all levels of government.

Historical Expenditure and Funding Trends

Exhibit 6-7 breaks out expenditures since 2002 by type. The largest percentage increases are related to debt service, as bond retirement expenditures grew at an average annual rate of 12.8 percent from 2002 to 2012, while interest on debt grew an average annual rate of 7.9 percent. Total highway expenditures grew by 5.0 percent per year over this period in nominal dollar terms, while capital outlay rose at an average annual rate of 4.4 percent with capital expenditures becoming a smaller share of total expenditures.

Exhibit 6-7 Expenditures for Highways by Type, All Units of Government, 2002–2012

	Highway Expenditures, Billions of Dollars						Annual Rate of Change
	2002	2004	2006	2008	2010	2012	2012/2002
Expenditure Type							
Capital Outlay	$68.2	$70.3	$80.2	$90.4	$100.0	$105.2	4.4%
Maintenance and Traffic Services	$33.2	$36.3	$40.8	$45.9	$46.3	$48.0	3.8%
Administration	$10.7	$12.7	$13.1	$17.8	$16.5	$16.0	4.1%
Highway Patrol and Safety	$11.7	$14.3	$14.7	$17.3	$16.8	$17.8	4.3%
Interest on Debt	$5.4	$5.8	$6.6	$8.5	$10.1	$11.6	7.9%
Total, Current Expenditures	**$129.1**	**$139.5**	**$155.5**	**$180.0**	**$189.7**	**$198.7**	4.4%
Bond Retirement	$6.8	$8.0	$8.1	$8.6	$14.6	$22.6	12.8%
Total, All Expenditures	**$135.9**	**$147.5**	**$163.5**	**$188.5**	**$204.3**	**$221.3**	5.0%

Source: Highway Statistics, various years, Tables HF-10A and HF-10.

Exhibit 6-8 shows that Federal funding for highways grew more slowly from 2002 to 2012 than did State or local funding. The Federal portion of total highway expenditures declined from 24.1 percent to 21.4 percent over this period, while the federally funded share of highway capital outlay declined from 46.1 percent to 43.1 percent.

Exhibit 6-8 Funding for Highways by Level of Government, 2002–2012

	Highway Funding, Billions of Dollars						Annual Rate of Change
	2002	2004	2006	2008	2010	2012	2012/2002
Capital Outlay							
Funded by Federal Government	$31.5	$30.8	$34.6	$37.6	$43.3	$45.3	3.7%
Funded by State or Local Governments	$36.7	$39.5	$45.6	$52.8	$56.7	$59.9	5.0%
Total	**$68.2**	**$70.3**	**$80.2**	**$90.4**	**$100.0**	**$105.2**	**4.4%**
Federal Share	**46.1%**	**43.8%**	**43.1%**	**41.6%**	**43.3%**	**43.1%**	
Total Expenditures							
Funded by Federal Government	$32.8	$33.1	$36.3	$39.8	$46.1	$47.4	3.7%
Funded by State Governments	$69.0	$72.8	$77.4	$96.6	$98.7	$105.8	4.4%
Funded by Local Governments	$34.1	$41.6	$49.8	$52.2	$59.5	$68.1	7.2%
Total	**$135.9**	**$147.5**	**$163.5**	**$188.5**	**$204.3**	**$221.3**	**5.0%**
Federal Share	**24.1%**	**22.4%**	**22.2%**	**21.1%**	**22.6%**	**21.4%**	

Source: Highway Statistics, various years, Tables HF-10A and HF-10.

The Federal expenditure figures for 2010 include $11.9 billion funded by the Recovery Act. By 2012, this figure had dropped to $3.0 billion as most Recovery Act projects had been completed. Federally funded highway expenditures grew by $1.3 billion from 2010 to 2012 (from $46.1 billion to $47.4 billion), indicating that cash-basis expenditures funded from other Federal sources increased by more than the decline in Recovery Act-funded expenditures over this period.

State funding for highways increased from $98.7 billion in 2010 to $105.8 billion in 2012. The 7.2 percent average annual increase in local government funding exceeded the growth rates for both Federal and State funding.

Constant Dollar Expenditures

The types of inputs of materials and labor associated with various types of highway expenditures significantly differ; for example, on a dollar-per-dollar basis, highway maintenance activities are generally more labor intensive than highway construction activities. This report uses different indices for converting nominal dollar highway spending to constant dollars for capital and noncapital expenditures. For constant-dollar conversions for highway capital expenditures, the Federal Highway Administration (FHWA) Composite Bid Price Index is used through 2006, the last year for which this index was produced. Capital expenditure conversions for subsequent years rely on a successor index, the FHWA National Highway Construction Cost Index. Constant-dollar conversions for other types of highway expenditures are based on the Bureau of Labor Statistics' Consumer Price Index.

> **Q&A**
>
> **How has the federally funded share of highway capital spending varied over time?**
>
> The federally funded share of highway capital outlay exceeded 50 percent each year from 1976 to 1986. Since then, this share has typically varied from 41 to 46 percent. In 1998, 1999, and 2007, however, it fell below 40 percent.
>
> Over the 10-year period from 2003 through 2012, the average of the federally funded shares for each year was 43.1 percent. This average matches the federally funded share in 2012.

Exhibit 6-9 illustrates the trends in cost indices used in the report, converted to a common base year of 2002. Over the 10-year period from 2002 to 2012, the Consumer Price Index increased by 27.6 percent, similar to the 25.0-percent rise in the combination of the Bid Price Index and Construction Cost Index. Within this 10-year period, however, the indices differed significantly;

Exhibit 6-9 Comparison of Inflation Indices (Converted to a 2002 Base Year), 2002–2012[1]

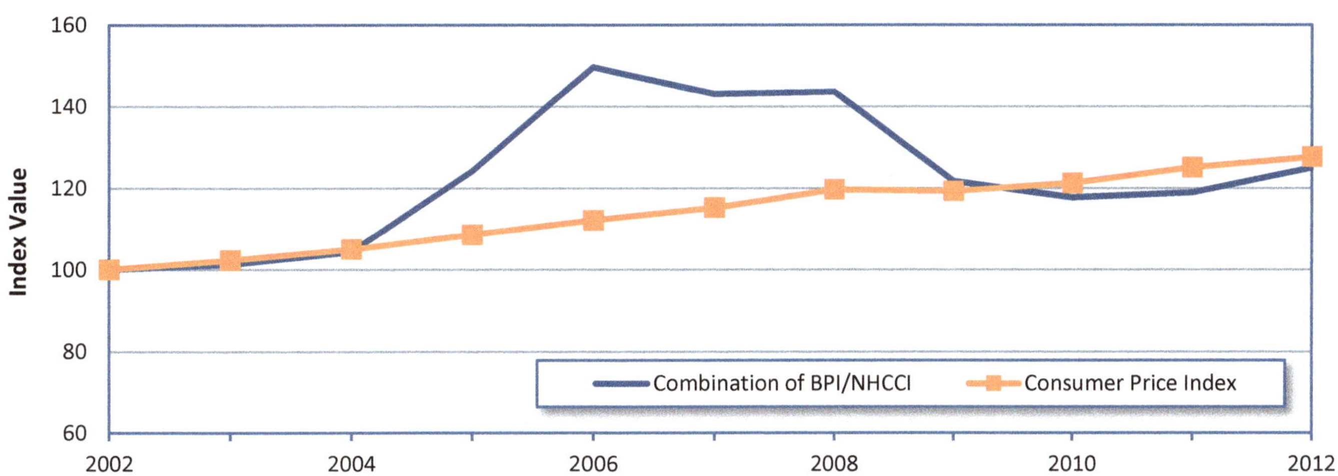

[1] *In order to facilitate comparisons of trends, each index was mathematically converted so that its value for the year 2002 would be equal to 100.*

Sources: FHWA Highway Statistics, various years, Table PT-1; http://www.bls.gov/cpi/.

for example, in the period between 2004 and 2006, sharp increases in the prices of materials such as steel, asphalt, and cement caused the Bid Price Index to increase by 43.3 percent, compared with a 6.7-percent increase in the Consumer Price Index. Although highway construction prices as measured by the Bid Price Index and the Construction Cost Index subsequently declined, the purchasing power of highway capital spending in 2006–2008 was significantly less than in 2009–2012. In other words, each dollar of highway capital outlay from 2009 to 2012 had the potential to have a bigger impact on system performance than was the case for each dollar spent from 2006 to 2008.

Exhibits 6-10 and *6-11* display time-series data on highway expenditures in both current (nominal) and constant (real) 2012 dollars. Although constant dollars for total highway expenditures have decreased periodically, they reached an all-time high in 2012. The same was not true for highway capital expenditures, which reached an all-time high in 2011, but dipped slightly in 2012.

Exhibit 6-10 Highway Capital, Noncapital, and Total Expenditures in Current and Constant 2012 Dollars, All Units of Government, 1992–2012[1]

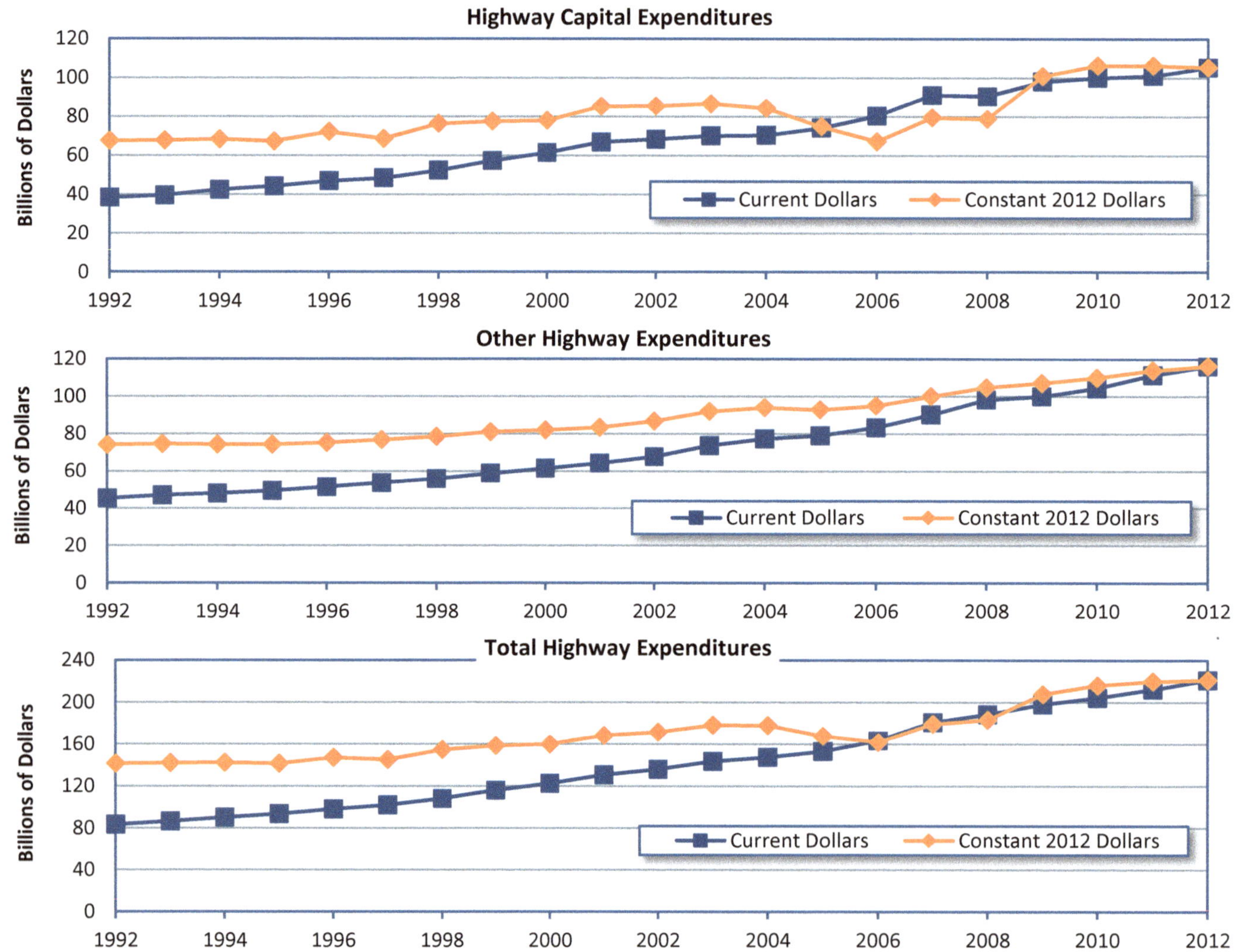

[1] *Constant dollar conversions for highway capital expenditures were made using the FHWA BPI through the year 2006 and the FHWA NHCCI in subsequent years. Constant dollar conversions for other types of highway spending were made using the Bureau of Labor Statistics CPI.*

Sources: Highway Statistics, various years, Tables HF-10A, HF-10, PT-1; http://www.bls.gov/cpi/.

For the 10-year period ending in 2012, highway capital spending increased at an average annual rate of 2.1 percent in constant-dollar terms, slightly below the 2.6-percent annual constant-dollar growth rate for total highway expenditures. From 2002 to 2012, federally funded highway expenditures increased at an average annual rate of 1.4 percent in constant-dollar terms; State and local constant-dollar expenditures grew much faster, rising by 2.9 percent per year on average.

The relative trends differ over the 20 years from 1992 to 2012. Over this period, highway capital expenditures and total highway expenditures both grew by 2.3 percent in constant-dollar terms. Federally funded highway capital spending increased by 2.4 percent annually in constant-dollar terms, increasing more quickly than combined State and local constant-dollar expenditures, which grew by only 2.2 percent per year.

Exhibit 6-11 Highway Expenditures Funded by Federal and Non-Federal Sources in Current and Constant 2012 Dollars, 1992–2012[1]

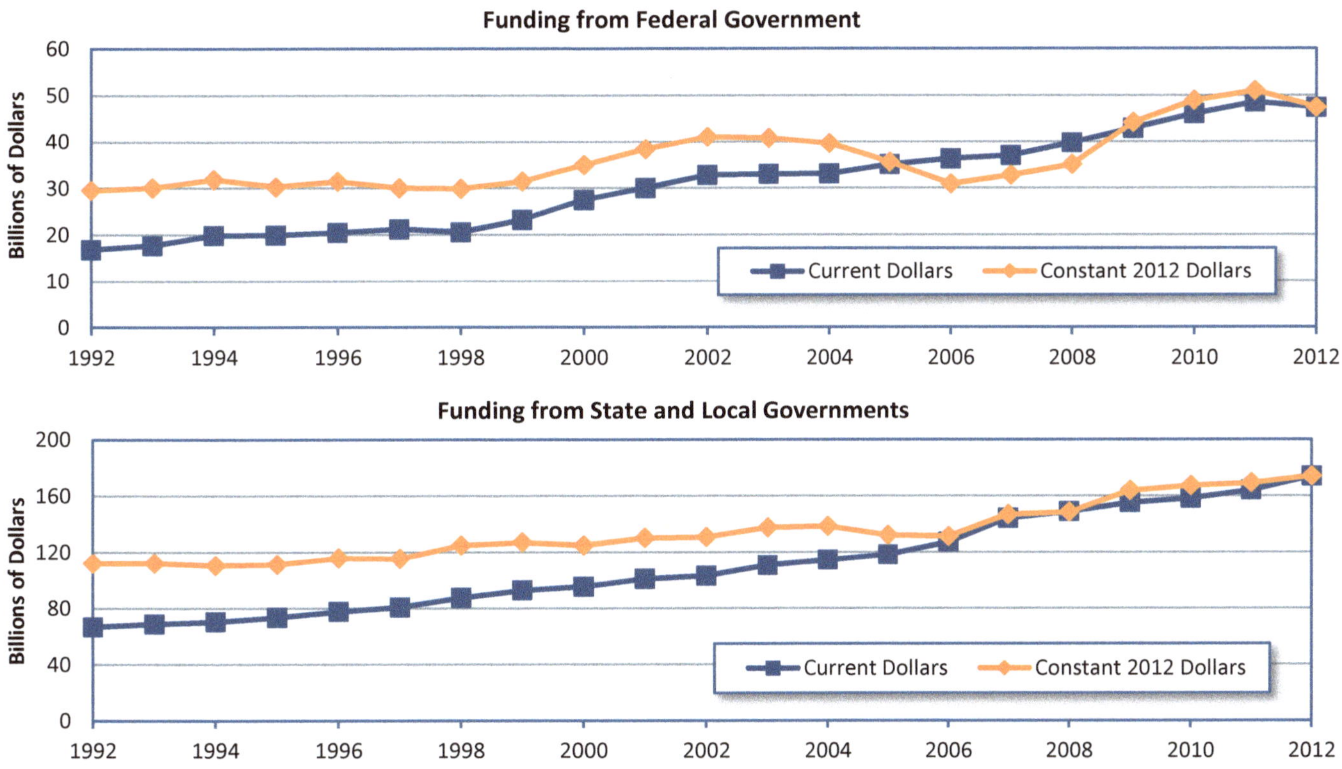

[1] Constant dollar conversions for highway capital expenditures were made using the FHWA BPI through the year 2006 and the FHWA NHCCI in subsequent years. Constant dollar conversions for other types of highway spending were made using the Bureau of Labor Statistics CPI.

Sources: Highway Statistics, various years, Tables HF-10A, HF-10, PT-1; http://www.bls.gov/cpi/.

The impact of the relatively high construction costs in 2006 to 2008 noted above in reference to *Exhibit 6-9* is evident in *Exhibits 6-10* and *6-11*, as the constant-dollar lines dip below the current dollar line in several of the line charts for those years. The difference is most pronounced in the highway capital expenditures chart in *Exhibit 6-10* and the funding from Federal government chart in *Exhibit 6-11*, as most Federal funds are used to support highway capital expenditures.

Highway Capital Outlay

States provide FHWA with detailed data on what they spend on arterials and collectors, classifying capital outlay on each functional system into 17 improvement types. Direct State expenditures on arterials and collectors totaled $65.8 billion in 2012, drawing on a combination of State revenues, transfers from the Federal government, and transfers from local governments. *Exhibit 6-12* illustrates the distribution of these expenditures by improvement type and shows how these

Exhibit 6-12 Highway Capital Outlay by Improvement Type, 2012

	Distribution of Capital Outlay, Billions of Dollars				
		System Expansion			
Type of Expenditure	System Rehabilitation	New Roads and Bridges	Existing Roads	System Enhancements	Total Outlay
Direct State Expenditures on Arterials and Collectors					
Right-of-Way		$1.5	$1.9		**$3.5**
Engineering	$4.6	$0.8	$1.0	$0.8	**$7.3**
New Construction		$5.5			**$5.5**
Relocation			$1.0		**$1.0**
Reconstruction—Added Capacity	$1.8		$4.1		**$5.9**
Reconstruction—No Added Capacity	$4.9				**$4.9**
Major Widening			$2.6		**$2.6**
Minor Widening	$1.0				**$1.0**
Restoration and Rehabilitation	$16.9				**$16.9**
Resurfacing	$0.0				**$0.0**
New Bridge		$0.7			**$0.7**
Bridge Replacement	$5.9				**$5.9**
Major Bridge Rehabilitation	$1.4				**$1.4**
Minor Bridge Work	$2.9				**$2.9**
Safety				$2.7	**$2.7**
Traffic Management/Engineering				$1.3	**$1.3**
Environmental and Other				$2.4	**$2.4**
Total, State Arterials and Collectors	**$39.4**	**$8.5**	**$10.6**	**$7.2**	**$65.8**
Total, Arterials and Collectors, All Jurisdictions (estimated)[1]					
Highways and Other	$35.4	$9.4	$13.0	$9.9	$67.6
Bridges	$12.5	$0.9			$13.4
Total, Arterials and Collectors	**$47.9**	**$10.2**	**$13.0**	**$9.9**	**$81.0**
Total Capital Outlay on All Systems (estimated)[1]					
Highways and Other	$45.7	$12.1	$14.0	$15.9	$87.7
Bridges	$16.4	$1.1			$17.5
Total, All Systems	**$62.1**	**$13.2**	**$14.0**	**$15.9**	**$105.2**
Percent of Total	**59.0%**	**12.6%**	**13.3%**	**15.1%**	**100.0%**

[1] Improvement type distribution was estimated based on State arterial and collector data.

Sources: Highway Statistics 2012, Table SF-12A, and unpublished FHWA data.

improvement types have been allocated among three broad categories: system rehabilitation, system expansion, and system enhancement. These broad categories are also used in Chapter 7 to discuss the components of future capital investment scenarios. These categories are defined as follows:

- **System rehabilitation:** capital improvements on existing roads and bridges intended to preserve the existing pavement and bridge infrastructure. These activities include reconstruction, resurfacing, pavement restoration or rehabilitation, widening of narrow lanes or shoulders, bridge replacement, and bridge rehabilitation. Also included is the portion of widening (lane addition) projects estimated for reconstructing or improving existing lanes. System rehabilitation does not include routine maintenance costs.
- **System expansion:** construction of new roads and new bridges and addition of new lanes to existing roads. Expansion includes all new construction, new bridge, and major widening, and most of the costs associated with reconstruction-added capacity, except for the portion of these expenditures estimated for improving existing lanes of a facility.
- **System enhancement:** safety enhancements, traffic operation improvements such as the installation of intelligent transportation systems, and environmental enhancements.

Exhibit 6-12 presents an estimated distribution of total highway capital outlay by all levels of government on all roads. Of the $105.2 billion in total highway capital outlay, an estimated $62.1 billion (59.0 percent) was used for system rehabilitation, $27.2 billion (25.9 percent) was used for system expansion, and $15.9 billion (15.1 percent) was used for system enhancement. These estimates are based primarily on State expenditure patterns on arterials and collectors, along with limited data from other sources. As shown in *Exhibit 6-12*, most types of highway capital improvement reported by States are assigned to one of these three broad categories; however, engineering is split among the three categories and reconstruction-added capacity is divided between system rehabilitation and system expansion.

Exhibit 6-13 shows the distribution of capital expenditures by type and functional system. In 2012, $30.1 billion was invested on rural arterials and collectors, with 64.5 percent directed to system

Exhibit 6-13 Distribution of Capital Outlay by Improvement Type and Functional System, 2012

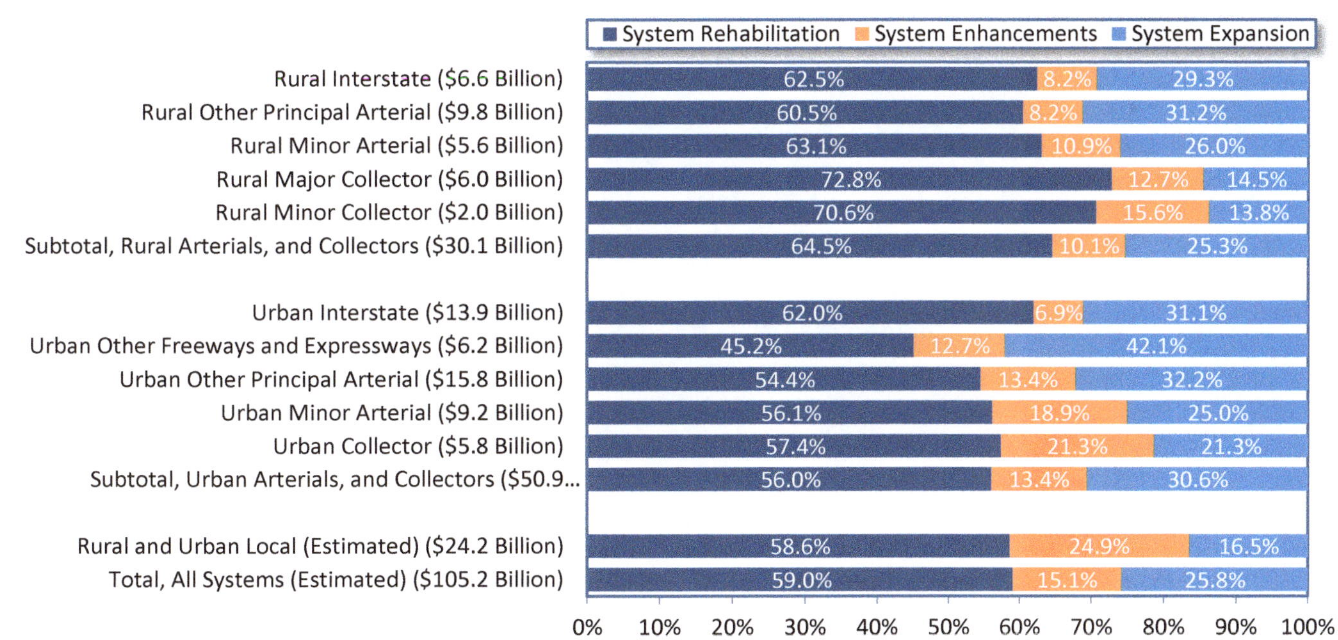

Sources: Highway Statistics 2012, Table SF-12A, and unpublished FHWA data.

rehabilitation and 25.3 percent to expansion; the remainder was directed to system enhancement. Capital outlays on urban arterials and collectors were $50.9 billion, of which 56.0 percent was for system rehabilitation and 30.6 percent was for system expansion. Among the individual functional systems, rural major collectors had the highest percentage of highway capital outlay directed to system rehabilitation (72.8 percent), while urban other freeways and expressways had the lowest percentage directed for that purpose (45.2 percent).

Exhibit 6-14, shows trends in capital outlays by improvement type from 2002 to 2012. Each year, a majority of capital outlays was directed to rehabilitation, reflecting the need to preserve the aging system. The share of total capital spending for system rehabilitation, however, rose dramatically between 2008 and 2010, from 51.1 percent to 60.5 percent. System rehabilitation expenditures increased from $46.2 billion to $60.5 billion, nearly 31 percent over the 2 years. This dramatic increase was partly driven by the Recovery Act; one of the Recovery Act's stated goals is to support jobs through construction expenditures, an aim best achieved by selecting projects that could be initiated and completed relatively quickly. This strategy led many States to direct a larger portion of their Recovery Act funding toward pavement improvement projects than they usually finance from regular Federal-aid funds in a typical year. Although most Recovery Act-funded projects had been completed before 2012, the overall share of highway capital spending directed to system preservation declined to just 59.0 percent, somewhat below the share in 2010, but still well above the share in 2008. This finding suggests that the shift toward system preservation beginning in 2008 could have been driven by factors in addition to the Recovery Act, and thus might represent the start of a long-term trend.

Exhibit 6-14 Capital Outlay on All Roads by Improvement Type, 2002–2012

	Capital Outlay, Billions of Dollars						Annual Rate of Change
Improvement Type	2002	2004	2006	2008	2010	2012	2012/2002
System Rehabilitation							
Highway	$25.5	$26.7	$31.0	$33.5	$43.4	$45.7	6.0%
Bridge	$10.7	$9.6	$10.3	$12.7	$17.0	$16.4	4.4%
Subtotal	**$36.2**	**$36.3**	**$41.3**	**$46.2**	**$60.5**	**$62.1**	**5.6%**
System Expansion							
Additions to Existing Roadways	$11.9	$12.1	$14.0	$15.7	$15.0	$14.0	1.6%
New Routes	$11.4	$12.6	$15.2	$16.1	$11.4	$12.1	0.6%
New Bridges	$1.1	$1.4	$1.2	$1.5	$0.9	$1.1	0.2%
Subtotal	**$24.4**	**$26.1**	**$30.4**	**$33.3**	**$27.4**	**$27.2**	**1.1%**
System Enhancements	**$7.6**	**$7.8**	**$8.5**	**$10.9**	**$12.2**	**$15.9**	**7.7%**
Total	**$68.2**	**$70.3**	**$80.2**	**$90.4**	**$100.0**	**$105.2**	**4.4%**
Percent of Total Capital Outlay							
System Rehabilitation	53.1%	51.7%	51.5%	51.1%	60.5%	59.0%	
System Expansion	35.8%	37.1%	37.9%	36.9%	27.4%	25.8%	
System Enhancements	11.1%	11.2%	10.6%	12.0%	12.2%	15.1%	

Sources: Highway Statistics 2010, Table SF-12A, and unpublished FHWA data.

From 2002 to 2012, system rehabilitation expenditures grew at an average annual rate of 5.6 percent. System expansion expenditures have increased more slowly—at an average annual rate of 1.1 percent. This slower expansion growth resulted in a decline in share of total capital outlays from 35.8 percent in 2002 to 25.8 percent in 2012. System enhancement expenditures have grown more quickly, rising from 11.1 percent of total capital outlays in 2002 to 15.1 percent in 2012.

> Q&A
>
> **How have constant dollar expenditures for different capital improvement types grown in recent years?**
>
> Total capital outlay by all levels of government grew at an average annual rate of 2.1 percent from 2002 to 2012 in constant-dollar terms. Constant-dollar system rehabilitation expenditures rose by 3.2 percent per year over this period, while system expansion expenditures declined by 1.2 percent annually when adjusted for inflation. Expenditures for system enhancements grew by 5.3 percent per year in constant-dollar terms from 2002 to 2012.

Capital Outlays on Federal-Aid Highways

As discussed in Chapter 2, "Federal-aid highways" includes all roads except those in functional classes that are generally ineligible for federal funding: rural minor, rural local, or urban local. *Exhibit 6-15* shows that total capital outlays on Federal-aid highways increased at an average annual rate of 3.9 percent from 2002 to 2012, rising to $79.0 billion in 2012.

The share of capital outlay on Federal-aid highways directed to system rehabilitation in 2012 was 58.9 percent, just below the comparable percentage for all roads of 59.0 percent. This pattern is consistent with that from 2002 to 2010 as well; in each year, the portion of Federal-aid highway capital outlay directed toward system rehabilitation and system enhancements was lower than the comparable shares for all roads, whereas the portion directed toward system expansion was higher than for all roads.

Exhibit 6-15 Capital Outlay on Federal-Aid Highways by Improvement Type, 2002–2012

	Capital Outlay, Billions of Dollars						Annual Rate of Change
Improvement Type	2002	2004	2006	2008	2010	2012	2012/2002
System Rehabilitation							
Highway	$19.6	$19.4	$22.9	$26.1	$33.1	$34.5	5.8%
Bridge	$8.3	$7.2	$7.7	$9.3	$12.5	$12.0	3.8%
Subtotal	**$27.9**	**$26.6**	**$30.6**	**$35.5**	**$45.6**	**$46.5**	**5.2%**
System Expansion							
Additions to Existing Roadways	$11.0	$11.6	$12.9	$14.3	$13.8	$12.8	1.5%
New Routes	$9.1	$9.8	$12.0	$12.8	$8.8	$9.3	0.2%
New Bridges	$0.9	$1.2	$0.9	$1.0	$0.7	$0.8	-0.4%
Subtotal	**$21.0**	**$22.6**	**$25.9**	**$28.1**	**$23.3**	**$22.9**	**0.9%**
System Enhancements	**$4.8**	**$5.0**	**$5.5**	**$6.4**	**$6.8**	**$9.6**	**7.1%**
Total	**$53.7**	**$54.2**	**$61.9**	**$70.0**	**$75.7**	**$79.0**	**3.9%**
Percent of Total Capital Outlay							
System Rehabilitation	52.0%	49.1%	49.3%	50.7%	60.3%	58.9%	
System Expansion	39.1%	41.6%	41.9%	40.1%	30.8%	29.0%	
System Enhancements	8.9%	9.3%	8.8%	9.2%	9.0%	12.1%	

Sources: Highway Statistics 2012, Table SF-12A, and unpublished FHWA data.

Capital Outlays on the National Highway System

The National Highway System (NHS) comprises roads essential to the Nation's economy, defense, and mobility, as described in Chapter 2. *Exhibit 6-16* shows that capital outlays for the NHS amounted to $44.6 billion in 2012, having grown at an average annual rate of 3.2 percent since 2002.

Exhibit 6-16 Capital Outlay on the National Highway System by Improvement Type, 2002–2012

	Capital Outlay, Billions of Dollars						Annual Rate of Change
Improvement Type	2002	2004	2006	2008	2010	2012	2012/2002
System Rehabilitation							
Highway	$10.6	$9.5	$12.3	$14.9	$19.9	$19.7	6.4%
Bridge	$4.5	$4.0	$4.3	$5.4	$7.4	$6.7	4.1%
Subtotal	**$15.1**	**$13.5**	**$16.6**	**$20.4**	**$27.3**	**$26.4**	**5.7%**
System Expansion							
Additions to Existing Roadways	$7.1	$7.1	$8.1	$9.2	$8.6	$8.0	1.2%
New Routes	$6.7	$6.8	$8.9	$8.6	$4.7	$5.6	-1.8%
New Bridges	$0.6	$0.9	$0.7	$0.6	$0.3	$0.5	-1.2%
Subtotal	**$14.5**	**$14.8**	**$17.7**	**$18.3**	**$13.7**	**$14.1**	**-0.2%**
System Enhancements	**$2.8**	**$2.8**	**$2.8**	**$3.3**	**$3.4**	**$4.0**	**3.6%**
Total	**$32.4**	**$31.1**	**$37.2**	**$42.0**	**$44.4**	**$44.6**	**3.2%**
Percent of Total Capital Outlay							
System Rehabilitation	46.7%	43.5%	44.7%	48.5%	61.6%	59.3%	
System Expansion	44.7%	47.6%	47.7%	43.7%	30.8%	31.7%	
System Enhancements	8.7%	8.9%	7.6%	7.8%	7.6%	9.0%	

Sources: Highway Statistics 2012, Table SF-12B, and unpublished FHWA data.

Between 2010 and 2012, system rehabilitation expenditures on the NHS declined from $27.3 billion to $26.4 billion, whereas system rehabilitation for all roads and for Federal-aid highways increased. Over the 10-year period beginning in 2002, system rehabilitation spending grew at an average annual rate of 5.7 percent, faster than the comparable growth rates for all roads (5.6 percent) or for Federal-aid highways (5.2 percent). System expansion expenditures on the NHS grew from $13.7 billion in 2010 to $14.1 billion in 2012, although from 2002 to 2012, NHS system expansion expenditures declined.

Capital Outlays on the Interstate System

Exhibit 6-17 shows that from 2010 to 2012, capital outlay increased by only 1.4 percent on the Interstate System, to $20.5 billion, well below the 5.2-percent increase observed for all roads. This increase is also much lower than the average annual increase in capital outlay for the Interstate System of 3.0 percent observed from 2002 and 2012.

The share of Interstate capital outlay directed to system rehabilitation in 2012 was 52.1 percent, higher than the comparable percentages for the NHS, Federal-aid highways, and all roads. This pattern is largely consistent with that from 2002 to 2010; the share of Interstate capital outlay directed to system rehabilitation was higher in each year from 2002 to 2010 than comparable percentages for the NHS or Federal-aid highways, although in some years it was lower than the comparable percentage for all roads. The share of Interstate capital outlay directed toward system enhancements was lower in each year from 2002 to 2012 than comparable percentages for all roads.

Exhibit 6-17 Capital Outlay on the Interstate System, by Improvement Type, 2002–2012

	Capital Outlay, Billions of Dollars						Annual Rate of Change
Improvement Type	**2002**	**2004**	**2006**	**2008**	**2010**	**2012**	**2012/2002**
System Rehabilitation							
Highway	$5.5	$4.7	$5.8	$7.5	$9.4	$8.9	4.9%
Bridge	$2.4	$2.3	$2.5	$3.3	$4.1	$3.8	4.6%
Subtotal	**$8.0**	**$7.0**	**$8.3**	**$10.8**	**$13.5**	**$12.7**	**4.8%**
System Expansion							
Additions to Existing Roadways	$3.2	$2.9	$3.2	$4.5	$3.5	$3.4	0.5%
New Routes	$2.5	$2.5	$3.5	$3.0	$1.7	$2.7	0.6%
New Bridges	$0.2	$0.2	$0.3	$0.3	$0.1	$0.2	0.3%
Subtotal	**$5.9**	**$5.6**	**$7.1**	**$7.8**	**$5.3**	**$6.3**	**0.6%**
System Enhancements	**$1.4**	**$1.1**	**$1.2**	**$1.4**	**$1.4**	**$1.5**	**0.5%**
Total	**$15.3**	**$13.7**	**$16.5**	**$20.0**	**$20.2**	**$20.5**	**3.0%**
Percent of Total Capital Outlay							
System Rehabilitation	52.1%	50.8%	49.9%	53.9%	66.7%	62.1%	
System Expansion	38.5%	40.9%	42.6%	38.9%	26.3%	30.5%	
System Enhancements	9.4%	8.3%	7.4%	7.1%	6.9%	7.3%	

Sources: Highway Statistics 2012, Table SF-12A, and unpublished FHWA data.

Project Finance

In recent years, State and local transportation agencies have adopted new ways of financing and delivering transportation projects. In the face of stagnating public revenues and demanding fiscal requirements, many jurisdictions are relying on innovative options such as public-private partnerships, Federal credit assistance, and other debt-financing tools. These strategies could enable financially strapped public agencies to deliver costly and complex infrastructure projects much earlier than would be possible through traditional mechanisms.

Public-Private Partnerships

Public-Private Partnerships (P3s) are contractual agreements between a public agency and a private entity that allow for greater private-sector participation in the delivery and financing of transportation projects. Typically, this participation involves the private sector's assuming additional project risks, such as design, finance, long-term operation, maintenance, or traffic and revenue. P3s are undertaken for a variety of purposes, including monetizing the value of existing assets, developing new transportation facilities, or rehabilitating or expanding existing facilities. Although P3s offer certain advantages, such as increased financing capacity and reduced upfront costs, the public sector still must identify a source of revenue for the project to provide a return to the private partner's investment and must ensure that the goals and interests of the public are adequately secured. Additional information on P3s is available at http://www.fhwa.dot.gov/ipd/p3/index.htm.

Public-Private Partnership Project: I-95 HOV/HOT Lanes

The I-95 Express Lanes are the second major step in creating a regional network of tolled managed lanes in Northern Virginia. The project consists of the development, design, finance, construction, maintenance, and operation of 29.4 miles of high occupancy vehicle (HOV)/high occupancy toll (HOT) lanes along the I-95 and I-395 corridor in Northern Virginia.

The project was divided into four segments:

- 8.3 miles of new construction – two-lane reversible (includes 7 new bridges),
- 7.0 miles of two-lane HOV conversion – two-lane reversible,
- 11.9 miles of two-lane HOV conversion – three-lane reversible, and
- 2.2 miles of two-lane HOV conversion – three-lane reversible (including connection to 495 Express Lanes at the Springfield Interchange).

The $922.6-million project was financed primarily through a Transportation Infrastructure and Finance Innovation Act loan, private activity bonds, and private equity. The concession agreement was finalized in 2012 between the Virginia Department of Transportation and 95 Express Lanes LLC, and the facility opened to traffic in late 2014.

Federal Credit Assistance

Federal credit assistance for highway improvements can take one of two forms: (1) loans, where project sponsors borrow Federal highway funds directly from a State DOT or the Federal government; and (2) credit enhancements, where a State DOT or the Federal government makes Federal funds available on a contingent (or standby) basis. Loans can provide the capital necessary to proceed with a project and reduce the amount of capital borrowed from other sources. Credit enhancement helps reduce risk to investors and thus allows project sponsors to borrow at lower interest rates. Loans also might serve a credit enhancement function by reducing the risk borne by other investors. Federal tools currently available to project sponsors include the Transportation Infrastructure and Finance Innovation Act (TIFIA) program, State Infrastructure Bank programs, and Section 129 loans.

The TIFIA Credit Program provides Federal credit assistance in the form of direct loans, loan guarantees, and standby lines of credit to finance surface transportation projects of national and regional significance. A TIFIA project must pledge repayment in whole or in part with dedicated revenue sources, such as tolls, user fees, special assessments (taxes), or other non-Federal sources. State Infrastructure Banks are State-run revolving funds that provide loans, credit enhancements, and other forms of nongrant assistance to surface transportation projects. State Infrastructure Banks can be capitalized with regularly apportioned Federal-aid funds. Section 129 loans allow States to lend apportioned Federal-aid highway funds to toll and nontoll projects generating dedicated revenue streams. Additional information on credit assistance tools is available at http://www.fhwa.dot.gov/ipd/finance/tools_programs/federal_credit_assistance/index.htm.

Debt Financing Tools

Some transportation projects are so large that their cost exceeds available current grant funding and tax receipts or would consume so much of these current funding sources as to delay many other planned projects. For this reason, State and local governments often seek financing for large projects through borrowing, which provides an immediate influx of cash to fund project construction costs. The borrower then retires the debt by making principal and interest payments over time. Tax-exempt municipal bonds, backed by future government revenues, are the most common method of borrowing by government agencies for transportation projects.

Two innovative debt instrument tools—GARVEES (Grant Anticipation Revenue Vehicles) and PABs (Private Activity Bonds) provide additional borrowing opportunities. A GARVEE is a debt-financing instrument—such as a bond, note, certificate, mortgage, lease, or other debt financing technique—that has a pledge of future Federal-aid funding. PABs are debt instruments issued by State or local governments on behalf of a private entity for highway and freight transfer projects, allowing a private project sponsor to benefit from the lower financing costs of tax-exempt municipal bonds. Additional information on Federal debt financing tools is available at http://www.fhwa.dot.gov/ipd/finance/tools_programs/federal_debt_financing/index.htm.

Washington State Route 520 Bridge

State Route (SR) 520 is one of two major east-west roadways crossing Lake Washington, located within the Seattle metropolitan area. It connects major population and employment centers between Seattle and the region's eastern suburbs.

The SR 520 Floating Bridge and Eastside plus West Approach Bridge Project included:

- Pontoon Construction Project – Construction of 33 bridge pontoons and a 55-acre site. Pontoon construction includes 21 longitudinal pontoons, 10 supplementary stability pontoons, and 2 cross pontoons;
- Floating Bridge and Landings Project – Construction of a new six-lane floating bridge across Lake Washington and removal of the existing floating bridge;
- Eastside Project – Widening of SR 520 and other corridor-wide improvements to complete a transit/high occupancy vehicle lane in each direction; and
- West Approach Bridge Project – Construction of a permanent west approach bridge structure to connect traffic from the bridge to an interchange on land and to complete the bicycle/pedestrian path from the east side to Seattle.

Financing for the bridge project included a $300-million Transportation Infrastructure and Finance Innovation Act loan and the use of $923 million from Grant Anticipation Revenue Vehicles.

Transit Finance

Transit funding comes from two major sources: public funds allocated by Federal, State, and local governments, and system-generated revenues earned from providing transit services. As shown in *Exhibit 6-18*, $58 billion was available for urban transit financing in 2012. Federal funding for transit includes fuel taxes dedicated to transit from the Mass Transit Account (MTA) of the Highway Trust Fund and undedicated taxes allocated from Federal general fund appropriations. State and local governments also provide funding for transit from their general fund appropriations and from fuel, income, sales, property, and other unspecified taxes, specific percentages of which can be dedicated to transit. These percentages vary considerably among taxing jurisdictions and by type of tax. Other public funds from toll revenues, general transportation funds, and other sources also might be used to fund transit. Passenger fares principally comprise system-generated revenues, although transit systems earn additional revenues from advertising and concessions, park-and-ride lots, investment income, and rental of excess property and equipment.

Exhibit 6-18 2012 Urban Revenue Sources for Transit Funding (Millions of Dollars)

	Federal	State	Local	Total	Percent
Public Funds	**10,859.4**	**12,697.9**	**18,951.9**	**42,509.2**	**73.3%**
General Fund	2,171.9	3,204.2	4,549.1	9,925.1	17.1%
Fuel Tax	8,687.5	909.8	190.4	9,787.6	16.9%
Income Tax		395.4	91.7	487.1	0.8%
Sales Tax		3,455.3	5,431.9	8,887.2	15.3%
Property Tax		10.4	651.1	661.4	1.1%
Other Dedicated Taxes		1,923.3	566.9	2,490.2	4.3%
Other Public Funds		2,799.7	7,470.9	10,270.6	17.7%
System-Generated Revenue				**15,451.2**	**26.7%**
Passenger Fares				13,608.4	23.5%
Other Revenue				1,842.8	3.2%
Total All Sources				**57,960.4**	**100.0%**

Source: National Transit Database.

Transit Finance

In 2012, the amount of funding available for urban transit financing was $58 billion. Transit funding comes from two major sources: (1) public funds that Federal, State, and local governments allocate and (2) revenues that transit systems generate by providing transit services. Federal funding for transit includes fuel taxes dedicated to transit from the Mass Transit Account of the Highway Trust Fund and undedicated taxes allocated from Federal general fund appropriations. Fuel taxes accounted for 77.3 percent of all Fund revenues for Fiscal Year 2012, which is similar to the share reported in the National Transit Database (80 percent). State and local governments also provide funding for transit from their general fund appropriations and from fuel, income, sales, property, and other unspecified taxes, specific percentages of which can be dedicated to transit. The percentages vary considerably among taxing jurisdictions and by type of tax. Other public funds from sources such as tolls and general transportation funds also can be used to fund transit. Passenger fares principally comprise system-generated revenues, although transit systems derive additional revenues from advertising and concessions, park-and-ride lots, investment income, and rental of excess property and equipment.

Level and Composition of Transit Funding

Exhibit 6-19 breaks down the sources of total urban transit funding. In 2012, public funds of $42.5 billion were available for urban transit, accounting for 73.3 percent of total transit funding. Of this amount, Federal funding was $10.9 billion or 25.5 percent of total public funding and 18.7 percent of all funding from both public and nonpublic sources. State funding was $12.7 billion, accounting for 29.9 percent of total public funds and 21.9 percent of all funding. Local jurisdictions provided the bulk of transit funds, $19 billion in 2012 or 44.6 percent of total public funds and 32.7 percent of all funding. System-generated revenues were $15.5 billion or 26.7 percent of all funding. During the American Recovery and Reinvestment Act (Recovery Act) years, 2009–2011, transit agencies reported annual expenditures averaging $17.0 billion. The infusion of $5.3 billion in Recovery Act funds during that period enabled the industry to maintain investment levels near the record 2008 funding level of $17.1 billion.

Exhibit 6-19 2012 Urban Transit Revenue Sources (Billions of Dollars)

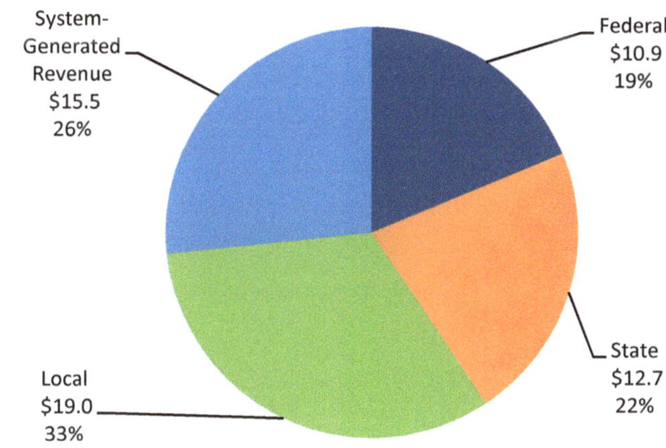

Source: National Transit Database.

Q&A How long has it been since excise tax revenue deposited into the Mass Transit Account exceeded expenditures?

The last time annual net receipts credited to the Mass Transit Account (MTA) of the Highway Trust Fund exceeded annual expenditures from the Highway Account was 2007. As shown in *Exhibit 6-20*, for 9 of the 11 years since 2002, total annual receipts to the MTA from excise taxes and other income (including amounts transferred from the Highway Account) have been lower than the annual expenditures from the MTA. The gap between MTA outlays and receipts increased by 40.5 percent from 2011 to 2012.

Exhibit 6-20 Mass Transit Account Receipts and Outlays, Fiscal Years 2002–2012

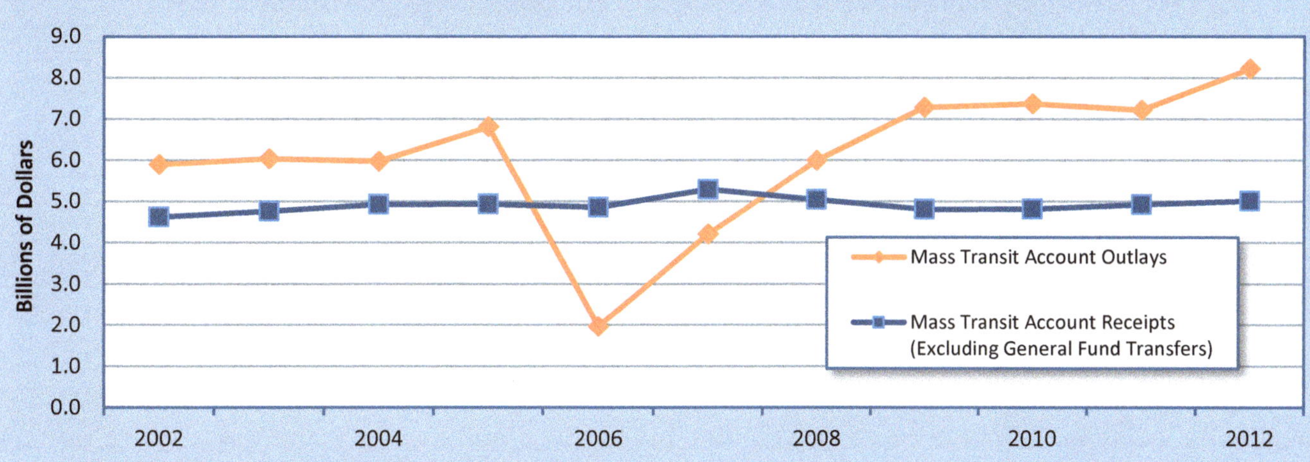

Sources: Highway Statistics, various years, Tables FE-210 and FE-10.

Federal Funding

Federal funding for transit comes from two sources: the general revenues of the U.S. government, and revenues generated from fuel taxes credited to the Highway Trust Fund's MTA. The largest part of the funding from the Highway Trust Fund is distributed by formula funding, which is legislatively defined. A smaller part is distributed competitively or at agency discretion.

General revenue sources include income taxes, corporate taxes, tariffs, fees, and other government income not required by statute to be accounted for in a separate fund. MTA, a trust fund for capital projects in transit, is generally the largest source of Federal funding for transit, although in 2009 Recovery Act funds from the general account surpassed the MTA contribution. *Exhibit 6-21* shows how Recovery Act funds were awarded in 2009, 2010, and 2011 compared to other Federal funding from the MTA and the General Fund. Of the funds authorized for transit grants in FTA's 2010 budget, 79.0 percent were derived from the MTA. Funding from the MTA in nominal dollars increased from $0.5 billion in 1983 to $8.3 billion in 2010.

Exhibit 6-21 Urban Recovery Act Funding Awards Compared to Other FTA Fund Awards

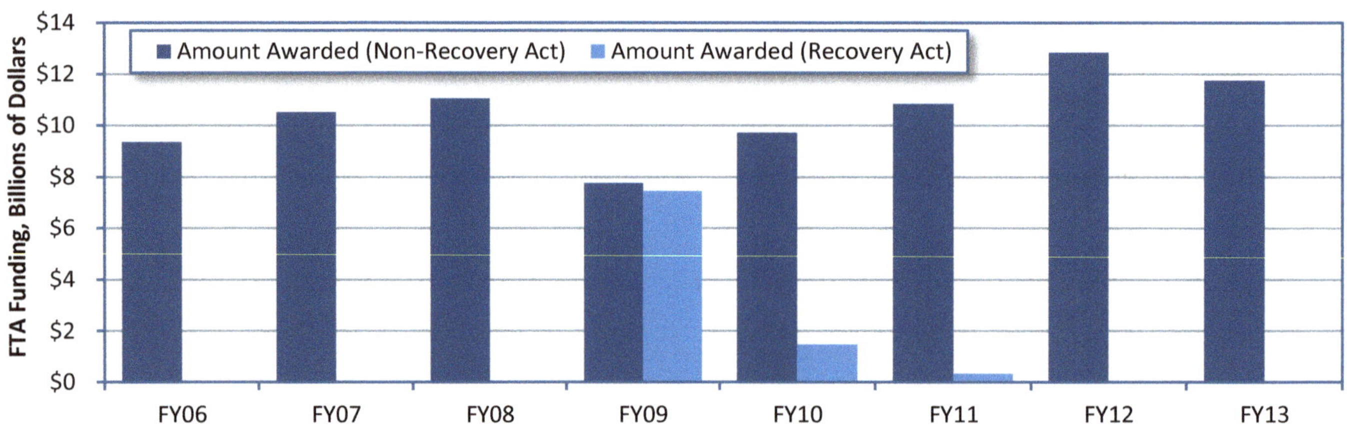

Source: Federal Transit Administration, Grants Data.

For the past 6 years, starting in 2008, the Highway Trust Fund has experienced shortfalls as authorized spending levels exceeded revenues. Revenues decreased due to several factors, including a reduction in fuel tax revenues caused by technological improvements in the manufacture of cars and other vehicles, which led to greater fuel efficiency, a drop in employment until early 2010 that led to reduced highway travel.

The Department of Homeland Security funds projects aimed at improving transit security. In 2012, the Department provided $87.5 million to transit service providers.

Since 1973, Federal statutes authorizing surface transportation have contained flexible funding provisions that enable transfers from certain highway funds to transit programs and vice versa. Transfers are subject to State and regional/local discretion, and priorities are established through statewide transportation planning processes. Forty-three States and the District of Columbia participate in the flexible funding program. Flexible funding transferred from highways to transit fluctuates from year to year and is drawn from several different sources.

The Surface Transportation Program is also the primary source of Federal Highway Administration (FHWA) funds that are "flexed" to the Federal Transit Administration (FTA) to pay for transit projects. Funding is at 80 percent of the Federal share and may be used for all capital and maintenance projects eligible for funds under current FTA programs. These funds may not be used for operating assistance.

FHWA's Congestion Mitigation and Air Quality (CMAQ) Improvement Program funds are another source of flexed funds used to support transit projects in air quality nonattainment areas. A CMAQ project must contribute to the attainment of the National Ambient Air Quality Standards by reducing air pollutant emissions from transportation sources. Public transportation projects can be funded through CMAQ, which also includes some provisions for transit operating assistance during project startup.

State and Local Funding

General funds and other dedicated public funds (such as vehicle licensing and registration fees, communications access fees, surcharges and taxes, lottery and casino receipts, and proceeds from property and asset sales) are important sources of funding for transit at both the State and local levels. State and local funding sources for transit are shown in *Exhibit 6-22*. Taxes, including fuel, sales, income, property, and other dedicated taxes, provide 43 percent of public funds for State and local sources. General funds provide 25 percent of transit funding, and other public funds provide the remaining 32 percent.

Exhibit 6-22 2012 Urban State and Local Sources of Transit Funding, Millions of Dollars

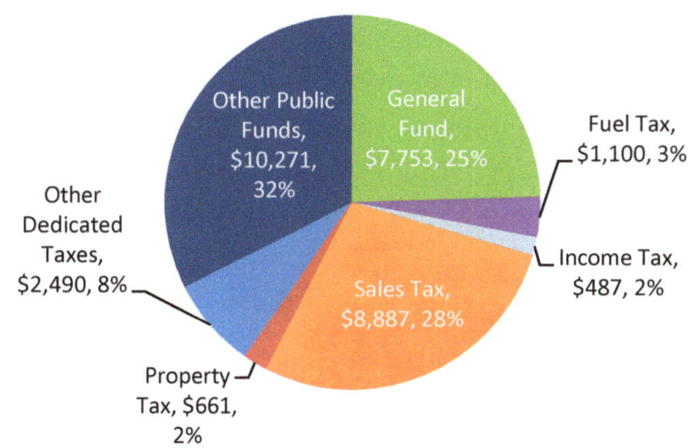

Source: National Transit Database.

System-Generated Funds

In 2012, system-generated funds were $15.5 billion and provided 26 percent of total transit funding. Passenger fares contributed $13.6 billion, accounting for 23.5 percent of total transit funds. These passenger fare figures do not include payments by State entities to transit systems that offset reduced transit fares for certain segments of the population, such as students and the elderly. These payments are included in the "other revenue" category.

Trends in Funding

Between 2002 and 2012, public funding for transit increased at an average annual rate of 3.5 percent, Federal funding increased at an average annual rate of 3.1 percent, and State and local funding increased at an average annual rate of 1.9 percent after adjusting for inflation (constant dollars). These data are presented in *Exhibit 6-23*.

Exhibit 6-23 Urban Funding for Transit by Government Jurisdiction, 2002–2012 (Constant Dollars)

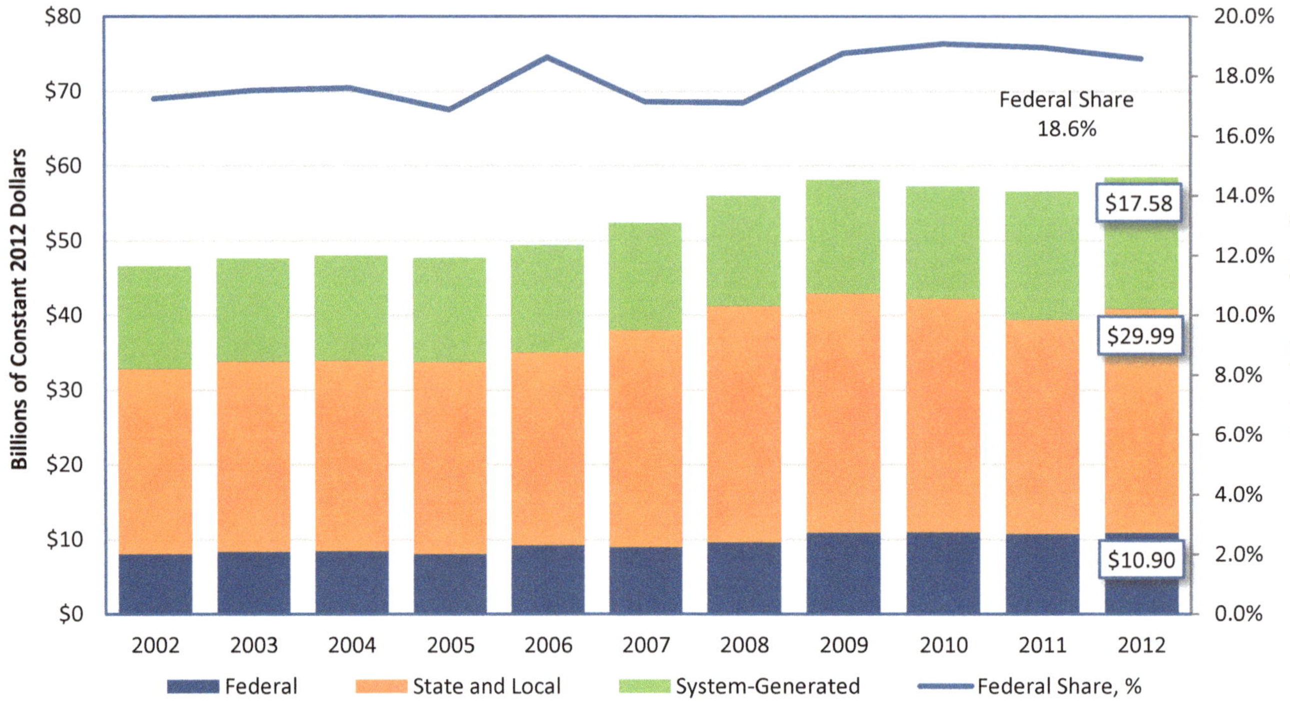

Source: National Transit Database.

Federal funding for transit, as a percentage of total funding for transit from Federal, State, and local sources combined, reached a peak of 42.9 percent in the late 1970s, and declined to near its present value by the early 1990s. State and local funding increased during this same period. *Exhibit 6-23* shows that, since 2002, the Federal government has provided between 16.9 and 19.5 percent of total funding for transit (including system-generated funds). In 2012, it provided 18.6 percent.

Funding in Current and Constant Dollars

Public funding for transit in current dollars and constant (adjusted for inflation) dollars since 1992 is presented in *Exhibit 6-24*. Total public funding for transit was $42.5 billion in 2012. In constant dollar terms, this amount was 3.2 percent lower than in 2010. Between 2010 and 2012, Federal funding increased from nearly $10.4 billion to $10.9 billion (4.8 percent) in current dollars. In constant dollars, however, this represents a 0.5-percent decrease in funding. From 2010 to 2012, in current dollars, State and local funding increased from $29.8 billion to $31.7 billion (6.4 percent). In constant dollars, this represents a 4.0-percent decrease in funding.

Federal funds directed to capital expenditures have increased 4.2 percent from 2002 to 2012, while capital funds applied to operating expenditures have increased 9.9 percent during the same period (current dollars). As indicated in *Exhibit 6-25*, $3.4 billion was applied to operating expenditures and $7.5 billion was applied to capital expenditures in 2012. More than half the operating expenditures were for preventive maintenance, which is reimbursed as a capital expense under FTA's 5307 grant program.

Exhibit 6-24 Urban Current and Constant Dollar Funding for Public Transportation, 1992–2012 (All Sources)[1]

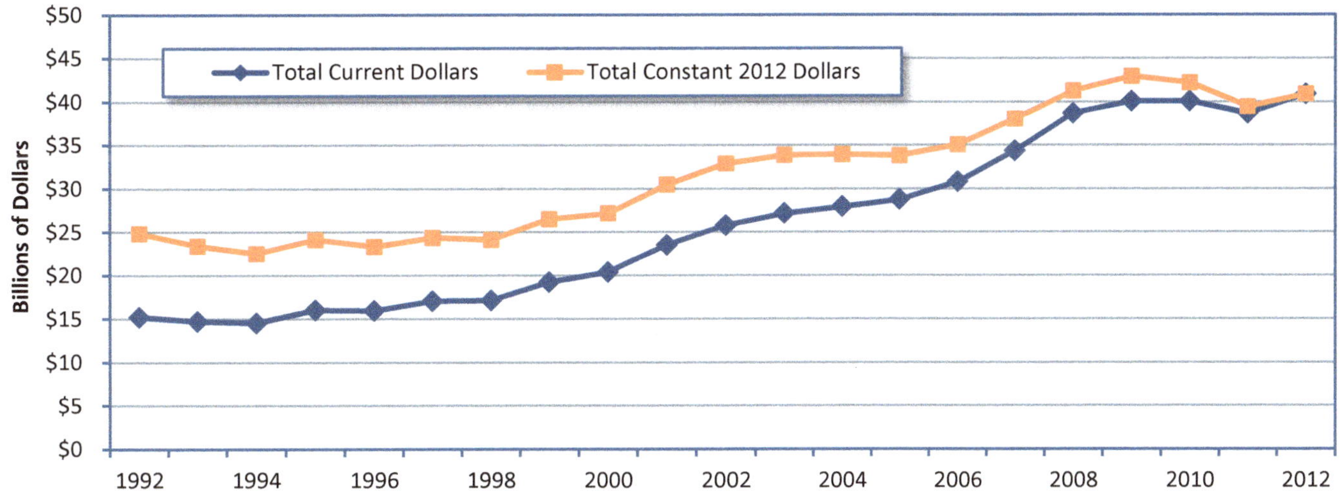

[1] *Constant dollar conversions were made using the U.S. Bureau of Labor Statistics CPI.*

Source: National Transit Database.

Exhibit 6-25 Urban Applications of Federal Funds for Transit Operating and Capital Expenditures, 2002–2012

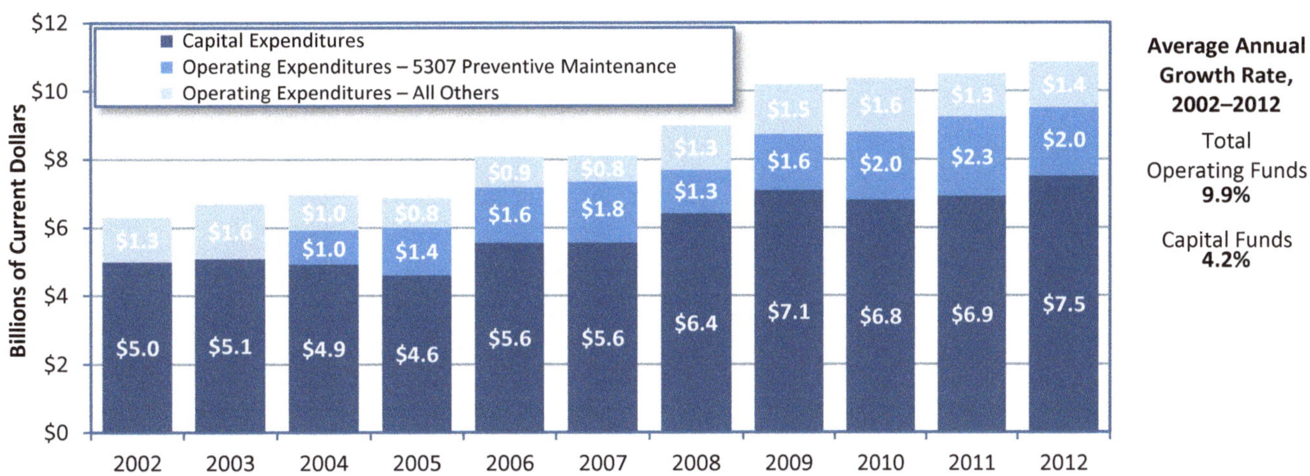

Source: National Transit Database.

Capital Funding and Expenditures

Funding for capital investments by transit operators in the United States comes primarily from public sources. A relatively small amount of private-sector funds for capital investment in transit projects is generated through innovative finance programs.

Capital investments include the design and construction of new transit systems, extensions of existing systems, and the modernization or replacement of existing assets. Capital investment expenditures can be for the acquisition, renovation, and repair of vehicles (e.g., buses, railcars, locomotives, and service vehicles) or fixed assets (e.g., fixed-guideway systems, terminals, stations, and maintenance and administrative facilities).

As shown in *Exhibit 6-26*, total public transit agency expenditures for capital investment were $16.9 billion in 2012. This expenditure accounted for 39.8 percent of total available funds for transit. Federal funds were $6.0 billion in 2012, 35.5 percent of total transit agency capital expenditures. State funds provided 11.9 percent and local funds provided 43.7 percent of total transit funding. Recovery Act funds provided the remaining 8.9 percent of revenues for agency capital expenditures.

Exhibit 6-26 Urban Sources of Funds for Transit Capital Expenditures, 2002–2012

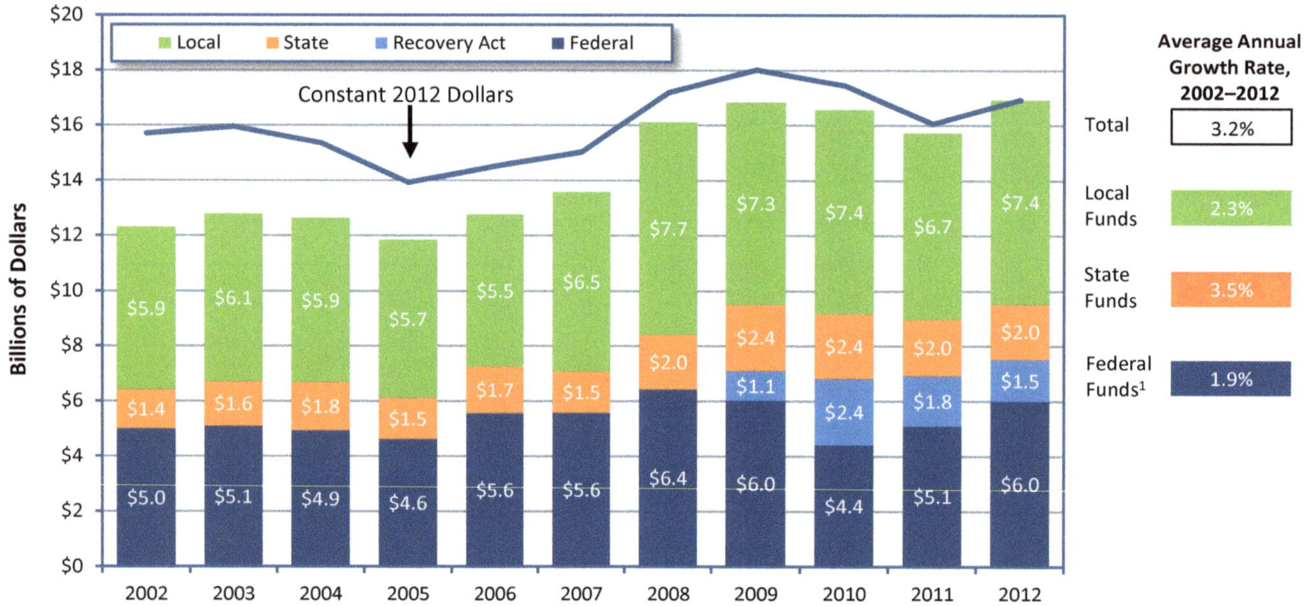

[1] *Growth rate shown for Federal Funds includes Recovery Act funds as well as other types of Federal funds.*

Source: National Transit Database.

From 2009 to 2011, substantial amounts of Recovery Act funds were expended, and non-Recovery Act Federal funds decreased compared to previous years. This decrease is not surprising, given the strict 2-year obligation limit specified for Recovery Act funds—these funds had to be used first due to their short availability period. In 2012, as most of the Recovery Act funds had been expended, expenditures using non-Recovery Act Federal funds reverted to pre-2009 levels.

As shown in *Exhibit 6-27*, rail modes require a higher percentage of total transit capital investment than fixed-route bus modes for two reasons: (1) the higher cost of building fixed guideways and rail stations and (2) fixed-route bus systems typically do not pay to build or maintain the roads on which they run. In 2012, $12.1 billion, or 72.1 percent of total transit capital expenditures, were invested in rail modes of transportation, compared with $4.7 billion, or 27.9 percent of the total, which was invested in nonrail modes. This investment distribution has been consistent over the past decade.

Exhibit 6-27 2012 Urban Transit Capital Expenditures by Mode and Type

	Rail Capital Expenditures (Millions of Dollars)						
Type	Commuter Rail	Heavy Rail	Light Rail	Hybrid Rail	Streetcar Rail	Other Rail[1]	Total Rail
Guideway	$1,398	$1,903	$2,370	$1	$68	$99	**$5,839**
Rolling Stock	$625	$249	$209	$0	$13	$12	**$1,108**
Systems	$171	$800	$132	$1	$1	$14	**$1,119**
Maintenance Facilities	$212	$355	$72	$0	$0	$1	**$641**
Stations	$301	$2,103	$391	$1	$3	$4	**$2,802**
Fare Revenue Collection Equipment	$9	$23	$14	$0	$0	$0	**$46**
Administrative Buildings	$8	$25	$2	$0	$0	$0	**$35**
Other Vehicles	$15	$28	$3	$0	$0	$4	**$50**
Other Capital Expenditures[2]	$72	$391	$21	$0	$1	$0	**$486**
Total	**$2,811**	**$5,877**	**$3,215**	**$3**	**$86**	**$135**	**$12,127**
Percent of Total	16.6%	34.7%	19.0%	0.0%	0.5%	0.8%	71.7%

	Nonrail Capital Expenditures (Millions of Dollars)							
Type	Bus	Bus Rapid Transit	Commuter Bus	Demand Response	Ferryboat	Vanpool	Trolley Bus	Total Nonrail
Guideway	$172	$28	$24	$0	$0	$0	$15	**$238**
Rolling Stock	$2,172	$13	$106	$157	$80	$33	$4	**$2,566**
Systems	$354	$0	$4	$23	$1	$0	$1	**$384**
Maintenance Facilities	$595	$0	$2	$18	$7	$0	$0	**$623**
Stations	$288	$7	$20	$3	$109	$0	$1	**$428**
Fare Revenue Collection Equipment	$62	$0	$0	$1	$1	$0	$1	**$65**
Administrative Buildings	$142	$0	$0	$25	$0	$0	$0	**$167**
Other Vehicles	$53	$0	$0	$1	$0	$0	$0	**$55**
Other Capital Expenditures[2]	$165	$5	$0	$6	$2	$0	$0	**$179**
Total	**$4,003**	**$53**	**$157**	**$234**	**$200**	**$33**	**$22**	**$4,703**
Percent of Total	23.7%	0.3%	0.9%	1.4%	1.2%	0.2%	0.1%	27.8%

Type	Total Expenditures (Millions of Dollars) for Rail and Nonrail Modes	Percent of Total
Guideway	**$6,077**	35.9%
Rolling Stock	**$3,674**	21.7%
Systems	**$1,503**	8.9%
Maintenance Facilities	**$1,264**	7.5%
Stations	**$3,230**	19.1%
Fare Revenue Collection Equipment	**$111**	0.7%
Administrative Buildings	**$202**	1.2%
Other Vehicles	**$104**	0.6%
Other Capital Expenditures[2]	**$665**	3.9%
Agencies operating less than 30 peak vehicles[3]	**$89**	0.5%
Total	**$16,919**	**100.0%**

[1] Includes Alaska railroad, monorail/automated guideway, cable car, and inclined plane.

[2] Capital expenditures not elsewhere included. These expenditures include furniture and equipment that are not an integral part of buildings and structures; they also include shelters, signs, and passenger amenities (e.g., benches) not in passenger stations.

[3] Agencies operating less than 30 peak vehicles do not report capital data by mode and type of expenditure.

Table does not include Público and demand response taxi.

Source: National Transit Database.

Fluctuations in the levels of capital investment in different types of transit assets reflect normal rehabilitation and replacement cycles and new investment. Capital investment expenditures have been reported to the National Transit Database only at the level of detail in *Exhibit 6-27* since 2002.

Total guideway investment was $6.1 billion in 2012, and total investment in systems was $1.5 billion. Guideway includes at-grade rail, elevated and subway structures, tunnels, bridges, track and power systems for all rail modes, and paved highway lanes dedicated to fixed-route buses. Investment in systems by transit operators includes groups of devices or objects forming a network, most notably for train control, signaling, and communications.

In 2012, total investment in vehicles, stations, and maintenance facilities was $3.7 billion, $3.2 billion, and $1.3 billion, respectively. Vehicles include the bodies and chassis of transit vehicles and their attached fixtures and appliances but do not include fare collection equipment and movement control equipment for revenue vehicles, such as radios. Stations include station buildings, platforms, shelters, parking and other forms of access, and crime prevention and security equipment at stations. Facilities include the purchase, construction, and rehabilitation of administrative and maintenance facilities. Facilities also include investment in building structures, climate control, parking, yard track, vehicle and facilities maintenance equipment, furniture, office equipment, and computer systems.

Other capital costs include those associated with general administration facilities, furniture, equipment that is not an integral part of buildings and structures, data processing equipment, and shelters located at on-street bus stops. Data processing equipment includes computers and peripheral devices for which the sole use is in data processing operations.

Q&A **How does FTA fund major transit construction projects?**

FTA provides funding for the design and construction of light rail, heavy rail, commuter rail, streetcar, bus rapid transit, and ferry projects through a competitive, discretionary grant program known as Capital Investment Grants. Title 49 U.S.C. Section 5309 provides funds for new transit systems and extensions to current systems and for capacity expansion projects on existing transit lines currently at or over capacity. These types of projects are known more commonly as "New Starts," "Small Starts," and "Core Capacity" projects.

To receive funds from the Capital Investment Grant program, the proposed project must emerge from the metropolitan or statewide planning process and proceed through a multiyear, multistep process outlined in law, which includes a detailed evaluation and rating of the project by FTA. FTA evaluates proposed projects based on financial criteria and project justification criteria as prescribed by statute.

Under current law, Capital Investment Grant funding may not exceed 80 percent of a project's total capital cost. Generally, however, the Capital Investment Grant program share of such projects averages about 50 percent due to the overwhelming demand for funds nationwide. Funds are typically provided over a multiyear period rather than all at once, due to the size of the projects and the size of the overall annual program funding level.

Most, but not all, major transit capital projects are constructed using Capital Investment Grant program funds. Some project sponsors choose to use other sources instead, such as FTA Urbanized Area Formula funds, FTA discretionary Ferry Program funds, and Transportation Investments Generating Economic Recovery funds from the Department of Transportation.

Exhibit 6-28 shows yearly capital expenditures for rehabilitation or expansion by mode. Rehabilitation expenses are those dollars used to replace service directly or to maintain existing service. Expansion expenses are those used to increase service. Example expansion expenses include procuring additional buses to create a new route, building a new rail line, or constructing an additional rail station on an existing rail line.

After adjusting for inflation (constant dollars), total capital expenditures from 2003 to 2012 have increased by an annual average of 5.7 percent. Although rehabilitation expenses over this period have increased modestly, service expansion investment, particularly in rail modes, has increased considerably. Average annual expenses for heavy rail expansion had the largest increase over this time, with an average annual expansion expense of 13.6 percent.

Exhibit 6-28 Capital Expenditures Applied by Rehabilitation and Expansion by Mode, 2003–2012 (Millions, Constant Dollars)

	2003	2004	2005	2006	2007	2008	2009	2010	2011	2012	Growth Rate[1]
Rail Rehabilitation	$5,391	$4,986	$5,092	$5,364	$5,990	$7,300	$6,993	$5,900	$5,731	$5,526	0.3%
Rail Expansion	$2,103	$2,278	$2,079	$2,733	$3,152	$4,227	$4,725	$5,428	$5,053	$6,601	13.6%
Rail Total	**$7,494**	**$7,264**	**$7,171**	**$8,097**	**$9,142**	**$11,527**	**$11,717**	**$11,328**	**$10,783**	**$12,127**	**5.5%**
Nonrail Rehabilitation	$2,366	$2,780	$2,536	$2,788	$2,656	$3,032	$3,590	$3,926	$4,012	$4,159	6.5%
Nonrail Expansion	$364	$346	$357	$312	$459	$527	$421	$478	$527	$544	4.6%
Non-Rail Total	**$2,730**	**$3,126**	**$2,893**	**$3,100**	**$3,115**	**$3,559**	**$4,011**	**$4,404**	**$4,539**	**$4,703**	**6.2%**
Total Rehabilitation	$7,757	$7,765	$7,628	$8,152	$8,646	$10,332	$10,583	$9,826	$9,742	$9,684	2.5%
Total Expansion	$2,466	$2,625	$2,436	$3,045	$3,611	$4,755	$5,146	$5,906	$5,580	$7,145	12.5%
Grand Total	**$10,223**	**$10,390**	**$10,064**	**$11,197**	**$12,257**	**$15,087**	**$15,728**	**$15,732**	**$15,323**	**$16,830**	**5.7%**

[1] Represents average annual growth rate (2012/2003).

Source: National Transit Database.

Operating Expenditures

Transit operating expenditures include wages, salaries, fuel, spare parts, preventive maintenance, support services, and certain leases used in providing transit service. As indicated in *Exhibit 6-29*, $41.4 billion was available for operating expenses in 2012, the Federal share of which decreased from the 2010 level of 9.4 percent to 8.1 percent. The largest share of Federal funds applied to operating expenditures comes from the Urbanized Area Formula Program (Title 49 U.S.C. Section 5307), which contributed 77 percent of all Federal funds. This program includes operating assistance for urbanized areas with populations less than 200,000 and capital funds eligible for operating assistance, such as preventive maintenance. Funds from the Recovery Act contributed 4 percent of Federal funds. The remaining 19 percent included FTA, Department of Transportation, and other Federal funds. The share generated from system revenues remained relatively stable, 37.5 percent in 2010 compared to 37.7 percent in 2012. The State share increased marginally from 25.0 percent in 2010 to 25.8 percent in 2012. The local share of operating expenditures was essentially unchanged from 28.2 percent in 2010 to 28.4 percent in 2012.

Operating Expenditures by Transit Mode

As shown in *Exhibit 6-30*, total transit operating expenditures were $37.6 billion in 2012. These expenditures increased at an average annual rate of 5.1 percent between 2002 and 2012 (in current dollars). Light rail and demand-response modes experienced the largest percentage increase in operating expenditures during this period. This increase is due to relatively greater investment in new light rail and demand-response capacity over the past 10 years.

Exhibit 6-29 Urban Sources of Funds for Transit Operating Expenditures, 2002–2012[1]

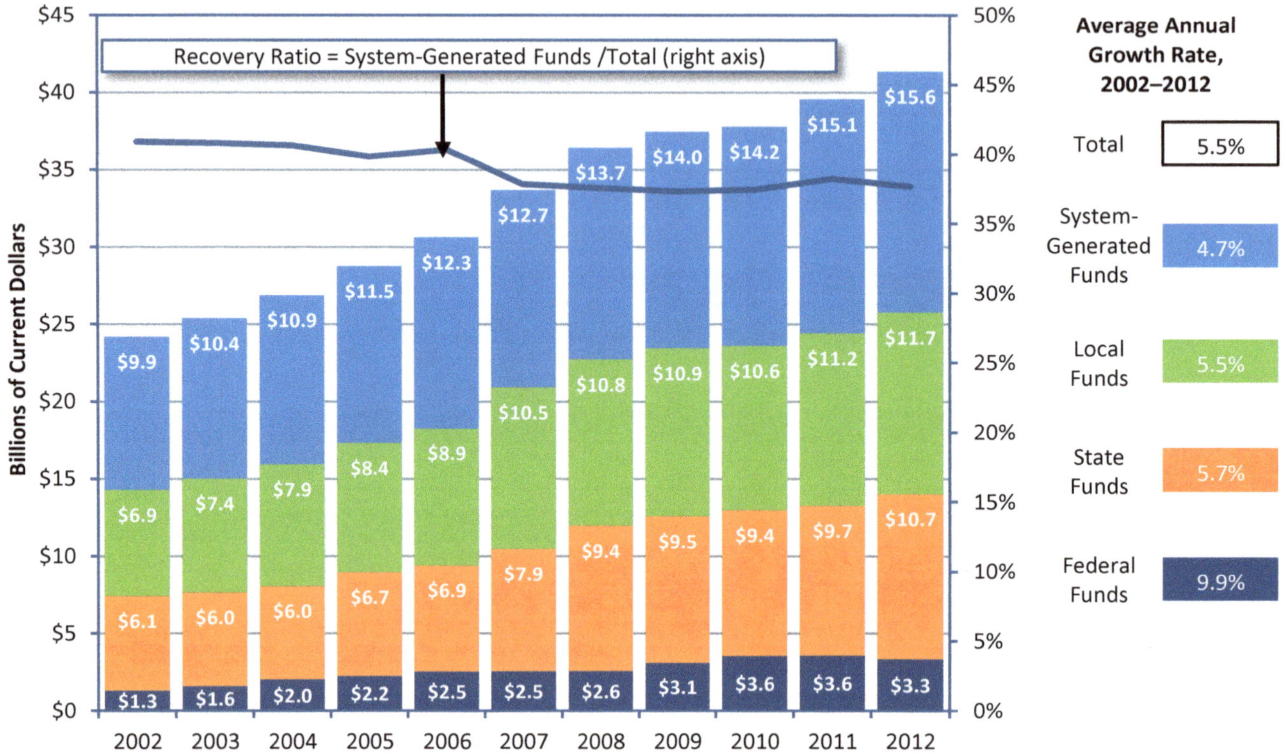

[1] *This chart includes reconciled funds that eliminate double counting and incorporates funds (fares and revenues accrued through a purchased transportation agreement) reported by sellers under contract to public agencies that are also reported by the buyers.*

Source: National Transit Database.

Exhibit 6-30 Urban Transit Operating Expenditures by Mode, 2002–2012[1]

	Expenditures, Millions of Current Dollars						
Year	Fixed-Route Bus	Heavy Rail	Commuter Rail	Light Rail	Demand Response	Other	Total
2002	$12,586	$4,267	$2,995	$778	$1,636	$643	**$22,905**
2003	$13,316	$4,446	$3,173	$754	$1,779	$718	**$24,185**
2004	$13,790	$4,734	$3,436	$826	$1,902	$739	**$25,427**
2005	$14,666	$5,145	$3,657	$978	$2,071	$721	**$27,238**
2006	$15,796	$5,287	$3,765	$1,070	$2,286	$820	**$29,025**
2007	$16,812	$5,888	$4,001	$1,163	$2,539	$901	**$31,304**
2008	$17,963	$6,129	$4,294	$1,259	$2,861	$975	**$33,479**
2009	$18,313	$6,311	$4,538	$1,393	$3,053	$1,030	**$34,638**
2010	$18,399	$6,370	$4,595	$1,499	$3,171	$1,037	**$35,071**
2011	$18,751	$6,669	$4,669	$1,559	$3,290	$1,071	**$36,009**
2012	$19,324	$6,982	$4,929	$1,683	$3,477	$1,162	**$37,556**
Percent of Total							
2002	54.9%	18.6%	13.1%	3.4%	7.1%	2.8%	**100.0%**
2012	51.5%	18.6%	13.1%	4.5%	9.3%	3.1%	**100.0%**

[1] This chart does not include reconciled funds.

Source: National Transit Database.

Operating Expenditures by Type of Cost

In 2012, $19.9 billion, or 53.0 percent of total transit operating expenditures, went toward vehicle operations. Smaller amounts were expended on maintenance and administration; these expenses, which have virtually been unchanged for the past several years, are broken down across cost categories in *Exhibit 6-31*.

Exhibit 6-31 Urban Operating Expenditures by Mode and Functions, 2012

	Distribution of Expenditures, Millions of Dollars				
Mode	Vehicle Operations	Vehicle Maintenance	Nonvehicle Maintenance	General Administration	Total
Fixed-Route Bus	$11,171	$3,728	$838	$2,951	**$18,938**
Heavy Rail	$3,008	$1,231	$1,768	$974	**$6,982**
Commuter Rail	$2,013	$1,223	$880	$812	**$4,929**
Demand Response	$2,080	$403	$91	$646	**$3,351**
Light Rail	$607	$327	$280	$272	**$1,486**
Ferry Boat	$349	$87	$38	$73	**$547**
Commuter Bus	$193	$67	$14	$61	**$349**
Trolley Bus	$123	$50	$19	$42	**$234**
Vanpool	$67	$21	$2	$81	**$172**
Other[1]	$62	$35	$27	$40	**$163**
Streetcar Rail	$56	$35	$12	$32	**$134**
Demand Response Taxi	$80	$15	$3	$26	**$126**
Hybrid Rail	$29	$9	$8	$17	**$63**
Publico	$31	$9	$0	$6	**$46**
Bus Rapid Transit	$28	$3	$3	$3	**$36**
Total	**$19,898**	**$7,241**	**$3,983**	**$6,035**	**$37,556**
Percent of All Modes	53.0%	19.3%	10.6%	16.1%	100.0%

[1] Includes Alaska railroad, cable car, inclined plane, and monorail/automated guideway.

Source: National Transit Database.

Exhibits 6-32 and *6-33* illustrate how road and rail operations have inherently different cost structures because, in most cases, roads are not paid for by the transit provider, but tracks are. A significantly higher percentage of expenditures for rail modes of transportation is classified as nonvehicle maintenance, corresponding to the repair and maintenance costs of fixed guideway systems.

Operating Expenditures per Vehicle Revenue Mile

Operating expenditures per vehicle revenue mile (VRM) is one measure of financial or cost efficiency. It shows the expense of operating a transit vehicle in revenue service. As shown in *Exhibit 6-34*, operating expenditures per VRM for all transit modes combined was $9.48 in 2012. The average annual increase in operating expenditures per VRM for all modes combined between 2002 and 2012 was 1.1 percent in constant dollars.

Exhibit 6-32 2012 Urban Rail Operating Expenditures by Type of Cost, Millions of Dollars

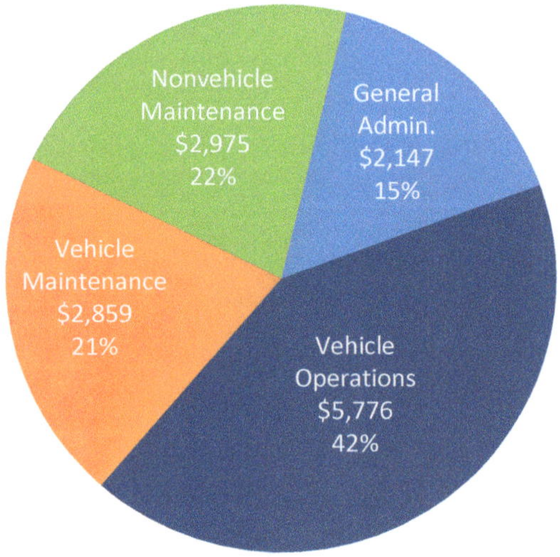

Source: National Transit Database.

Exhibit 6-33 2012 Urban Nonrail Operating Expenditures by Type of Cost, Millions of Dollars

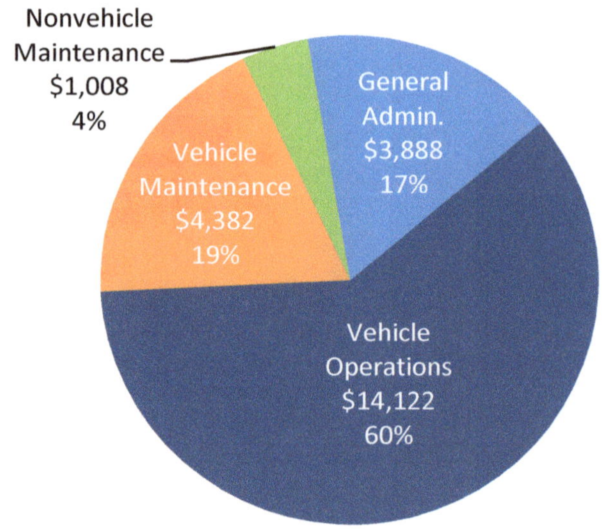

Source: National Transit Database.

Exhibit 6-34 Urban Operating Expenditures per Vehicle Revenue Mile, 2002–2012 (Constant Dollars)

Year	Fixed-Route Bus[1]	Heavy Rail	Commuter Rail	Light Rail	Demand Response	Other[2]	Total
2002	$8.62	$9.02	$14.75	$16.57	$3.97	$7.13	$8.53
2003	$8.83	$9.07	$15.11	$15.29	$4.08	$7.94	$8.68
2004	$8.89	$9.21	$15.54	$15.07	$4.12	$6.33	$8.71
2005	$9.15	$9.64	$15.52	$16.93	$4.11	$5.48	$8.89
2006	$9.42	$9.50	$14.94	$16.70	$4.29	$5.84	$8.33
2007	$9.64	$10.21	$14.93	$15.63	$4.36	$5.72	$9.20
2008	$9.79	$9.97	$14.82	$15.54	$4.43	$5.22	$9.17
2009	$9.95	$10.13	$15.55	$16.78	$4.51	$4.86	$9.30
2010	$10.11	$10.36	$15.38	$17.23	$4.65	$4.73	$9.42
2011	$10.14	$10.70	$15.31	$16.87	$4.53	$4.47	$9.39
2012	$10.21	$10.94	$15.51	$17.04	$4.58	$4.56	$9.48
Average Annual Rate of Change							
2012/2002	1.7%	1.9%	0.5%	0.3%	1.4%	-4.4%	1.1%

[1] Note that annual changes in operating expense per capacity-equivalent VRM and unadjusted fixed-route bus operating expenditures are consistent with those shown in *Exhibit 6-31*.

[2] Includes monorail/automated guideway, Alaska railroad, cable car, ferryboat, inclined plane, Público, trolleybus, and vanpool.

Source: National Transit Database.

Exhibit 6-35 shows average per-mile fares and costs for the Nation's 10 largest transit agencies since 2002. After adjusting for inflation (constant dollars), fares per mile over this period increased 6.6 percent, while the average cost per mile increased 6.3 percent. These increases resulted in a 0.3-percent increase in the "fare recovery ratio," which is the percentage of operating costs covered by passenger fares. The 2012 fare recovery ratio for these 10 agencies was 38.0 percent. These 10 agencies are all rail agencies, and rail systems tend to have lower operating costs per mile. Therefore, this fare recovery ratio is higher than would be found for most fixed-route bus or demand-response operations.

Exhibit 6-35 Urban Average Fares and Operating Costs per Mile—Top 10 Transit Agencies, 2002–2012 (Constant Dollars)

Top 10 Systems[1]	2002	2003	2004	2005	2006	2007	2008	2009	2010	2011	2012	% Increase 2002–2012	% Increase Average Annual
Average Fare per Mile	$2.57	$2.63	$2.88	$3.08	$3.38	$3.58	$3.81	$3.80	$4.10	$4.67	$4.90	90%	6.6%
Average Operating Cost per Mile	$6.99	$7.34	$7.85	$8.53	$9.23	$10.11	$10.48	$10.72	$11.38	$12.27	$12.90	85%	6.3%
Average Recovery Ratio	36.8%	35.9%	36.6%	36.1%	36.6%	35.4%	36.4%	35.4%	36.1%	38.1%	38.0%	3%	0.3%

[1] Includes Metropolitan Transit Authority New York City, Chicago Transit Authority, Los Angeles County Metropolitan Transportation Authority, Washington Metropolitan Area Transit Authority, Massachusetts Bay Transportation Authority, Southeastern Pennsylvania Transportation Authority, New Jersey Transit Corporation, San Francisco Municipal Railway, Metropolitan Atlanta Rapid Transit Authority, and Maryland Transit Administration.

Source: National Transit Database.

As shown in *Exhibit 6-36*, analysis of the National Transit Database reports for the top 10 agencies in urbanized areas with greater than 1 million in population shows that the growth in operating expenses is led by the cost of fringe benefits, which have been increasing at a rate of 2.3 percent per year above inflation (constant dollars) since 2002. By comparison, average salaries at these 10 agencies decreased at an inflation-adjusted rate of 0.9 percent per year in that period.

Exhibit 6-36 Urban Growth in Operating Costs—Top 10 Agencies, 2002–2012 (Constant Dollars), Directly Operated Services[1]

	Average Cost per Mile, Constant Dollars											% Increase	
Cost Component	2002	2003	2004	2005	2006	2007	2008	2009	2010	2011	2012	2002–2012	Average Annual
Salaries	$5.53	$5.47	$5.40	$5.26	$5.29	$5.37	$4.98	$5.02	$5.04	$5.10	$5.03	-9%	-0.9%
Fringe Benefits	$3.29	$3.46	$3.59	$3.65	$3.71	$3.96	$3.53	$3.72	$3.91	$4.08	$4.13	26%	2.3%
Labor Cost	$8.82	$8.93	$8.99	$8.91	$9.01	$9.34	$8.51	$8.74	$8.95	$9.18	$9.16	4%	0.4%

[1] Includes Metropolitan Transit Authority New York City, Chicago Transit Authority, Los Angeles County Metropolitan Transportation Authority, Washington Metropolitan Area Transit Authority, Massachusetts Bay Transportation Authority, Southeastern Pennsylvania Transportation Authority, New Jersey Transit Corporation, San Francisco Municipal Railway, Metropolitan Atlanta Rapid Transit Authority, and Maryland Transit Administration. Includes directly operated services only as labor data are not reported for purchased transportation services in the NTD.

Source: National Transit Database.

Operating expenditures per capacity-equivalent VRM are a better measure of comparing cost efficiency among modes than operating expenditures per VRM because they adjust for passenger-carrying capacities. As demonstrated in *Exhibit 6-37*, rail systems are more cost efficient in providing service than nonrail systems, once investment in rail infrastructure has been completed.

Based on operating costs alone, heavy rail is the most efficient at providing transit service, and demand-response systems are the least efficient. Annual changes in operating expense per capacity-equivalent VRM are not comparable across modes because average capacities for all vehicle types are adjusted separately each year based on reported fleet averages.

Exhibit 6-37 Urban Operating Expenditures per Capacity-Equivalent Vehicle Revenue Mile by Mode, 2002–2012 (Constant Dollars)

Year	Fixed-Route Bus	Heavy Rail	Commuter Rail	Light Rail	Demand Response	Other[1]	Total
2002	$8.62	$3.82	$6.33	$6.57	$22.08	$10.76	$6.78
2003	$8.83	$3.66	$5.93	$5.68	$22.65	$11.95	$6.85
2004	$8.89	$3.72	$6.10	$5.60	$24.22	$11.07	$6.90
2005	$9.15	$3.88	$5.07	$6.15	$24.78	$10.18	$7.06
2006	$9.42	$3.82	$4.88	$6.06	$25.86	$11.29	$7.17
2007	$9.64	$4.13	$4.91	$5.75	$25.99	$11.09	$7.14
2008	$9.79	$4.03	$4.87	$5.72	$26.44	$13.77	$7.22
2009	$9.95	$4.82	$5.66	$6.70	$27.36	$12.16	$7.95
2010	$10.11	$4.19	$5.63	$6.27	$26.83	$11.75	$7.62
2011	$10.14	$4.32	$5.48	$5.87	$25.95	$11.58	$7.61
2012	$10.20	$4.42	$5.55	$5.93	$26.27	$11.73	$7.70
Compound Annual Growth Rate							
2012/2002	1.7%	1.5%	-1.3%	-1.0%	1.8%	0.9%	1.3%

[1] Includes monorail/automated guideway, Alaska railroad, cable car, ferryboat, inclined plane, Público, trolleybus, and vanpool.

Source: National Transit Database.

Operating Expenditures per Passenger Mile

Operating expense per passenger mile is an indicator of the cost effectiveness of providing a transit service. It shows the relationship between service inputs as expressed by operating expenses and service consumption as measured in passenger miles traveled. Operating expenditures per passenger mile for all transit modes combined increased at an average annual rate of 0.7 percent between 2002 and 2012 (from $0.64 to $0.68). These data are shown in *Exhibit 6-38.*

Farebox Recovery Ratios

The farebox recovery ratio represents farebox revenues as a percentage of total transit operating costs. It measures users' contributions to the variable cost of providing transit services and is influenced by the number of riders, fare structure, and rider profile. Low regular fares, high availability and use of discounted fares, and high transfer rates tend to result in lower farebox recovery ratios. Farebox recovery ratios for 2004 to 2012 are provided in *Exhibit 6-39*. The average farebox recovery ratio over this period for all transit modes combined was 35.0 percent; heavy rail had the highest average farebox recovery ratio at 61.1 percent. Farebox recovery ratios for total costs are not provided because capital investment costs are not evenly distributed across years. Rail modes have farebox recovery ratios for total costs that are significantly lower than for operating costs alone because of these modes' high level of capital costs.

Exhibit 6-38 Urban Operating Expenditures per Passenger Mile, 2002–2012 (Constant Dollars)

Year	Fixed-Route Bus	Heavy Rail	Commuter Rail	Light Rail	Demand Response	Other[1]	Total
2002	$0.82	$0.40	$0.40	$0.69	$3.21	$0.70	**$0.64**
2003	$0.86	$0.41	$0.41	$0.69	$3.22	$0.70	**$0.66**
2004	$0.89	$0.40	$0.43	$0.68	$3.28	$0.64	**$0.67**
2005	$0.89	$0.42	$0.45	$0.68	$3.30	$0.62	**$0.68**
2006	$0.88	$0.41	$0.41	$0.65	$3.46	$0.66	**$0.67**
2007	$0.91	$0.40	$0.40	$0.67	$3.61	$0.66	**$0.67**
2008	$0.90	$0.39	$0.42	$0.64	$3.61	$0.61	**$0.66**
2009	$0.93	$0.40	$0.44	$0.68	$3.71	$0.62	**$0.69**
2010	$0.94	$0.41	$0.45	$0.73	$3.82	$0.60	**$0.70**
2011	$0.93	$0.39	$0.42	$0.67	$3.82	$0.58	**$0.68**
2012	$0.91	$0.40	$0.44	$0.68	$3.92	$0.58	**$0.68**
Compounded Annual Growth Rate							
	1.1%	0.0%	1.0%	-0.3%	2.0%	-1.9%	**0.7%**

[1] Includes monorail/automated guideway, Alaska railroad, cable car, ferryboat, inclined plane, Público, trolleybus, and vanpool.

Source: National Transit Database.

Exhibit 6-39 Urban Farebox Recovery Ratio by Mode, 2004–2012[1]

Year	Fixed-Route Bus	Heavy Rail	Commuter Rail	Light Rail	Demand Response	Other[2]	Total
2004	27.9%	61.3%	47.0%	26.2%	9.6%	36.2%	**35.5%**
2005	27.6%	58.4%	47.2%	25.4%	9.5%	12.6%	**34.8%**
2006	26.6%	60.9%	49.4%	27.4%	9.3%	34.3%	**34.8%**
2007	26.6%	56.8%	49.5%	26.6%	8.2%	35.3%	**34.0%**
2008	26.3%	59.4%	50.3%	29.3%	7.5%	32.7%	**34.1%**
2009	26.7%	60.2%	47.9%	28.0%	7.8%	34.9%	**34.2%**
2010	26.7%	62.3%	48.5%	27.5%	7.9%	37.0%	**34.7%**
2011	27.9%	66.0%	52.0%	28.9%	7.3%	37.9%	**36.5%**
2012	28.2%	64.6%	51.8%	29.0%	7.7%	40.4%	**36.6%**
Average	**27.2%**	**61.1%**	**49.3%**	**27.6%**	**8.3%**	**33.5%**	**35.0%**

[1] Note that the ratios presented in this exhibit were calculated differently from the ratios presented in the 2008 C&P Report; therefore, they are not totally comparable. The ratios presented here were calculated using data from NTD data table 26, "Fares per Passenger and Recovery Ratio" (available at www.ntdprogram.gov/ntdprogram/data.htm).

[2] Includes monorail/automated guideway, Alaska railroad, cable car, ferryboat, inclined plane, Público, trolleybus, and vanpool.

Source: National Transit Database.

Rural Transit

The Federal government has contributed to the funding of transit in rural areas (i.e., areas with populations less than 50,000) since 1978. These rural areas are estimated to account for approximately 36 percent of the U.S. population and 38 percent of the transit-dependent population.

Funding for rural transit is currently provided through Section 5311, the Rural Formula Grant Program. Rural transit funding increased substantially with passage of the Transportation Equity

Act for the 21st Century (TEA-21) and continued to increase under the Safe, Accountable, Flexible, Efficient Transportation Equity Act: A Legacy for Users (SAFETEA-LU). Federal funding for rural transit was $240 million in the last year of TEA-21, Fiscal Year 2004, and reached $569 million in the last year of SAFETEA-LU, Fiscal Year 2012. States may transfer additional funds to rural transit from highway projects or formula transit funds for small urbanized areas.

Federal funds constitute the bulk of capital funds applied to rural transit. As shown in *Exhibit 6-40*, Federal funds accounted for 80 percent of rural transit capital budgets. This amount of funding is a historic record and was due to funds being made available from the Recovery Act, which accounted for 30 percent of the total capital investment. The two main sources of non-Recovery Act Federal funds are the FTA Capital Program (Section 5309) and Urbanized Area Formula Funds (49 U.S.C. Section 5311). These two sources combined contributed 38.3 percent of non-Recovery Act funds. The other 11.7 percent originated from other FTA programs and Federal sources.

Exhibit 6-40 Rural Transit Funding Sources for Capital Expenditures, 2012

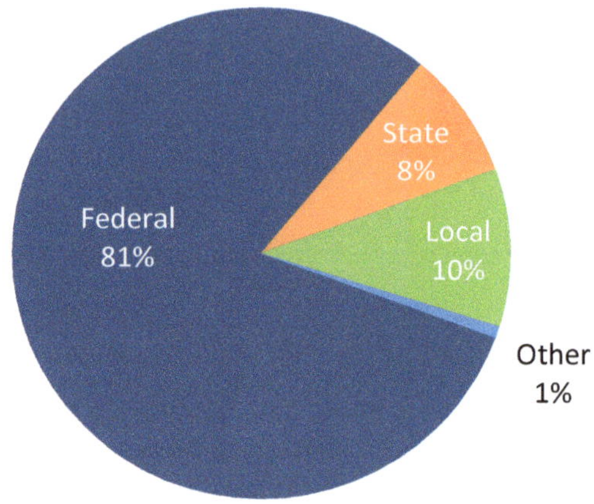

Source: N

As shown in *Exhibit 6-41*, 33 percent of rural transit authorities' operating budgets come from Federal funds. State and local governments cover 41 percent of their rural transit operating budgets through a combination of dedicated State and local taxes, appropriations from State general revenues, and allocations from other city and county funds. Contract revenue, defined as reimbursement from a private entity (profit or nonprofit) for the provision of transit service, accounts for 18 percent of operating budgets for rural transit. Fares accounted for only 8 percent, close to the average farebox recovery rate for demand-response service (which constitutes most of rural transit). In 2012, the total value of rural transit operating budgets reported to the National Transit Database was $1.37 billion.

Exhibit 6-41 Rural Transit Funding Sources for Operating Expenditures, 2012

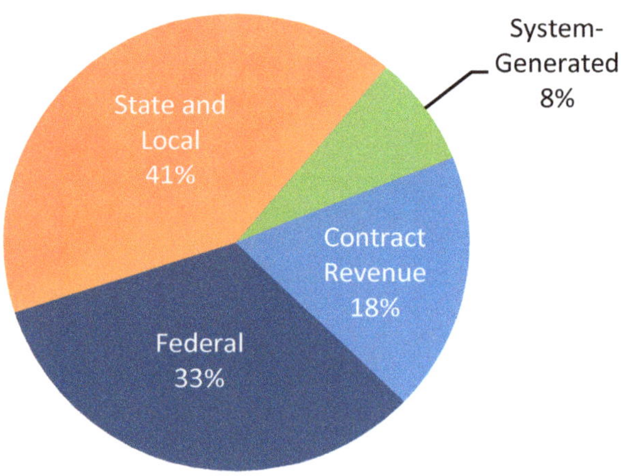

Source: National Transit Database.

part II

Investment/Performance Analysis

Introduction

To support the development and evaluation of transportation policies and programs, Chapters 7 through 10 present and analyze general scenarios for future capital investment in highways, bridges, and transit. In each of these 20-year scenarios, the investment level is an estimate of the spending that would be required to achieve a certain level of infrastructure performance. **The scenarios do not address how much different levels of government might contribute to funding the investment, nor do they directly address the potential contributions of different public or private revenue sources**.

The four investment-related chapters in Part II measure investment levels in constant 2012 dollars, except where noted otherwise, and include the following analyses:

Chapter 7, **Potential Capital Investment Impacts**, analyzes the projected impacts of alternative levels of future investment on measures of physical condition, operational performance, and benefits to system users. Each alternative pertains to investment from 2013 through 2032 and is presented as an annual average level of investment and as the constant annual percentage rate of increase or decrease in investment that would produce that annual average.

Chapter 8, **Selected Capital Investment Scenarios**, examines several scenarios distilled from the investment alternatives considered in Chapter 7. Some of the scenarios are oriented around maintaining different aspects of system condition and performance or achieving a specified minimum level of performance, while others link to broader measures of system user benefits. The scenarios included in this chapter are intended to be illustrative and do not represent comprehensive alternative transportation policies; the U.S. Department of Transportation (DOT) does not endorse any scenario as a target level of investment.

Chapter 9, **Supplemental Scenario Analysis**, explores some of the implications of the scenarios presented in Chapter 8 and contains some additional policy-oriented analyses addressing issues not covered in Chapters 7 and 8. As part of this analysis, highway projections from previous editions of the C&P report are compared with actual outcomes to elucidate the value and limitations of the projections presented in this edition.

Chapter 10, **Sensitivity Analysis**, explores the impacts on scenario projections by varying some of the key assumptions. The investment scenario projections in this report are developed using models that evaluate current system condition and operational performance and make 20-year projections based on assumptions about future travel growth and a variety of engineering and economic variables. The accuracy of these projections depends, in large part, on the realism of these assumptions. Since the future rate of growth in transit travel is uncertain, Chapter 7 considers alternative high and low values for this parameter. Chapter 10 likewise varies the assumed rate of growth in highway travel and the values assumed for the discount rate, the value of travel time savings, and other assumed parameters. Other sources of uncertainty in the modeling procedures are discussed below.

Unlike Chapters 1 through 6, which largely present highway and transit statistics drawn from other sources, the investment scenario projections presented in these chapters (and the models used to create the projections) were developed exclusively for the C&P report. The procedures for developing the investment scenario estimates have evolved over time to incorporate recent research, new data sources, and improved estimation techniques. These procedures are described more fully in Appendices A (Highways), B (Bridges), and C (Transit).

The combination of engineering and economic analysis in this part of the C&P report is consistent with the movement of transportation agencies toward asset and performance management, value engineering, and greater consideration of cost effectiveness in decision-making. The economic approach to transportation investment is discussed at the end of this section.

Capital Investment Scenarios

The projections for the 20-year capital investment scenarios shown in this report reflect complex technical analyses that attempt to predict the impact that capital investment might have on the future conditions and performance of the transportation system. These scenarios are illustrative, and DOT does not endorse any of them as a target level of investment. Where practical, supplemental information is included to describe the impacts of other possible investment levels.

This report does not attempt to address issues of cost responsibility. The investment scenarios predict the impact that particular levels of combined Federal, State, local, and private investment might have on the overall conditions and performance of highways, bridges, and transit. Although Chapter 6 provides information on what portion of highway investment has come from different levels of government in the past, the report makes no specific recommendations about what these portions, or that from the private sector, should be in the future.

The system condition and performance projections in this report's capital investment scenarios represent what could be achievable assuming a particular level of investment, rather than what would be achieved. The models used to develop the projections generally assume that, when funding is constrained, the benefit-cost ratio (BCR) establishes the order of precedence among potential capital projects, with projects having higher BCRs selected first. In actual practice, the BCR generally omits some types of benefits and

State Use of Benefit-Cost Analysis

DOT recently issued a report to Congress, *Use of Benefit-Cost Analysis by State Departments of Transportation*, in response to a requirement in Senate Report 113-182 accompanying the Transportation and Housing and Urban Development, and Related Agencies Appropriations Bill, 2015.

The study revealed that the extent to which State DOTs use benefit-cost analysis continues to vary significantly among States, project types, and planning stages. The quality of benefit-cost analysis also varies, as it is affected by availability of data and appropriate baselines for comparison, benefit definitions, and accuracy of traffic demand forecasts. State DOTs face institutional, resource, and technical challenges in conducting benefit-cost analysis. Potential strategies to address these challenges include outreach and communication, technical training, and provision of assistance in methodological issues.

costs because of difficulties in valuing them monetarily, and these other benefits and costs can and do affect project selection. In addition, actual project selection can be guided by political or other considerations outside benefit-cost analysis.

Highway and Bridge Investment Scenarios

Projections for future conditions and performance under alternative potential levels of investment are developed independently for highways and bridges in Chapter 7 using separate models and techniques, and then combined for selected investment scenarios in Chapter 8. Investments in bridge repair, rehabilitation, and replacement are modeled by the National Bridge Investment Analysis System (NBIAS); those in capacity expansion and the highway resurfacing and reconstruction component of system rehabilitation are modeled by the Highway Economic Requirements System (HERS). Although HERS was primarily designed to analyze highway segments, it also factors in the costs of expanding bridges and other structures when determining whether to add lanes to a highway segment. Some elements of highway investment spending are modeled by neither HERS nor NBIAS. Chapter 8 factors these elements into the investment levels associated with each scenario using scaling procedures external to the models. The scenario investment levels are estimates of the amount of future capital spending required to meet the performance goals specified in the scenarios.

Chapter 8 uses consistent performance criteria to create separate but parallel investment scenarios for all Federal-aid highways, the National Highway System, and the Interstate System. Corresponding scenarios are also presented for all roads system wide, but projections for these scenarios are less reliable because data coverage is more limited off the Federal-aid highways. Although the NBIAS database includes information on all bridges, the Highway Performance Monitoring System (HPMS) database, on which the HERS model relies, includes detailed information only on Federal-aid highways; for the scenarios based on all roads, non-model-based estimates must be generated for roads functionally classified as rural minor collectors, rural local, or urban local.

The **Sustain 2012 Spending scenario** projects the potential impacts of sustaining capital spending at 2012 base-year levels in constant-dollar terms over the 20-year period 2013 through 2032. The **Maintain Conditions and Performance scenario** assumes that combined highway capital investment by all levels of government gradually changes in constant-dollar terms over 20 years to the point at which selected performance indicators in 2032 are maintained at their 2012 base-year levels. For this edition of the C&P report, the HERS component of the scenario is defined as the lowest level of investments required to at least maintain each of two performance indicators—average pavement roughness and average delay per vehicle mile traveled (VMT)—at their base-year level or better. For the NBIAS component, the benchmark performance indicator is the percentage of deck area on deficient bridges.

The investment levels for the **Improve Conditions and Performance scenario** are determined by identifying the highest rate of annual spending growth for which potentially cost-beneficial highway and bridge improvements can be identified. This scenario represents an "investment ceiling" above which further investment would not be cost-beneficial, even if available funding were unlimited. The portion of this scenario directed toward addressing engineering deficiencies on pavements and bridges is described as the **State of Good Repair benchmark**.

Transit Investment Scenarios

The transit section of Chapter 7 evaluates the impact of varying levels of capital investment on various measures of condition and performance, while the transit section of Chapter 8 provides a more in-depth analysis of specific investment scenarios.

The **Sustain 2012 Spending scenario** projects the potential impacts of sustaining preservation and expansion spending at 2012 base-year levels in constant-dollar terms over the 20-year period of 2013 through 2032. The scenario applies benefit-cost analysis to prioritize investments within this constrained budget target.

The **State of Good Repair benchmark** projects the level of investment needed to bring all assets to a state of good repair over the next 20 years, defined as asset condition ratings of 2.5 or higher on a 5-point scale (Chapter 3 discusses these ratings). This scenario does not apply a benefit-cost test and focuses solely on the preservation of existing assets.

The **Low-Growth and High-Growth scenarios** each add a system expansion component to the system preservation needs associated with the State of Good Repair benchmark. The goal of these scenarios is to preserve existing assets and expand the transit asset base to support projected ridership growth over 20 years based on forecasts linked to the average annual growth experienced between 1997 and 2012. The Low-Growth scenario projects ridership growth at 0.5 percent per year less than the historic trend, while the High-Growth scenario incorporates a more extensive expansion of the existing transit asset base to support ridership growth at 0.5 percent per year above the historic trend. Both scenarios incorporate a benefit-cost test for evaluating potential investments; thus, their system preservation components are somewhat smaller than the level identified in the State of Good Repair benchmark.

Comparisons between Report Editions

When comparing capital investment scenarios presented in different editions of the C&P report, several considerations should be taken into account.

Scenario definitions have been modified over time. Between the 2013 C&P Report and the current edition, the target performance indicators in the **Maintain Conditions and Performance scenarios** have changed. In the 2013 edition, the indicator for investments modeled by HERS was the average between the investment levels required to maintain, alternatively, average pavement roughness or average congestion delay per VMT. For the investments modeled by NBIAS, the target performance indicator in the 2013 edition was the average sufficiency rating for bridges.

Before the 2013 edition, the scenarios in the C&P report for highway and bridge investment assumed that VMT would grow as forecast by the States for HPMS. The 2013 edition added an alternative set of scenarios that projected aggregate growth in VMT at the 15-year historic trend rate. This change made the highway and bridge investment scenarios more comparable to the transit investment scenarios, which have included an alternative trend-based forecast for ridership growth since the 2010 edition. In the current report edition, however, all scenarios for highway and bridge investment assume aggregate growth in VMT at the rate forecast by an econometric model recently developed for FHWA. This forecast was judged more realistic than the aggregate growth rate based on the forecasts the States submit to HPMS, which evidence presented in Chapter 9 suggests has been over-predicting in recent years.

The base year of the analysis advances two years between successive editions of this biennial report. During this period, changes in many real-world factors can affect the investment scenario estimates. Among these factors are construction costs and other prices, conditions and performance of the highway and transit systems, expansion of the system asset base, and changes in technology (such as improvements in motor vehicle fuel economy). Although relevant to all scenarios, this issue is particularly significant for scenarios aimed at maintaining base-year conditions. Comparability across C&P report editions is also limited by changes over time in the analytical tools and data sets used in generating the scenarios.

The Economic Approach to Transportation Investment Analysis

The methods and assumptions used to analyze future highway, bridge, and transit investment scenarios are continuously evolving. Since the beginning of the highway report series in 1968, enhancements to the highway investment scenarios have resulted from innovations in analytical methods, new data and evidence, and changes in transportation planning objectives. Estimates of future requirements for highway investment, as reported in the *1968 National Highway Needs Report to Congress*, began as a combined "wish list" of State highway "needs." As the focus of national highway investment changed from system expansion to management of the existing system during the 1970s, national engineering standards were defined and applied to identify system deficiencies, and the investments necessary to remedy these deficiencies were estimated. By the end of the decade, a comprehensive database, the HPMS, had been developed to enable monitoring of highway system conditions and performance nationwide.

In the early 1980s, a sophisticated simulation model, the HPMS Analytical Process (HPMS-AP), became available to evaluate the impact of alternative investment strategies on system conditions and performance. The procedures used in HPMS-AP were based on engineering principles. Engineering standards were applied to determine which system attributes were considered deficient, and improvement option packages were developed using standard engineering countermeasures for given deficiencies, but without consideration of comparative economic benefits and costs.

In 1988, the Federal Highway Administration embarked on a long-term research and development effort to produce an alternative simulation procedure combining engineering principles with economic analysis. The product of this effort, the HERS model, was first used to develop one of the two highway investment scenarios presented in the 1995 C&P Report. In subsequent reports, HERS has been used to develop all the highway investment scenarios.

Executive Order 12893, "Principles for Federal Infrastructure Investments," issued on January 26, 1994, directs that Federal infrastructure investments should be based on a systematic analysis of expected benefits and costs. This order provided additional momentum for the shift toward developing analytical tools that incorporate economic analysis into the evaluation of investment requirements.

In the 1997 C&P Report, the Federal Transit Administration introduced the Transit Economics Requirements Model (TERM), which was used to develop both of the transit investment scenarios. TERM incorporates benefit-cost analysis into its determination of transit investment levels.

The 2002 C&P Report incorporated economic analysis into bridge investment modeling for the first time with the introduction of NBIAS.

The Economic Approach in Theory and Practice

The economic approach to transportation investment entails analysis and comparison of benefits and costs. Investments that yield benefits for which the values exceed their costs increase societal welfare and are thus considered "economically efficient," or "cost-beneficial." For such analysis to be reliable, it must adequately consider the range of possible benefits and costs and the range of possible investment alternatives.

Which Benefits and Costs Should Be Considered?

A comprehensive benefit-cost analysis of a transportation investment considers all impacts of potential significance for society and values them in monetary terms, to the extent feasible. For some types of impacts, monetary valuation is facilitated by the existence of observable market prices. Such prices are generally available for inputs to the provision of transportation infrastructure, such as concrete for building highways or buses purchased for a transit system. The same is true for some types of benefits from transportation investments, such as savings in business travel time, which are conventionally valued at a measure of average hourly labor cost of the travelers.

For some other types of impacts for which market prices are not directly observable, monetary values can be reasonably inferred from behavior or expressed preferences. In this category are savings in non-business travel time and reductions in risk of crash-related fatality or other injury. As discussed in Chapter 10 (under "Value of a Statistical Life"), what is inferred is the amount that people typically would be willing to pay per unit of improvement, for example, per hour of non-business travel time saved. These values are combined with estimates of the magnitude of the improvement (or, as may happen, deterioration).

For other impacts, monetary valuation may not be possible because of problems with reliably estimating the magnitude of the improvement, placing a monetary value on the improvement, or both. Even when possible, reliable monetary valuation may require time and effort that would be out of proportion to the likely importance of the impact concerned. Benefit-cost analyses of transportation investments thus typically will omit valuing certain impacts that are difficult to monetize but, nevertheless, could be of interest.

The models used in this report—HERS, NBIAS, and TERM—each omit various types of investment impacts from their benefit-cost analyses. To some extent, this omission reflects the national coverage of their primary databases. Although consistent with this report's focus on the Nation's highways and transit systems, such broad geographic coverage requires some sacrifice of detail to stay within feasible budgets for data collection. In the future, technological progress in data collection and growing demand for data for performance management systems for transportation infrastructure likely will yield national databases that are more comprehensive and of better quality.

In addition, DOT will continue to explore other avenues for addressing impacts not captured by the suite of models used for the C&P report. One approach is to have the models represent impacts in ways that are sufficiently simplified to demand no more data than are available. This approach was taken to represent within HERS the impacts of traffic disruptions resulting from road construction. Another approach that DOT will continue to explore for the C&P report is to supplement the findings from HERS, NBIAS, and TERM with evidence from other sources. This approach could elucidate various environmental, health, and community impacts of highway and transit investments. Examples include environmental impacts of increased water runoff from highway pavements, barrier effects of highways for human and animal populations, health benefits from the additional walking activity when travelers use transit rather than cars, and other impacts related to livability. Another effect the DOT models do not consider, but which could be significant for some transportation investments, is the boost to economic competitiveness that results when travel times among competing producers are lessened. Faced with stiffer competition from rivals in other locations, producers may become more efficient and lower prices.

What Alternatives Should Be Analyzed?

Benefit-cost analyses of transportation investments need to include a sufficiently broad range of investment alternatives to be able to identify which is optimal. For transit and highway projects, this can entail consideration of cross-modal alternatives. Transit and highway projects can be complements, as when the addition of high-occupancy toll lanes to a freeway allows for new or improved bus express services; they can also be substitutes, as when construction of a light rail line lessens the demand for travel on a parallel freeway. In contrast, HERS and TERM each focus on investment in just one mode, and to incorporate a cross-modal perspective properly would require a major investment of time and resources, entailing major changes to the benefit-cost methodologies and the addition of considerable detail to the supporting databases. (As was noted above, the models' databases necessarily sacrifice detail to make national-level coverage feasible). For the foreseeable future, the best way to address this deficiency in future editions of the C&P report likely would be through review of evidence obtained from more regionally focused analyses using other modeling frameworks. Opportunities for future development of HERS, TERM, and NBIAS, including efforts to allow feedback between the models, were discussed in Appendix D of the 2013 C&P Report.

Beyond related cross-modal investment possibilities, economic evaluations of investments in highways or transit should also attempt to consider related public choices, such as policies for travel demand management and local zoning, or investment in other infrastructure. Several previous editions of the C&P report presented HERS modeling of highway investment combined with system-wide highway congestion pricing. Although the results indicated that pricing could substantially reduce the amount of highway investment that would be cost-beneficial, a review of the methodology in 2010 revealed significant limitations, which reflected in part the lack of transportation network detail in the HPMS database. The decision to exclude such modeling from the 2013 and current editions of the C&P report also took into account that the results would have been unlikely to differ from those reported previously, and that system-wide congestion pricing has yet to gain widespread public support.

A more limited form of congestion pricing is tolling on designated express lanes within a full access-controlled highway. When the tolling includes a discount or exemption for high-occupancy vehicles, such facilities are termed HOT (High Occupancy Toll) lanes. Over the past three decades, tolled express lanes have been implemented in urban areas across the United States and have been gaining popular support. Future versions of the HERS model could include a capability to analyze the costs and benefits of tolled express lanes and their effects on investment needs.

Measurement of Costs and Benefits in "Constant Dollars"

Benefit-cost analyses normally measure all benefits and costs in "constant dollars," that is, at the prices prevailing in some base year, typically near the year when the analysis is released. Future price changes can be difficult to forecast, and benefits and costs measured in base-year prices are more comprehensible.

In the simplest form of constant-dollar measurement, conversion of any quantity to a dollar value is done at that quantity's base-year price. Future savings in gallons of gasoline, for example, are monetized at the average price per gallon of gasoline in the base year (with the price possibly measured net of excise tax, as in HERS). This approach, still quite common in benefit-cost analysis, was the general practice in pre-2008 editions of the C&P report. It assumes any future inflation will change all prices in equal proportion, so that the ratios among prices will remain constant at their base-year levels. With relative prices constant, whether a benefit-cost analysis uses actual base-year prices or those prices are inflated uniformly at a projected rate of inflation is purely a presentational issue.

An alternative approach is warranted when significant changes in the relative price of a quantity important to the analysis can be predicted with sufficient confidence. What constitutes sufficient confidence is a judgment call, but some predictions carry official weight. The Energy Information Administration's *Annual Energy Outlook* forecasts changes in motor-fuel prices relative to the consumer price index (CPI) 25 years out. Starting with the 2008 C&P Report, the highway investment scenarios have incorporated these CPI-deflated forecasts. Since the 2010 edition, the C&P report also has incorporated CPI-deflated forecasts of the marginal damage cost of CO_2 emissions. Values for the marginal damage are those recommended by a Federal interagency working group for use in regulatory impact analysis. For this edition of the C&P report, the values are taken from the 2013 update to these recommendations, which specify values for each year between 2010 and 2050.[1]

In this C&P report, the HERS modeling of highway investment also factors in the future growth in values of travel time savings. Such growth could be expected to result from future increases in average incomes: Notwithstanding periods of relative stagnation, the real incomes of Americans have increased over the long term and evidence indicates that people value their travel time more highly as their incomes increase. The growth rate assumed for the HERS modeling, 1.2 percent per year, was that stipulated in 2014 guidance for DOT on valuation of travel time savings for analyses with a base year of 2012. This assumption is a significant departure from the HERS analysis presented in the 2013 C&P Report, where growth in the value of travel time savings was included only as a sensitivity test and no growth was factored into the main scenarios.

Notwithstanding allowances for likely changes in prices relative to the consumer price index, the analysis in this report can be considered to measure benefits and costs in constant 2012 dollars. Office of Management and Budget guidance on benefit-cost analysis defines "real or constant dollar values" as follows: "Economic units measured in terms of constant purchasing power. A real value is not affected by general price inflation. Real values can be estimated by deflating nominal values with a general price index, such as the implicit deflator for Gross Domestic Product or the Consumer Price Index."[2]

Uncertainty in Transportation Investment Modeling

The three investment analysis models used in this report are deterministic, not probabilistic: They provide a single projected value of total investment for a given scenario rather than a range of likely values. As a result, only general statements can be made about the element of uncertainty in these projections, based on the characteristics of the process used to develop them; specific information about confidence intervals cannot be developed. As was indicated above, the analysis in Chapter 10 of this edition of the C&P report enables statements about the sensitivity of the scenario projections to variation in the underlying parameters (e.g., discount rates, value of time saved, statistical value of lives saved). As much as possible, the range of variation considered in these tests corresponds to the range considered plausible in the corresponding research literature or to ranges recommended in authoritative guidance. The sensitivity tests address only some of the elements of uncertainty in the scenario projections. In some cases, the uncertainty extends beyond the value of a model parameter to the entire specification of the equations in which the parameters are embedded.

The relative level of uncertainty differs among the various projections made in this report. As already noted, the projections for all roads system wide are less reliable than those for Federal-aid highways. In addition, the projections for absolute levels of condition and performance indicators entail more uncertainty than the differences among these levels according to an assumed level of investment. For example, if speed limits were increased nationwide in the future, contrary to the HERS modeling assumption of no change from the base-year speed limits, this might significantly reduce the accuracy performance of the model's projections for average speed. At the same time, projections of how the amount of future investments in highways affects average speed could be relatively accurate. Although investments in highway capacity expansion increase average speed, the increase will occur primarily under conditions of congestion when average speeds can be well below even the current speed limit. Under such conditions, an increase in the speed limit might have a negligible effect on the congestion reduction benefits of adding lanes.

---

[1] Interagency Working Group on Social Cost of Carbon. May 2013. *Technical Update of the Social Cost of Carbon for Regulatory Impact Analysis under Executive Order 12866*, https://www.whitehouse.gov/sites/default/files/omb/inforeg/social_cost_of_carbon_for_ria_2013_update.pdf.

[2] OMB Circular No. A-94 Revised, http://www.whitehouse.gov/omb/circulars_a094.

chapter 7

Potential Capital Investment Impacts

Potential Highway Capital Investment Impacts

The analyses presented in this section use a common set of assumptions to derive relationships between alternative levels of future highway capital investment and various measures of future highway and bridge conditions and performance. A subsequent section in this chapter provides comparable information for different types and levels of potential future transit investments.

This section examines the types of investment within the scopes of the Highway Economic Requirements System (HERS) and the National Bridge Investment Analysis System (NBIAS) and lays the foundation for the capital investment scenarios for highways presented in Chapter 8. The accuracy of the projections for highway investments in this chapter depends on the validity of the technical assumptions underlying the analysis, some of which are explored in the sensitivity analysis in Chapter 10. The analyses presented in this section make no explicit assumptions regarding how future investment in highways might be funded.

Types of Capital Spending Projected by HERS and NBIAS

The types of investments HERS and NBIAS evaluate can be related to the system of highway functional classification introduced in Chapter 2 and to the broad categories of capital improvements introduced in Chapter 6 (system rehabilitation, system expansion, and system enhancement). NBIAS relies on the National Bridge Inventory (NBI) database, which covers bridges on all highway functional classes, and evaluates improvements that generally fall within the system rehabilitation category.

HERS evaluates pavement improvements—resurfacing or reconstruction—and highway widening; the types of improvements included in these categories roughly correspond to system rehabilitation and system expansion as described in Chapter 6. In estimating the per-mile costs of widening improvements, HERS recognizes a typical number of bridges and other structures that would need modification. Thus, the estimates from HERS are considered to represent system expansion costs for both highways and bridges. Coverage of the HERS analysis is limited, however, to Federal-aid highways, as the Highway Performance Monitoring System (HPMS) sample does not include data for rural minor collectors, rural local roads, or urban local roads.

The term "nonmodeled spending" refers in this report to spending on highway and bridge capital improvements that are not evaluated in HERS or NBIAS; such spending is not included in the analyses presented in this chapter, but the capital investment scenarios presented in Chapter 8 are adjusted to account for them. Nonmodeled spending includes capital improvements on highway classes omitted from the HPMS sample and hence the HERS model. The development of the future investment scenarios for the highway system as a whole thus required supplementary estimation outside the HERS modeling process.

Nonmodeled spending also includes types of capital expenditures classified in Chapter 6 as system enhancements, which neither HERS nor NBIAS currently evaluate. Although HERS incorporates assumptions about future operations investments, the capital components of which would be classified as system enhancements, the model does not directly evaluate the need for these deployments. In addition, HERS does not identify specific safety-oriented investment opportunities, but instead considers the ancillary safety impacts of capital investments that are directed primarily toward system rehabilitation or capacity expansion. This limitation of the model owes to the HPMS database's containing no information on the locations of crashes and safety devices such as guardrails or rumble strips.

Q&A **How closely do the types of capital improvements modeled in HERS and NBIAS correspond to the specific capital improvement type categories presented in Chapter 6?**

Exhibit 6-12 (see Chapter 6) provides a crosswalk between a series of specific capital improvement types for which data are routinely collected from the States and three major summary categories: system rehabilitation, system expansion, and system enhancement. The types of improvements covered by HERS and NBIAS are assumed to correspond with the system rehabilitation and system expansion categories. As in *Exhibit 6-12*, HERS splits spending on "reconstruction with added capacity" among these categories.

For some of the detailed categories in *Exhibit 6-12*, the assumed correspondence is close overall but not exact. In particular, the extent to which HERS covers construction of new roads and bridges is ambiguous. Although not directly modeled in HERS, such investments are often motivated by a desire to alleviate congestion on existing facilities in a corridor, and thus would be captured indirectly by the HERS analysis in the form of additional normal-cost or high-cost lanes. As described in Appendix A, the costs per mile assumed in HERS for high-cost lanes are based on typical costs of tunneling, double-decking, or building parallel routes, depending on the functional class and area population size for the section being analyzed. To the extent that investments in the "new construction" and "new bridge" improvement types identified in Chapter 6 are motivated by desires to encourage economic development or accomplish other goals aside from the reduction of congestion on the existing highway network, such investments would not be captured in the HERS analysis.

Some other comparability issues include:

- Some of the relocation expenditures identified in *Exhibit 6-12* could be motivated by considerations beyond those reflected in the curve and grade rating data that HERS uses in computing the benefits of horizontal and vertical realignments.
- The bridge expenditures that *Exhibit 6-12* counts as system rehabilitation could include work on bridge approaches and ancillary improvements that NBIAS does not model.
- HERS and NBIAS are assumed not to capture improvements that count as system enhancement spending, including the spending on the "safety" category in *Exhibit 6-12*. Some safety deficiencies, however, might be addressed as part of broader pavement and capacity improvements modeled in HERS.
- The HERS operations preprocessor described in Appendix A includes capital investments in operations equipment and technology that would fall under the definition of the "traffic management/engineering" improvement type in Chapter 6. These investments are counted among the nonmodeled system enhancements because they are not evaluated within the benefit-cost framework that HERS applies to system preservation and expansion investments.

Exhibit 7-1 shows that, systemwide in 2012, highway capital spending was $105.2 billion, of which $57.4 billion was for the types of improvement that HERS models and $16.4 billion was for the types of improvement NBIAS models. The other $31.4 billion, which was for nonmodeled highway capital spending, was divided about evenly between system enhancement expenditures and capital improvements to classes of highways not reported in HPMS.

Exhibit 7-1 Distribution of 2012 Capital Expenditures by Investment Type

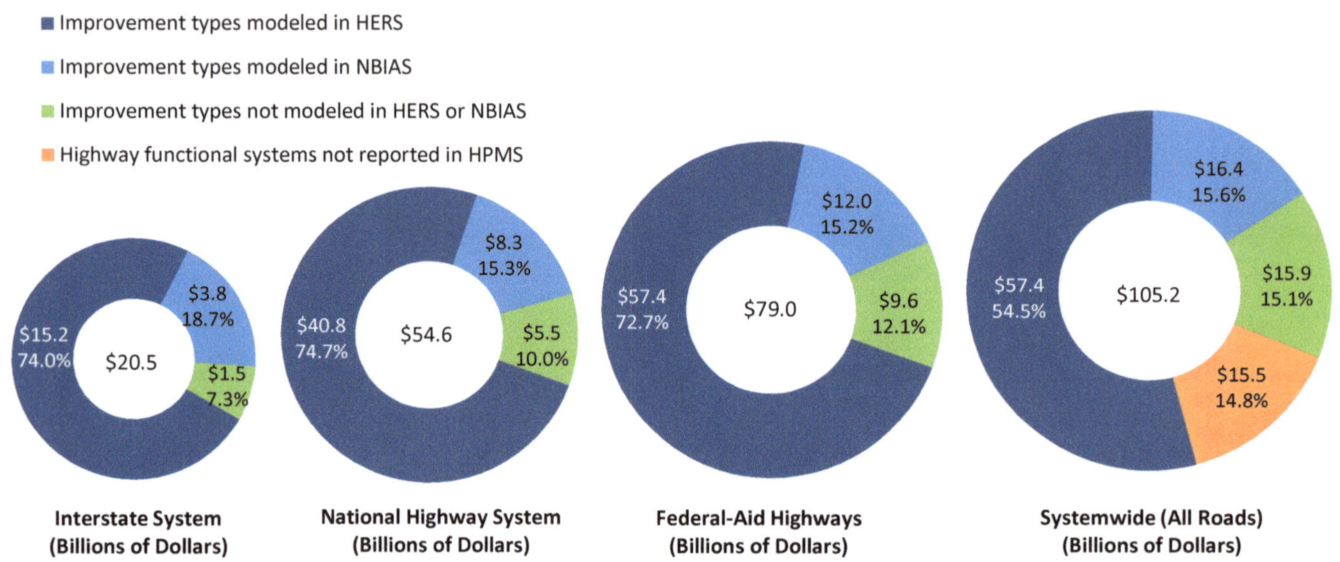

Source: Highway Statistics 2012 (Table SF-12A) and unpublished FHWA data.

Because the HPMS sample data are available only for Federal-aid highways, the percentage of capital improvements classified as nonmodeled spending is lower for Federal-aid highways than is the case systemwide. Of the $79.0 billion spent by all levels of government on capital improvements to Federal-aid highways in 2012, 72.7 percent was within the scope of HERS, 15.2 percent was within the scope of NBIAS, and 12.1 percent was for spending captured by neither. The percentage distribution differs somewhat for the Interstate System, with a slightly higher share within the scope of HERS and NBIAS (74.0 percent and 18.7 percent, respectively) and a smaller share captured by neither (7.3 percent).

Of note is that the statistics presented in this chapter and in Chapter 8 relating to future National Highway System (NHS) investment are based on an estimate of how the NHS will look after its expansion pursuant to MAP-21, rather than as the system existed in 2012. Although the 2012 HPMS sample data incorporate the MAP-21-driven expansion of the NHS, the 2012 NBI data do not reflect the expanded NHS. As indicated in Chapter 6, combined highway capital spending by all levels of government on the NHS in 2012 totaled $44.6 billion. The NHS capital spending figure of $54.6 billion referenced in *Exhibit 7-1* includes amounts spent on other principal arterials, as much of this mileage was be added to the NHS by MAP-21.

Treatment of Traffic Growth

For the HERS analysis in this report, growth in vehicle miles traveled (VMT) is based on two primary inputs: HPMS section-level forecasts of future annual average daily traffic that States provide and a national-level forecast developed from a new FHWA model. The national-level forecast serves as a control, which the sum of the forecast section-level changes in VMT must match. To match the national-level control, the section-level forecasts are scaled proportionally. For this report, the sum of the section-level forecasts yielded an aggregate average annual VMT growth rate of 1.42 percent that exceeded the national-level forecast of 1.04 percent per year, and thus were scaled proportionally downward to match the national-level forecast. Chapter 9 discusses the national-level forecast and reviews the accuracy of VMT projections in previous C&P Reports.

The national-level forecast includes separate VMT growth rates for light-duty vehicles, single-unit trucks, and combination trucks; these separate growth rates were applied in the HERS analysis. VMT in light-duty vehicles is forecast to grow at 0.92 percent per year. VMT for heavy-duty vehicles is forecast to grow at a rate more than twice that for light-duty vehicles (2.15 percent per year for single-unit trucks and 2.12 percent per year for combination trucks). The higher rate of forecast VMT growth for heavy-duty vehicles reflects a close relationship between heavy-vehicle VMT and economic output (GDP or gross domestic product). Economic factors (e.g., sensitivity of VMT demand to income and fuel prices) also influence the forecast of light-duty VMT, but to a weaker extent than the influence on heavy-duty vehicles. The difference in projected VMT growth rates for heavy-duty and light-duty vehicles reflects the direct role of freight transportation in facilitating the production and sale of outputs measured within GDP; increases in income associated with GDP growth do not influence light-duty VMT to the same degree.

The procedures used for estimating traffic growth in the NBIAS analysis presented in this report are similar to those used for HERS. For NBIAS, these forecasts build off bridge-level forecasts of future average daily traffic that States provide in the NBI. The sum of the bridge-level forecasts yielded an aggregate growth rate of 1.46 percent per year; growth rates for individual bridges were adjusted downward to match the 1.04 percent control total from the national-level VMT forecast model referenced above. Unlike the HERS analysis, the NBIAS analysis applied the same growth rate to all vehicle classes, as NBIAS is not currently equipped to handle separate growth rates by vehicle type.

An underlying assumption applied in both HERS and NBIAS is that VMT will grow linearly (so that 1/20th of the additional VMT is added each year), rather than geometrically (i.e., at a constant annual rate). With linear growth, the annual rate of growth gradually declines over the forecast period. Estimated VMT growth rates within each highway investment scenario deviate from the FHWA forecast values due to estimated changes in user travel cost, as discussed in the following section.

In previous reports, the State-reported travel growth forecasts in the HPMS and the NBI were applied directly (i.e., they were not scaled to match a national-level control). Chapter 10 considers an alternative in which VMT grows consistently with the State-reported forecasts.

Alternative Levels of Future Capital Investment Analyzed

Both the HERS and NBIAS analyses presented in this chapter assume that capital investment within the scopes of the models will grow over 20 years at a constant annual percentage rate, which could be positive, negative, or zero. Because future levels are measured in constant 2012 dollars, the rates of growth are real (inflation-adjusted). This "ramped" approach to analyzing alternative investment levels was introduced in the 2008 C&P Report. Analyses for previous editions either assumed a fixed amount would be spent in each year or set funding levels based on benefit-cost ratios, which tended to "front-load" the investment within the 20-year analysis period. Chapter 9 includes an analysis of the impacts on conditions and performance of these alternative timing patterns of investments and presents an example of how the ramping approach influences year-by-year funding levels for some of the highway investment scenarios presented in Chapter 8.

This chapter quantifies potential highway and bridge system outcomes under various assumptions about the rate of ramped investment growth. The particular investment levels were selected from among the results of a much larger number of model simulations. Each investment level presented corresponds to a particular target outcome, such as funding all potential capital improvements with a benefit-cost ratio above a certain threshold or attaining a certain level of performance for highways or bridges. Although each selected rate of change has some specific analytical significance, the analyses presented in this chapter do not constitute complete investment scenarios, but rather form the building blocks for such scenarios, which are presented in Chapter 8.

Highway Economic Requirements System

Simulations conducted with HERS provide the basis for this report's analysis of investment in highway resurfacing and reconstruction and for highway and bridge capacity expansion. HERS uses incremental benefit-cost analysis to evaluate highway improvements based on data from HPMS. HPMS includes State-supplied information on current roadway characteristics, conditions, and performance and anticipated future travel growth for a nationwide sample of more than 120,000 highway sections. HERS analyzes individual sample sections only as a step toward providing results at the national level; the model does not provide definitive improvement recommendations for individual sections.

HERS simulations begin with evaluations of the current state of the highway system using data from the HPMS sample. These data provide information on pavements, roadway geometry, traffic volume and composition (percentage of trucks), and other characteristics of the sampled highway sections. For sections with one or more identified deficiencies, the model then considers potential improvements, including resurfacing, reconstruction, alignment improvements, and widening or adding travel lanes. HERS selects the improvement (or combination of improvements) with the greatest net benefits, with benefits defined as reductions in direct highway user costs, agency costs for road maintenance, and societal costs from vehicle emissions of greenhouse gases and other pollutants. (The model uses estimates of emission costs that include damage to property and human health and, for greenhouse gases, other potential impacts such as loss of outdoor recreation amenities.) The model allocates investment funding only to those sections for which at least one potential improvement is projected to produce benefits exceeding construction costs.

HERS normally considers highway conditions and performance over a period of 20 years from the base ("current") year—the most recent year for which HPMS data are available. This analysis period is divided into four equal funding periods. After analyzing the first funding period, HERS updates the database to reflect the projected outcomes of the first period, including the effects of the selected highway improvements. The updated database is then used to analyze conditions and performance in the second period, the database is updated again, and so on through the fourth and last period. Appendix A contains a detailed description of the project selection and implementation process HERS uses.

Operations Strategies

Since the 2004 C&P Report, HERS has considered the impacts of certain types of highway operational improvements that feature intelligent transportation systems (ITS). The operations strategies HERS currently evaluates are:

- **Freeway management:** ramp metering, electronic roadway monitoring, variable message signs, integrated corridor management, variable speed limits, queue warning systems, lane controls.
- **Incident management:** detection, verification, response.
- **Arterial management:** upgraded signal control, electronic monitoring, variable message signs.
- **Traveler information:** 511 systems, advanced in-vehicle navigation systems with real-time traveler information.

Appendix A describes these strategies in detail and their treatment in HERS. Of importance to note is that HERS does not analyze the benefits and costs of these investments, nor does it directly analyze tradeoffs between them and the pavement improvements and widening options the model also considers. Instead, a separate preprocessor estimates the impacts of these operations strategies on the performance of highway sections where they are deployed. The analyses presented in this chapter assume a package of investments that continue existing deployment trends, and a sensitivity analysis presented in Chapter 10 considers the impacts of a more aggressive deployment pattern. HERS does not currently model applications of various developing vehicle-to-vehicle and vehicle-to-infrastructure communications because reliably predicting the impacts and patterns of their deployment is premature.

Q&A **How will Vehicle-to-Vehicle (V2V) and Vehicle-to-Infrastructure (V2I) communications potentially impact future investment needs?**

Cellular, Wi-Fi, and other dedicated short-range communication technologies are expanding the possibilities for a Connected Vehicle Environment. Communications among vehicles on the road (V2V)—and between these vehicles and infrastructure (V2I)—hold promise for substantial reductions in crashes and vehicle emissions and for enhanced mobility through more efficient management and operations of transportation systems. Adding to this potential are rapid advances in vehicle automation. For example, under advanced speed harmonization, vehicle speed would adjust automatically to speed limits that vary based on road, traffic, and weather conditions (an existing V2I application).

Additional examples of connectivity applications include blind spot monitoring/lane change warning, smart parking, forward collision warning, do-not-pass warning, curve speed warning, red light violation warning, transit pedestrian warning, cooperative adaptive cruise control, braking assist, and dynamic lane closure management.

Reaching the full potential of connected vehicles will require investment, coordination, and partnership with public and private entities. As development and implementation of connected vehicle applications proceed, additional information should make possible their representation in HERS. Research efforts by FHWA, Federal Transit Administration (FTA), National Highway Traffic Safety Administration (NHTSA), American Association of State Highway and Transportation Officials (AASHTO), and others that will measure benefits and costs of these applications include: (1) Applications for the Environment: Real-Time Information Synthesis Program; (2) AASHTO Connected Vehicle Field Infrastructure Footprint Analysis; (3) Connected and Automated Vehicle Benefit Cost Analysis; and (4) Measuring Local, Regional and Statewide Economic Development Associated with the Connected Vehicle program.

Travel Demand Elasticity

A key feature of the HERS economic analysis is the influence of the cost of travel on the demand for travel. HERS represents this relationship as a travel demand elasticity that relates demand, measured by VMT, to changes in the average user cost of travel that result from either: (1) changes in highway conditions and performance as measured by travel delay, pavement condition, and crash costs, relative to base year levels; the elasticity mechanism reduces travel demand when these changes are for the worse (e.g., an increase in travel delay) and increase travel demand when they are improvements (e.g., better pavement condition); or (2) deviations from the price projections built into the baseline demand forecasts. This report considers the latter deviations only in Chapter 10, where one of the sensitivity tests alters the projections for motor fuel prices.

HERS also allows the induced demand predicted through the elasticity mechanism to influence the cost of travel to highway users. On congested sections of highway, the initial congestion relief afforded by an increase in capacity will reduce the average user cost per VMT, which in turn will stimulate demand for travel; this increased demand, in turn, will reverse some of the initial congestion relief. The elasticity feature operates likewise with respect to improvements in pavement quality by allowing for induced traffic that adds to pavement wear. (Conversely, an initial increase in user costs can start a causal chain with effects in the opposite direction.) By capturing these offsets to initial impacts on highway user costs, HERS can estimate the net impacts.

Impacts of Federal-Aid Highway Investments Modeled by HERS

The HERS analysis for this edition of the C&P report starts with an evaluation of the state of Federal-aid highways in 2012—the base year. *Exhibit 7-1* shows that capital spending on the types of improvements modeled in HERS for these highways in the base year was $57.4 billion (total highway capital spending was $105.2 billion). The analysis continues by considering the potential impacts on system performance of raising or lowering the amount of investment within the scope of HERS at various annual rates over 20 years. Spending in any year is measured in constant 2012 dollars, so that spending and its rate of growth are both measured in real, rather than nominal, terms. Chapter 9 includes an illustration of how future spending levels could be converted from real to nominal dollar levels under alternative assumptions about the future inflation rate.

Selection of Investment Levels for Analysis

Exhibit 7-2 introduces the six investment levels presented in the next several exhibits to illuminate the relationship between the levels of investment modeled in HERS and the future conditions and performance of Federal-aid highways.

The highest level of spending shown in *Exhibit 7-2* corresponds to the annual growth rate in real spending (2.53 percent) associated with attaining a minimum benefit-cost ratio (BCR) of 1.0 over the 20-year analysis period. As explained in the introduction to Part II of this report, HERS ranks potential projects in order of BCR and implements them until the funding constraint is reached. The lowest BCR among the projects selected, the "marginal BCR," varies across the four funding periods, and HERS refers to the lowest of these values across the funding periods as the "minimum BCR." The attainment of a minimum BCR of 1.0 can be interpreted as having gradually implemented all potentially cost-beneficial projects (BCR ≥ 1.0) over 20 years. The "Improve C&P" reference in *Exhibit 7-2* signifies that this level of investment feeds into the Improve Conditions and Performance scenario presented in Chapter 8.

Another funding level shown in *Exhibit 7-2* represents the annual growth rate in real spending geared toward matching a specific level of performance in 2032; an average annual growth rate of −1.16 percent is projected to be adequate to allow average pavement roughness as measured by the International Roughness Index (IRI) in 2032 to match the level in 2012 (see discussion of IRI in Chapter 3) and for average delay to be at least as low in 2032 as it was in 2012. This "Maintain C&P" reference in *Exhibit 7-2* signifies that this level of investment feeds into the Maintain Conditions and Performance scenario, also presented in Chapter 8.

The remaining four of the six funding levels shown in *Exhibit 7-2* represent a range of annual growth rates in real highway spending above, at, and below 2012 funding (2, 1, 0, and −1 percent). The "2012 Spending" reference in *Exhibit 7-2* for the 0.00-percent growth rate row signifies that this level of spending feeds into the Sustain 2012 Spending scenario presented in Chapter 8.

Exhibit 7-2 HERS Annual Investment Levels Analyzed for Federal-Aid Highways

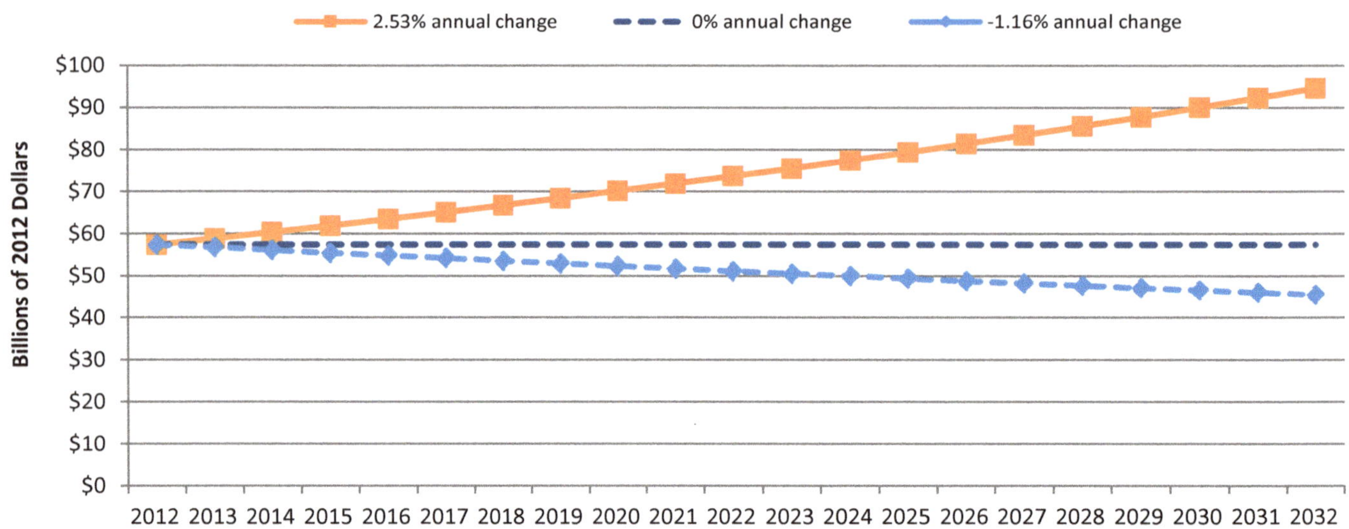

Annual Percent Change in HERS Capital Spending	Spending Modeled in HERS (Billions of 2012 Dollars): Cumulative 5-Year 2013 Through 2017	5-Year 2018 Through 2022	5-Year 2023 Through 2027	5-Year 2028 Through 2032	20-Year 2013 Through 2032	Average Annual Over 20 Years: Total HERS Spending[1]	System Rehabilitation Spending[2]	System Expansion Spending[2]	Link to Chapter 8 Scenario
2.53%	$309	$351	$397	$450	$1,507	**$75.4**	$45.4	$30.0	Improve C&P
2.00%	$305	$336	$371	$410	$1,422	**$71.1**	$43.1	$28.0	
1.00%	$296	$311	$326	$343	$1,276	**$63.8**	$38.7	$25.1	
0.00%	$287	$287	$287	$287	$1,147	**$57.4**	$35.0	$22.4	**2012 Spending**
-1.00%	$278	$265	$252	$239	$1,034	**$51.7**	$31.5	$20.2	
-1.16%	$277	$261	$247	$233	$1,017	**$50.9**	$30.9	$19.9	Maintain C&P

[1] The amounts shown represent the average annual investment over 20 years that would occur if annual investment grows in constant dollar terms by the percentage shown in each row of the first column.

[2] HERS splits its available budget between system rehabilitation and system expansion based on the mix of spending it finds to be most cost-beneficial, which varies by funding level.

Source: Highway Economic Requirements System.

The portion of each investment level that HERS directs to system rehabilitation versus system expansion is significant, as these types of investments have varying degrees of influence on different performance measures. Investment in system rehabilitation (ranging from $30.9 billion to $45.4 billion across reported investment levels) tends to have a stronger influence on physical condition measures such as pavement ride quality. Investment in system expansion (ranging from $19.9 billion to $30.0 billion across reported investment levels) has a more pronounced impact on operational performance measures such as delay.

Q&A **How large is the investment backlog estimated by HERS?**

The investment backlog represents all improvements that could be economically justified for immediate implementation, based solely on the current conditions and operational performance of the highway system (without regard to potential future increases in vehicle miles traveled or potential future physical deterioration of pavements).

HERS does not routinely produce rolling backlog figures over time as an output, but is equipped to do special analyses to identify the base-year backlog. To determine which action items to include in the backlog, HERS evaluates the current state of each highway section before projecting the effects of future travel growth on congestion and pavement deterioration. Any potential improvement that would correct an existing pavement or capacity deficiency and that has a benefit-cost ratio greater than or equal to 1.0 is considered part of the current highway investment backlog.

HERS estimates the size of the backlog as $463.1 billion for Federal-aid highways, stated in constant 2012 dollars. The estimated backlog for the Interstate System is $105.1 billion; adding other principal arterials produces an estimated backlog of $281.2 billion for the expanded NHS. The investment levels associated with a minimum benefit-cost ratio of 1.0 presented in this chapter would fully eliminate this backlog and address other deficiencies that arise over the next 20 years, when doing so might be cost-beneficial.

Of note is that these figures reflect only a subset of the total highway investment backlog; they do not include the types of capital improvements modeled in NBIAS (presented later in this chapter) or the types of capital improvements not currently modeled in HERS or NBIAS. Chapter 8 presents an estimate of the combined backlog for all types of improvements (see *Exhibit 8-4*).

Investment Levels and BCRs by Funding Period

Exhibit 7-2 illustrates how the six alternative funding growth rates for Federal-aid highways that were selected for further analysis in this chapter would translate into cumulative spending in 5-year intervals (corresponding to 5-year analysis periods used in HERS). The portions of these investment levels relating to system rehabilitation and system expansion are also identified, as the former would be expected to have a greater impact on measures of physical conditions such as IRI, while the latter would be expected to have a greater impact on measures of operational performance, such as user delay.

As shown in *Exhibit 7-2*, achieving a minimum BCR of 1.0 is estimated to require $1.507 trillion over the analysis period. Achieving a minimum BCR of 1.0 would necessitate an increase in spending of $360 billion over the analysis period relative to a scenario in which 2012 spending levels were maintained from 2012 through 2032.

Exhibit 7-3 illustrates the marginal benefit-cost ratios (i.e., the lowest benefit-cost ratio among the improvements selected within a funding period) associated with the six alternative funding levels. *Exhibit 7-3* also provides the minimum benefit-cost ratios across all funding periods (which is identical to the lowest marginal benefit-cost ratio) and the average benefit-cost ratios across all funding periods (i.e., the total level of benefits of all improvements divided by the total cost of all improvements). For positive growth rates in spending levels, the marginal BCR declines over time, reflecting the tendency in HERS to implement the most worthwhile improvements first; the minimum BCR over the entire 20-year analysis period, shown in the last column, equals the marginal BCR in the last 5-year period. Conversely, for negative (and zero) growth rates in spending levels, the minimum BCR equals the marginal BCR in the third 5-year period. This pattern reflects the impacts of funding constraints; the relative scarcity of funding toward the end of the analysis period is inadequate to keep pace with newly emerging needs, limiting the range of needs that can be addressed.

Exhibit 7-3 Minimum and Average Benefit-Cost Ratios (BCRs) for Different Possible Funding Levels on Federal-Aid Highways

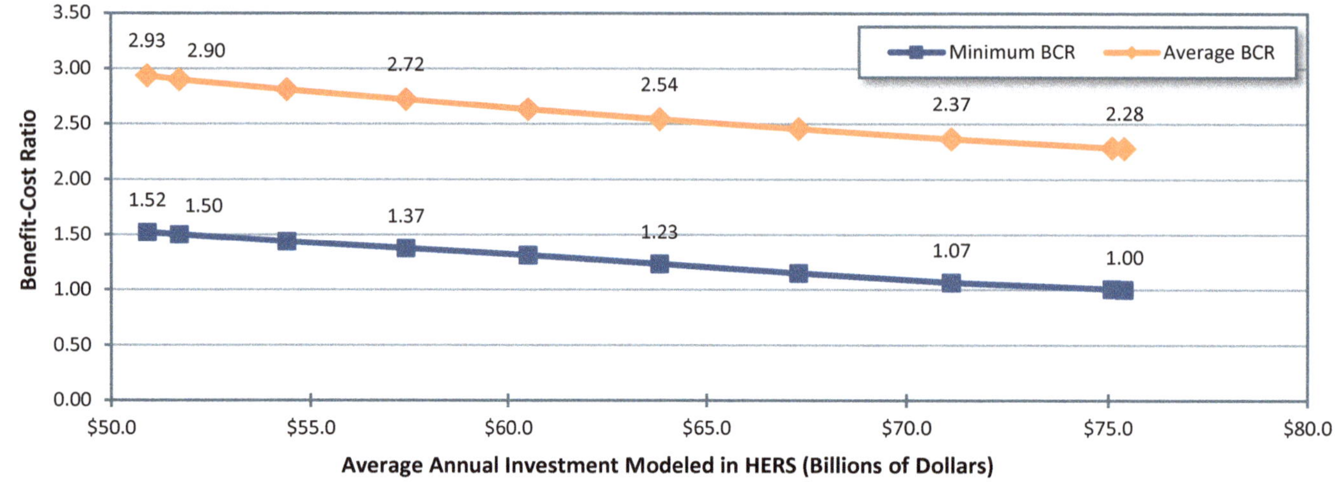

HERS-Modeled Investment on Federal-Aid Highways		Benefit-Cost Ratios[1]						
			Marginal BCR[3]					
Average Annual Investment[2] (Billions of 2012 Dollars)	Average Annual Percent Change vs. 2012	Average BCR 20-Year 2013 Through 2032	5-Year 2013 Through 2017	5-Year 2018 Through 2022	5-Year 2023 Through 2027	5-Year 2028 Through 2032	Minimum BCR 20-Year 2013 Through 2032	Link to Chapter 8 Scenario
$75.4	2.53%	2.28	1.80	1.24	1.08	**1.00**	**1.00**	**Improve C&P**
$71.1	2.00%	2.37	1.82	1.27	1.14	**1.07**	**1.07**	
$63.8	1.00%	2.54	1.86	1.35	1.25	**1.23**	**1.23**	
$57.4	0.00%	2.72	1.90	1.43	**1.37**	1.40	**1.37**	2012 Spending
$51.7	-1.00%	2.90	1.94	1.52	**1.50**	1.59	**1.50**	
$50.9	-1.16%	2.93	1.95	1.53	**1.52**	1.62	**1.52**	Maintain C&P

[1] As HERS ranks potential improvements by their estimated BCRs and assumes that the improvements with the highest BCRs will be implemented first (up until the point where the available budget specified is exhausted), the minimum and average BCRs will naturally tend to decline as the level of investment analyzed rises.

[2] The amounts shown represent the average annual investment over 20 years that would occur if annual investment grows in constant dollar terms by the percentage shown in each row of the first column.

[3] The marginal BCR represents the lowest benefit-cost ratio for any project implemented during the period identified at the level of funding shown. The minimum BCRs, indicated by bold font, are the smallest of the marginal BCRs across the funding periods.

Source: Highway Economic Requirements System.

Further evident in *Exhibit 7-3* is the inverse relationship between the minimum BCR and the level of investment. At any given level of average annual investment, the average BCR always exceeds the marginal BCR. For example, at the highest level of investment considered, an average annual investment level of $75.4 billion, the average BCR of 2.28 exceeds the minimum BCR of 1.00.

Impact of Future Investment on Highway Pavement Ride Quality

The primary measure of highway physical condition in HPMS is pavement ride quality as measured by IRI (defined in Chapter 3). The HERS analysis presented in this report focuses on VMT-weighted IRI values; the average IRI values shown thus reflect the pavement ride quality experienced on a typical mile traveled. *Exhibit 7-4* shows how the projection for the average IRI on Federal-aid highways in 2032 varies with the portion of investment that HERS allocates to system rehabilitation, as identified in *Exhibit 7-2*; system rehabilitation is more significant than investment in system expansion in influencing average pavement ride quality. The levels of system rehabilitation analyzed range from an average annual investment level of $30.9 billion (which feeds the Maintain Conditions and Performance scenario in Chapter 8) to an average annual investment level of $45.4 billion (which feeds the Improve Conditions and Performance scenario in Chapter 8).

Exhibit 7-4 Projected 2032 Pavement Ride Quality Indicators on Federal-Aid Highways Compared with 2012 for Different Possible Funding Levels

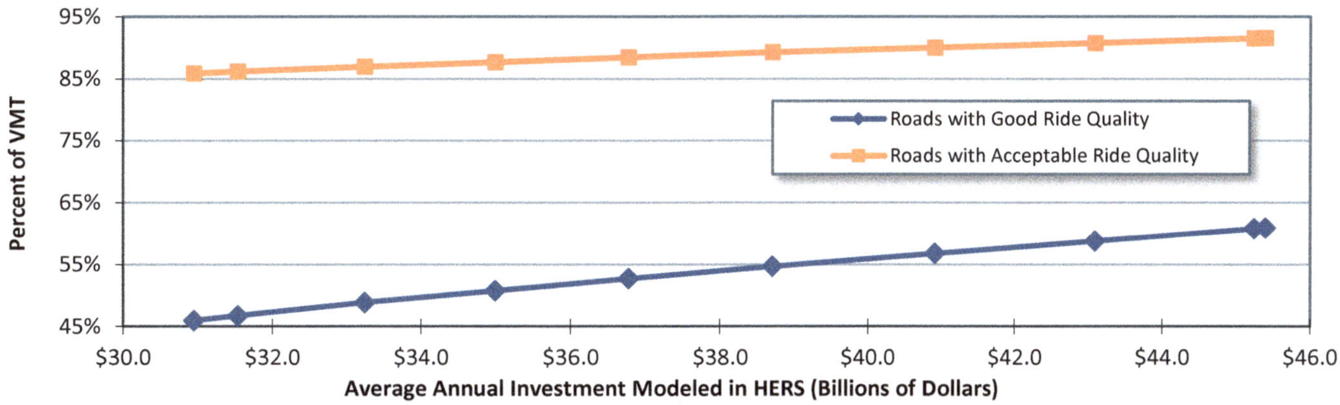

HERS-Modeled Capital Investment	Projected 2032 Condition Measures on Federal-Aid Highways[1,2]				
	Percent of VMT on Roads With Ride Quality of:		Average IRI (VMT-Weighted)		
Average Annual for System Rehabilitation (Billions of 2012 Dollars)[2]	Good (IRI<95)[3]	Acceptable (IRI<=170)[3]	Inches Per Mile	Change Relative to Base Year	Link to Chapter 8 Scenario
$45.4	60.9%	91.5%	100.7	-14.0%	Improve C&P
$43.1	58.8%	90.8%	102.9	-12.1%	
$38.7	54.7%	89.3%	107.4	-8.3%	
$35.0	50.8%	87.7%	111.8	-4.5%	2012 Spending
$31.5	46.7%	86.2%	116.3	-0.7%	
$30.9	46.0%	85.9%	117.1	**0.0%**	**Maintain C&P**
Base Year Values:	**44.9%**	**83.3%**	**117.1**		

[1] The HERS model relies on information from the HPMS sample section database, which is limited to those portions of the road network that are generally eligible for Federal funding (i.e., "Federal-aid highways") and excludes roads classified as rural minor collectors, rural local, and urban local.

[2] The amounts shown represent only the portion of HERS-modeled spending directed toward system rehabilitation, rather than system expansion. Other types of spending can affect these indicators as well.

[3] As discussed in Chapter 3, IRI values of 95 and 170 inches per mile, respectively, are the thresholds associated with "good" and "acceptable" ride quality.

Source: Highway Economic Requirements System.

For all investment levels presented in *Exhibit 7-4*, pavements on Federal-aid highways are projected to be smoother on average in 2032 than in 2012, with the exception of the lowest investment level, which matches the average base-year pavement condition exactly. VMT-weighted average IRI decreases by up to 14 percent across alternatives (from 117.1 to 100.7).

Exhibit 7-4 also shows the HERS projections for the percentage of travel occurring on pavements with ride quality that would be rated good or acceptable based on the IRI thresholds set in Chapter 3. Under all circumstances represented in the exhibit, the 2032 projection for the percentage of travel occurring on pavements with good ride quality exceeds the 44.9 percent that occurred in 2012; the improvement in the share of pavements with good ride quality increases roughly linearly with spending. The projections for 2032 range from 60.9 percent at the highest level of investment modeled (an average annual investment level for system rehabilitation of $45.4 billion) to 46.0 percent at the lowest level of investment (an average annual investment level for system rehabilitation of $30.9 billion).

In all the circumstances considered, *Exhibit 7-4* reveals increases relative to the base-year level of 83.3 percent in the proportion of travel occurring on pavements with ride quality rated as acceptable. The projection for 2032 ranges from 91.5 percent at the highest level of investment modeled to 85.9 percent at the lowest. When no change from the 2012 level of investment is modeled, 87.7 percent of travel in 2032 in the forecast traffic growth case is projected to occur on pavements with acceptable ride quality. As noted in Chapter 3, the IRI threshold of 170 used to identify acceptable ride quality was originally set to measure performance on the NHS and might not be fully applicable to non-NHS routes, which tend to have lower travel volumes and speeds.

Q&A

Why does HERS predict smaller improvements to pavement quality in this report compared to previous analyses?

Two primary factors limit the extent to which future highway investment results in improvements in projected future pavement quality in this report relative to previous analyses. First, the rate of forecast growth in vehicle miles traveled is lower in this analysis than in previous analyses, resulting in the selection of fewer projects that generate improved pavement quality through surface widening. That is, the estimated benefits of widening lanes (with concurrent increases in pavement quality) are reduced relative to the past, because there are fewer projected users to whom benefits would accrue. Second, changes to the pavement model in HERS better reflect the effects of aging road infrastructure and the challenges associated with maintaining pavement quality over time. In particular, revisions to HERS have decreased the rate at which pavement quality is assumed to decline, dampening the estimated benefits of surface rehabilitation projects.

Impact of Future Investment on Highway Operational Performance

Exhibit 7-5 shows the HERS projections for the impact of investment levels on average speed and traveler delay. *Exhibit 7-5* splits out the portion of that investment that HERS programs for system expansion (such as widening existing highways or building new routes in existing corridors), which tend to reduce congestion delay more than spending on system rehabilitation. The levels of system expansion analyzed range from an average annual investment level of $19.9 billion (which feeds the Maintain Conditions and Performance scenario in Chapter 8) to an average annual investment level of $30.0 billion (which feeds the Improve Conditions and Performance scenario in Chapter 8).

Exhibit 7-5 Projected Changes in 2032 Highway Travel Delay and Speed on Federal-Aid Highways Compared with Base Year for Different Possible Funding Levels

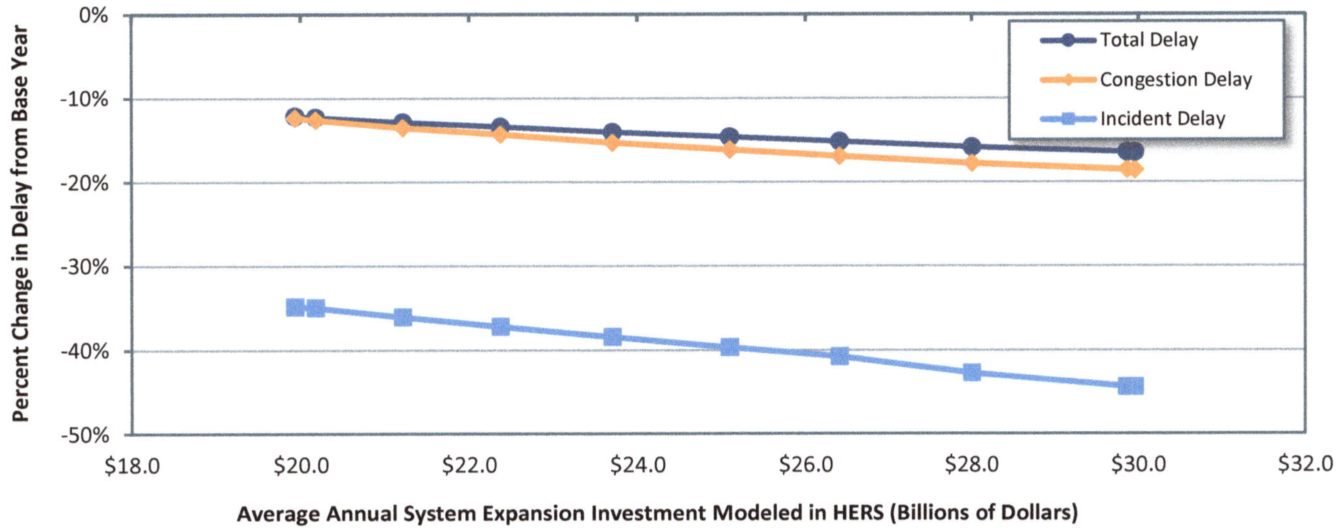

HERS-Modeled Capital Investment	Projected 2032 Performance Measures on Federal-Aid Highways					Link to Chapter 8 Scenario
			Percent Change Relative to Baseline			
Average Annual System Expansion (Billions of 2012 Dollars)[1]	Average Speed in 2030 (mph)	Annual Hours of Delay per Vehicle[2]	Total Delay per VMT	Congestion Delay per VMT	Incident Delay per VMT	
$30.0	44.3	46.0	-16.5%	-18.6%	-44.4%	Improve C&P
$28.0	44.3	46.3	-15.9%	-17.8%	-42.8%	
$25.1	44.2	47.0	-14.7%	-16.2%	-39.7%	
$22.4	44.0	47.6	-13.4%	-14.4%	-37.2%	2012 Spending
$20.2	43.9	48.2	-12.4%	-12.7%	-35.0%	
$19.9	43.9	48.3	-12.2%	-12.4%	-34.9%	Maintain C&P
Base Year Values:	42.3	55.0				

[1] The amounts shown represent only the portion of HERS-modeled spending directed toward system expansion rather than system rehabilitation. Other types of spending can affect these indicators as well.

[2] The values shown were computed by multiplying HERS estimates of average delay per VMT by 11,707, the average VMT per registered vehicle in 2012. HERS does not forecast changes in VMT per vehicle over time. The HERS delay figures include delay attributable to stop signs and signals as well as delay resulting from congestion and incidents.

Source: Highway Economic Requirements System; Highway Statistics 2013, Table VM-1.

As noted above, HERS assumes the continuation of existing trends in the deployment of certain system management and operations strategies. Among these strategies are several that can be expected to mitigate delay associated with isolated incidents more than the delay associated with recurring congestion ("congestion delay"), such as freeway incident management programs. In line with this, *Exhibit 7-5* shows the amount of incident delay decreasing strongly relative to congestion delay over the period 2012–2032. HERS projects incident delay per VMT on Federal-aid highways to decrease by between 34.9 percent (in the Maintain Conditions and Performance alternative) and 44.4 percent (in the Improve Conditions and Performance alternative) between

2012 and 2032. The results in *Exhibit 7-5* also reveal investment within the scope of HERS to be a potent instrument for reducing congestion delay. HERS projects congestion delay to decrease by between 12.4 percent and 16.5 percent.

Q&A **Why does HERS predict larger reductions in delay in this report compared to previous analyses?**

The strong tendency for delay costs to fall is driven by multiple factors. The relatively low forecast growth rate in vehicle miles traveled (VMT) reduces upward pressure on delay compared to previous analyses. Likewise, lower forecast VMT growth enables developments in intelligent transportation systems to mitigate delay more effectively as VMT increases. Improvements in data quality related to obstacles to implementing widening projects improve the ability of HERS to identify economically beneficial projects that add capacity.

Notably, changes to the pavement model have tended to reduce the estimated benefits of pavement improvements, leading to an increased selection rate for projects that add capacity at lower investment levels.

Across all scenarios presented in *Exhibit 7-5*, annual delay per vehicle in 2032 is lower than the 2012 level (55 hours), with reductions in delay ranging from 6.7 hours in the lowest level of investment analyzed to 9.0 hours in the highest. The projected reductions in delay are associated with relatively small variations in average vehicle speed, ranging from 43.9 miles per hour to 44.3 miles per hour, compared to the 2012 level of 42.3 miles per hour.

Some traffic basics are important to keep in mind when interpreting these results. In addition to congestion and incident delay, some delay inevitably results from traffic control devices. For this reason, and because traffic congestion occurs only at certain places and times, *Exhibit 7-5* shows the variation in investment level as having less impact on projections for total delay and average speed than on the projections for congestion and incident delay. In addition, although the impacts of additional investment on average speed are proportionally small, these impacts apply to a vast amount of travel; hence, the associated savings in user cost are not necessarily small relative to the cost of the investment.

Impact of Future Investment on Highway User Costs

In HERS, the benefits from highway improvements are the reductions in highway user costs, agency costs, and societal costs of vehicle emissions. In measuring the highway user costs, the model includes the costs of travel time, vehicle operation, and crashes.

Exhibit 7-6 shows the projected changes from 2012 to 2032 in average user cost of travel on Federal-aid highways by cost component. For Federal-aid highways, HERS estimates that user costs—the costs of travel time, vehicle operation, and crashes—averaged $1.283 per mile traveled in 2012.

Average user cost per VMT is projected to increase at a lower rate at the spending level HERS indicates would be needed to fund all cost-beneficial projects (averaging $75.4 billion annually); under this spending level, average user cost per mile of VMT in 2032 is projected to be $1.382, or 7.7 percent higher than in 2012. Average user cost per VMT is projected to increase between 2012 and 2032 by 8.7 percent and 9.2 percent under the assumptions that real annual spending remains at the base-year level (average annual growth rate of 0.0 percent) or, alternatively, decreases annually at the rate geared toward maintaining average pavement roughness (1.16 percent).

Exhibit 7-6 Projected 2032 Average Total User Costs on Federal-Aid Highways Compared with Base Year for Different Possible Funding Levels

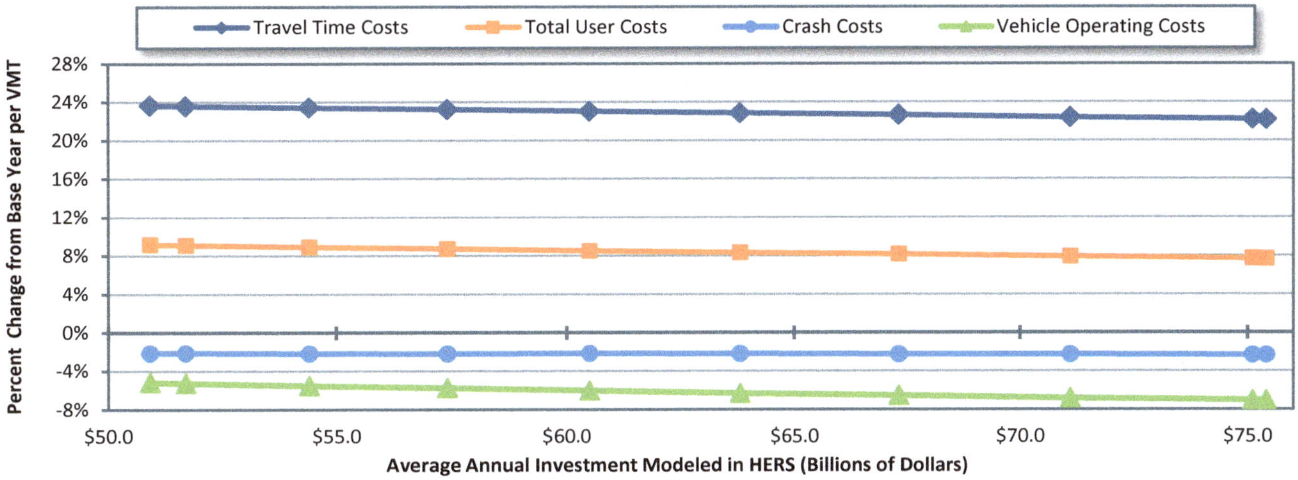

HERS-Modeled Investment On Federal-Aid Highways		Projected 2032 Performance Measures on Federal-Aid Highways					
			Percent Change Relative to Baseline Average per VMT				
Average Annual (Billions of 2012 Dollars)	Average Annual Change vs. 2012	Average Total User Costs ($/VMT)	Total User Costs	Travel Time Costs	Vehicle Operating Costs	Crash Costs	Link to Chapter 8 Scenario
$75.4	2.53%	$1.381	7.7%	22.1%	-7.1%	-2.3%	Improve C&P
$71.1	2.00%	$1.384	7.9%	22.4%	-6.8%	-2.3%	
$63.8	1.00%	$1.389	8.3%	22.8%	-6.3%	-2.2%	
$57.4	**0.00%**	$1.394	8.7%	23.2%	-5.8%	-2.2%	**2012 Spending**
$51.7	-1.00%	$1.399	9.1%	23.6%	-5.3%	-2.1%	
$50.9	-1.16%	$1.400	9.2%	23.6%	-5.2%	-2.1%	Maintain C&P
	Base Year Values:	$1.283					

Source: Highway Economic Requirements System.

The cost of crashes is the user cost component with the lowest absolute sensitivity to the assumed level of highway investment, which as an annual average varies between $50.9 billion (which feeds the Maintain Conditions and Performance scenario in Chapter 8) and $75.4 billion (which feeds the Improve Conditions and Performance scenario in Chapter 8). Crash costs in 2032 are projected to be between 2.1 percent and 2.3 percent lower than in 2012.

The levels of spending in each scenario are limited to the types of improvements that HERS evaluates, which are basically system rehabilitation and expansion. Because HPMS lacks detailed information on the current location and characteristics of safety-related features (e.g., guardrail, rumble strips, roundabouts, yellow change intervals at signals), safety-focused investments are not evaluated. Thus, the findings presented in *Exhibit 7-6* establish nothing about how such investments affect highway safety.

Crash costs also form the smallest of the three components of highway user costs. For 2012 travel on Federal-aid highways, HERS estimates the breakdown by cost component to be crash cost, 14.0 percent; travel time cost, 48.3 percent, and vehicle operating cost, 37.8 percent. Research under way to update the vehicle operating cost equations in HERS (see Appendix A) might alter the split among these costs somewhat, but crash costs will remain a small component. Although highway trips always consume traveler time and resources for vehicle operation, only a small fraction involves crashes. In addition, most crashes are non-catastrophic: Particularly on urban highways, many crashes involve only damage to property with no injuries.

The projections for travel time costs are less sensitive to the assumed level of investment than are the projections for vehicle operating costs. The projected 2012–2032 change in travel time cost per VMT ranges from an increase of 22.1 percent at the highest level of assumed investment to an increase of 23.6 percent at the lowest. These projections indicate that investing at the highest level rather than the lowest level would reduce the time cost of travel per VMT in 2032 by 1.2 percent, saving travelers hundreds of millions of hours per year in aggregate. The projected impacts on travel time costs in this report differ from the corresponding projected impacts in the 2013 C&P Report, which projected a small decrease in travel time cost under high levels of investment. This distinction was driven by assumptions about increasing real travel time costs in future years, as noted previously; the revisions incorporate projected increases in real income, which is a central input to estimated values of travel time savings.

Q&A **What are the monetized national-level impacts implied by the changes in average user costs projected by HERS?**

Exhibit 7-6 presents measures of average user costs per vehicle mile traveled (VMT), rather than projections of aggregate, national-level user costs. To identify monetized impacts of changes in investment levels on national-level user costs, national VMT in 2032 can be multiplied by differences in average user costs across investment levels. At the highest level of investment (an annual average of $75.4 billion), average total user costs are projected to be $1.381 per VMT. Average total user costs at the highest level of investment represent decreases in average total user costs of $0.013 per VMT when spending is held at the base-year level ($57.4 billion per year) and $0.019 per VMT at the lowest level of investment (an annual average of $50.9 billion).

Investing at the highest level is projected to result in a decrease in total user costs in 2032 of between $59.6 billion and $60.0 billion relative to the lowest level of investment, depending on the measure of projected VMT specified in the calculation (i.e., the choice of projected VMT among investment levels). Investing at the highest level is projected to result in a decrease in total user costs in 2032 of $40.9 billion relative to investing at the base-year level, for the projected VMT when investing at the lowest level.

Approximately half the projected national-level impacts on average user costs can be attributed to impacts on vehicle operating costs. At the highest investment level, average vehicle operating costs per VMT in 2032 are projected to be $0.009 lower than under the lowest investment level and $0.006 lower than when spending is held at the base-year level. Investing at the highest level is projected to result in a decrease in total vehicle operating costs in 2032 of $28.3 relative to the lowest level of investment, based on projected VMT for the lowest investment level in 2032. Investing at the highest level is projected to result in a decrease in total vehicle operating costs in 2032 of $18.9 billion relative to investing at the base-year level, based on projected VMT for the lowest investment level in 2032.

Impact on Vehicle Operating Costs

Exhibit 7-7 presents projections for vehicle operating costs per VMT, including separate values for four-tire vehicles (light-duty vehicles) and trucks (heavy-duty vehicles). The projected 2012–2032 change in vehicle operating costs per VMT ranges from a decrease of 5.2 percent at the lowest level of assumed investment (from $0.485 to $0.460 per VMT) to a decrease of 7.0 percent at the highest (from $0.485 to $0.451 per VMT). These projections indicate that investing at the highest level rather than at the lowest level would reduce the operating cost of travel per VMT in 2032 by 2.0 percent (from $0.460 to $0.451 per VMT).

Exhibit 7-7 Projected 2032 Vehicle Operating Costs on Federal-Aid Highways Compared with Base Year for Different Possible Funding Levels

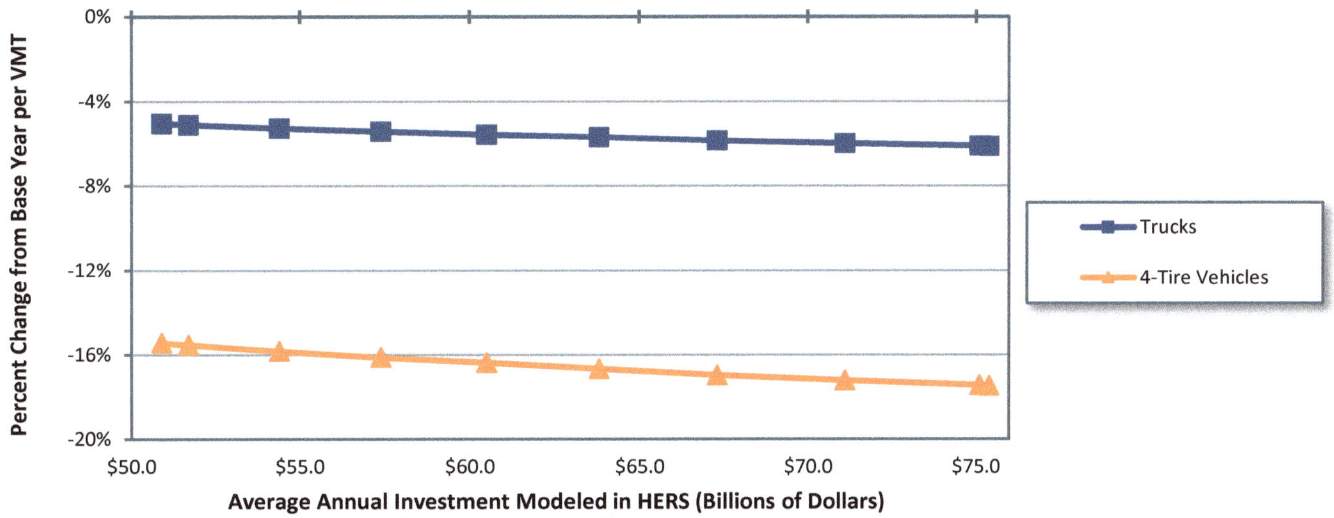

HERS-Modeled Investment on Federal-Aid Highways		Projected 2032 Performance Measures on Federal-Aid Highways					
		Average Vehicle Operating Costs			Percent Change Relative to Baseline		
Average Annual Investment (Billions of 2012 Dollars)	Average Annual Percent Change vs. 2012	All Vehicles ($/VMT)	4-Tire Vehicles ($/VMT)	Trucks ($/VMT)	4-Tire Vehicles	Trucks	Link to Chapter 8 Scenario
$75.4	2.53%	$0.451	$0.342	$1.087	-17.5%	-6.1%	Improve C&P
$71.1	2.00%	$0.452	$0.343	$1.089	-17.2%	-6.0%	
$63.8	1.00%	$0.454	$0.346	$1.092	-16.7%	-5.7%	
$57.4	**0.00%**	$0.457	$0.348	$1.096	-16.1%	-5.4%	**2012 Spending**
$51.7	-1.00%	$0.459	$0.350	$1.099	-15.5%	-5.1%	
$50.9	-1.16%	$0.460	$0.351	$1.100	-15.5%	-5.1%	Maintain C&P
	Base Year Values:	$0.485					

Source: Highway Economic Requirements System.

The projected impacts on vehicle operating costs are larger for four-tire vehicles than for trucks when compared to both the 2012 values and the adjusted baseline. When comparing the vehicle operating cost projections to the adjusted baseline, the magnitudes of the impacts are much larger; isolating the effects of future highway investment reveals that vehicle operating costs per mile are projected to decline by between 15.5 percent and 17.5 percent for four-tire vehicles, and by between 5.1 percent and 6.1 percent for trucks from 2012 to 2032.

The projected reductions in vehicle operating costs per VMT are driven by projected increases in fuel efficiency across the analysis horizon. The assumed paths of fuel efficiency are based on projections from the Energy Information Administration's *Annual Energy Outlook 2014*. The average price of gasoline is assumed to decrease between 2012 and 2032 by 4.7 percent relative to the consumer price index, while the average price of diesel fuel is assumed to increase by 9.2 percent relative to the consumer price index. The projected changes in fuel prices are added to the fuel cost savings that would result from the improvements in vehicle energy efficiency that the Energy Information Administration projects for this same period; these changes are represented in HERS as increases in average miles per gallon (mpg) of 54.6 percent for light-duty vehicles, 53.9 percent for two-axle trucks, and 15.5 percent for trucks with three or more axles. These projections incorporate the effect of increases in Corporate Average Fuel Economy (CAFE) standards and U.S. Environmental Protection Agency (EPA) standards for emissions of greenhouse gases by automobiles and light trucks through model year 2025. The projections also account for new standards for fuel efficiency and greenhouse gas emissions for medium- and heavy-duty trucks through model year 2018 adopted by the U.S. Department of Transportation and EPA.

Q&A What changes in CAFE standards have recently been adopted, and what impacts are these changes expected to have?

On May 7, 2010, the National Highway Traffic Safety Administration (NHTSA) and U.S. Environmental Protection Agency (EPA) jointly adopted Corporate Average Fuel Economy (CAFE) and carbon dioxide (CO_2) emission standards for cars and light trucks produced during model years 2012 through 2016. In combination with NHTSA's previous actions, this rule raised required fleet-average fuel economy levels for cars from 27.5 miles per gallon (mpg) in model year 2010 to 37.8 mpg for model year 2016, and those for light trucks from 23.5 mpg in 2010 to 28.8 mpg for 2016. On August 28, 2012, the two agencies adopted new rules that further increased CAFE standards for model year 2021 to 46.1 to 46.8 mpg for automobiles and to 32.6 to 33.3 mpg for light trucks; this most recent action also established tentative CAFE standards for model year 2025 of 55.3 to 56.2 mpg for cars and 39.3 to 40.3 mpg for light trucks. All of the adopted and tentative CAFE standards apply to the vehicle fleet as a whole, and are minimum standards for the vehicle fleet.

The impacts of these standards on the fuel economy of the overall vehicle fleet will continue to grow for many years beyond 2025, as new vehicles meeting the higher fuel economy requirements gradually replace older, less fuel-efficient vehicles. In announcing the most recent increases in CAFE standards, NHTSA estimated that the cumulative effects of its actions would be to save more than 500 billion gallons of fuel and to reduce CO_2 emissions by 6 billion metric tons over the lifetimes of cars and light trucks produced in 2011 through 2025. The agency also estimated that its standards would save the Nation's drivers more than $1.7 trillion in fuel costs over these vehicles' lifetimes.

In 2011, NHTSA and EPA also established new fuel efficiency and CO_2 emission standards for medium- and heavy-duty trucks produced from 2014 through 2018. These standards are expected to reduce fuel consumption by an additional 22 billion gallons, while further reducing CO_2 emissions by nearly 270 million metric tons.

Impact of Future Investment on Future VMT

As discussed above, the travel demand elasticity features in HERS modify future VMT growth for each HPMS sample section based on changes to highway user costs. In the absence of information to the contrary, previous C&P reports assumed that the HPMS forecasts represented the level of travel that would occur if user costs did not change. Because the baseline VMT forecasts used in this report are tied to a specific VMT forecasting model with known inputs, this assumption was changed. For this report, HERS was programmed to assume that the baseline projections of future VMT already accounted for anticipated independent changes in user cost component values.

In computing the impact of user cost changes on future VMT growth on an HPMS sample section, HERS compares projected highway user costs against assumed user costs that would have occurred had the physical conditions or operating performance on that highway section remained unchanged. This concept is illustrated in *Exhibit 7-8*. Based on the 2012 values assigned to various user cost components (e.g., value of travel time per hour, fuel prices, fuel efficiency, truck travel as a percentage of total travel), HERS computes baseline 2012 user costs at $1.283 per mile. If the 2032 values assigned to those same user cost components were applied in 2012, however, HERS would compute 2012 user costs to be $1.437 per mile. This "adjusted baseline" is the relevant point of comparison when examining the impact of user cost changes on VMT.

Exhibit 7-8 Projected 2032 User Costs and VMT on Federal-Aid Highways Compared with Base Year for Different Possible Funding Levels

HERS-Modeled Investment on Federal-Aid Highways		Projected 2032 Indicators on Federal-Aid Highways					
		Average Total User Costs[1]			Projected VMT[2]		
			Percent Change				
Average Annual Investment (Billions of 2012 Dollars)	Average Annual Percent Change vs. 2012	($/VMT)	vs. Actual 2012	vs. Adjusted Baseline	Trillions of VMT	Annual Percent Change vs. 2012	Link to Chapter 8 Scenario
$75.4	2.53%	$1.381	7.7%	-3.9%	3.160	1.15%	Improve C&P
$71.1	2.00%	$1.384	7.9%	-3.7%	3.157	1.15%	
$63.8	1.00%	$1.389	8.3%	-3.3%	3.151	1.14%	
$57.4	**0.00%**	$1.394	8.7%	-3.0%	3.145	1.13%	**2012 Spending**
$51.7	-1.00%	$1.399	9.1%	-2.6%	3.140	1.12%	
$50.9	-1.16%	$1.400	9.2%	-2.6%	3.139	1.12%	Maintain C&P
	Base Year Values:	$1.283			2.513	1.04%	
	Adjusted Baseline:	$1.437					

[1] The computation of user costs includes several components (value of travel time per hour, fuel prices, fuel efficiency, truck travel as a percent of total travel, etc.) that are assumed to change over time independently of future highway investment. The adjusted baseline applies the parameter values for 2032 to the data for 2012 so that changes in user costs attributable to future highway investment can be identified.

[2] The operation of the travel demand elasticity features in HERS cause future VMT growth to be influenced by future changes in average user costs per VMT. For this report, the model was set to assume that the baseline projections of future VMT already took into account anticipated independent future changes in user cost component values; hence, it is the changes versus the adjusted baseline user costs that are relevant. Since the percentage change in adjusted total user costs declined for each of the investment levels identified, the annual projected VMT growth was higher than the 1.04-percent baseline projection in all cases.

Source: Highway Economic Requirements System.

Although user costs are projected to increase in absolute terms from 2012 to 2032, they are projected to decline relative to the adjusted baseline by between 2.6 percent (at the lowest level of investment analyzed) and 3.9 percent (at the highest level of investment analyzed in 2032). Because the percentage change in adjusted total user costs declined for each investment level identified, the effective annual projected VMT growth associated with each investment level was higher than the 1.04 percent baseline projection in all cases, ranging from 1.12 percent to 1.15 percent.

Impacts of NHS Investments Modeled by HERS

As described in Chapter 2, the NHS includes the Interstate System and other routes most critical to national defense, mobility, and commerce. As noted earlier, the NHS analyses presented in this section are based on an estimate of what the NHS will look like after its expansion pursuant to MAP-21, rather than the system as it existed in 2012.

This section examines the impacts that investment on NHS roads could have on future NHS conditions and performance, independently of spending on other Federal-aid highways. The analysis presented in this section centers on HERS runs that used a database consisting only of NHS roads. This process differs from that used in previous reports, in which the levels of future investment in NHS roads were extracted from analyses that compared potential investments across a database of all Federal-aid highways. The estimated annual growth rates of investment levels for the NHS are different from those of Federal-aid highways above, because the trade-offs among costs and corresponding benefits of potential improvements are not identical across the two sets of roadways. The investment levels presented in this section were selected by applying the operational constraints used in the analysis of all Federal-aid roads (e.g., average annual spending growth rates, minimum BCR, maintaining pavement roughness, and average delay at the base-year level) to the NHS-specific database.

Impact of Future Investment on NHS User Costs and VMT

Exhibit 7-9 presents the projected impacts of NHS investment on VMT and total average user costs on NHS roads in 2032. Average user costs are projected to be lower in 2032 than for the adjusted baseline ($1.367 per VMT) for all investment levels presented. When increasing spending gradually over 20 years to implement all cost-beneficial projects (the highest level of investment, an annual average of $53.0 billion), average total user costs are projected to be 5.0 percent lower ($1.299 per VMT) than in 2012. At the lowest level of investment presented (an annual average of $36.5 billion), average total user costs are projected to be 3.4 percent lower ($1.320 per VMT) than in 2012.

Projected VMT growth on NHS roads is relatively insensitive to the range of investment levels presented in *Exhibit 7-9*. At the highest level of investment presented in *Exhibit 7-9* (an annual average of $53.0 billion), VMT is projected to grow at an average annual rate of 1.18 percent from 2012 to 2032 (2.071 trillion VMT in 2032 versus 1.638 trillion VMT in 2012). At the lowest level of investment presented in *Exhibit 7-9* (an annual average of $36.5 billion), VMT is projected to grow at an average annual rate of 1.14 percent from 2012 to 2032 (2.056 trillion VMT in 2032 versus 1.638 trillion VMT in 2012).

Exhibit 7-9 HERS Investment Levels Analyzed for the National Highway System and Projected Minimum Benefit-Cost Ratios, User Costs, and Vehicle Miles Traveled

HERS-Modeled Investment On the NHS				Projected NHS Indicators			
Average Annual Percent Change vs. 2012	Average Annual Over 20 Years						
	Total HERS Spending[1]	System Rehabilitation Spending[2]	System Expansion Spending[2]	Minimum BCR 20-Year 2013 through 2032[3]	Average 2032 Total User Costs ($/VMT)[4]	Projected 2032 VMT (Trillions)[5]	Link to Chapter 8 Scenario
2.52%	**$53.0**	$29.8	$23.2	**1.00**	$1.299	2.071	**Improve C&P**
2.00%	**$50.0**	$28.4	$21.7	1.06	$1.302	2.068	
1.00%	**$44.9**	$25.4	$19.5	1.20	$1.308	2.064	
0.00%	**$40.4**	$22.8	$17.6	1.29	$1.314	2.060	**2012 Spending**
-0.38%	**$38.9**	$21.9	$17.0	1.33	$1.316	2.059	**Maintain C&P**
-1.00%	**$36.5**	$20.4	$16.0	1.39	$1.320	2.056	
				Base Year Values:	$1.222	1.638	
				Adjusted Baseline:	$1.367		

[1] The amounts shown represent the average annual investment over 20 years that would occur if annual investment grows in constant dollar terms by the percentage shown in each row of the first column.

[2] HERS splits its available budget between system rehabilitation and system expansion based on the mix of spending it finds to be most cost-beneficial which varies by funding level.

[3] As HERS ranks potential improvements by their estimated BCRs and assumes that the improvements with the highest BCRs will be implemented first (up until the point where the available budget specified is exhausted), the minimum BCR will naturally tend to decline as the level of investment analyzed rises.

[4] The computation of user costs includes several components (value of travel time per hour, fuel prices, fuel efficiency, truck travel as a percent of total travel, etc.) that are assumed to change over time independently of future highway investment. The adjusted baseline applies the parameter values for 2032 to the data for 2012, so that changes in user costs attributable to future highway investment can be identified.

[5] The operation of the travel demand elasticity features in HERS cause future VMT growth to be influenced by future changes in average user costs per VMT. For this report, the model was set to assume that the baseline projections of future VMT already took into account anticipated independent future changes in user cost component values; hence, it is the changes versus the adjusted baseline user costs that are relevant.

Source: Highway Economic Requirements System.

Across the investment levels presented in *Exhibit 7-9*, HERS allocates between $20.4 billion and $29.8 billion in average annual spending on NHS roads to system rehabilitation and between $16.0 billion and $23.2 billion in average annual spending on NHS roads to system expansion.

Impact of Future Investment on NHS Travel Times and Travel Time Costs

Exhibit 7-10 presents the projections of NHS averages for time-related indicators of performance, along with the spending amount that HERS programs for NHS expansion projects (which have stronger effects on time-related indicators of performance than preservation projects have). For all investment levels presented in *Exhibit 7-10*, average travel speed in 2032 exceeds average travel speed in 2012 (48.3 miles per hour). The range of average travel speeds is narrow across the investment levels. At the lowest level of investment in system expansion presented in *Exhibit 7-10* (an annual average of $16.0 billion), the average travel speed in 2032 is projected to be 51.1 miles per hour. At the highest level of investment in system expansion presented in *Exhibit 7-10* (an annual average of $23.2 billion), the average travel speed in 2032 is projected to be 51.8 miles per hour.

Exhibit 7-10 Projected Changes in 2032 Highway Speed, Travel Delay, and Travel Time Costs on the National Highway System Compared with Base Year for Different Possible Funding Levels

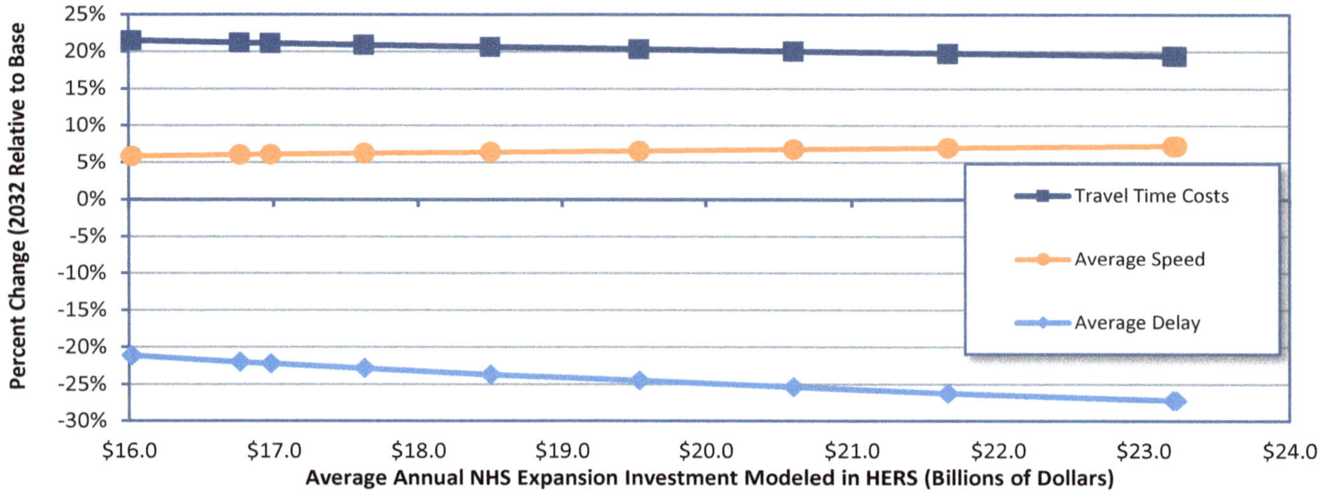

HERS-Modeled Investment on the NHS	Projected 2032 Performance Measures on the NHS				Link to Chapter 8 Scenario
		Percent Change Relative to Baseline			
Average Annual for System Expansion (Billions of 2012 Dollars)[1]	Average Speed (mph)	Average Speed	Average Delay per VMT	Travel Time Costs per VMT[2]	
$23.2	51.8	7.2%	-27.2%	19.4%	Improve C&P
$21.7	51.7	7.0%	-26.3%	19.8%	
$19.5	51.5	6.6%	-24.5%	20.3%	
$17.6	51.3	6.2%	-22.9%	20.9%	2012 Spending
$17.0	51.2	6.1%	-22.2%	21.1%	Maintain C&P
$16.0	51.1	5.8%	-21.2%	21.5%	
Base Year Values:	48.3				

[1] The amounts shown represent only the portion of HERS-modeled spending directed toward system expansion, rather than system rehabilitation. Other types of spending can affect these indicators as well.

[2] Travel time costs are affected by an assumption that the value of time will increase by 1.2 percent in real terms each year. Hence, costs would rise even if travel time remained constant.

Source: Highway Economic Requirements System; Highway Statistics 2013, Table VM-1.

The global increase in average travel speed across investment levels corresponds with large decreases in average delay per VMT across investment levels. At the highest level of investment in system expansion presented in *Exhibit 7-10*, average delay per VMT in 2032 is projected to be 27.2 percent lower than in 2012. At the lowest level of investment in system expansion presented in *Exhibit 7-10*, average delay per VMT in 2032 is projected to be 21.2 percent lower than in 2012.

Due to increases in the value of time from 2012 to 2032, the projected increases in average travel speed do not correspond to decreases in travel time costs per VMT. Travel time costs per VMT in 2032 are projected to increase across the investment levels presented. Travel time costs per VMT in 2032 are projected to increase by 19.4 percent relative to 2012 at the highest investment level and to increase by 21.5 percent at the lowest level of investment.

Impact of Future Investment on NHS Pavement Ride Quality

Exhibit 7-11 shows the portion of modeled NHS spending that HERS allocates to rehabilitation projects (which influence average pavement quality more than expansion projects do). The projected average pavement roughness of NHS roads is sensitive to the level of investment on NHS roads. At the highest level of investment presented in *Exhibit 7-11* (an annual average of $29.8 billion allocated to system rehabilitation), the model projects average pavement roughness on the NHS to be 12.0 percent lower in 2032 than in 2012. At the lowest level of investment presented in *Exhibit 7-11* (an annual average of $20.4 billion allocated to system rehabilitation), the model projects average pavement roughness on the NHS to be 2.5 percent higher in 2032 than in 2012.

Exhibit 7-11 Projected 2032 Pavement Ride Quality Indicators on the National Highway System Compared with 2012 for Different Possible Funding Levels

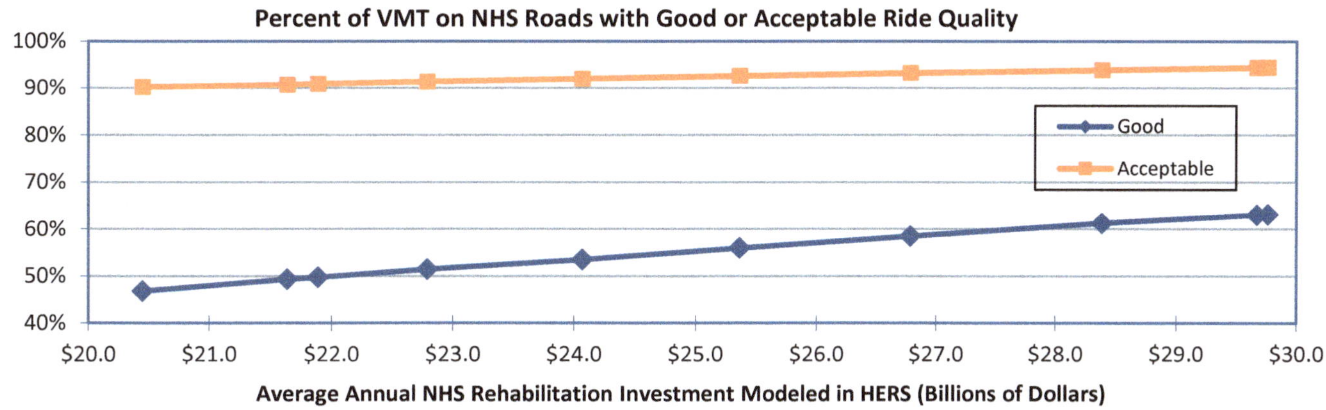

HERS-Modeled Investment on the NHS	Projected 2032 Condition Measures on the NHS[1]				Link to Chapter 8 Scenario
Average Annual for System Rehabilitation (Billions of 2012 Dollars)[2]	Percent of VMT on Roads With Ride Quality of: Good (IRI<95)	Percent of VMT on Roads With Ride Quality of: Acceptable (IRI<=170)	Average IRI (VMT-Weighted): Inches Per Mile	Average IRI (VMT-Weighted): Change Relative to Base Year	
$29.8	63.1%	94.4%	94.6	-12.0%	Improve C&P
$28.4	61.2%	93.8%	96.5	-10.2%	
$25.4	55.9%	92.6%	101.5	-5.6%	
$22.8	51.4%	91.4%	105.8	-1.6%	2012 Spending
$21.9	49.7%	90.9%	107.5	**0.0%**	**Maintain C&P**
$20.4	46.8%	90.2%	110.2	2.5%	
Base Year Values:	57.1%	89.0%	107.5		

[1] As discussed in Chapter 3, IRI values of 95 and 170 inches per mile, respectively, are the thresholds associated with "good" and "acceptable" pavement ride quality on the NHS.

[2] The amounts shown represent only the portion of HERS-modeled spending directed toward system rehabilitation, rather than system expansion. Other types of spending can affect these indicators as well.

Source: Highway Economic Requirements System.

At the highest level of investment presented in *Exhibit 7-11*, the model projects that pavements with an IRI below 95, which was the criterion in Chapter 3 for rating ride quality as "good," will carry 63.1 percent of the VMT on the NHS, up from the 57.1 percent estimated for 2012. At this investment level, the average IRI of the system would be 94.6, achieving the classification of providing good ride quality at the aggregate level. Furthermore, at the highest level of investment presented in *Exhibit 7-11*, HERS projects that 94.4 percent of VMT on the NHS would be on roads with an IRI at or below 170, which was the criterion in Chapter 3 for rating ride quality as "acceptable." This projection represents an improvement of 5.4 percentage points in the share of NHS roads with acceptable ride quality relative to the base year (89.0 percent of NHS roads with acceptable ride quality).

At the lowest level of investment presented in *Exhibit 7-11*, the model projects that pavements with an IRI below 95 will carry 46.8 percent of the VMT on the NHS, down from the 57.1 percent estimated for 2012. At this investment level, the average IRI of the system would increase to 110.2, which fails to achieve the classification of providing good ride quality at the aggregate level. The share of NHS roads with acceptable ride quality is projected to increase slightly by 2032 at the lowest level of investment presented in *Exhibit 7-11*; HERS projects that 90.2 percent of VMT on the NHS would be on roads with an IRI at or below 170, which is slightly higher than the share of NHS roads with acceptable ride quality in 2012 (89.0 percent of NHS roads with acceptable ride quality).

Based on these modeling results, additional investment to bring the percentage of NHS VMT on roads with "good" or "acceptable" ride quality closer to 100 percent would be economically inefficient, as the costs would exceed the benefits. A key factor leading to this result is that some improvements are not cost-beneficial until IRI rises above the threshold for acceptable ride quality by a sufficient margin. Thus, for some roads with an IRI above 170, improvements would not generate benefits exceeding costs. A further restriction in achieving a state in which all roads have an IRI at or below 170 is that, at any given point, some pavements will be under construction.

Impacts of Interstate System Investments Modeled by HERS

The Interstate System, unlike the broader NHS of which it is a part, has standard design and signage requirements, making it the most recognizable subset of the highway network. This section examines the impacts that investment in the Interstate System could have on future Interstate System conditions and performance, independently of spending on other Federal-aid highways. The analysis presented in this section centers on HERS runs that used a database consisting only of Interstate System roads. This process differs from that used in previous reports, in which the levels of future investment in the Interstate System were extracted from analyses that compared potential investments across a database of all Federal-aid highways.

The Interstate investment levels presented in this section were selected by applying the operational constraints used in the analysis of all Federal-aid roads (e.g., average annual spending growth rates, minimum BCR, maintaining pavement roughness, and average delay at the base-year level) to the Interstate System-specific database.

Impact of Future Investment on Interstate User Costs and VMT

Exhibit 7-12 presents the projected impacts of highway investment on VMT and total average user costs on Interstate roads in 2032, along with the amount that HERS allocates to Interstate projects. Average user costs are projected to be lower in 2032 than the adjusted baseline ($1.267 per VMT) for all investment levels presented. At the highest level of investment presented in *Exhibit 7-12* (an annual average of $23.7 billion), average total user costs are projected to be 4.9 percent lower ($1.205 per VMT) than in 2012. At the lowest level of investment presented (an annual average of $13.7 billion), average total user costs are projected to be 2.1 percent lower ($1.241 per VMT) than in 2012.

Exhibit 7-12 HERS Investment Levels Analyzed for the Interstate System and Projected Minimum Benefit-Cost Ratios, User Costs, and Vehicle Miles Traveled

HERS-Modeled Investment On the Interstate System				Projected Interstate Indicators			
Average Annual Percent Change vs. 2012	Average Annual Over 20 Years						
	Total HERS Spending[1]	System Rehabilitation Spending[2]	System Expansion Spending[2]	Minimum BCR 20-Year 2013 through 2032[3]	Average 2032 Total User Costs ($/VMT)[4]	Projected 2032 VMT (Trillions)[5]	Link to Chapter 8 Scenario
4.08%	**$23.7**	$12.7	$11.0	**1.00**	$1.205	0.926	**Improve C&P**
4.00%	**$23.5**	$12.6	$10.9	1.01	$1.205	0.926	
3.00%	**$21.0**	$11.4	$9.6	1.12	$1.212	0.923	
2.00%	**$18.8**	$10.3	$8.5	1.25	$1.219	0.921	
1.74%	**$18.3**	$10.1	$8.2	1.26	$1.222	0.920	**Maintain C&P**
1.00%	**$16.9**	$9.3	$7.6	1.37	$1.226	0.919	
0.00%	**$15.2**	$8.3	$6.8	1.54	$1.234	0.917	**2012 Spending**
-1.00%	**$13.7**	$7.6	$6.1	1.66	$1.241	0.915	
				Base Year Values:	**$1.129**	**0.728**	
				Adjusted Baseline:	**$1.267**		

[1] The amounts shown represent the average annual investment over 20 years that would occur if annual investment grows in constant dollar terms by the percentage shown in each row of the first column.

[2] HERS splits its available budget between system rehabilitation and system expansion based on the mix of spending it finds to be most cost-beneficial which varies by funding level.

[3] As HERS ranks potential improvements by their estimated BCRs and assumes that the improvements with the highest BCRs will be implemented first (up until the point where the available budget specified is exhausted), the minimum BCR will naturally tend to decline as the level of investment analyzed rises.

[4] The computation of user costs includes several components (value of travel time per hour, fuel prices, fuel efficiency, truck travel as a percent of total travel, etc.) that are assumed to change over time independent of future highway investment. The adjusted baseline applies the parameter values for 2032 to the data for 2012 so that changes in user costs attributable to future highway investment can be identified.

[5] The operation of the travel demand elasticity features in HERS cause future VMT growth to be influenced by future changes in average user costs per VMT. For this report, the model was set to assume that the baseline projections of future VMT already took into account anticipated independent future changes in user cost component values; hence, it is the changes versus the adjusted baseline user costs that are relevant.

Source: Highway Economic Requirements System.

Projected VMT growth on Interstate roads is relatively insensitive to the range of investment levels presented in *Exhibit 7-12*. At the highest level of investment presented in *Exhibit 7-12* (an annual average of $23.7 billion), VMT is projected to grow at an average annual rate of 1.21

percent from 2012 to 2032 (926 billion VMT in 2032 versus 728 billion VMT in 2012). At the lowest level of investment presented in *Exhibit 7-12* (an annual average of $13.7 billion), VMT is projected to grow at an average annual rate of 1.15 percent from 2012 to 2032 (915 billion VMT in 2032 versus 728 billion VMT in 2012).

Across the investment levels presented in *Exhibit 7-12*, HERS allocates between $7.6 billion and $12.7 billion in average annual spending on Interstate roads to system rehabilitation, and between $6.1 billion and $11.0 billion in average annual spending on Interstate roads to system expansion.

Impact of Future Investment on Interstate System Travel Times and Travel Costs

Exhibit 7-13 presents the projections of Interstate System averages for time-related indicators of performance, along with the amount that HERS programs for Interstate System expansion projects (which have a relatively large impact on travel time). Across all investment levels presented in *Exhibit 7-13*, average speed on the Interstate System is projected to be higher than its 2012 level (61.6 miles per hour) in 2032. At the highest level of investment presented in *Exhibit 7-13* (average annual investment in system expansion of $11.0 billion), average Interstate highway travel speed is projected to be 8.9 percent higher (67.1 miles per hour) in 2032 than in 2012. At the lowest level of investment presented in *Exhibit 7-13* (average annual investment in system expansion of $6.1 billion), average Interstate highway travel speed is projected to be 5.3 percent higher (64.9 miles per hour) in 2032 than in 2012.

The global increase in average travel speed across investment levels corresponds with large decreases in average delay per VMT across investment levels. At the highest level of investment presented in *Exhibit 7-13*, average delay per VMT in 2032 is projected to be 49.0 percent lower than in 2012. At the lowest level of investment presented in *Exhibit 7-13*, average delay per VMT in 2032 is projected to be 26.4 percent lower than in 2012.

The projected impacts on travel delay across investment levels are much greater for Interstates than for other portions of Federal-aid highways. This result suggests the presence of a large scope of congestion-related benefits that could be achieved through investments in Interstate highway improvements.

Due to increases in the value of time from 2012 to 2032, the projected increases in average travel speed do not correspond to decreases in travel time costs per VMT. Travel time costs per VMT in 2032 are projected to increase across all investment levels. Travel time costs per VMT in 2032 are projected to increase by 18.7 percent relative to 2012 at the highest level of investment presented in *Exhibit 7-13* and by 24.3 percent at the lowest level of investment.

The ranges of average travel speeds and, in turn, travel time cost impacts across investment levels are larger for Interstate highways than for the NHS. This result indicates that outcomes related to travel speed and travel time on Interstate highways are more sensitive to the level of investment than corresponding outcomes on the NHS overall.

Exhibit 7-13 Projected Changes in 2032 Highway Speed, Travel Delay, and Travel Time Costs on the Interstate System Compared with Base Year for Different Possible Funding Levels

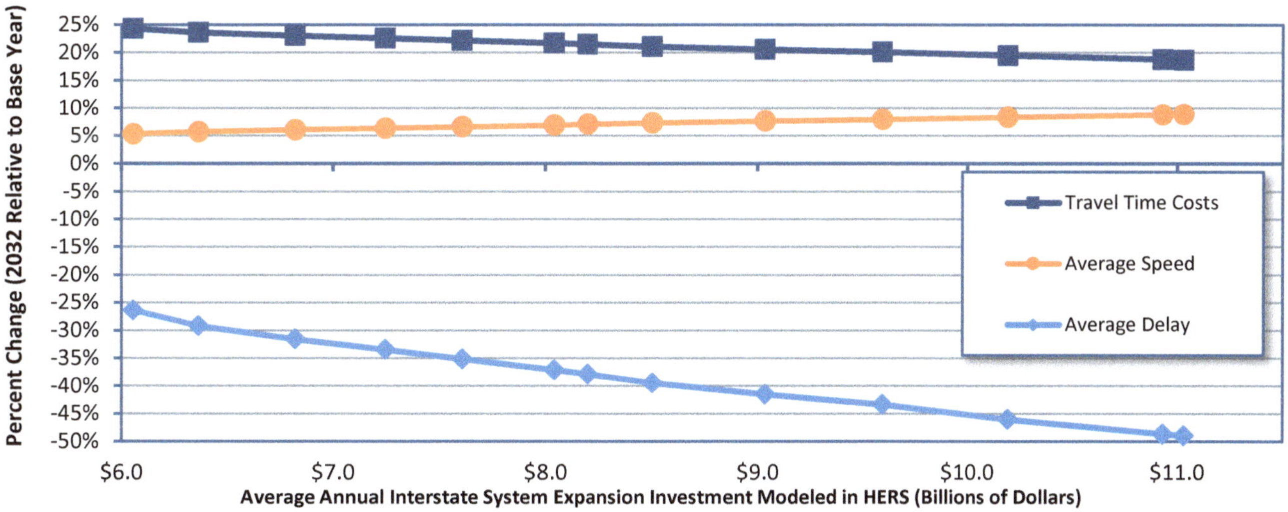

HERS-Modeled Investment on Interstate Highways	Projected 2032 Performance Measures on Interstate Highways				
		Percent Change Relative to Baseline			
Average Annual for System Expansion (Billions of 2012 Dollars)[1]	Average Speed (mph)	Average Speed	Average Delay per VMT	Travel Time Costs per VMT[2]	Link to Chapter 8 Scenario
$11.0	67.1	8.9%	-49.0%	18.7%	Improve C&P
$10.9	67.1	8.8%	-48.7%	18.8%	
$9.6	66.5	8.0%	-43.4%	20.1%	
$8.5	66.2	7.3%	-39.5%	21.1%	
$8.2	66.0	7.1%	-37.9%	21.5%	Maintain C&P
$7.6	65.7	6.6%	-35.2%	22.2%	
$6.8	65.4	6.1%	-31.6%	23.1%	2012 Spending
$6.1	64.9	5.3%	-26.4%	24.3%	
Base Year Values:	61.6				

[1] The amounts shown represent only the portion of HERS-modeled spending directed toward system expansion, rather than system rehabilitation. Other types of spending can affect these indicators as well.

[2] Travel time costs are affected by an assumption that the value of time will increase by 1.2 percent in real terms each year; hence, costs would rise even if travel time remained constant.

Source: Highway Economic Requirements System; Highway Statistics 2013, Table VM-1.

Impact of Future Investment on Interstate Pavement Ride Quality

Exhibit 7-14 shows the sub-portions of modeled Interstate System spending that HERS allocates to rehabilitation projects (which influence average pavement quality more than expansion projects do). The projected average pavement roughness of NHS roads is sensitive to the level of investment on Interstate System roads. At the highest level of investment presented in *Exhibit 7-14* (an annual average of $12.7 billion allocated to system rehabilitation), the model projects average pavement roughness on the Interstate System to be 9.3 percent lower in 2032 than in 2012. At the lowest level of investment presented in *Exhibit 7-14* (an annual average of $7.6 billion allocated to system rehabilitation), the model projects average pavement roughness on the Interstate System to be 11.4 percent higher in 2032 than in 2012.

Exhibit 7-14 Projected 2032 Pavement Ride Quality Indicators on the Interstate System Compared with 2012 for Different Possible Funding Levels

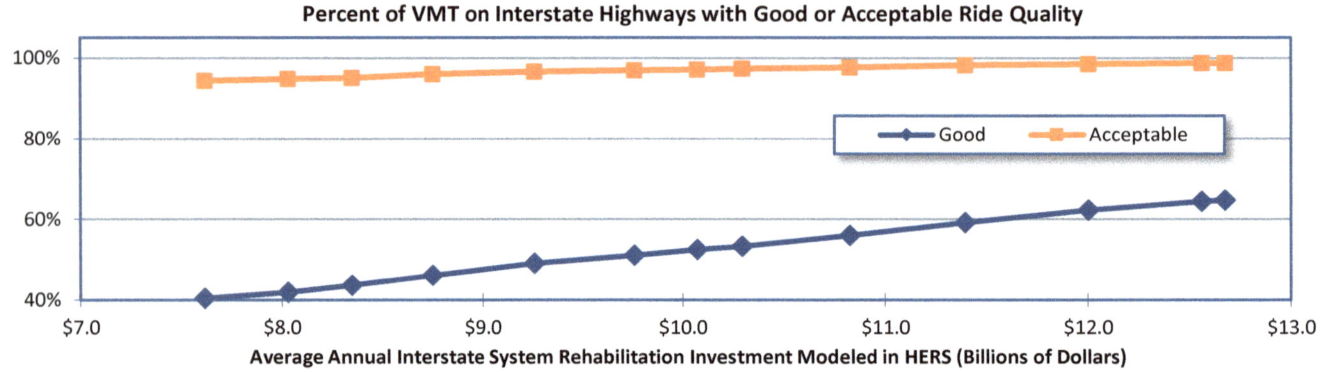

HERS-Modeled Investment on Interstate Highways	Projected 2032 Condition Measures Interstate Highways[1]				Link to Chapter 8 Scenario
	Percent of VMT on Roads with Ride Quality of:		Average IRI (VMT-Weighted)		
Average Annual for System Rehabilitation (Billions of 2012 Dollars)[2]	Good (IRI<95)	Acceptable (IRI<=170)	Inches Per Mile	Change Relative to Base Year	
$12.7	64.7%	98.7%	85.6	-9.3%	Improve C&P
$12.6	64.4%	98.7%	85.8	-9.1%	
$11.4	59.1%	98.2%	89.4	-5.3%	
$10.3	53.3%	97.3%	93.6	-0.8%	
$10.1	52.5%	97.1%	94.4	**0.0%**	**Maintain C&P**
$9.3	49.1%	96.6%	97.2	3.0%	
$8.3	43.7%	95.1%	101.9	7.9%	2012 Spending
$7.6	40.4%	94.3%	105.2	11.4%	
Base Year Values:	**66.8%**	**95.2%**	**94.4**		

[1] As discussed in Chapter 3, IRI values of 95 and 170 inches per mile, respectively, are the thresholds associated with "good" and "acceptable" pavement ride quality on the NHS.

[2] The amounts shown represent only the portion of HERS-modeled spending directed toward system rehabilitation, rather than system expansion. Other types of spending can affect these indicators as well.

Source: Highway Economic Requirements System.

Across all investment levels presented in *Exhibit 7-14*, the model projects that the share of pavements with an IRI below 95, which was the criterion in Chapter 3 for rating ride quality as "good," would be below the corresponding share in 2012 (66.8 percent). The share of VMT on Interstate highways with an IRI below 95 in 2032 is highly sensitive to investment levels. At the highest level of investment presented in *Exhibit 7-14*, 64.7 percent of VMT on Interstate highways is projected to be on roads with an IRI below 95 (a decrease of 2.1 percentage points relative to the base year). At the lowest level of investment presented in *Exhibit 7-14*, 40.4 percent of VMT on Interstate highways is projected to be on roads with an IRI below 95 (a decrease of 26.4 percentage points relative to the base year).

The share of Interstate pavements with an IRI at or below 170, which was the criterion in Chapter 3 for rating ride quality as "acceptable," is projected to increase from the corresponding share in

2012 (95.2 percent) at the highest level of investment presented in *Exhibit 7-14* (98.7 percent). At the lowest level of investment, the share of Interstate pavements with an IRI at or below 170 is slightly below the 2012 level (94.3 percent).

Based on these modeling results, additional investment to increase the percentage of VMT on Interstate highways with "good" quality would be economically inefficient, as the costs would exceed the benefits; however, increasing the percentage of VMT on Interstate highways with "acceptable" ride quality is warranted. A key factor leading to this result is that some improvements are not cost-beneficial until IRI rises above 170 (or even higher). Thus, the HERS analysis tended to focus on improving Interstate roads to reach or maintain "acceptable" status, while forgoing non-cost-beneficial improvements that would achieve or maintain "good" ride quality on some roads.

A related limiting factor is that the Interstate-wide average road quality in the base year is relatively high, with an average IRI below 95, and with 95.2 percent of Interstate roads at "acceptable" road quality. Thus, not only is the model constrained by a relatively small subset of roads for which surface rehabilitation would be cost-beneficial, but also the model confirms that allocating funding to alternative projects can be optimal, provided Interstate pavement quality remains at or near "acceptable" on unimproved surfaces.

National Bridge Investment Analysis System

The scenario estimates relating to bridge repair and replacement shown in this report are derived primarily from NBIAS. NBIAS can synthesize element-level data from the general condition ratings reported for individual bridges in the NBI. The analyses presented in this report are based on synthesized element-level data. Examples of bridge elements include the bridge deck, a steel girder used for supporting the deck, a concrete pier cap on which girders are placed, a concrete column used for supporting the pier cap, or a bridge railing.

NBIAS uses a probabilistic approach to model bridge deterioration for each synthesized bridge element. It relies on a set of transition probabilities to project the likelihood that an element will deteriorate from one condition state to another over a given period. This information, along with details on the cost of maintenance, repair and rehabilitation (MR&R) actions, is used to predict lifecycle costs of maintaining existing bridges, and to develop MR&R policies specifying what MR&R action to perform based on the existing condition of a bridge element. Another key input to the model is the overall objective assumed for MR&R policies. The State of Good Repair strategy, although the most aggressive of the available MR&R policies, generates results more consistent with agency practices and recent trends in bridge conditions compared to the other three strategies evaluated (see Appendix B). Therefore, the State of Good Repair strategy has been adopted for use in the baseline analyses presented in this chapter and in Chapter 8.

The State of Good Repair strategy aims to improve all bridges to good condition that can be sustained through ongoing investment. MR&R investment is front loaded under the State of Good Repair strategy, as large MR&R investments are required in the early years of the forecast period to improve bridge conditions, while smaller MR&R investments are needed in the later years to sustain bridge conditions. Replacement of a bridge is recommended if a bridge evaluation results in lower life-cycle costs as compared to the recommended MR&R work.

To estimate functional improvement needs, NBIAS applies a set of improvement standards and costs to each bridge in the NBI. The system then identifies potential improvements—such as widening existing bridge lanes, raising bridges to increase vertical clearances, and strengthening bridges to increase load-carrying capacity—and evaluates their potential benefits and costs. NBIAS evaluates potential bridge replacements by comparing their benefits and costs with what could be achieved through MR&R work alone. Appendix B discusses NBIAS in detail.

In using NBIAS to project conditions and performance of the Nation's bridges over 20 years, this section considers the alternatives of continuing to invest in bridge rehabilitation at the 2012 level (in constant dollars) and at higher or lower levels. The expenditures modeled pertain only to bridge system rehabilitation; expenditures associated with bridge system expansion are modeled separately as part of the capacity expansion analysis in HERS. (The NBIAS-modeled investments presented here should be considered as additive to the HERS-modeled investments presented above; each capital investment scenario presented in Chapter 8 combines one HERS analysis with one NBIAS analysis and makes adjustments to account for non-modeled spending.)

Performance Measures

NBIAS incorporates engineering criteria to evaluate bridge deficiencies at the level of individual bridge elements, and computes an initial value for the cost of a set of corrective actions that would address all such element-level deficiencies. NBIAS projects the deterioration of the individual bridge elements for future years, which determines the timing, type, and cost of any needed future corrective actions. Of note is that these corrective actions are not limited to bridges rated as structurally deficient or functionally obsolete (see Chapter 3). Instead, the model considers potential actions on all bridges, which allows the software to address element-level deficiencies before they trigger a deficiency rating for the overall bridge (i.e., the bridge as a whole is classified as structurally deficient), consistent with sound principles of asset management.

Most previous editions of the C&P report used the economic bridge investment backlog as the sole indicator of bridge system performance. For this edition of the C&P report, four metrics are presented to provide a more comprehensive view of bridge performance:

- Percentage Structurally Deficient by Deck Area
- Total Percentage Deficient by Deck Area (used in computing the Maintain Conditions and Performance scenario in Chapter 8)
- Average Health Index
- Economic Investment Backlog (used in computing the Improve Conditions and Performance scenario in Chapter 8)

The Percent Structurally Deficient by Deck Area metric indicates the amount of deck area on bridges classified as structurally deficient. Total Percent Deficient by Deck Area metric is the amount of deck area on bridges classified as structurally deficient and functionally obsolete. The Health Index metric is a ranking system (0–100) for bridge elements typically used in the context of decision-making for bridge preventive maintenance. Although the condition state of a bridge element is categorical, it is useful to consider an element's condition at a given time as a point along a continuous timeline with 100 percent in the best state to 0 percent in the worst state. The Health Index merely indicates where the element is along this continuum.

Q&A **Why are functionally obsolete bridges not represented as separate items in the exhibits in this chapter?**

Although included in the total deficient bridge figures, functionally obsolete bridges are intentionally not featured in the exhibits; NBIAS can model some improvements that address functional obsolescence, but it currently does not consider replacing bridges with wider bridges having more through lanes (these types of capacity expansions are instead modeled in HERS). Under this limitation, the percentage of functional obsolete bridges in NBIAS might not consistently decrease as investment increases. As discussed in Chapter 3, if a bridge is both structurally deficient and functionally obsolete, it is classified as structurally deficient. Hence, at higher levels of investment, NBIAS might address structural deficiencies but be unable to address functional obsolescence, causing the percentage of functionally obsolete bridges to increase.

To aggregate the element-level result to the bridge level (i.e., assign a value for the Health Index), a weight is assigned to each element according to the economic consequences of its failure, and then an average of all the weighted elements is calculated. Thus, an element for which a failure has relatively little economic effect, such as a railing, would receive less weight than an element for which a failure might result in closing the bridge, such as a girder. In general, the lower the Health Index is, the higher the priority for rehabilitation or maintenance of the structure, although other factors also are instrumental in determining priority of work on bridges.

The Economic Investment Backlog metric represents the combined cost of all corrective actions for which NBIAS estimates implementation would be cost-beneficial. Consistent with the HERS analysis, implementing all cost-beneficial corrective actions in NBIAS would not necessarily mean that no structurally deficient or functionally obsolete bridges would remain; rather, implementing all cost-beneficial corrective actions in NBIAS would indicate that it would not be cost-beneficial to take any further corrective actions. As noted above, these actions extend to all bridges, not just those rated as structurally deficient or functionally obsolete.

Impacts of Systemwide Investments Modeled by NBIAS

As referenced in Chapter 6, of the $105.2 billion invested in highways in 2012, $16.4 billion was used for bridge system rehabilitation. For investments of the types modeled by NBIAS, *Exhibit 7-15* shows how the total amount invested over the 20-year analysis period influences the bridge performance levels projected for the final year, 2032. If spending were sustained at its 2012 level in constant dollar terms ($16.4 billion, the investment level feeding the Sustain 2012 Spending scenario presented in Chapter 8), projected performance for 2032 would improve relative to 2012 for each performance measure considered. The share of bridges classified as structurally deficient, weighted by deck area, would decrease from 8.2 percent to 2.9 percent. The average Health Index would rise from 92.0 to 95.1. The Economic Investment Backlog would decrease by 83.5 percent relative to its 2012 level of $123.1 billion.

Exhibit 7-15 Projected Impact of Alternative Investment Levels on 2032 Bridge Condition Indicators for All Bridges

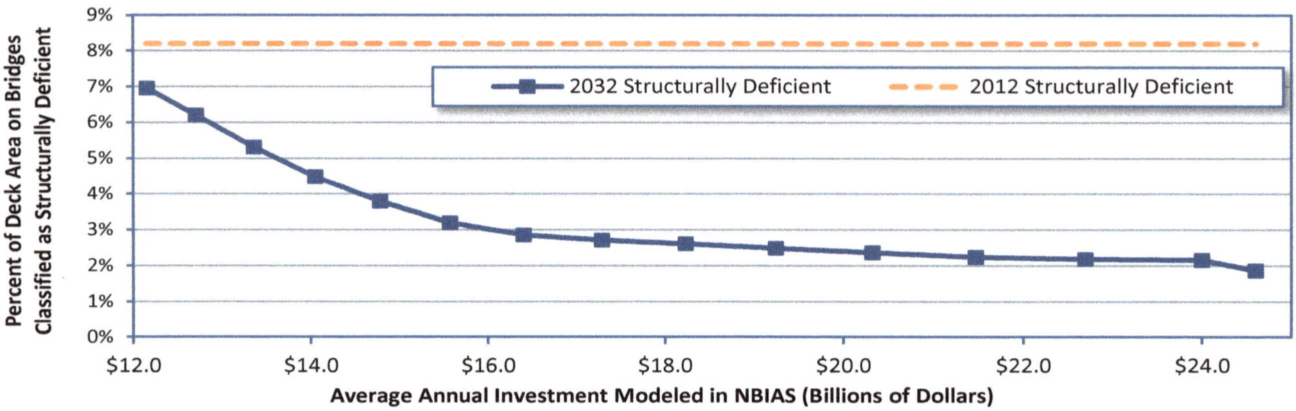

NBIAS-Modeled Investment on All Bridges		Projected 2032 Condition Indicators—All Bridges				
Average Annual Investment[1] (Billions of 2012 Dollars)	Average Annual Percent Change vs. 2012	Percent Structurally Deficient By Deck Area	Total Percent Deficient By Deck Area	Health Index	Economic Investment Backlog[1] (Billions of 2012 Dollars)	Link to Chapter 8 Scenario
$24.6	3.72%	1.9%	21.2%	95.4	**$0.0**	Improve C&P
$22.7	3.00%	2.2%	21.7%	95.3	$5.5	
$20.3	2.00%	2.4%	22.0%	95.2	$10.3	
$18.2	1.00%	2.6%	22.4%	95.2	$15.0	
$16.4	**0.00%**	2.9%	22.9%	95.1	$20.3	2012 Spending
$14.8	-1.00%	3.8%	23.9%	94.7	$32.9	
$13.4	-2.00%	5.3%	25.2%	94.1	$50.5	
$12.2	-2.95%	7.0%	**26.7%**	93.4	$67.6	Maintain C&P
2012 Baseline Values:		8.2%	26.7%	92.0	$123.1	

[1] The amounts shown do not reflect system expansion needs; the bridge components of such needs are addressed as part of the HERS model analysis.

Source: National Bridge Investment Analysis System.

The highest level of spending shown in *Exhibit 7-15* averages $24.6 billion per year (this feeds the Improve Conditions and Performance scenario in Chapter 8). This level of investment is projected to reduce the Percent Structurally Deficient by Deck Area to 1.9 percent and to eliminate the Economic Investment Backlog for bridges by 2032. This indicates that the model does not find that completely eliminating structural deficiencies would be cost-beneficial at any single point in time. In some cases, the model recommends that corrective actions be deferred and, in other cases, estimates that the benefits of replacing a bridge would be outweighed by its costs (suggesting that it should eventually be closed, diverting traffic to other available crossings).

Exhibit 7-15 also indicates that the average annual bridge investment could be reduced from the 2012 level while maintaining bridge performance. An average annual spending decline of 2.95 percent to an average annual investment level of $12.2 billion would still be sufficient to maintain the Total Percent Deficient by Deck Area at its 2012 level through 2032. At this level of investment, the deck area-weighted share of bridges classified as structurally deficient is projected to drop (improve), the average Health Index is projected to rise (improve), and the Economic Investment Backlog is projected to shrink (improve).

Q&A **Why does the economic backlog estimated by NBIAS differ from bridge backlog figures estimated by some other organizations?**

One major reason for such differences is that the backlog estimated by NBIAS is not intended to constitute a complete bridge investment estimate backlog. The NBIAS figures relate only to investment needs associated with the condition of existing structures, and not capacity expansion needs. The backlog HERS estimates includes estimates of capacity-related needs for highways and bridges combined.

Some estimates of bridge backlog produced by other organizations do attempt to combine estimates of needs relating to bridge capacity with those relating to existing structures.

Impacts of Federal-Aid Highway Investments Modeled by NBIAS

For bridges on Federal-aid highways, *Exhibit 7-16* compares performance projections for 2032 at various levels of investment with measured performance in 2012. If spending on the types of improvements modeled in NBIAS were sustained at the 2012 level of $12.0 billion (in constant dollars), performance is projected to improve. The Percent Structurally Deficient by Deck Area would decrease from 7.5 percent to 3.3 percent and the average Health Index would rise from 92.0 to 94.8. The Economic Investment Backlog would decrease by 73.4 percent from its 2012 level of $105.8 billion.

If spending declined by 1.77 percent per year to an average annual investment level of $10.0 billion, NBIAS projects Total Percent Deficient by Deck Area would be the same in 2032 as in 2012. The remaining metrics would improve with the Economic Investment Backlog showing the largest change from its 2012 level, a reduction of 47.2 percent by 2032. If spending increased by 4.39 percent per year to an average annual level of $19.5 billion, the Economic Investment Backlog would fall to zero by 2032.

Exhibit 7-16 Projected Impact of Alternative Investment Levels on 2032 Bridge Condition Indicators for Bridges on Federal-Aid Highways

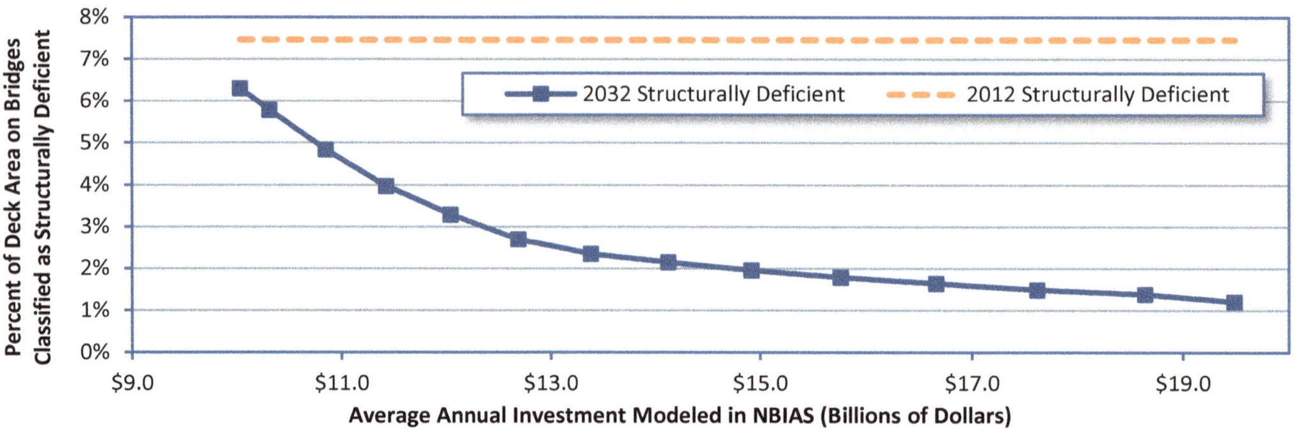

NBIAS-Modeled Investment On Federal-Aid Bridges		Projected 2032 Condition Indicators Bridges on Federal-Aid Highways				
Average Annual Investment[1] (Billions of 2012 Dollars)	Average Annual Percent Change vs. 2012	Percent Structurally Deficient By Deck Area	Total Percent Deficient By Deck Area	Health Index	Economic Investment Backlog[1] (Billions of 2012 Dollars)	Link to Chapter 8 Scenario
$19.5	4.39%	1.2%	21.1%	95.4	**$0.0**	Improve C&P
$18.6	4.00%	1.4%	21.4%	95.4	$2.8	
$16.7	3.00%	1.6%	21.8%	95.3	$7.0	
$14.9	2.00%	2.0%	22.3%	95.2	$12.4	
$13.4	1.00%	2.3%	22.9%	95.2	$17.2	
$12.0	**0.00%**	3.3%	24.0%	94.8	$28.1	2012 Spending
$10.8	-1.00%	4.8%	25.4%	94.2	$43.5	
$10.0	-1.77%	6.3%	**26.6%**	93.5	$55.9	Maintain C&P
2012 Baseline Values:		7.5%	26.6%	92.0	$105.8	

[1] The amounts shown do not reflect system expansion needs; the bridge components of such needs are addressed as part of the HERS model analysis.

Source: National Bridge Investment Analysis System.

Impacts of NHS Investments Modeled by NBIAS

The impact of various funding levels on the performance of the bridges on the NHS is shown in *Exhibit 7-17*. If spending on types of improvements modeled in NBIAS on NHS bridges were sustained at the 2012 level of $8.3 billion in constant dollar terms, projected performance for 2032, as measured by the level of Total Percent Deficient by Deck Area, would decrease from 26.9 percent to 25.4 percent. The Percent Structurally Deficient by Deck Area would decrease from 7.1 percent to 3.1 percent, the average Health Index would increase from 92.0 to 94.9, and the Economic Investment Backlog would decrease by 74.3 percent.

Exhibit 7-17 Projected Impact of Alternative Investment Levels on 2032 Bridge Condition Indicators for Bridges on the National Highway System

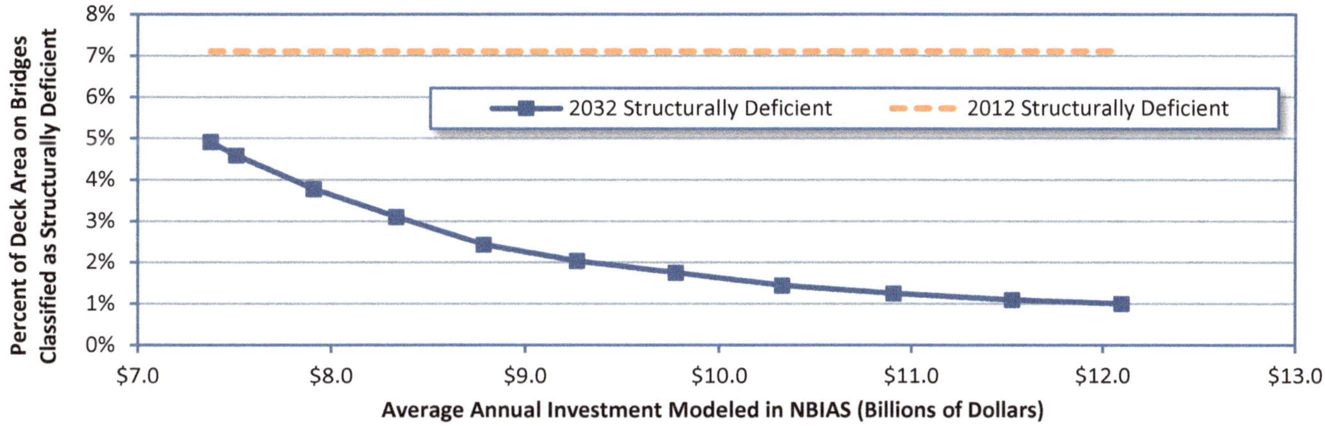

NBIAS-Modeled Investment on NHS Bridges		Projected 2032 Condition Indicators—NHS Bridges				
Average Annual Investment[1] (Billions of 2012 Dollars)	Average Annual Percent Change vs. 2012	Percent Structurally Deficient By Deck Area	Total Percent Deficient By Deck Area	Health Index	Economic Investment Backlog[1] (Billions of 2012 Dollars)	Link to Chapter 8 Scenario
$12.1	3.43%	1.0%	23.1%	95.5	**$0.0**	Improve C&P
$11.5	3.00%	1.1%	23.2%	95.4	$1.4	
$10.3	2.00%	1.4%	23.7%	95.4	$5.4	
$9.3	1.00%	2.0%	24.4%	95.3	$10.2	
$8.3	**0.00%**	3.1%	25.4%	94.9	$19.1	2012 Spending
$7.5	-1.00%	4.6%	26.6%	94.3	$30.8	
$7.4	-1.17%	4.9%	**26.9%**	94.1	$32.7	Maintain C&P
2012 Baseline Values:		7.1%	26.9%	92.0	$74.2	

[1] The amounts shown do not reflect system expansion needs; the bridge components of such needs are addressed as part of the HERS model analysis.

Source: National Bridge Investment Analysis System.

A 3.43-percent annual increase in spending to an average annual investment level of $12.1 billion would reduce the Economic Investment Backlog to zero by 2032. The Percent Structurally Deficient by Deck Area would decrease to 1.0 percent from 7.1 percent in 2012. The Total Percent Deficient by Deck Area would decrease from 26.9 in 2012 to 23.1 percent in 2032 and the average Health Index would increase from 92.0 to 95.5 during the same period. A decline in spending by 1.17 percent per year to an average annual investment level of $7.4 billion would reduce the Economic Investment Backlog in 2032 by 55.9 percent from the level in 2012 (from $74.2 billion to $32.7 billion).

Impacts of Interstate Investments Modeled by NBIAS

Exhibit 7-18 shows the impact of varying funding levels on the performance of bridges on the Interstate System. If spending on types of improvements modeled in NBIAS on Interstate bridges were sustained at the 2012 level of $3.8 billion in constant dollar terms, the Total Percent

Deficient by Deck Area would increase from 28.5 percent to 29.2 percent by 2032. Projected performance for 2032 would improve for the other metrics relative to 2012: the Percent Structurally Deficient by Deck Area would decrease from 7.1 percent in 2012 to 6.6 percent in 2032; the average Health Index would rise from 91.6 to 93.6; and the Economic Investment Backlog would decrease by 44.5 percent to $22.3 billion relative to the 2012 level of $40.2 billion.

Exhibit 7-18 Projected Impact of Alternative Investment Levels on 2032 Bridge Condition Indicators for Interstate Bridges

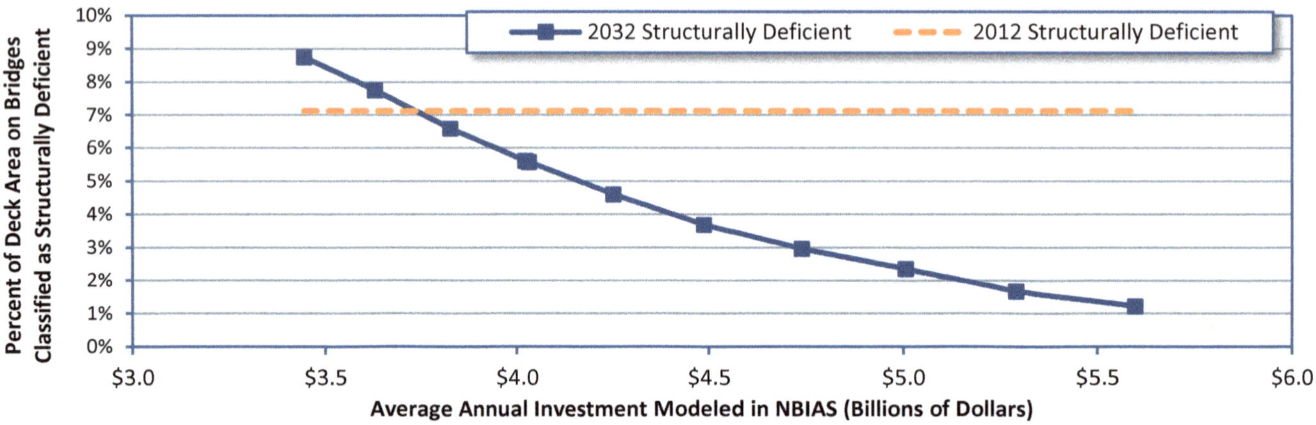

NBIAS-Modeled Investment On Interstate Bridges		Projected 2032 Condition Indicators - Interstate Bridges				
Average Annual Investment[1] (Billions of 2012 Dollars)	Average Annual Percent Change vs. 2012	Percent Structurally Deficient By Deck Area	Total Percent Deficient By Deck Area	Health Index	Economic Investment Backlog[1] (Billions of 2012 Dollars)	Link to Chapter 8 Scenario
$5.8	3.77%	1.0%	24.7%	95.4	**$0.0**	Improve C&P
$5.3	3.00%	1.7%	25.4%	95.3	$3.1	
$4.7	2.00%	3.0%	26.7%	95.1	$7.9	
$4.3	1.00%	4.6%	27.7%	94.4	$15.1	
$4.0	0.48%	5.6%	**28.5%**	94.0	$18.9	Maintain C&P
$3.8	**0.00%**	6.6%	29.2%	93.6	$22.3	2012 Spending
$3.4	-1.00%	8.7%	31.0%	92.6	$28.8	
2012 Baseline Values:		**7.1%**	**28.5%**	**91.6**	**$40.2**	

[1] The amounts shown do not reflect system expansion needs; the bridge components of such needs are addressed as part of the HERS model analysis.

Source: National Bridge Investment Analysis System.

A spending increase of 3.77 percent per year to an average annual level of $5.8 billion is estimated to be sufficient to reduce the Economic Investment Backlog to zero by 2032, decrease the Percent Structurally Deficient by Deck Area to 1.0 percent, increase the average Health Index to 95.4, and reduce the Total Percent Deficient by Deck Area to 24.7 percent.

Potential Transit Capital Investment Impacts

This section examines how different types and levels of annual capital investments would likely affect transit system condition and performance by 2032. It begins with an overview of the types of capital spending projected by the Federal Transit Administration's (FTA's) Transit Economic Requirements Model (TERM), the primary analysis tool used to assess transit investment needs and impacts in Part II of this report. The section then examines how variations in the level of annual capital spending are likely to affect future transit conditions and performance—both at the national level and for urbanized areas (UZAs) with populations greater than 1 million.

Types of Capital Spending Projected by TERM

TERM is an analysis tool that uses algorithms based on engineering and economic concepts to forecast total capital investment needs for the U.S. transit industry through a 20-year time horizon. Specifically, TERM is designed to forecast the following types of investment needs:

- Preservation: The level of investment in the rehabilitation and replacement of existing transit capital assets required to attain specific investment goals (e.g., to attain a state of good repair [SGR]) subject to potentially limited capital funding.
- Expansion: The level of investment in the expansion of transit fleets, facilities, and rail networks required to support projected growth in transit demand (i.e., to maintain performance at current levels as demand for service increases).

TERM includes a benefit-cost test that is applied to expansion scenarios to determine which investments are cost effective and which are not. TERM reports investment costs only for investments that pass the test. The SGR Benchmark, described in Chapter 8, uses a zero-growth assumption and turns off the cost-benefit test. It estimates the cost of maintaining what is currently in service as an analytical exercise and is not considered a realistic scenario.

The data used to support TERM's needs estimates are derived from a variety of sources—including fleet investment and transit performance data obtained from the National Transit Database (NTD), asset inventory data provided by local transit agencies (at FTA's request), and historical rates of ridership growth calculated by region, agency size, and mode. The Low-Growth scenario is 0.5 percent less than the historical trend rate in growth while the High-Growth scenario is 0.5 percent more than the historical trend rate in growth. Appendix C contains a detailed description of the analysis methodology TERM uses, and Chapter 9 provides additional detail on the growth rates.

Preservation Investments

TERM estimates current and future preservation investment needs by first assessing the age and current condition of the Nation's existing stock of transit assets (the results of this analysis were presented in Chapter 3 of this report). TERM then uses this information to assess both current reinvestment needs (i.e., the reinvestment backlog) and the expected level of ongoing investment required to meet the life-cycle needs of the Nation's transit assets over the next 20 years, including all required rehabilitation and replacement activities.

Condition-Based Reinvestment

Rather than relying on age alone in assessing the timing and cost of current and future reinvestment activities, TERM uses a set of empirical asset deterioration curves that estimate asset condition (both current and future) as a function of asset type, age, past rehabilitation activities, and, depending on asset type, past maintenance and utilization levels. An asset's estimated condition at the start of each year over the 20-year forecast horizon determines the timing of specific rehabilitation and replacement activities. Asset condition declines as the asset ages, triggering reinvestment events at different levels of deterioration and ultimately leading to outright replacement.

Financial Constraints, the Investment Backlog, and Future Conditions

TERM is designed to estimate investment needs with or without annual capital funding constraints. When run without funding constraints, TERM estimates the total level of investment required to complete all rehabilitation and replacement needs the model identifies at the time those investment needs come due (hence, with unconstrained analyses after any initial deferred investment is addressed, investment backlog is not appreciable). In contrast, when TERM is run in a financially constrained mode, sufficient funding might not be available to cover the reinvestment needs of all assets. In this case, some reinvestment activities would be deferred until sufficient funds become available. The lack of funds to address all reinvestment needs for some or all of the 20 years of the model forecast results in varying levels of investment backlog during this period. Most analyses presented in this chapter were completed using funding constraints. Similarly, TERM's ability to estimate asset conditions—both current and future—allows for assessment of how future asset conditions are likely to improve or decline given varying levels of capital reinvestment. Finally, note that TERM's benefit-cost analysis is used to determine the order in which reinvestment activities are completed when funding capacity is limited, with investments having the highest benefit-cost ratios addressed first.

Expansion Investments

In addition to ongoing reinvestment in existing assets, most transit agencies also invest in the expansion of their vehicle fleets, maintenance facilities, fixed guideway, and other assets. Investments in expansion assets can be considered as serving two distinct purposes. First, the demand for transit services typically increases over time in line with population growth, employment, and other factors. To maintain current levels of performance in the face of expanding demand, transit operators must similarly expand the capacity of their services (e.g., by increasing the number of vehicles in their fleets). Failure to accommodate this demand would result in increased vehicle crowding, increased dwell times at passenger stops, and decreased operating speeds for existing services. Second, transit operators also invest in expansion projects with the aim of improving current service performance. Such improvements include capital expansion projects (e.g., a new light rail segment) to reduce vehicle crowding or increase average operating speeds. TERM is designed to assess investment needs and impacts for both types of expansion investments.

Expansion Investments: Maintain Performance

To assess the level of investment required to maintain existing service quality, TERM estimates the rate of growth in transit vehicle fleets required to maintain current vehicle occupancy levels given the projected growth rate in transit passenger miles. In addition to assessing the level of investment in new fleet vehicles required to support this growth, TERM also forecasts investments in the expansion of other assets needed to support projected fleet growth, including bus maintenance facilities and, in the case of rail systems, additional investment in guideway, track work, stations, maintenance facilities, train control, and traction power systems. Asset expansion investment needs are assessed on a mode-by-mode basis for all agencies reporting to NTD. Cost-benefit constraints, however, prevent TERM from investing in asset expansion for those agency modes having lower ridership (per vehicle) than the national average.

Expansion Investments: Improve Performance

In previous editions of the C&P report, TERM was used to estimate the level of investment required to improve current transit performance by (1) reducing crowding in higher-utilization transit systems, and (2) expanding existing investment in rail to improve average operating speeds in urbanized areas having average operating speeds (across all transit modes) well below the national average. For this edition, the impact of increased investment on system performance is assessed by developing TERM scenarios where the rate of investment in transit asset expansion exceeds the projected rate of growth in transit passenger miles. This difference between the rate of asset expansion and actual growth in travel demand represents projected long-term reductions in in-vehicle crowding and potential increases in average operating speed.

Recent Investment in Transit Preservation and Expansion

Exhibit 7-19 shows the broad composition of the 2012 spending by U.S. transit agencies on capital projects that correspond to the investment types TERM models. Of the total spending of $17.1 billion, $10.0 billion or 58.5 percent was devoted to preserving existing assets, and the rest was spent on expansion investments.

As expected, preservation and expansion spending were concentrated in the large urban systems. In combination, urbanized areas with populations greater than 1 million in 2012 accounted for 87.6 percent of preservation spending and 92.5 percent of expansion spending. Smaller urbanized areas and rural areas accounted for the rest. Although preservation and expansion spending for rural systems is small relative to that for large urban systems, rural transit service has been growing rapidly. Every State and four U.S. Territories provide some form of rural transit service in low-density areas improving the accessibility for Americans living in these areas.

Exhibit 7-19 2012 Transit Capital Expenditures[1]

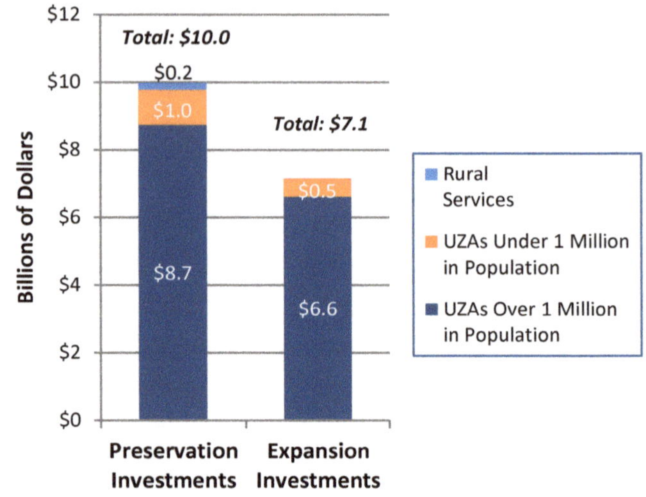

[1] Numbers may not sum to total due to rounding.

Source: National Transit Database.

Impacts of Systemwide Investments Modeled by TERM

This section uses TERM analyses to assess how various levels of investment in the preservation and expansion of the Nation's transit asset base can be expected to influence transit conditions and performance over the next 20 years. A key objective here is to place a broad range of potential future investment levels—and the consequences of those levels of investment—within the context of both the current expenditures on transit preservation and expansion and some potential investment goals (e.g., attainment of an SGR within 20 years). More specifically, these analyses consider the impact of different levels of transit capital expenditures on the following:

- Preservation Investments—Average condition rating of U.S. transit assets and SGR backlog
- Expansion Investments—Additional ridership (boardings) capacity.

Each analysis is completed first at the national level (the remainder of this section) and then repeated (in the following section) for two different segments of urbanized areas, including the following:

- Urbanized areas with populations greater than 1 million
- All other urbanized areas and rural areas with existing transit services.

Impact of Preservation Investments on Transit Backlog and Conditions

This subsection considers the expected impact of varying levels of aggregate capital reinvestment by all levels of government on the future investment backlog and physical condition (as of 2032) for the Nation's existing stock of transit assets.

Transit Backlog

The 2010 Conditions and Performance Report introduced the concept of reinvestment backlog as an indication of the amount of near-term investment needed to replace assets that are past their expected useful lifetime. Reinvestment backlog focuses attention on assets that are in the worst condition rather than on the average condition of all assets, which is reported below and had been the primary measure in previous editions. This additional perspective is needed because average condition has become less meaningful in the current environment with high levels of investment in new assets for transit system expansion. Investment backlog is a measure of the need for investment in infrastructure preservation. TERM estimates that investment backlog is $89.8 billion (see Chapter 8).

Exhibit 7-20 focuses on the impact of future spending levels on this investment backlog. Specifically, *Exhibit 7-20* presents the estimated impact of differing levels of annual capital reinvestment on the expected size of the investment backlog in 2032. Here the investment backlog is defined as the level of investment required to bring all of the Nation's assets to an SGR. This includes replacing those assets that currently exceed their useful lives (the $89.8 billion) and completing all major rehabilitation activities and replacing assets that will exceed their useful lives during the analysis period. If future reinvestment rates are insufficient to address these ongoing reinvestment needs as they arise, the size of the backlog will increase over time. Reinvestment at a rate above that required to address new needs as they arise will ultimately result in elimination of the existing backlog.

As shown in *Exhibit 7-20*, TERM analysis suggests that the current rate of capital reinvestment of $9.8 billion is insufficient to keep pace with ongoing rehabilitation and replacement needs and, if maintained over the next 20 years, would result in a reinvestment backlog of roughly $122.1 billion by 2032. In contrast, increasing the annual rate of reinvestment to an average of $17.0 billion would eliminate the backlog by 2032. The annual level of reinvestment would need to be increased to roughly $11.4 billion just to maintain the backlog at roughly its current size.

Transit Conditions

Exhibit 7-21 presents the estimated impact of various levels of annual rehabilitation and replacement investments on the average physical condition of all existing assets nationwide as of 2032. The exhibit shows ongoing improvements to the overall condition of the Nation's existing transit asset base from increasing levels of transit capital reinvestment. Of special note is that average condition provides a measure of asset conditions taken together. Hence, even though overall conditions improve with additional expenditures, the condition of some individual assets is expected to continue to deteriorate (given the length of asset lives and the timing of their replacement cycles) while the condition of other assets improves. The value of the aggregate measure lies in providing an overall, single measure of asset conditions. Moreover, given the relationship between asset condition and asset reliability, any general improvement in overall asset conditions also can be associated with related improvements to service quality, reliability, and possibly safety.

Exhibit 7-20 Impact of Preservation Investment on 2032 Transit State of Good Repair Backlog in All Urbanized and Rural Areas

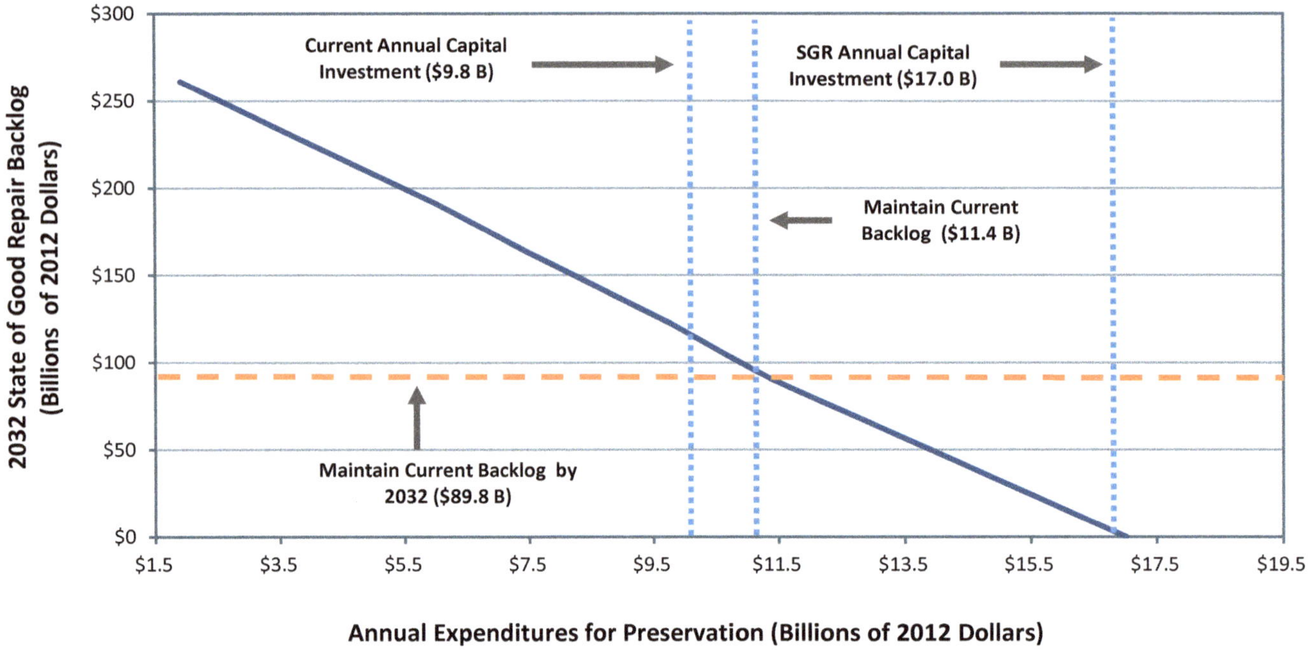

Average Annual Investment (Billions of 2012 Dollars)	Average Annual Percent Change vs. 2012	Average Condition Rating in 2032[1]	Backlog in 2032 (Billions of 2012 Dollars)[2]	Percent Change From Current Backlog	Funding Level Description
$17.0	5.7%	3.20	$0.0	-100%	SGR (unconstrained, replace at 2.50)
$11.4	1.5%	3.17	$89.8	0%	Maintain current backlog
$9.8	0.0%	3.10	$122.1	36%	2012 capital expenditures (sustain 2012 spending)
$7.5	-2.7%	3.03	$162.5	81%	Reduce 2.5 percent[3]
$6.0	-5.3%	2.96	$191.1	113%	Reduce 5 percent[3]
$3.8	-11.2%	2.86	$228.2	154%	Reduce 10 percent[3]
$1.9	-23.7%	2.80	$260.9	191%	Reduce 20 percent[3]

[1] For this report, assets are considered past their useful lives once their estimated condition in TERM falls below condition 2.50.

[2] Data points depicted in the chart might not correspond exactly to data presented in the associated table due to rounding of Average Annual Investment (Billions of 2012 Dollars) amounts.

[3] Funding is reduced from current level by the percentage identified every year for 20 years.

Source: Transit Economic Requirements Model.

The table portion of *Exhibit 7-21* presents the same investment and average condition information as in the chart. This table also presents the impact of reinvestment on asset conditions for five key transit asset categories (i.e., guideway and track, facilities, systems, stations, and vehicles) and the average annual percentage change in constant dollar funding from 2012 levels to achieve each projected condition level.

Exhibit 7-21 Impact of Preservation Investment on 2032 Transit Conditions in All Urbanized and Rural Areas[1,2]

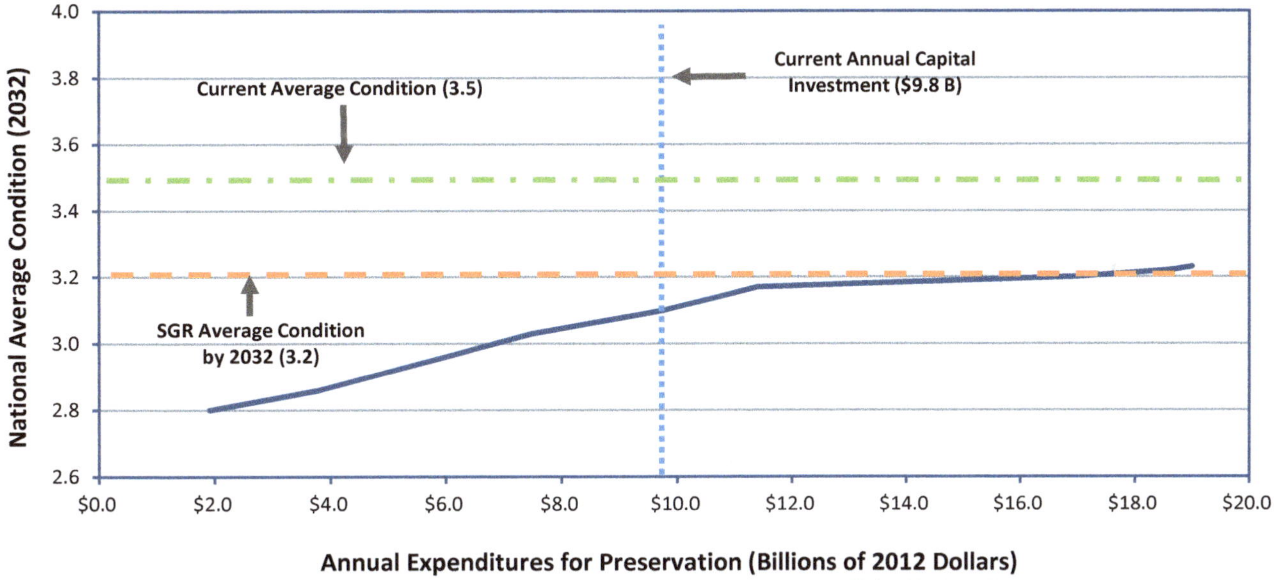

Average Annual Investment (Billions of 2012 Dollars) Total Capital Outlay	Average Annual Percent Change vs. 2012	Average Transit Conditions in 2032						Funding Level Description
		Asset Categories						
		Guideway	Facilities	Systems	Stations	Vehicles	All Transit Assets[3]	
$19.0	7.8%	2.77	3.19	3.55	3.64	3.38	3.23	Unconstrained, replace at 3.00
$18.6	6.5%	2.76	3.19	3.55	3.64	3.38	3.22	Unconstrained, replace at 2.75
$17.0	5.8%	2.71	3.19	3.55	3.64	3.38	3.20	SGR (unconstrained, replace at 2.50)
$11.4	1.6%	2.67	2.94	3.58	3.63	3.36	3.17	Maintain current backlog
$9.8	0.0%	2.64	2.61	3.48	3.52	3.40	3.10	2012 capital expenditures
$7.5	-2.7%	2.55	2.61	3.35	3.48	3.34	3.03	Reduce 2.5 percent
$6.0	-5.3%	2.49	2.61	3.12	3.46	3.28	2.96	Reduce 5 percent
$3.8	-11.2%	2.44	2.61	2.77	3.45	3.05	2.86	Reduce 10 percent
$1.9	-23.7%	2.40	2.61	2.67	3.45	2.80	2.80	Reduce 20 percent

[1] The conditions of individual transit assets are estimated using TERM's asset decay curves, which estimate asset conditions on a scale of 5 (excellent) through 1 (poor), as described earlier in this chapter and in Appendix C to this report. The average national condition is the weighted average of the condition of all assets nationwide, weighted by the estimated replacement cost of each asset.

[2] This preservation analysis is intended to consider reinvestment needs only for existing transit assets (as of 2012), not for expansion assets to be added to the existing capital stock in future years.

[3] Data points depicted in the chart might not correspond exactly to data presented in the associated table due to rounding of Average Annual Investment (Billions of 2012 Dollars) amounts.

Source: Transit Economic Requirements Model.

Further review of *Exhibit 7-21* reveals several observations. First, note that none of the selected reinvestment rates presented (including the current level of reinvestment, which was $9.8 billion in 2012) is sufficient to maintain aggregate conditions at or near the current national average condition rating of 3.5. Even the highest reinvestment rate presented here of $19.0 billion annually (replacement at condition rating 3.0), which is an aggressive reinvestment rate, is not sufficient to maintain aggregate conditions at current levels. A primary factor driving this result is the ongoing expansion investment in new rail systems over the past several decades. Although this expansion investment has tended to maintain or even increase the average condition rating of assets nationwide (despite the ongoing deterioration of older assets), it also has resulted in an average condition rating that is not sustainable in the long term (i.e., without including the influence of further expansion investments or replacing assets at an unreasonably early age). Second, note that reinvestment at roughly $17.0 billion annually is required to attain an SGR condition by 2032 and that this level of reinvestment is estimated to yield an average condition value of roughly 3.20 by that year. Given the definition of the SGR Benchmark (described in more detail in Chapter 8), which seeks to eliminate the existing investment backlog and then address all subsequent rehabilitation and replacement activities "on time" thereafter, the 3.20 value could be considered representative of the expected long-term average condition of a well-maintained and financially unconstrained national transit system. Hence, an average condition rating of roughly 3.20 represents a more reasonable long-term condition target for existing transit infrastructure than the current aggregate rating of 3.5.

Another observation is that a significant level of reinvestment is required to alter the estimated 2032 average condition measure by a point or more. This result is also driven in part by a large proportion of transit assets with expected useful lives of 80 years or more that will not require significant reinvestment over the 20-year period of this analysis (regardless of the level of reinvestment). These assets tend to contribute a high weighting in the average condition measure, making the measure somewhat insensitive to the rate of reinvestment (note that a high proportion of reinvestment activity is focused on the replacement of those assets with relatively shorter useful lives, such as vehicles).

Finally, TERM prioritizes asset needs based on five criteria (condition, reliability, safety, riders impacted, and operations and maintenance cost impacts) with condition having the highest weighting. Replacement and rehabilitation investments are both subject to this same prioritization scoring. Replacement needs tend to score higher, however, as they tend to reflect the needs of assets that are in poorer condition than those assets requiring rehabilitation. Therefore, rehabilitation needs tend not to be addressed until most (but far from all) replacement needs are addressed. TERM currently predicts improvement in asset condition only following a replacement. Hence, expenditures past approximately $11.8 billion on the chart increase total cost as rehabilitation projects are added, but these projects do not contribute to an increase in condition ratings.

Expansion Investments and Transit Ridership

Although capital spending on preservation primarily benefits the physical condition of existing transit assets, expansion investments are typically undertaken to expand the asset base to accommodate projected growth in ridership and potentially to improve service performance for existing transit system users.

Exhibit 7-22 shows the relationship between aggregated annual capital spending by all levels of government on expansion investments and the additional number of annual passenger boardings that transit systems would be able to support by 2032. More precisely, this chart presents the level of expansion investment required to ensure that transit vehicle occupancy rates are maintained at current levels over the next two decades for a broad range of the potential rates of growth in transit passenger miles traveled. As the upward sloping curve of the chart indicates, higher levels of investment are required to support greater numbers of additional riders at a constant level of service. If investment levels are insufficient to support the projected growth in ridership fully, vehicle occupancy rates will tend to increase, leading to increased crowding on high-utilization systems and potentially leading to increased dwell times at stops, reduced average operating speeds, and increased rates of vehicle wear. Conversely, if the rate of transit capacity expansion exceeds the actual rate of ridership growth, occupancy rates will tend to decline and service performance would likely improve.

Exhibit 7-22 New Ridership Supported in 2032 by Expansion Investments in All Urbanized and Rural Areas[1]

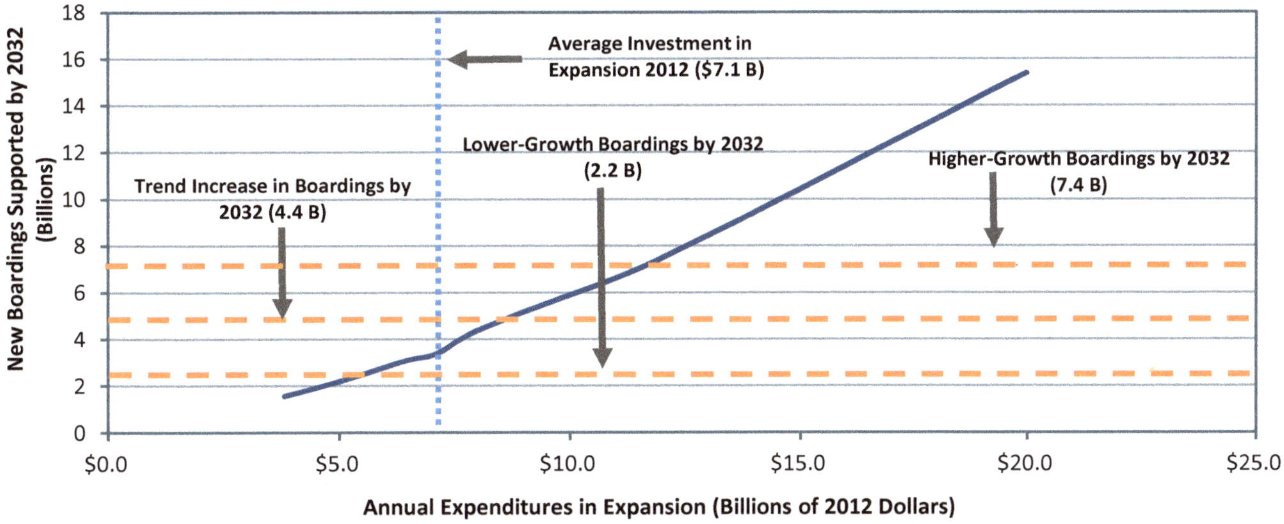

Average Annual Investment (Billions of 2012 Dollars)	Average Annual Percent Change vs. 2012	Total New Boardings by 2032: New Riders Supported (Billions of Annual Boardings)	Total New Boardings by 2032: Average Annual Growth in Boardings[2]	Funding Level Description
$20.0	9.3%	15.4	4.6%	Highest-growth scenario (+1.5%)
$11.9	4.9%	7.4	2.7%	Higher-growth scenario (+1.0%)
$9.9	3.2%	5.8	2.2%	High-growth scenario (+0.5%)
$8.0	1.2%	4.4	1.7%	15-year historic growth rate trend
$7.1	0.0%	3.4	1.5%	2012 expansion expenditures
$6.4	-1.0%	3.1	1.3%	Low-growth scenario (-0.5%)
$5.1	-3.5%	2.2	1.0%	Lower-growth scenario (-1.0%)
$3.8	-6.9%	1.6	0.7%	Lowest-growth Scenario (-1.5%)

[1] TERM assesses expansion needs at the agency-mode level subject to (1) current vehicle occupancy rates at the agency-mode level and (2) expected transit PMT growth at the UZA level (hence, all agency modes within a given UZA are subject to the same transit PMT growth rate). Note, however, that TERM does not generate expansion needs estimates for agency modes that have occupancy rates that are well below the national average for that mode.

[2] As compared with total urban ridership in 2012; only includes increases covered by investments passing TERM's benefit-cost test.

Source: Transit Economic Requirements Model.

The findings presented in *Exhibit 7-22* suggest the following trends. First, the recent rate of investment in asset expansion ($7.1 billion in 2012) could support roughly 3.4 billion additional boardings by 2032 (approximately a 1.5-percent annual growth in ridership). Assuming that the actual rate of ridership growth is close to the trend rate of growth for the past 15 years, an average capital investment of $8.0 billion annually in transit expansion would be required over the next 20 years to support an additional 4.4 billion annual boardings—again after excluding expansion investments that do not pass TERM's benefit-cost test. Hence, although the existing levels of transit capital expansion investment might be sufficient to maintain current service performance (i.e., vehicle occupancy rates), if ridership growth is relatively low (1 percent average annual growth in boardings), the corresponding average annual level of investment ($5.1 billion) is roughly two-thirds of what is required to support a level of ridership growth consistent with that experienced over the most recent 15-year period.

Impacts of Urbanized Area Investments Modeled by TERM

The remainder of this chapter focuses on how different levels of annual capital investment in the U.S. transit infrastructure affect urbanized areas with dissimilar transit investment needs. Specifically, this section explores the impact of capital expenditures by transit agencies sorted into two distinct UZA groupings: (1) the urbanized areas with populations greater than 1 million and (2) all other urbanized and rural areas with existing transit services.

Urbanized Areas Over 1 Million in Population

The Nation's largest urbanized areas own and operate most of the Nation's existing transit assets. These urbanized areas also typically have the highest levels of investment in older rail assets.

In 2012, transit agencies operating in urbanized areas with populations greater than 1 million spent $15.4 billion on capital projects. This expenditure consisted of $8.7 billion on preservation investments intended to rehabilitate or replace existing assets and $6.6 billion on expansion investments designed to increase service capacity. The following is a discussion of the transit asset preservation and expansion needs of these urbanized areas with populations greater than 1 million.

Preservation Investments

As shown in *Exhibit 7-23*, the 2012 level of capital reinvestment for the largest urbanized areas—$8.7 billion—is insufficient to keep pace with ongoing rehabilitation and replacement needs. Further, maintaining this reinvestment amount over the next 20 years would result in a larger SGR backlog of roughly $120.5 billion by 2032 compared with the current $88.6 billion backlog. In contrast, increasing the rate of reinvestment to an annual average of roughly $15.7 billion would eliminate the entire backlog by 2032. The annual level of reinvestment would need to be increased to roughly $10.2 billion to maintain the backlog at about its current size.

Exhibit 7-23 Impact of Preservation Investment on 2032 Transit State of Good Repair Backlog in Urbanized Areas with Population over 1 Million

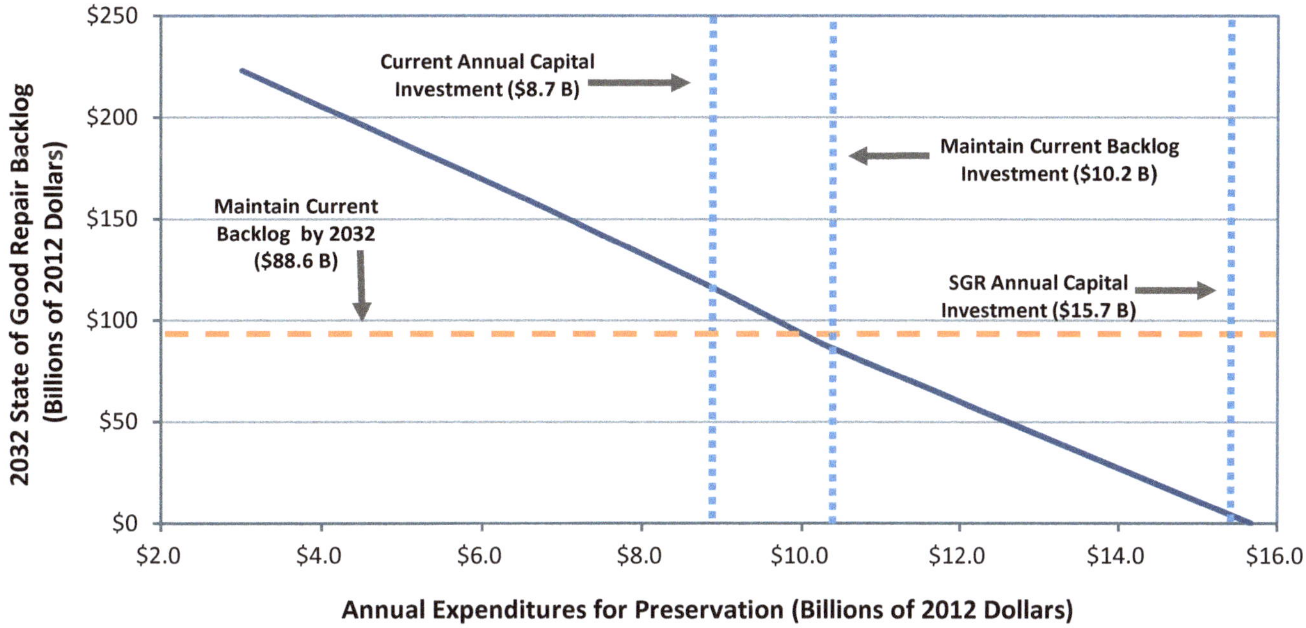

Average Annual Investment (Billions of 2012 Dollars)	Average Annual Percent Change vs. 2012	Replacement Condition[1]	Average Condition Rating in 2032	Backlog in 2032 (Billions of 2012 Dollars)[2]	Funding Level Description
$15.7	5.5%	2.50	3.19	$0.0	SGR (unconstrained, replace at 2.50)
$10.2	1.6%	2.50	3.16	$88.6	Maintain current backlog
$8.7	0.0%	2.50	3.09	$120.5	2012 capital expenditures (sustain 2012 spending)
$6.6	-2.9%	2.50	3.02	$159.3	Reduce 2.5 percent
$5.0	-6.1%	2.50	2.95	$186.6	Reduce 5 percent
$3.0	-13.1%	2.50	2.85	$223.2	Reduce 10 percent

[1] For this report, assets are considered past their useful lives once their estimated condition in TERM falls below condition 2.50.

[2] Data points depicted in the chart might not correspond exactly to data presented in the associated table due to rounding of Average Annual Investment (Billions of 2012 Dollars) amounts.

Source: Transit Economic Requirements Model.

Exhibit 7-24 shows the estimated impact of varying levels of preservation investments on the future condition of existing transit assets located in urbanized areas with populations greater than 1 million. As was shown in *Exhibit 7-21* covering the entire industry, this chart clearly indicates that, due to significant recent investments in long-lived expansion assets, the current average condition rating for transit assets located in the largest urbanized areas is not sustainable in the long term without replacing assets on an aggressive schedule (i.e., replacement at or before condition rating 3.0). At the same time, the 2012 level of reinvestment ($8.7 billion) is less than that required to attain an SGR ($15.7 billion), with the latter supporting a more sustainable long-term average condition rating of roughly 3.19.

Exhibit 7-24 Impact of Level of Preservation Investment on 2032 Transit Conditions in Urbanized Areas with Population over 1 Million[1,2]

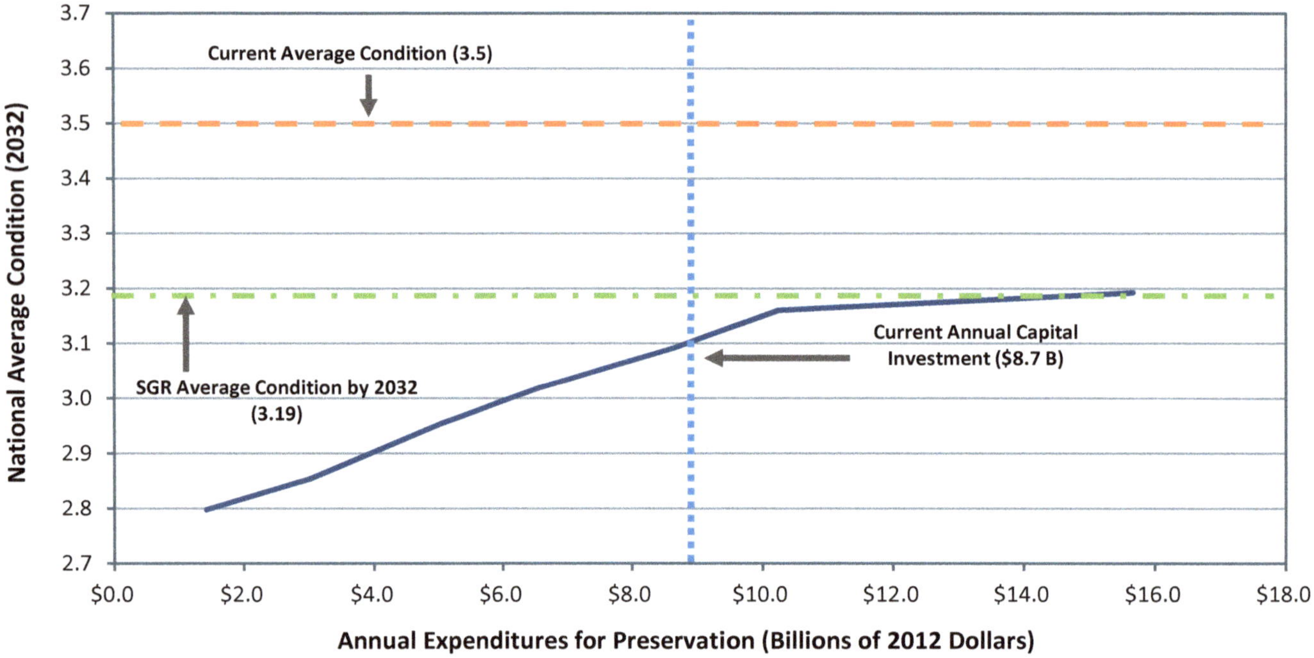

Average Annual Investment (Billions of 2012 Dollars) Total Capital Outlay	Average Annual Percent Change vs. 2012	Average Transit Conditions in 2032 — Asset Categories: Guideway	Facilities	Systems	Stations	Vehicles	All Transit Assets[3]	Funding Level Description
$15.7	5.6%	2.68	3.19	3.55	3.64	3.38	3.19	SGR (unconstrained condition, replace at 2.50)
$10.2	1.6%	2.64	2.97	3.58	3.63	3.36	3.16	Maintain current backlog
$8.7	0.0%	2.61	2.62	3.49	3.52	3.39	3.09	2012 capital expenditures (maintain current spending)
$6.6	-2.8%	2.53	2.62	3.36	3.48	3.33	3.02	Reduce 2.5 percent
$5.0	-5.9%	2.47	2.62	3.12	3.46	3.28	2.95	Reduce 5 percent
$3.0	-12.9%	2.42	2.62	2.77	3.45	3.03	2.85	Reduce 10 percent
$1.4	-29.0%	2.38	2.62	2.67	3.45	2.80	2.80	Reduce 20 percent

[1] The conditions of individual transit assets are estimated using TERM's asset decay curves, which estimate asset conditions on a scale of 5 (excellent) through 1 (poor), as described earlier in this chapter and in Appendix C to this report. The average national condition is the weighted average of the condition of all assets nationwide, weighted by the estimated replacement cost of each asset.

[2] This preservation analysis is intended to consider reinvestment needs only for existing transit assets (as of 2012), not for expansion assets to be added to the existing capital stock in future years.

[3] Data points depicted in the chart might not correspond exactly to data presented in the associated table due to rounding of Average Annual Investment (Billions of 2012 Dollars) amounts.

Source: Transit Economic Requirements Model.

Expansion Investments

Although urbanized areas with populations greater than 1 million tend to be cities with slower rates of increase in population and transit ridership (e.g., Boston, Philadelphia, and Chicago), this group also includes urbanized areas expected to experience relatively high rates of growth in transit boardings and passenger miles traveled over the next two decades, including Los Angeles, Atlanta, and Seattle. Given the high numbers of existing riders and transit capacity in these higher-growth large urbanized areas, they will require significant increases in expansion investments to maintain current service performance during this period.

Exhibit 7-25 presents estimates of expansion investment level required to support varying levels of growth in transit demand while maintaining current performance levels (as measured by vehicle capacity utilization) for these large urbanized areas. Note that the 2012 level of investment for these urbanized areas ($6.6 billion) was more than that required to support the rate of increase in transit demand as projected by the Low-Growth scenario (0.5 percent below the trend rate of increase as experienced in recent years) but well short of that required to support a high rate of growth (0.5 percent above the trend rate of increase as experienced in recent years).

Other Urbanized and Rural Areas

The following analysis considers the combined preservation and expansion needs of urbanized areas with populations less than 1 million and those of all rural areas with existing transit service. This diverse group therefore includes numerous mid-sized and small urbanized and rural transit operators offering only bus or para-transit services, or both.

In 2012, transit agencies operating outside of the largest urbanized areas spent $1.7 billion on capital projects, with $1.2 billion on preservation intended to rehabilitate or replace existing assets and $0.5 billion on expansion designed to increase service capacity. The following is a discussion of the transit asset preservation and expansion needs of transit agencies in these areas.

Preservation Investments

As shown in *Exhibit 7-26*, the 2012 level of capital reinvestment of $1.1 billion for rural areas and smaller urbanized areas is insufficient to keep pace with ongoing rehabilitation and replacement needs. If maintained over the next 20 years, this rate of investment would result in a larger SGR backlog of roughly $1.7 billion by 2032, as compared with the current backlog of $1.3 billion for this group. In contrast, increasing the rate of reinvestment to an annual average of roughly $1.3 billion would eliminate the entire backlog by 2032. The annual level of reinvestment would need to be increased to roughly $1.2 billion annually to maintain the backlog at about its current size.

Exhibit 7-25 New Ridership Supported in 2032 by Expansion Investments in Urbanized Areas with Population over 1 Million[1]

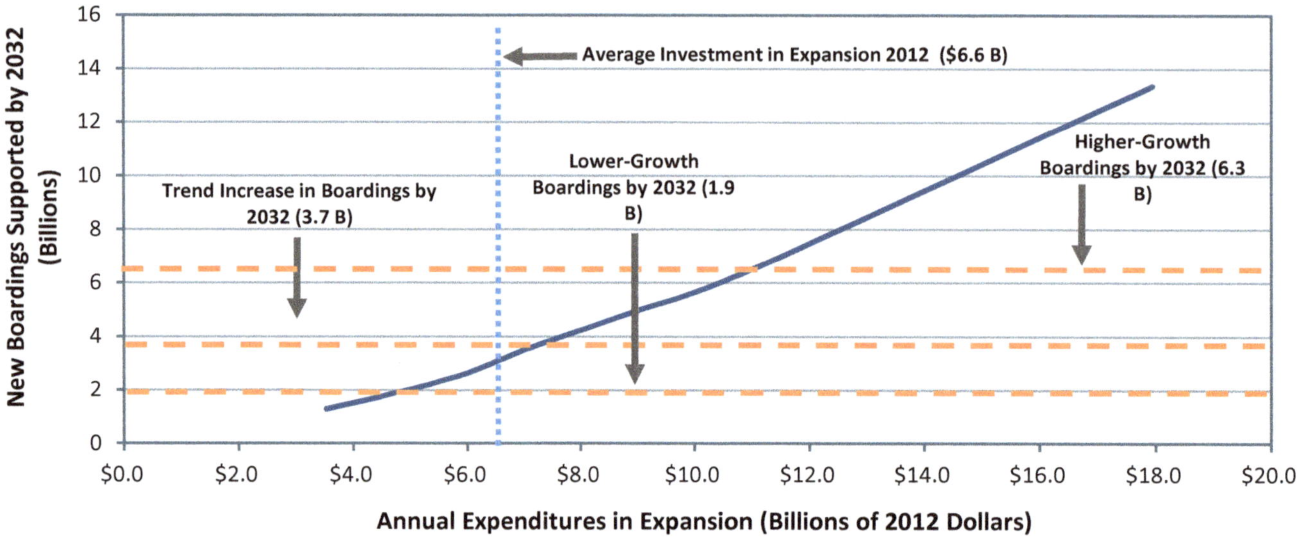

Average Annual Investment (Billions of 2012 Dollars)	Average Annual Percent Change vs. 2012	Total New Boardings by 2032: New Riders Supported (Billions of Annual Boardings)[2]	Total New Boardings by 2032: Average Annual Growth in Boardings[3]	Funding Level Description
$17.9	9.9%	13.3	4.5%	Highest-growth scenario (+1.5%)
$10.8	5.5%	6.3	2.6%	Higher-growth scenario (+1.0%)
$9.0	3.8%	4.9	2.1%	High-growth scenario (+0.5%)
$7.3	1.9%	3.7	1.7%	15-year historic growth rate trend
$6.6	0.0%	3.1	1.4%	2012 expansion expenditures
$5.9	-0.2%	2.6	1.2%	Low-growth scenario (-0.5%)
$4.7	-2.7%	1.9	0.9%	Lower-growth scenario (-1.0%)
$3.5	-6.0%	1.3	0.6%	Lowest-growth scenario (-1.5%)

[1] TERM assesses expansion needs at the agency-mode level subject to (1) current vehicle occupancy rates at the agency-mode level and (2) expected transit PMT growth at the UZA level (hence, all agency modes within a given UZA are subject to the same transit PMT growth rate). Note, however, that TERM does not generate expansion needs estimates for agency modes that have occupancy rates that are well below the national average for that mode.

[2] Data points depicted in the chart might not correspond exactly to data presented in the associated table due to rounding of Average Annual Investment (Billions of 2012 Dollars) amounts.

[3] As compared with total urban ridership in 2012; only includes increases covered by investments passing TERM's benefit-cost test.

Source: Transit Economic Requirements Model.

Exhibit 7-26 Impact of Preservation Investment on 2032 Transit State of Good Repair Backlog in Urbanized Areas with Population under 1 Million and Rural Areas

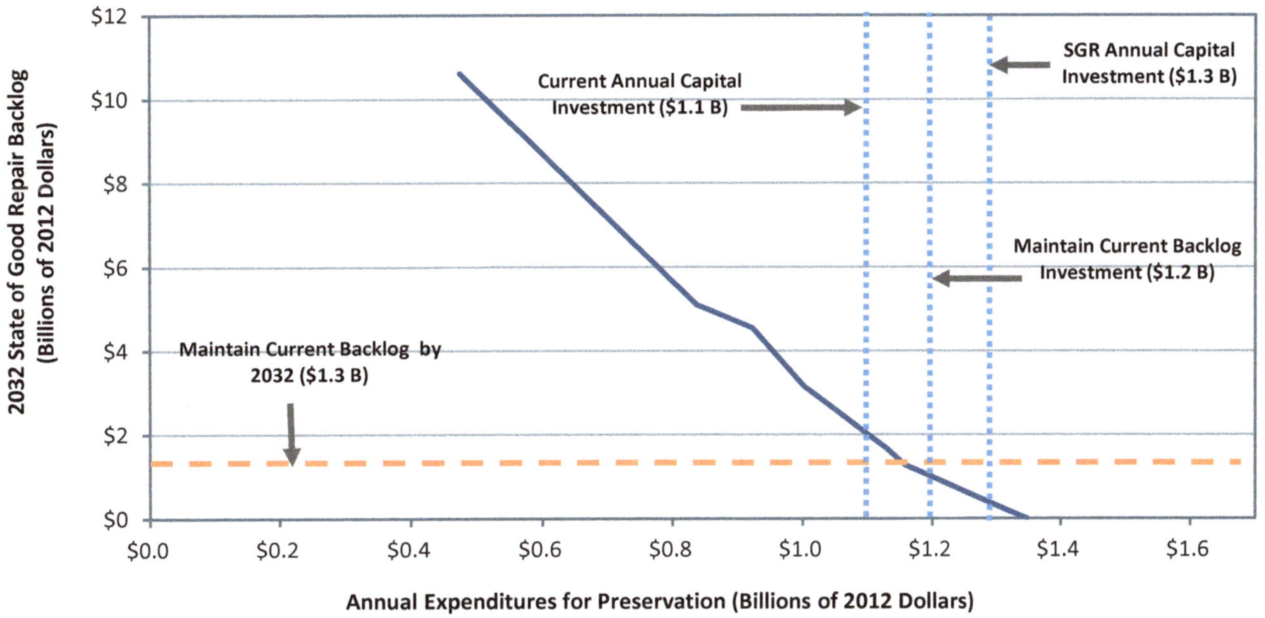

Average Annual Investment (Billions of 2012 Dollars)	Average Annual Percent Change vs. 2012	Replacement Condition[1]	Average Condition Rating in 2032	Backlog in 2032 (Billions of 2012 Dollars)[2]	Funding Level Description
$1.3	1.7%	2.50	3.64	$0.0	SGR (unconstrained, replace at 2.50)
$1.2	0.3%	2.50	3.65	$1.3	Maintain current backlog
$1.1	0.0%	2.50	3.63	$1.7	2012 capital expenditures (sustain 2012 spending)
$1.0	-1.2%	2.50	3.44	$3.1	Reduce 2.5 percent
$0.9	-2.1%	2.50	3.33	$4.5	Reduce 5 percent
$0.8	-3.1%	2.50	3.30	$5.1	Reduce 10 percent
$0.5	-10.3%	2.50	3.06	$10.6	Reduce 20 percent

[1] For this report, assets are considered past their useful lives once their estimated condition in TERM falls below condition 2.50.

[2] Data points depicted in the chart might not correspond exactly to data presented in the associated table due to rounding of Average Annual Investment (Billions of 2012 Dollars) amounts.

Source: Transit Economic Requirements Model.

Exhibit 7-27 shows the estimated impact of varying levels of preservation investments on the future condition of existing transit assets located in urbanized areas with populations less than 1 million and in rural areas. As was shown in *Exhibit 7-24* for the largest urbanized areas, this chart also indicates that the current average condition rating for transit assets in these smaller urbanized and rural areas is not sustainable in the long term without replacing assets on an aggressive schedule (i.e., replacement at or before condition rating 3.0). At the same time, the 2012 level of reinvestment ($1.1 billion) is less than that required to attain an SGR ($1.3 billion), with the latter supporting a more sustainable long-term average condition rating of roughly 3.63.

Exhibit 7-27 Impact of Preservation Investment on 2032 Transit Conditions in Urbanized Areas with Population under 1 Million and Rural Areas[1,2]

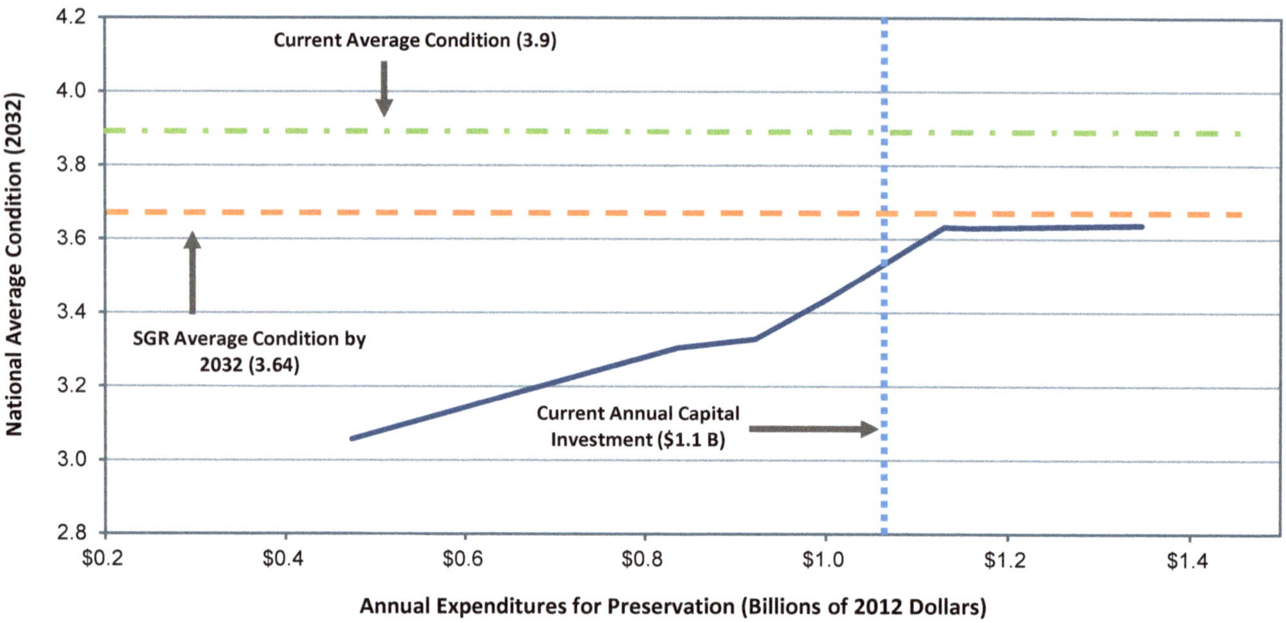

Average Annual Investment (Billions of 2012 Dollars) Total Capital Outlay	Average Annual Percent Change vs. 2012	Average Transit Conditions in 2032 — Asset Categories: Guideway	Facilities	Systems	Stations	Vehicles	All Transit Assets[3]	Funding Level Description
$1.3	1.4%	4.29	3.18	3.47	4.00	3.36	3.64	SGR (unconstrained, replace at 2.50)
$1.2	0.6%	4.22	2.57	3.40	3.18	3.42	3.63	Maintain current backlog
$1.1	0.0%	4.35	2.57	3.42	3.84	3.42	3.63	2012 capital expenditures (maintain current spending)
$1.0	-1.2%	3.89	2.57	3.02	3.84	3.35	3.44	Reduce 2.5 percent
$0.9	-2.4%	3.63	2.57	3.00	3.61	3.29	3.33	Reduce 5 percent
$0.8	-3.6%	3.63	2.57	2.99	3.61	3.25	3.30	Reduce 10 percent
$0.5	-9.4%	3.63	2.57	2.99	3.61	2.81	3.06	Reduce 20 percent

[1] The conditions of individual transit assets are estimated using TERM's asset decay curves, which estimate asset conditions on a scale of 5 (excellent) through 1 (poor), as described earlier in this chapter and in Appendix C to this report. The average national condition is the weighted average of the condition of all assets nationwide, weighted by the estimated replacement cost of each asset.

[2] This preservation analysis is intended to consider reinvestment needs only for existing transit assets (as of 2012), not for expansion assets to be added to the existing capital stock in future years.

[3] Data points depicted in the chart might not correspond exactly to data presented in the associated table due to rounding of Average Annual Investment (Billions of 2012 Dollars) amounts.

Source: Transit Economic Requirements Model.

Expansion Investments

Although the urbanized and rural areas in this group represent fewer riders and a smaller existing transit asset base, these areas are also expected to have a higher projected rate of increase in transit ridership.

Exhibit 7-28 presents estimates of the level of expansion investment required to support varying levels of growth in transit demand while maintaining current performance levels (as measured by transit passenger miles per peak vehicle) for the smaller urbanized and all rural areas. Note that the 2012 level of investment for these areas ($0.5 billion) was the same as that required to support the rate of increase in transit demand as projected by the Low-Growth trend and less than the High-Growth trend rate of increase as experienced over the past several years. Such investments should yield improvements in transit performance in these urbanized areas and help promote transit-led urban development in urbanized areas subject to above average rates of population and transit growth.

Exhibit 7-28 New Ridership Supported in 2032 by Expansion Investments in Urbanized Areas with Population under 1 Million and Rural Areas[1]

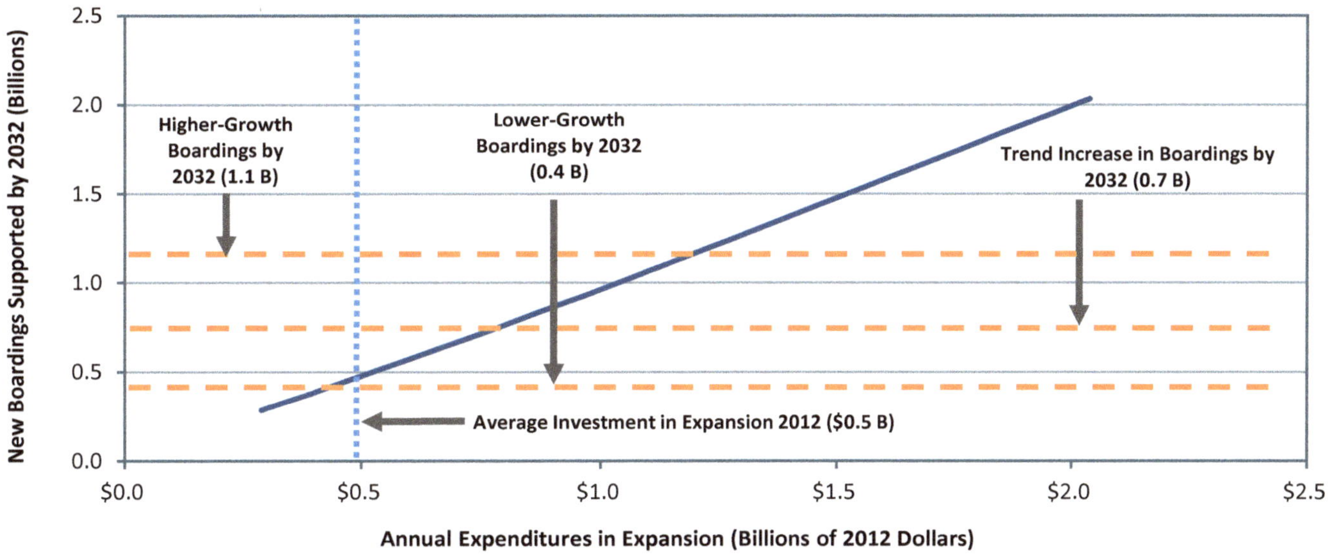

Average Annual Investment (Billions of 2012 Dollars)	Average Annual Percent Change vs. 2012	Total New Boardings by 2032: New Riders Supported (Billions of Annual Boardings)[2]	Total New Boardings by 2032: Average Annual Growth in Boardings[3]	Funding Level Description
$2.0	12.4%	2.0	5.2%	Highest-growth scenario (+1.5%)
$1.1	7.2%	1.1	3.3%	Higher-growth scenario (+1.0%)
$0.9	5.4%	0.9	2.8%	High-growth scenario (+0.5%)
$0.7	3.4%	0.7	2.3%	15-year historic growth rate trend
$0.5	0.8%	0.5	1.9%	Low-growth scenario (-0.5%)
$0.4	-2.0%	0.4	1.5%	Lower-growth scenario (-1.0%)
$0.3	-5.9%	0.3	1.1%	Lowest-growth scenario (-1.5%)

[1] TERM assesses expansion needs at the agency-mode level subject to (1) current vehicle occupancy rates at the agency-mode level and (2) expected transit PMT growth at the UZA level (hence, all agency modes within a given UZA are subject to the same transit PMT growth rate). Note, however, that TERM does not generate expansion needs estimates for agency modes that have occupancy rates that are well below the national average for that mode.

[2] Data points depicted in the chart might not correspond exactly to data presented in the associated table due to rounding of Average Annual Investment (Billions of 2012 Dollars) amounts.

[3] As compared with total urban ridership in 2012; only includes increases covered by investments passing TERM's benefit-cost test.

Source: Transit Economic Requirements Model.

chapter 8

Selected Capital Investment Scenarios

Selected Highway Capital Investment Scenarios

This section presents future investment scenarios that build on the Chapter 7 analyses of alternative levels of future investment in highways and bridges. Each scenario includes projections for system conditions and performance based on simulations with the Highway Economic Requirements System (HERS) and National Bridge Investment Analysis System (NBIAS). Each scenario scales up the total amount of simulated investment to account for capital improvements (highway and bridge investments) that are beyond the scopes of the models. Later in this chapter, transit investment scenarios are explored that, like those of this section, start with 2012 as the base year and cover the 20-year period through 2032. **All scenarios are illustrative, and none is endorsed as a target level of funding.**

Supplemental analyses relating to these scenarios, including comparisons with the investment levels presented for comparable scenarios in previous C&P reports, are the subject of Chapter 9. A series of sensitivity analyses that explore the implications of alternative technical assumptions for the scenario investment levels is presented in Chapter 10. The Introduction to Part II provides essential background information relating to the technical limitations of the analysis, which are discussed further in the appendices.

Scenarios Selected for Analysis

This section examines three scenarios (described in *Exhibit 8-1*) based on capital investment by all levels of government combined. What portion should be funded by the Federal government, State governments, local governments, or the private sector is beyond the scope of this report. Analyses were conducted first for the entire road network (titled "All Roads" in the exhibits) and then separately for Federal-aid highways, the National Highway System (NHS), and the Interstate System (these subsets of the road network are explained in Chapter 2). Each scenario pairs an assumed level of total investment in the types of improvements HERS models with an assumed level of investment in the types of improvements NBIAS models; these levels are drawn from those considered in Chapter 7. Together, the scopes of HERS and NBIAS cover spending on highway expansion and pavement improvements on Federal-aid highways (HERS) and spending on bridge rehabilitation on all highways (NBIAS). In the absence of data required for other types of highway and bridge investment (those not modeled in HERS or NBIAS), each scenario simply assumes that the percentage of highway and bridge investment spent on nonmodeled investments remains at the 2012 percentage. Percentage shares from 2012 also serve as a way to distribute the amount of nonmodeled investment among the component categories: pavement spending on non-Federal-aid highways, system expansion spending on non-Federal-aid highways, and system enhancement spending (which includes safety enhancements, operational improvements, and environmental projects) on all roads.

Exhibit 8-1 Capital Investment Scenarios for Highways and Bridges and Derivation of Components

Scenario Component	Sustain 2012 Spending Scenario	Maintain Conditions and Performance Scenario	Improve Conditions and Performance Scenario	State of Good Repair Benchmark
HERS-Derived	Sustain spending on types of capital improvements modeled in HERS at 2012 levels in constant dollar terms over next 20 years.	Set spending at the lowest level at which (1) projected average IRI in 2032 matches (or is better than) the value in 2012 and (2) projected average delay per VMT in 2032 matches (or is better than) the value in 2012.	Set spending at the level sufficient to gradually fund all cost-beneficial potential projects (i.e., those with a BCR greater than or equal to 1.0) over 20 years.	Subset of Improve Conditions and Performance scenario; includes spending on system rehabilitation, excludes spending on system capacity.
NBIAS-Derived	Sustain spending on types of capital improvements modeled in NBIAS at 2012 levels in constant dollar terms over the next 20 years.	Set spending at the level at which the projected percentage of deck area on deficient (structurally deficient or functionally obsolete) bridges in 2032 matches that in 2012.	Set spending at the level sufficient to gradually fund all cost-beneficial potential projects over 20 years.	Includes all NBIAS-derived spending included in the Improve Conditions and Performance scenario.
Other (Nonmodeled)	Sustain spending on types of capital improvements not modeled in HERS or NBIAS at 2012 levels in constant dollar terms over the next 20 years.	Set spending at the level necessary so that the nonmodeled share of total highway and bridge investment will remain the same as in 2012.	Set spending at the level necessary so that the nonmodeled share of total highway and bridge investment will remain the same as in 2012.	Subset of Improve Conditions and Performance scenario; includes spending on system rehabilitation, excludes spending on system capacity and system enhancement.

The projections for conditions and performance in each scenario are estimates of what could be achieved with a given level of investment assuming an economically driven approach to project selection (the project selection method is explained in Chapter 7). The projections do not necessarily represent what would be achieved given current decision-making practices. Consequently, comparing the relative conditions and performance outcomes across the different scenarios might be more illuminating than focusing on the specific projections for each scenario individually.

Q&A **What is the Federal share of highway capital spending?**

The Federal share of total capital spending on highways was 43.1 percent in 2012. Over the past 20 years, the share has ranged from a low of 37.1 percent (1998) to a high of 46.1 percent (2002). The remainder of capital spending is funded by States, local governments, and the private sector. Due to data limitations, however, separately identifying the shares for those funding sources is not possible.

Q&A **How do the definitions of the selected scenarios presented in this report compare to those presented in the 2013 C&P Report?**

As the base year of the analysis for this report is 2012 rather than 2010, the Sustain 2012 Spending scenario replaces the Sustain 2010 Spending scenario analyzed in the 2013 C&P Report. The names and definitions of the Improve Conditions and Performance scenario and the State of Good Repair benchmark are unchanged.

The Maintain Conditions and Performance scenario is similar in concept to the comparable scenario in the 2013 C&P Report, in that it attempts to maintain selected performance measures at their base-year levels through the end of the 20-year analysis period; however, the target measures have been modified. The NBIAS-derived component of the scenario targets the percentage of total bridge deck area that is on bridges classified as deficient (structurally deficient or functionally obsolete), whereas in the 2013 C&P Report, the target was the average bridge sufficiency rating. The HERS-derived component of this scenario used for the current edition is defined as the lowest investment level that is sufficient to maintain the current average IRI (International Roughness Index) and current average delay. In practice, this approach results in one of these target measures maintaining its current level and the other improving somewhat over 20 years. This approach differs from the method used in the 2013 C&P Report, which used the average of the investment level estimated to be sufficient to maintain average IRI and the investment level estimated to be sufficient to maintain average delay.

At the systemwide level, using the criteria from the 2013 C&P Report would have produced an average annual investment level of $55.4 billion, or 38.4 percent less than the $89.9 billion for the Maintain Conditions and Performance scenario shown in *Exhibit 8-2*. This significant difference is attributable to changes to HERS and the Highway Performance Monitoring System database referenced in Chapter 7 and Appendix A. Following the integration of new pavement performance models, highway-capacity estimation formulas, new pavement distress data, new widening feasibility data, and less-aggressive forecasts of future highway demand, HERS now estimates relatively higher benefit-cost ratios for expansion projects and relatively lower benefit-cost ratios for resurfacing and reconstruction projects than was the case for the 2013 C&P Report. Thus, if projects are implemented in order of benefit-cost ratio (from high to low), HERS now finds that maintaining average delay is significantly cheaper than maintaining average pavement condition. Defining the Maintain Conditions and Performance scenario in the way it was used in the 2013 C&P Report would have resulted in average pavement condition worsening, which is inconsistent with the current emphasis on achieving a state of good repair.

Scenario Derivation and Associated Spending Levels

Future spending levels by scenario, summarized in *Exhibit 8-2*, are stated in constant 2012 dollars. (Chapter 9 illustrates how to convert these real-dollar values into nominal [future dollar] values that factor in inflation beyond 2012.) The modeling on which the scenarios are based (which is presented in Chapter 7) assumes that spending grows at an annual percentage rate that is constant over the 20-year analysis period, but which differs between the types of investments modeled by HERS and those modeled by NBIAS. (The average annual investment levels are determined by summing the amounts expended for each year from 2013 through 2032 under the scenario and dividing by 20.)

The application of the four illustrative scenarios to different highway systems produces the subscenarios displayed as columns in *Exhibit 8-2*. The goal of the subscenario is fulfilled for the particular highway system named, but does not necessarily force any subsystems to meet the scenario's goal individually. For example, the subscenario for Federal-aid highways in the Sustain 2012 Spending scenario fixes average annual spending on those highways at actual 2012 spending without likewise forcing the portions of that spending directed to the NHS or the Interstate System to match their 2012 levels. Differences between the level of investment for the subsystems and the corresponding base-year amounts arise because HERS and NBIAS rely on benefit-cost principles to allocate spending flexibly among potential improvements within their scope.

Exhibit 8-2 Summary of Average Annual Investment Levels by Scenario

Scenario and Comparison Parameter	All Roads	Federal-Aid Highways	NHS	Interstate System
Sustain 2012 Spending Scenario				
Average annual investment (billions of 2012 dollars), for 2013 through 2032	**$105.2**	**$79.0**	**$54.6**	**$20.5**
Maintain Conditions and Performance Scenario				
Average annual investment (billions of 2012 dollars), for 2013 through 2032	**$89.9**	**$69.3**	**$51.7**	**$24.1**
Percent difference relative to 2012 spending	-14.6%	-12.3%	-5.2%	17.4%
Annual spending increase needed to support scenario investment level[1]	-1.52%	-1.26%	-0.51%	1.50%
Improve Conditions and Performance Scenario				
Average annual investment (billions of 2012 dollars), for 2013 through 2032	**$142.5**	**$107.9**	**$72.9**	**$31.8**
Percent difference relative to 2012 spending	35.5%	36.6%	33.7%	55.2%
Annual spending increase needed to support scenario investment level[1]	2.81%	2.89%	2.68%	4.02%
State of Good Repair Benchmark				
Average annual investment (billions of 2012 dollars), for 2013 through 2032	**$85.3**	**$64.9**	**$42.2**	**$18.4**

[1] This percentage represents the annual percent change for each year relative to 2012 that would be required to achieve the average annual funding level specified for the scenario in constant dollar terms. Additional increases in nominal dollar terms would be needed to offset the impact of future inflation. Negative values indicate that the average annual investment level associated with the scenario is lower than 2012 spending.

Sources: Highway Economic Requirements System and National Bridge Investment Analysis System.

The Sustain 2012 Spending scenario, which fixes average annual investment to actual 2012 levels, results in average annual investment of $105.2 billion for all roads, of which $79.0 billion is for Federal-aid Highways, $54.6 billion is for NHS, and $20.5 billion is for the Interstate System.

The Maintain Conditions and Performance scenario uses average pavement roughness as measured by the International Roughness Index (IRI) and average delay per vehicle miles traveled (VMT) (both modeled in HERS) as the measures of overall highway conditions and performance that it seeks to maintain at 2012 levels. The scenario uses the percentage of total bridge deck area on bridges classified as deficient (structurally deficient or functionally obsolete, as modeled in NBIAS) as the measure of bridge conditions it seeks to maintain. Chapter 3 explains these metrics. Both HERS and NBIAS, used to simulate the scenarios, are designed to determine the investment program that minimizes the cost of achieving the scenario goal. Because HERS assumes that projects will be implemented in order of their benefit-cost ratios, the levels of investment that maintain each key highway measure (IRI and average delay per VMT) differ; consequently, this scenario incorporates the higher of those two levels. Because it is focused on *overall* conditions and performance, this scenario might sometimes entail improvement and sometimes deterioration in average conditions and performance on subsets of some networks. For example, when the scenario relates to maintaining average conditions and performance on Federal-aid highways, it could entail improvement to the Interstate System.

For the entire road network overall and specifically for Federal-aid highways and the NHS, the average amount of investment needed annually to maintain conditions and performance is less than actual 2012 spending. For all roads, the average annual investment level of $89.9 billion for the Maintain Conditions and Performance scenario is 14.6 percent lower than the actual 2012 capital spending of $105.2 billion. The goals of the scenario could be achieved even if capital spending declined by 1.52 percent per year over 20 years in constant dollar terms. Similar percentage decreases are evident in the scenarios for Federal-aid highways (12.3 percent) and the NHS (5.2 percent).

Q&A What are the benefit-cost ratios associated with each spending scenario?

By design, the Improve Conditions and Performance scenario gradually increases funding over 20 years to implement all projects that have a benefit-cost ratio (BCR) greater than 1.0. For the Sustain 2012 Spending scenario, the amount of funding was sufficient to fund all projects with a BCR of 1.37 or greater (the minimum BCR for the Federal-aid highway subscenario was identical, as HERS only evaluates Federal-aid highways). For the Sustain 2012 Spending subscenarios focused on the NHS and Interstate System, the corresponding minimum BCR values were 1.29 and 1.54, respectively. For the Maintain Conditions and Performance scenario, the minimum BCR is 1.52 for all roads and Federal-aid highways, 1.33 for the NHS, and 1.26 for the Interstate System.

In contrast, the level of investment needed to maintain conditions and performance for the Interstate System is estimated to be 17.4 percent higher than the amount of investment directed to that system in 2012. The reasons for this result are twofold. First, spending on rehabilitation projects for the Interstate System has grown more slowly than for other subsets of the highway network (see Chapter 6), resulting in a relatively larger backlog of rehabilitation projects. Second, the Interstate System is aging and reconstruction needs likely will rise over time.

Targeting investment at a level projected to maintain base-year conditions and performance makes sense only if one is satisfied with that level of performance. The analyses reflected in the Improve Conditions and Performance scenario suggest that an economically driven approach to investment that funds all cost-beneficial improvements would substantially increase real spending on highways and bridges above base-year levels. The annual percentage increase in investment associated with implementing all cost-beneficial capital improvements is 2.81 percent for all roads, 2.89 percent for Federal-aid highways, 2.68 percent for the NHS, and 4.02 percent for the Interstate System. These levels of spending represent investment ceilings above which investing would not be cost-beneficial, even if available funding were unlimited. The average annual spending in this scenario exceeds the 2012 levels by 35.5 percent for all roads, 36.6 percent for Federal-aid highways, 33.7 percent for the NHS, and 55.2 percent for the Interstate System. For all roads, the average annual spending amount to implement all cost-beneficial investments fully is estimated to be $142.5 billion—or $2.9 trillion for the 20-year period—stated in constant 2012 dollars.

The State of Good Repair benchmark represents the portion of average annual spending that the Improve Conditions and Performance scenario allocates to system rehabilitation investments. Put at $85.3 billion in *Exhibit 8-2* for all roads, this benchmark represents the amount of cost-beneficial investment identified for rehabilitating existing pavements and bridges. In determining the size of this benchmark, HERS and NBIAS screen out through benefit-cost analysis any assets that might have outlived their original purpose, rather than automatically reinvest in all assets in perpetuity. With national consensus lacking on exactly what constitutes a "state of good repair" for the various transportation assets, alternative benchmarks with different objectives could be equally valid from a technical perspective.

How does the State of Good Repair benchmark compare to comparable spending in 2012?

The average annual investment level for the State of Good Repair benchmark for all roads is $85.3 billion. That value is 37.3 percent higher than the $62.1 billion all levels of government spent in 2012 for all roads on system rehabilitation. The $64.9-billion State of Good Repair benchmark value for Federal-aid highways is 39.5 percent higher than the comparable 2012 spending—$46.5 billion. The $42.2-billion State of Good Repair benchmark estimate for the NHS is 33.2 percent higher than the estimated $31.6 billion spent on roads included on the NHS (following its expansion under MAP-21) for system rehabilitation. The $18.4-billion State of Good Repair benchmark value for Federal-aid highways is 44.8 percent higher than comparable 2012 spending ($12.7 billion).

The sources of the estimates of average annual investment levels are presented in *Exhibit 8-3*. The HERS-derived component, which accounts for most of the total investment in each scenario, represents spending on pavement rehabilitation and capacity expansion on Federal-aid highways.

Exhibit 8-3 Source of Estimates of Highway Investment Scenarios, by Model

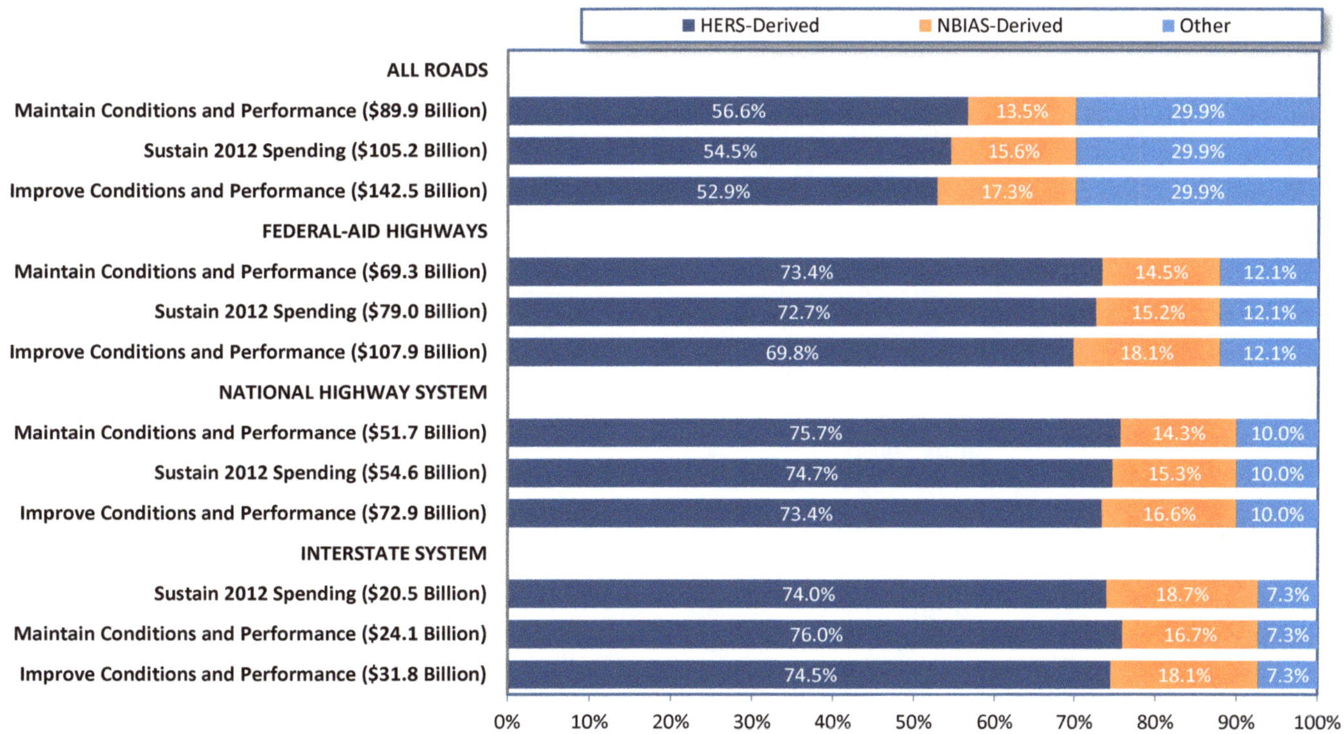

Sources: Highway Economic Requirements System and National Bridge Investment Analysis System.

The NBIAS-derived component represents rehabilitation spending on all bridges, including those not on Federal-aid highways. Nonmodeled spending, which accounted for 29.9 percent of total investment in 2012, is assumed to comprise the same share in all systemwide scenarios. Similarly, nonmodeled spending ("other" in *Exhibit 8-3*) is held constant across all scenarios at 12.1 percent for Federal-aid highways, at 10.0 percent for the NHS, and at 7.3 percent for the Interstate System.

Highway and Bridge Investment Backlog

Exhibit 8-4 presents an estimate of the 2012 backlog for the types of capital improvements modeled in HERS and NBIAS, plus an adjustment factor for nonmodeled capital improvement types. The investment backlog represents all highway and bridge improvements that could be economically justified for immediate implementation, based solely on the current conditions and operational performance of the highway system (without regard to potential future increases in VMT or potential future physical deterioration of infrastructure assets). Conceptually, this backlog represents a subset of the investment levels reflected in the Improve Conditions and Performance scenario, which addresses the existing backlog plus additional projected pavement, bridge, and capacity needs that might arise over the next 20 years.

Exhibit 8-4 Estimated Highway and Bridge Investment Backlog as of 2012

	Billions of 2012 Dollars[1]						
	System Rehabilitation			System	System		Percent
System Component	Highway	Bridge	Total	Expansion	Enhancement	Total	of Total
Federal-aid highways—rural	$94.2	$32.7	$126.9	$15.6	*$21.7*	$164.2	19.6%
Federal-aid highways—urban	$235.8	$73.1	$308.9	$117.5	*$54.2*	$480.6	57.5%
Federal-aid highways—total	$330.0	$105.8	$435.8	$133.1	*$75.9*	$644.8	77.1%
Non-Federal-aid highways	*$89.5*	$17.3	$106.8	*$33.9*	*$50.4*	$191.2	22.9%
All Roads	**$419.5**	**$123.1**	**$542.6**	**$167.0**	**$126.4**	**$836.0**	**100.0%**
Interstate System	$62.2	$40.2	$102.3	$42.9	*$11.5*	$156.8	18.8%
National Highway System	$184.1	$74.2	$258.3	$97.1	*$39.6*	$394.9	47.2%

[1] Italicized values are estimates for those system components and capital improvement types not modeled in HERS or NBIAS, such as system enhancements and pavement and expansion improvements to roads functionally classified as rural minor collector, rural local, or urban local for which Highway Performance Monitoring System data are not available to support a HERS analysis.

Sources: Highway Economic Requirements System and National Bridge Investment Analysis System.

Of the estimated $836.0-billion total backlog, approximately $156.8 billion (18.8 percent) is for the Interstate System, $394.9 billion (47.2 percent) is for the NHS, and $644.8 billion (77.1 percent) is for Federal-aid highways.

Approximately 64.9 percent ($542.6 billion) of the total backlog is attributable to system rehabilitation needs, 20.0 percent ($167.0 billion) is for system expansion, and 15.1 percent ($126.4 billion) for system enhancement. The share of the total backlog attributable to system rehabilitation is roughly similar across all highway systems.

The $836.0-billion estimated backlog is heavily weighted toward urban areas; approximately 57.5 percent of this total is attributable to Federal-aid highways in urban areas. As noted in Chapter 3, average pavement ride quality on Federal-aid highways is worse in urban areas than in rural areas; urban areas also face relatively greater problems with congestion and functionally obsolete bridges than do rural areas. Very little of the backlog spending (just 1.9 percent) is targeted toward system expansion on rural Federal-aid highways.

Scenario Spending Patterns and Conditions and Performance Projections

Systemwide Scenarios

The systemwide distribution of spending among improvement types for each scenario is shown in *Exhibit 8-5*. In the Improve Conditions and Performance scenario, annual spending on highway and bridge rehabilitation averages $85.3 billion, considerably more than the $62.1 billion of such spending in 2012 identified in Chapter 6. This result suggests that achieving a state of good repair on the Nation's highways would require either a significant increase in overall highway and bridge investment or a significant redirection of investment from other types of improvements toward system rehabilitation.

Exhibit 8-5 Systemwide Highway Capital Investment Scenarios for 2013 Through 2032: Distribution by Capital Improvement Type Compared with Actual 2012 Spending

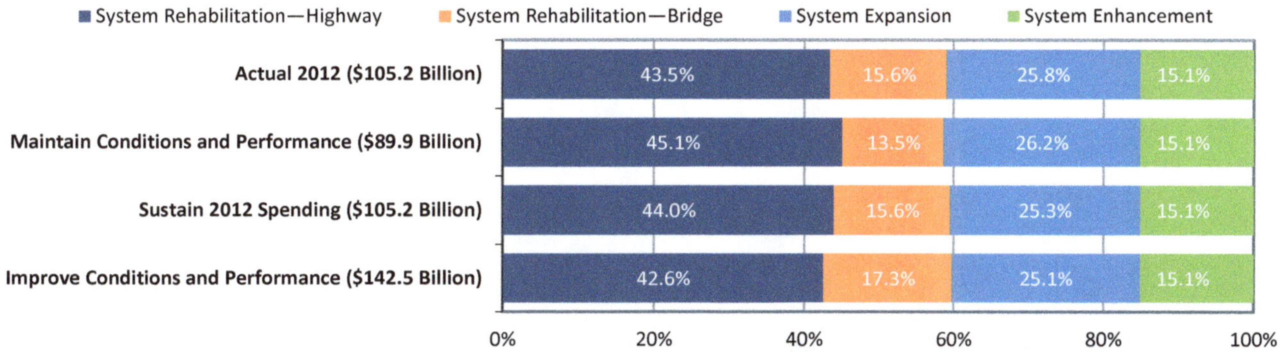

Capital Improvement Type	Actual 2012 Spending Distribution	Sustain 2012 Spending Scenario	Maintain Conditions & Performance Scenario	Improve Conditions & Performance Scenario
Average Annual Distribution by Capital Improvement Type (Billions of 2012 Dollars)				
System rehabilitation—highway	$45.7	$46.3	$40.6	$60.7
System rehabilitation—bridge	$16.4	$16.4	$12.2	$24.6
System rehabilitation—total	$62.1	$62.7	$52.7	$85.3
System expansion	$27.2	$26.6	$23.6	$35.7
System enhancement	$15.9	$15.9	$13.6	$21.5
Total, All Improvement Types	**$105.2**	**$105.2**	**$89.9**	**$142.5**

Sources: Highway Economic Requirements System and National Bridge Investment Analysis System.

Exhibit 8-5 compares the distributions from each scenario for investment spending by improvement type with the actual distribution of capital spending in 2012. At first glance, the proportional splits between improvement types roughly match the current 2012 spending by improvement type. Of importance to note, however, is that each percentage point change represents an approximate $1-billion shift in spending. Comparing the Sustain 2012 Spending scenario to the Actual 2012 Spending scenario, HERS modeling results support less spending on system expansion and more spending on highway rehabilitation than actually occur. At the higher levels of spending implied by the Improve Conditions and Performance scenario, the modeling results suggest spending relatively more on bridge system rehabilitation and relatively less on highway system rehabilitation and system expansion.

Exhibit 8-6 presents conditions and performance indicators for systemwide scenarios. (This information also can be found in various tables in Chapter 7). Because HERS considers only Federal-aid highways, the indicators for the Federal-aid highway scenarios are presented in place of indicators for all roads in *Exhibit 8-6*. These results are discussed more fully in the Federal-aid highway section below. In contrast, NBIAS considers bridges on all roads and provides several indicators to describe conditions and performance.

Exhibit 8-6 Systemwide Highway Capital Investment Scenarios for 2013 Through 2032: Projected Impacts on Selected Highway Performance Measures

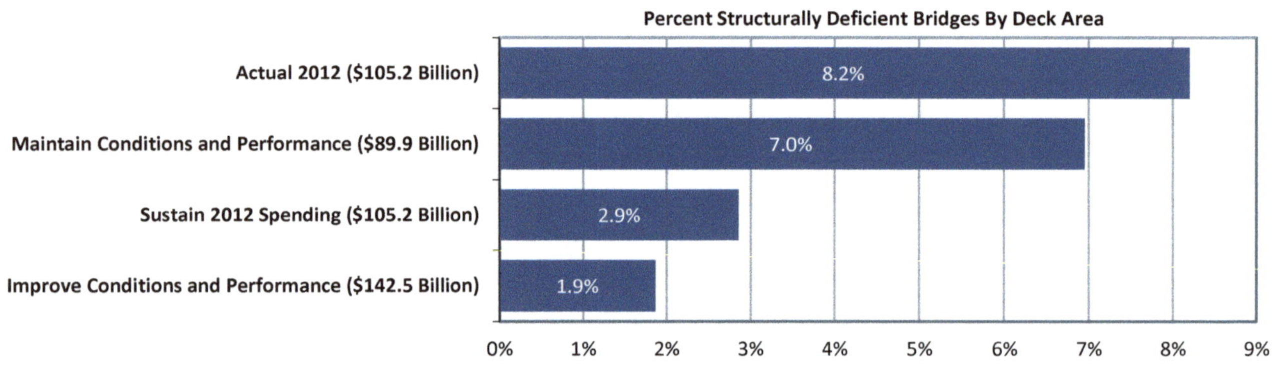

Highway Performance Measure	Actual 2012 Values	Sustain 2012 Spending Scenario	Maintain Conditions & Performance Scenario	Improve Conditions & Performance Scenario
Projected 2032 Values for Selected NBIAS Indicators (for Which Lower Numbers are Better)				
Percent structurally deficient bridges by deck area	8.2%	2.9%	7.0%	1.9%
Total percent deficient bridges by deck area	**26.7%**	22.9%	**26.7%**	21.2%
Economic bridge investment backlog (billions of 2012 dollars)	$123.1	$20.3	$67.6	**$0.0**
Projected 2032 Values for Selected HERS Indicators (for Which Higher Numbers are Better)				
Percent of VMT on roads with good ride quality[1]	44.9%	50.8%	46.0%	60.9%
Percent of VMT on roads with acceptable ride quality[1]	83.3%	87.7%	85.9%	91.5%
Projected Changes by 2032 Relative to 2012 for Selected HERS Indicators (for Which Negative Numbers are Better)				
Percent change in average IRI (VMT-weighted)[1]	0.0%	-4.5%	**0.0%**	-14.0%
Percent change in average delay per VMT[1]	0.0%	-13.4%	-12.2%	-16.5%

[1] The HERS indicators shown apply only to Federal-aid highways as HPMS sample data are not available for rural minor collectors, rural local, or urban local roads.

Sources: Highway Economic Requirements System and National Bridge Investment Analysis System.

Under the Sustain 2012 Spending scenario, the economic bridge investment backlog would drop from $123.1 billion in 2012 to $20.3 billion in 2032 and total percentage of bridges by deck area that are deficient would drop from 26.7 percent to 22.9 percent. The percentage of VMT on roads with good ride quality would rise from 44.9 percent to 50.8 percent and the average IRI would improve by 4.5 percent, while the average delay per VMT would fall by 13.4 percent.

The cells shaded blue in *Exhibit 8-6* (and similar exhibits that follow) are the values that define the scenarios. For the Maintain Conditions and Performance scenario, the cell showing that 26.7 percent of bridges (as measured by deck area) in 2032 would be deficient is shaded blue as it matches the actual value in 2012 (the goal of that scenario is to set funding to a level sufficient to maintain bridge conditions at their 2012 level). The cell showing that the average change in VMT-weighted IRI is 0.0 percent also is shaded blue, showing that this metric is unchanged relative to the actual 2012 value. Under the Maintain Conditions and Performance scenario, the economic bridge investment backlog would be $67.6 billion in 2032.

For the Improve Conditions and Performance scenario, the cell showing $0.0 in economic bridge investment backlog is shaded blue because the target of that scenario is to set spending at a level that would fund all cost-beneficial projects (thus eliminating the backlog). Under the Improve Conditions and Performance scenario, the percentage of bridges (measured by deck area) that are structurally deficient is projected to drop from 8.2 percent in 2012 to 1.9 percent in 2032. The total percentage of deficient bridges (including structurally deficient or functionally obsolete bridges) by deck area would drop from 26.7 percent in 2012 to 21.2 percent under the Improve Conditions and Performance scenario. (Of note is that this statistic understates the likely reduction in functionally obsolete bridges under this scenario, as it only captures improvements modeled in NBIAS and thus does not reflect the potential impact that system expansion investments modeled in HERS might have on addressing functionally obsolete bridges by replacing them with wider bridges). The Improve Conditions and Performance scenario also would eliminate the total economic bridge investment backlog, which totaled $123.1 in 2012.

Federal-Aid Highway Scenarios

For the scenarios that focus on Federal-aid highways, the average annual investment totals and the breakdown of those funds by type of investment are shown in *Exhibit 8-7*. The Maintain Conditions and Performance scenario involves a $9.7-billion reduction in spending on Federal-aid highways and the resulting lower level of spending would be allocated to different types of improvements in roughly the same proportion as actual 2012 spending. The Improve Conditions and Performance scenario involves an increase of $28.9 billion in spending per year, on average. At this higher level of spending, relatively more spending is directed toward rehabilitation and relatively less toward system expansion. System rehabilitation received 58.9 percent of funds in 2012 but would receive 60.1 percent under the Improve Conditions and Performance scenario.

Exhibit 8-8 presents conditions and performance indicators for the Federal-aid highways scenarios. Regarding performance indicators for roads, in 2012, the percentage of all VMT on roads in the Federal-aid highway system with good ride quality was 44.9 percent. That indicator would reach 60.9 percent under the Improve Conditions and Performance scenario. The Improve Conditions and Performance scenario raises the percentage of VMT on roads with acceptable ride quality to 91.5 percent from the 2012 value of 83.3 percent. The average VMT-weighted IRI (for

Exhibit 8-7 Federal-Aid Highway Capital Investment Scenarios for 2013 Through 2032: Distribution by Capital Improvement Type Compared with Actual 2012 Spending

Capital Improvement Type	Actual 2012 Spending Distribution	Sustain 2012 Spending Scenario	Maintain Conditions & Performance Scenario	Improve Conditions & Performance Scenario
Average Annual Distribution by Capital Improvement Type (Billions of 2012 Dollars)				
System rehabilitation—highway	$34.5	$35.0	$30.9	$45.4
System rehabilitation—bridge	$12.0	$12.0	$10.0	$19.5
System rehabilitation—total	$46.5	$47.0	$41.0	$64.9
System expansion	$22.9	$22.4	$19.9	$30.0
System enhancement	$9.6	$9.6	$8.4	$13.1
Total, all improvement types	**$79.0**	**$79.0**	**$69.3**	**$107.9**
Percent Distribution by Capital Improvement Type				
System rehabilitation	58.9%	59.6%	59.1%	60.1%
System expansion	29.0%	28.3%	28.8%	27.8%
System enhancement	12.1%	12.1%	12.1%	12.1%

Sources: Highway Economic Requirements System and National Bridge Investment Analysis System.

Exhibit 8-8 Federal-Aid Highway Capital Investment Scenarios for 2013 Through 2032: Projected Impacts on Selected Highway Performance Measures

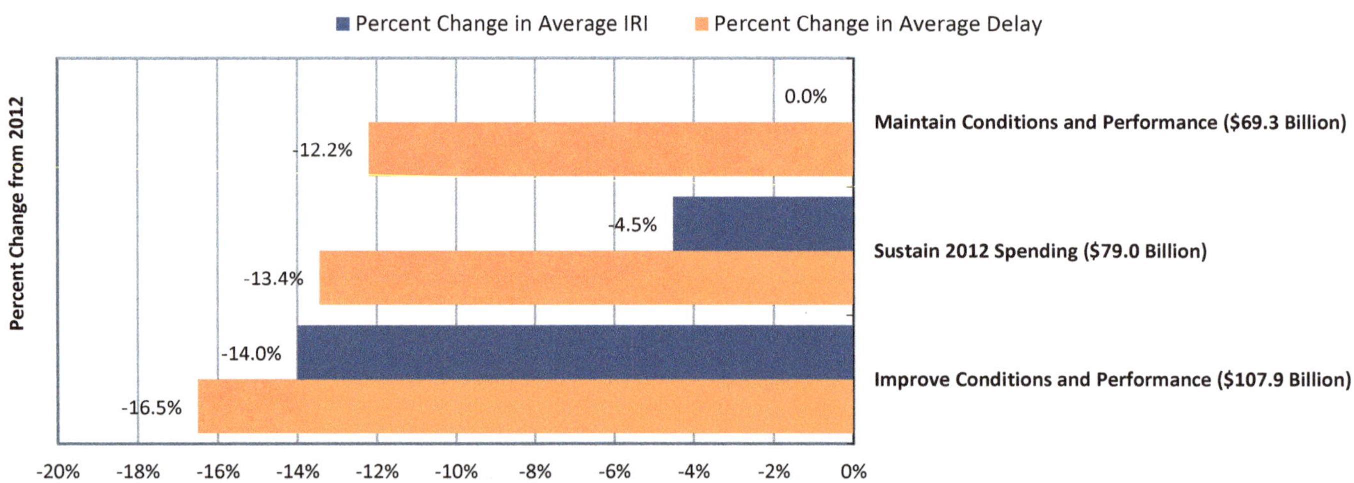

Highway Performance Measure	Actual 2012 Values	Sustain 2012 Spending Scenario	Maintain Conditions & Performance Scenario	Improve Conditions & Performance Scenario
Projected 2032 Values for Selected NBIAS Indicators (for Which Lower Numbers are Better)				
Percent structurally deficient by deck area	7.5%	3.3%	6.3%	1.2%
Total percent deficient bridges by deck area	**26.6%**	24.0%	**26.6%**	21.1%
Economic bridge investment backlog (billions of 2012 dollars)	$105.8	$28.1	$55.9	**$0.0**
Projected 2032 Values for Selected HERS Indicators (for Which Higher Numbers are Better)				
Percent of VMT on roads with good ride quality	44.9%	50.8%	46.0%	60.9%
Percent of VMT on roads with acceptable ride quality	83.3%	87.7%	85.9%	91.5%
Projected Changes by 2032 Relative to 2012 for Selected HERS Indicators (for Which Negative Numbers are Better)				
Percent change in average IRI (VMT-weighted)	0.0%	-4.5%	**0.0%**	-14.0%
Percent change in average delay per VMT	0.0%	-13.4%	-12.2%	-16.5%

Sources: Highway Economic Requirements System and National Bridge Investment Analysis System.

which lower numbers are better) would improve by 14.0 percent from its 2012 value under the Improve Conditions and Improvement scenario. The average delay per VMT would improve by 16.5 percent from its 2012 value under the Improve Conditions and Performance scenario, by 12.2 percent under the Maintain Conditions and Performance and by 13.4 percent in the Sustain 2012 Spending scenario. The reason average delay per VMT improves while spending remains constant or decreases is that capacity expansion projects tend to yield relatively high benefit-cost ratios, thus HERS opts to fund those projects, even when only limited funding is available. This level of forecast improvement in average delay per VMT is a departure from the findings of the 2013 C&P Report. This difference is due to changes in the modeling approach, discussed in Chapter 7, and because the forecast growth in VMT for this analysis is significantly lower than the growth rates used in the previous analysis. The 2013 C&P Report provides results for assumed annual VMT growth rates of 1.85 percent and 1.36 percent, while this current analysis assumes a VMT growth rate of 1.04 percent.

Under the Improve Conditions and Performance scenario, the percentage of bridges in the Federal-aid highway network (measured by deck area) that are structurally deficient is projected to drop from 7.5 percent in 2012 to 1.2 percent in 2032. The total percentage classified as deficient decreases from 26.6 percent in 2012 to 21.1 percent under the Improve Conditions and Performance scenario. The Improve Conditions and Performance scenario also eliminates the total economic bridge investment backlog, which was $105.8 billion in 2012. Under the Sustain 2012 Spending scenario, the bridge backlog drops to $28.1 billion, while the Maintain Conditions and Performance scenario results in a $55.9-billion backlog in 2032.

Spending by Improvement Type and Highway Functional Class

Exhibit 8-9 presents the distribution by improvement type and highway functional class for the Improve Conditions and Performance scenario compared to actual 2012 spending for Federal-aid highways. Moving to a finer level of detail in the analysis tends to reduce the reliability of simulation results from HERS and NBIAS, so the results presented in this exhibit should be viewed with caution. Nevertheless, the patterns strongly suggest certain directions in which spending patterns would need to change for scenario goals to be achieved. The scenarios can feature shifts in spending across highway functional classes and in highway spending between rehabilitation and expansion because the modeling frameworks determine allocations through benefit-cost optimization.

The Improve Conditions and Performance scenario shows that using a benefit-cost framework for project selection would dramatically shift spending away from rural roads and toward urban roads. Spending on rural roads would decrease by 1.2 percent from actual 2012 spending to $27.7 billion, while spending on urban roads would increase 57.5 percent to $80.2 billion.

The reduced spending on rural roads derives entirely from decreases in system expansion spending, which is reduced by 64.0 percent compared to actual 2012 spending. This indicates that HERS finds that sustaining spending in rural expansion at current levels over 20 years would not be cost-beneficial. In contrast, the Improve Conditions and Performance scenario suggests that a 75.5-percent increase in funding for system expansion of urban roads would be cost-beneficial.

Exhibit 8-9 Improve Conditions and Performance Scenario for Federal-Aid Highways: Distribution of Average Annual Investment for 2013 Through 2032 Compared with Actual 2012 Spending by Functional Class and Improvement Type

Average Annual National Investment on Federal-Aid Highways (Billions of 2012 Dollars)						
	System Rehabilitation			System Expansion	System Enhancement	Total
Functional Class	Highway	Bridge	Total			
Rural Arterials and Major Collectors						
Interstate	$4.7	$1.5	$6.2	$0.7	$0.7	$7.7
Other principal arterial	$5.0	$1.0	$6.0	$0.8	$1.1	$8.0
Minor arterial	$3.1	$0.9	$4.0	$0.5	$0.8	$5.3
Major collector	$3.6	$1.6	$5.2	$0.6	$1.0	$6.8
Subtotal	**$16.4**	**$5.0**	**$21.4**	**$2.6**	**$3.7**	**$27.7**
Urban Arterials and Collectors						
Interstate	$7.4	$4.7	$12.2	$9.9	$1.3	$23.4
Other freeway and expressway	$3.9	$1.8	$5.7	$7.3	$1.1	$14.1
Other principal arterial	$8.2	$3.5	$11.7	$4.3	$2.9	$18.9
Minor arterial	$6.4	$3.2	$9.6	$4.0	$2.4	$15.9
Collector	$3.1	$1.3	$4.4	$1.8	$1.7	$7.9
Subtotal	**$29.0**	**$14.5**	**$43.5**	**$27.3**	**$9.3**	**$80.2**
Total, Federal-aid highways[1]	**$45.4**	**$19.5**	**$64.9**	**$30.0**	**$13.1**	**$107.9**

Percent Above Actual 2012 Capital Spending on Federal-Aid Highways by All Levels of Government Combined						
	System Rehabilitation			System Expansion	System Enhancement	Total
Functional Class	Highway	Bridge	Total			
Rural Arterials and Major Collectors						
Interstate	28.8%	241.9%	51.3%	-62.4%	36.6%	16.8%
Other principal arterial	-4.9%	42.4%	0.7%	-72.4%	36.6%	-19.2%
Minor arterial	14.1%	3.8%	11.6%	-65.7%	36.6%	-5.8%
Major collector	6.4%	58.4%	18.3%	-35.4%	36.6%	12.8%
Subtotal	**9.3%**	**65.4%**	**18.7%**	**-64.0%**	**36.6%**	**-1.2%**
Urban Arterials and Collectors						
Interstate	42.1%	39.7%	41.1%	129.7%	36.6%	68.4%
Other freeway and expressway	83.9%	159.3%	103.0%	180.1%	36.6%	127.0%
Other principal arterial	39.7%	28.0%	36.0%	-16.2%	36.6%	19.3%
Minor arterial	71.0%	122.4%	85.3%	73.0%	36.6%	73.1%
Collector	23.6%	62.7%	32.7%	46.8%	36.6%	36.6%
Subtotal	**49.0%**	**60.7%**	**52.7%**	**75.5%**	**36.6%**	**57.5%**
Total, Federal-aid highways[1]	**31.8%**	**61.9%**	**39.5%**	**30.7%**	**36.6%**	**36.6%**

[1] The term "Federal-aid highways" refers to those portions of the road network that are generally eligible for Federal funding. Roads functionally classified as rural minor collectors, rural local, and urban local are excluded, although some types of Federal program funds can be used on such facilities.

Sources: Highway Economic Requirements System and National Bridge Investment Analysis System.

Spending on system rehabilitation for rural roads increases 18.7 percent in the Improve Conditions and Performance scenario compared to actual 2012 spending, but that increase is significantly lower than the 52.7-percent increase in spending for system rehabilitation needed for urban roads. Bridges on both urban and rural roads require substantial system rehabilitation spending, however, to achieve the goals of the scenario. The Improve Conditions and Performance scenario calls for 65.4-percent and 60.7-percent increases in system rehabilitation spending over actual 2012 spending for rural and urban bridges respectively.

The Improve Conditions and Performance scenario suggests that the largest funding gaps (in percentage terms) are for bridge rehabilitation on the rural portion of the Interstate System (241.9 percent), system expansion for urban freeways and expressways (180.1 percent), bridge rehabilitation on urban freeways and expressways (159.3 percent), and expansion of the urban portion of the Interstate System (129.7 percent).

Scenarios for the National Highway System and the Interstate System

Parallel to the analysis for the Federal-aid highways, *Exhibit 8-10* presents the scenarios for the NHS, and *Exhibit 8-11* presents the scenarios for the Interstate System. The results from these scenarios are derived in the same way, and the only spending component that is not modeled is system enhancements. System enhancements in 2012 accounted for slightly smaller shares of spending on the NHS and Interstate System than on all Federal-aid highways. Comparison of these scenarios with the Federal-aid highway scenarios in *Exhibit 8-7* reveals several patterns of interest:

- For both the NHS and the Interstate System, the Sustain 2012 Spending scenario suggests that some of the current spending should be shifted from system rehabilitation to system expansion. The suggested shift from rehabilitation to expansion is even more pronounced for the Maintain Conditions and Performance scenarios.
- The Improve Conditions and Performance scenario suggests that, with increased funding for the Interstate System, proportionally more funding should be directed at system expansion than at lower levels of funding.
- The Improve Conditions and Performance scenario suggests that the Interstate System is most in need of system expansion capital spending compared to the NHS or Federal-aid highways. The percentage of funding for system expansion is 27.8 percent for Federal-aid highways, 32.2 percent for the NHS, and 34.7 percent for the Interstate System.
- The Improve Conditions and Performance scenario results in more substantial road improvements (for all measures except pavement roughness) for the NHS and Interstate System than for Federal-aid highways as a whole. The average delay per VMT is reduced by 49.0 percent over 2012 conditions for the Interstate System, by 27.2 percent for the NHS, and by 16.5 percent for Federal-aid highways under the Improve Conditions and Performance scenario. The percentage of VMT on roads with acceptable ride quality is 98.7 percent for the Interstate System, 94.4 percent for the NHS, and 91.5 percent for Federal-aid highways under the Improve Conditions and Performance scenario.
- The Improve Conditions and Performance scenario results in more substantial pavement roughness for Federal-aid highways as a whole than for the Interstate System and the NHS. The percentage change in VMT-weighted average IRI is −14.0 percent for Federal-aid highways, −12.0 percent for NHS, and −9.3 percent for the Interstate System.
- In 2012, 7.1 percent of bridges (measured by deck area) on the Interstate System and the NHS were structurally deficient. The Improve Conditions and Performance scenario reduces that percentage by 6.1 percentage points to 1.0 percent for both the Interstate System and the NHS.

Exhibit 8-10 National Highway System Capital Investment Scenarios for 2013 Through 2032: Distribution by Capital Improvement Type and Projected Impacts on Selected Highway Performance Measures

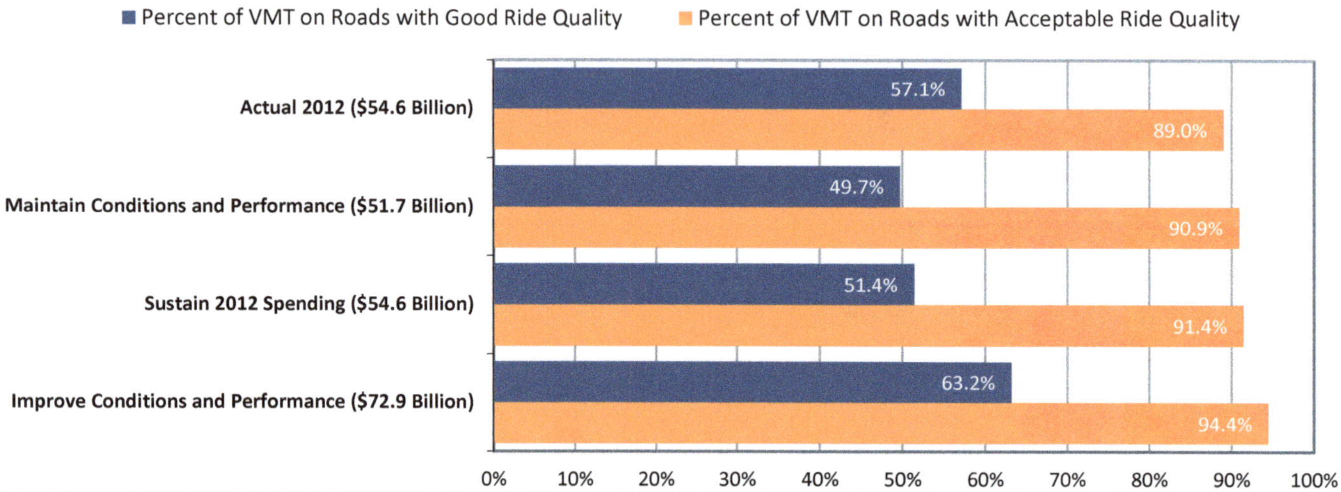

Capital Improvement Type	Actual 2012 Values	Sustain 2012 Spending Scenario	Maintain Conditions & Performance Scenario	Improve Conditions & Performance Scenario
Distribution by Capital Improvement Type, Average Annual (Billions of Base Year Dollars)				
System rehabilitation—highway	$23.3	$23.0	$22.1	$30.1
System rehabilitation—bridge	$8.3	$8.3	$7.4	$12.1
System rehabilitation—total	$31.6	$31.3	$29.4	$42.2
System expansion	$17.4	$17.8	$17.1	$23.5
System enhancement	$5.5	$5.5	$5.2	$7.3
Total, all improvement types	**$54.6**	**$54.6**	**$51.7**	**$72.9**
Percent Distribution by Capital Improvement Type				
System rehabilitation	58.0%	57.4%	56.9%	57.8%
System expansion	32.0%	32.6%	33.1%	32.2%
System enhancement	10.0%	10.0%	10.0%	10.0%
Projected 2032 Values for Selected NBIAS Indicators (for Which Lower Numbers Are Better)				
Percent structurally deficient by deck area	7.1%	3.1%	4.9%	1.0%
Total percent deficient bridges by deck area	**26.9%**	25.4%	**26.9%**	23.1%
Economic bridge investment backlog (billions of 2012 dollars)	$74.2	$19.1	$32.7	**$0.0**
Projected 2032 Values for Selected HERS Indicators (for Which Higher Numbers Are Better)				
Percent of VMT on roads with good ride quality	57.1%	51.4%	49.7%	63.2%
Percent of VMT on roads with acceptable ride quality	89.0%	91.4%	90.9%	94.4%
Projected Changes by 2032 Relative to 2012 for Selected HERS Indicators (for Which Negative Numbers Are Better)				
Percent change in average IRI (VMT-weighted)	0.0%	-1.6%	**0.0%**	-12.0%
Percent change in average delay per VMT	0.0%	-22.9%	-22.2%	-27.2%

Sources: Highway Economic Requirements System and National Bridge Investment Analysis System.

Exhibit 8-11 Interstate System Capital Investment Scenarios for 2013 Through 2032: Distribution by Capital Improvement Type and Projected Impacts on Selected Highway Performance Measures

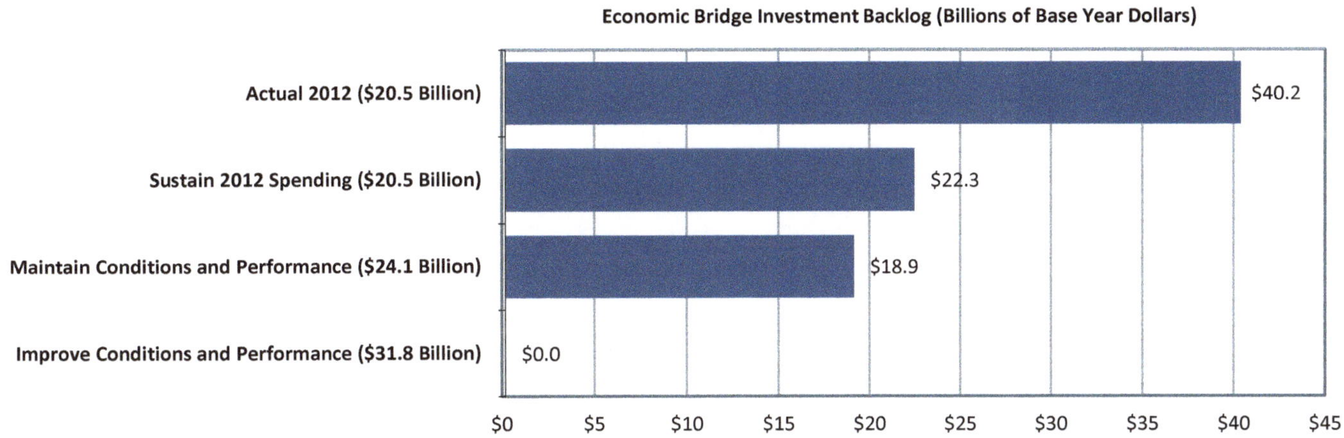

Capital Improvement Type	Actual 2012 Values	Sustain 2012 Spending Scenario	Maintain Conditions & Performance Scenario	Improve Conditions & Performance Scenario
Distribution by Capital Improvement Type, Average Annual (Billions of Base Year Dollars)				
System rehabilitation—highway	$8.9	$8.3	$10.1	$12.7
System rehabilitation—bridge	$3.8	$3.8	$4.0	$5.8
System rehabilitation—total	$12.7	$12.2	$14.1	$18.4
System expansion	$6.3	$6.8	$8.2	$11.0
System enhancement	$1.5	$1.5	$1.8	$2.3
Total, all improvement types	**$20.5**	**$20.5**	**$24.1**	**$31.8**
Percent Distribution by Capital Improvement Type				
System rehabilitation	62.1%	59.4%	58.6%	58.0%
System expansion	30.5%	33.3%	34.1%	34.7%
System enhancement	7.3%	7.3%	7.3%	7.3%
Projected 2032 Values for Selected NBIAS Indicators (for Which Lower Numbers Are Better)				
Percent structurally deficient by deck area	7.1%	6.6%	5.6%	1.0%
Total percent deficient bridges by deck area	**28.5%**	29.2%	**28.5%**	24.7%
Economic bridge investment backlog (billions of 2012 dollars)	$40.2	$22.3	$18.9	**$0.0**
Projected 2032 Values for Selected HERS Indicators (for Which Higher Numbers Are Better)				
Percent of VMT on roads with good ride quality	66.8%	43.7%	52.5%	64.7%
Percent of VMT on roads with acceptable ride quality	95.2%	95.1%	97.1%	98.7%
Projected Changes by 2032 Relative to 2012 for Selected HERS Indicators (for Which Negative Numbers Are Better)				
Percent change in average IRI (VMT-weighted)	0.0%	7.9%	**0.0%**	-9.3%
Percent change in average delay per VMT	0.0%	-31.6%	-37.9%	-49.0%

Sources: Highway Economic Requirements System and National Bridge Investment Analysis System.

Selected Transit Capital Investment Scenarios

Chapter 7 considered the impacts of varying levels of capital investment on transit conditions and performance. This chapter provides in-depth analysis of four specific investment scenarios, as outlined below in *Exhibit 8-12*. The Sustain 2012 Spending scenario assesses the impact of sustaining current expenditure levels on asset conditions and system performance over the next 20 years. Given that current expenditure rates are generally less than are required to maintain current condition and performance levels, this scenario reflects the magnitude of the expected declines in condition and performance should current capital investment rates be maintained. The State of Good Repair (SGR) Benchmark considers the level of investment required to eliminate the existing capital investment backlog and the condition and performance impacts of doing so. In contrast to the other scenarios considered here, the SGR Benchmark considers only the preservation needs of existing transit assets (it does not consider expansion requirements). Moreover, the SGR Benchmark does not require investments to pass the Transit Economic Requirements Model's (TERM's) benefit-cost test. Hence, it brings all assets to an SGR regardless of TERM's assessment of whether reinvestment is warranted. Finally, the Low-Growth and High-Growth scenarios both assess the required levels of reinvestment to (1) preserve existing transit assets at a condition rating of 2.5 or higher and (2) expand transit service capacity to support differing levels of ridership growth while passing TERM's benefit-cost test.

Exhibit 8-12 Capital Investment Scenarios for Transit

	SGR Benchmark	Sustain 2012 Spending Scenario	Low-Growth Scenario	High-Growth Scenario
Description	Level of investment to attain and maintain SGR over next 20 years (no assessment of expansion needs)	Sustain preservation and expansion spending at current levels over next 20 years	Preserve existing assets and expand asset base to support historical rate of ridership growth less 0.5% (1.3% between 1997 and 2012)	Preserve existing assets and expand asset base to support historical rate of ridership growth plus 0.5% (2.2% between 1997 and 2012)
Objective	Requirements to attain SGR (as defined by assets in condition 2.5 or better)	Assess impact of constrained funding on condition, SGR backlog, and ridership capacity	Assess unconstrained preservation and capacity expansion needs assuming low ridership growth	Assess unconstrained preservation and capacity expansion needs assuming high ridership growth
Apply Benefit-Cost Test?	No	Yes[1]	Yes	Yes
Preservation?	Yes[2]	Yes[2]	Yes[2]	Yes[2]
Expansion?	No	Yes	Yes	Yes

[1] To prioritize investments under constrained funding.
[2] Replace at condition 2.5.

TERM's estimates for capital expansion needs in the Low- and High-Growth scenarios are driven by the projected growth in passenger miles traveled (PMT). For this C&P report, Federal Transit Administration (FTA) has applied a new methodology for estimating growth in PMT that is considered more accurate and provides greater consistency between the Low- and High-Growth scenarios.

In prior years, PMT projections obtained from metropolitan planning organizations (MPOs) drove the Low-Growth scenario. Specially, PMT growth projections at the urbanized area (UZA) level were obtained from MPOs representing the Nation's 30 largest UZAs along with a sample of projections for MPOs representing smaller UZAs (population less than 1 million). These projections then were used to estimate transit capital expansion needs for the Low-Growth scenario. UZA growth rates for smaller UZAs not included in the sample were based on an average for UZAs of comparable size and region of the country. In contrast, the High-Growth scenario was driven by the historical (compound average annual) trend in rate of growth, also at the UZA level, based on data from the National Transit Database (NTD) for the most recent 15-year period.

For this C&P report, the Low- and High-Growth scenarios use a common, consistent approach that better reflects differences in PMT growth by mode. Specifically, these scenarios are now based on the trend rate of growth in PMT, calculated as the compound average annual PMT growth by FTA region, UZA stratum, and mode over the most recent 15-year period. For example, all bus operators located in the same FTA region in UZAs of the same population stratum are assigned the same growth rate. Use of the 10 FTA regions captures regional differences in PMT growth, while use of population strata (greater than 1 million; 1 million to 500,000; 500,000 to 250,000; and less than 250,000) captures differences in urban area size. Perhaps more importantly, the revised approach now recognizes differences in PMT growth trends by transit mode. Over the past decade, the rate of PMT growth has differed markedly across transit modes: highest for heavy rail, vanpool, and demand response and low to flat for motor bus. These differences are now accounted for in the expansion need projections for the Low- and High-Growth scenarios.

Exhibit 8-13 summarizes the analysis results for each scenario. Note that each scenario presented in *Exhibit 8-13* imposes the same asset condition replacement threshold (i.e., assets are replaced at condition rating 2.5 when budget is sufficient) when assessing transit reinvestment needs. Hence, the differences in the total preservation expenditure amounts across each scenario primarily reflect the impact of either (1) an imposed budget constraint (Sustain 2012 Spending scenario) or (2) application of TERM's benefit-cost test (the SGR Benchmark does not apply the benefit-cost test). A brief review of *Exhibit 8-13* reveals the following:

- **SGR Benchmark:** The level of expenditures required to attain and maintain an SGR over the upcoming 20 years, which would cover preservation needs but excludes expansion investments, is 8.6 percent higher than that currently expended on asset preservation and expansion combined.
- **Sustain 2012 Spending Scenario:** Total spending under this scenario is well below that of the other scenarios, indicating that sustaining recent spending levels is insufficient to attain the investment objectives of the SGR Benchmark, the Low-Growth scenario, or the High-Growth scenario. This result suggests future increases in the size of the SGR backlog and a likely increase in the number of transit riders per peak vehicle—including an increased incidence of crowding—in the absence of increased expenditures.

- **Low and High-Growth Scenarios:** The level of investment to address expected preservation and expansion needs is estimated to be roughly 46 to 69 percent higher than currently expended by the Nation's transit operators. Preservation and expansion needs are highest for UZAs exceeding 1 million in population.

Exhibit 8-13 Annual Average Cost by Investment Scenario, 2013 through 2032[1]

Mode, Purpose, and Asset Type	SGR Benchmark	Sustain 2012 Spending Scenario	Low-Growth Scenario	High-Growth Scenario
Urbanized Areas Over 1 Million in Population[2]				
Nonrail[3]				
Preservation	$4.1	$2.9	$3.7	$3.8
Expansion	NA	$0.4	$0.4	$1.1
Subtotal Nonrail4	**$4.1**	**$3.3**	**$4.1**	**$4.9**
Rail				
Preservation	$11.5	$5.8	$11.4	$11.5
Expansion	NA	$6.1	$5.5	$7.9
Subtotal Rail4	**$11.5**	**$11.9**	**$16.9**	**$19.3**
Total, Over 1 Million[4]	**$15.7**	**$15.1**	**$21.1**	**$24.2**
Urbanized Areas Under 1 Million in Population and Rural				
Nonrail[3]				
Preservation	$1.2	$1.1	$1.1	$1.1
Expansion	NA	$0.5	$0.5	$0.9
Subtotal Nonrail4	**$1.2**	**$1.6**	**$1.7**	**$2.0**
Rail				
Preservation	$0.2	$0.1	$0.1	$0.2
Expansion	NA	$0.03	$0.03	$0.04
Subtotal Rail4	**$0.2**	**$0.1**	**$0.2**	**$0.2**
Total, Under 1 Million and Rural[4]	**$1.3**	**$1.7**	**$1.8**	**$2.2**
Total[4]	$17.0	$16.8	$22.9	$26.4

[1] The average annual costs shown reflect investment over the 20-year period immediately following the end of the 2012 base year.

[2] Includes 42 different urbanized areas.

[3] Includes buses, vans, and other (including ferryboats).

[4] Note that totals may not sum due to rounding.

Source: Transit Economic Requirements Model.

The following subsections present more details on the assessments for each scenario and the SGR Benchmark.

Sustain 2012 Spending Scenario

In 2012, as reported to NTD by transit agencies, transit operators spent $17.1 billion on capital projects (see *Exhibit 7-20* and the corresponding discussion in Chapter 7). Of this amount, $10.0 billion was dedicated to preserving existing assets, while the remaining $7.1 billion was dedicated to investing in asset expansion—to support ongoing ridership growth and to improve service performance. This Sustain 2012 Spending scenario considers the expected impact on the long-term physical condition and service performance of the Nation's transit infrastructure if these 2012 expenditure levels were to be sustained in constant dollar terms through 2032. Similar to the discussion in Chapter 7, the analysis considers the impacts of asset-preservation investments separately from those of asset expansion.

Transit Investment Scenarios (*Exhibits 8-12* and *8-13*)

The Sustain 2012 Spending scenario assesses the impact of sustaining current expenditure levels on asset conditions and system performance over the next 20 years. Current expenditure rates are generally less than those required to maintain current condition and performance levels. This scenario therefore reflects the magnitude of the expected declines in condition and performance at current capital investment rates. The State of Good Repair (SGR) Benchmark considers the level of investment required to eliminate the existing capital investment backlog and the condition and performance impacts of doing so. In contrast to the other scenarios considered here, the SGR Benchmark considers only the preservation needs of existing transit assets (not expansion requirements). Moreover, the SGR Benchmark does not require investments to pass the Transit Economic Requirements Model's (TERM's) benefit-cost test. Hence, it brings all assets to an SGR regardless of TERM's assessment of whether reinvestment is warranted. Finally, both the Low-Growth and High-Growth scenarios assess the required levels of reinvestment to (1) preserve existing transit assets at a condition rating of 2.5 or higher and (2) expand transit service capacity to support differing levels of ridership growth while passing TERM's benefit-cost test.

- **Sustain 2012 Spending Scenario:** Total spending under this scenario is well below that of the other needs-based scenarios, indicating that sustaining recent spending levels is insufficient to attain the investment objectives of the SGR Benchmark, the Low-Growth scenario, or the High-Growth scenario. This finding suggests future increases in the size of the SGR backlog and a likely increase in the number of transit riders per peak vehicle—including an increased incidence of crowding—in the absence of increased expenditures.
- **SGR Benchmark:** The level of expenditures required to attain and maintain an SGR over the next 20 years—which covers preservation needs but excludes any expenditures on expansion investments—is 8.6 percent higher than that currently expended on asset preservation and expansion combined.
- **Low- and High-Growth Scenarios:** The level of investment to address expected preservation and expansion needs is estimated to be roughly 46 percent to 69 percent higher than the Nation's transit operators currently expend. Preservation and expansion needs are highest for urbanized areas with populations greater than 1 million.

Capital Expenditures for 2012: As reported to NTD, the level of transit capital expenditures peaked in 2009 at $16.8 billion, experienced a slight decrease in 2011 to $15.6 billion, and increased again in 2012 to $16.8 billion (see *Exhibit 8-14*). Although the annual transit capital expenditures averaged $14.7 billion from 2004 to 2012, expenditures averaged $16.4 billion in the most recent 5 years of NTD reporting. Furthermore, even though capital expenditures for preservation purposes in 2012 decreased $0.2 billion relative to prior-year levels, capital expenditures for expansion purposes increased $1.4 billion in 2012.

TERM's Funding Allocation: The following analysis of the Sustain 2012 Spending scenario relies on TERM's allocation of 2012-level preservation and expansion expenditures to the Nation's existing transit operators, their modes, and their assets over the upcoming 20 years, as depicted in *Exhibit 8-15*. As with other TERM analyses involving the allocation of constrained transit funds, TERM allocates limited funds based on the results of the model's benefit-cost analysis, which ranks potential investments based on their assessed benefit-cost ratios (with the highest-ranked investments funded first). Note that this TERM benefit-cost-based allocation of funding between assets and modes could differ from the allocation that local agencies might actually pursue, assuming that total spending is sustained at current levels over 20 years.

Exhibit 8-14 Annual Transit Capital Expenditures, 2004–2012

	(Billions of Current-Year Dollars)			(Billions of Constant 2012 Dollars)		
Year	Preservation	Expansion	Total	Preservation	Expansion	Total
2004	$9.4	$3.2	$12.6	$11.5	$3.9	$15.3
2005	$9.0	$2.9	$11.8	$10.5	$3.4	$13.9
2006	$9.3	$3.5	$12.8	$10.6	$3.9	$14.5
2007	$9.6	$4.0	$13.6	$10.6	$4.4	$15.0
2008	$11.0	$5.1	$16.1	$11.8	$5.4	$17.2
2009	$11.3	$5.5	$16.8	$12.1	$5.9	$18.0
2010	$10.3	$6.2	$16.6	$10.9	$6.5	$17.4
2011	$9.9	$5.7	$15.6	$10.1	$5.8	$16.0
2012	$9.7	$7.1	$16.8	$9.7	$7.1	$16.8
Average[1]	**$10.0**	**$4.8**	**$14.7**	**$10.9**	**$5.2**	**$16.0**

[1] Reflects the average expenditures over the nine-year period starting in 2004 and ending in 2012.

Source: National Transit Database.

Preservation Investments

As noted above, transit operators spent an estimated $10.0 billion in 2012 rehabilitating and replacing existing transit infrastructure. Based on current TERM analyses, this level of reinvestment is less than that required to address the anticipated reinvestment needs of the Nation's existing transit assets. If sustained over the forecasted 20 years, this level would result in an overall decline in the condition of existing transit assets and an increase in the size of the investment backlog.

For example, *Exhibit 8-16* presents the projected increase in the proportion of existing assets that exceeds their useful life by asset category from 2012 to 2032. Given the benefit-cost-based prioritization TERM imposes for this scenario, the proportion of existing assets that exceeds their useful life is projected to undergo a near-continuous increase across each asset category. This condition projection uses TERM's benefit-cost test to prioritize rehabilitation and replacement investments in this scenario. Specifically, for each investment period in the forecast, TERM ranks all proposed investment activities based on their assessed benefit-cost ratios (highest to lowest). TERM then invests

Exhibit 8-15 Sustain 2012 Spending Scenario: Average Annual Investment by Asset Type, 2013 through 2032

	Investment Category		
	Preservation	Expansion	Total
Asset Type	(Billions of 2012 Dollars)		
Rail			
Guideway Elements	$1.8	$1.3	$3.1
Facilities	$0.0	$0.2	$0.2
Systems	$2.4	$0.3	$2.7
Stations	$0.3	$0.8	$1.1
Vehicles	$1.5	$2.1	$3.6
Other Project Costs	$0.0	$1.4	$1.4
Subtotal Rail[1]	**$6.0**	**$6.1**	**$12.0**
Subtotal UZAs Over 1 Million[1]	**$5.8**	**$6.1**	**$11.9**
Subtotal UZAs Under 1 Million and Rural[1]	**$0.2**	**$0.0**	**$0.2**
Nonrail			
Guideway Elements	$0.0	$0.0	$0.0
Facilities	$0.0	$0.1	$0.1
Systems	$0.2	$0.0	$0.2
Stations	$0.0	$0.0	$0.0
Vehicles	$3.8	$0.8	$4.6
Other Project Costs	$0.0	$0.0	$0.0
Subtotal Nonrail[1]	**$4.0**	**$0.9**	**$4.9**
Subtotal UZAs Over 1 Million[1]	**$2.9**	**$0.5**	**$3.4**
Subtotal UZAs Under 1 Million and Rural[1]	**$1.0**	**$0.5**	**$1.5**
Total[1]	$10.0	$7.1	$17.1
Total UZAs Over 1 Million[1]	$8.7	$6.6	$15.3
Total UZAs Under 1 Million and Rural[1]	$1.2	$0.5	$1.7

[1] Note that totals may not sum due to rounding.

Source: Transit Economic Requirements Model and FTA staff estimates.

in the highest-ranked projects for each period until the available funding for the period is exhausted. Apparent here is that TERM investment priorities favor vehicle investments (as do those of most transit agencies because reinvesting in vehicles is important for reliability, safety, and operations and maintenance and patrons physically interact with them). Between 2015 and 2025, TERM invests in vehicles that rate highly on several investment criteria, and the vehicle over-age forecast for this period stays flat. (Investments not addressed in the current period as a result of the funding constraint are then deferred until the following period.) Also, given that the proportion of over-age assets is projected to increase for all asset categories under this prioritization, any reprioritization to favor reinvestment in one asset category over another clearly would accelerate the rate of increase of the remaining categories. Note that these over-age assets tend to deliver the lowest-quality transit service to system users (e.g., these assets have the highest likelihood of in-service failures). Due to changes in the asset inventory, the assessed reinvestment needs for stations, facilities, and guideway, as presented in this C&P report, are both higher and more critical (i.e., in poorer condition) than those presented in the 2013 C&P Report, whereas reinvestment needs for vehicles are fairly similar. This higher and more critical need creates greater competition for limited funds (recall that the sustained funding scenario is financially constrained) with less funding available for vehicles over the 20-year model run. Hence, the percentage of over-age vehicles is higher over the 20-year forecast period for this C&P report than for the 2013 C&P Report.

Exhibit 8-16 Sustain 2012 Spending Scenario: Over-Age Forecast by Asset Category, 2012–2032

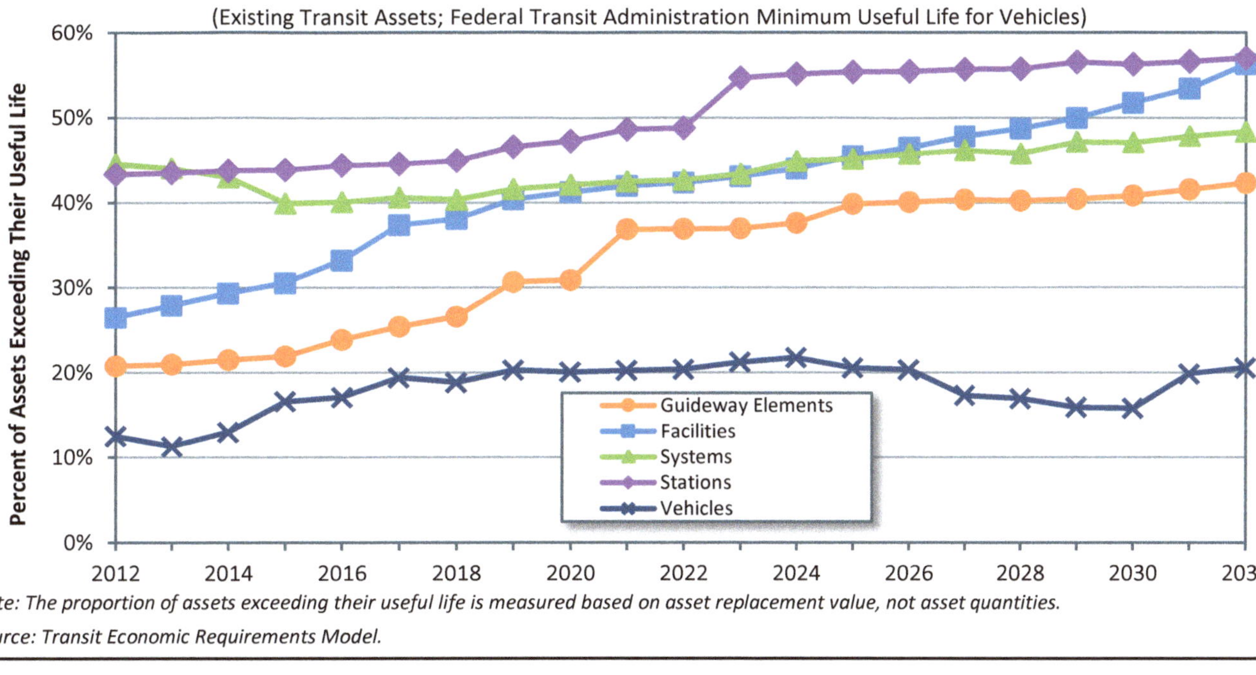

Note: The proportion of assets exceeding their useful life is measured based on asset replacement value, not asset quantities.

Source: Transit Economic Requirements Model.

Finally, *Exhibit 8-17* presents the projected change in the size of the investment backlog if reinvestment levels are sustained at the 2012 level of $10.0 billion, in constant dollar terms. As described in Chapter 7, the investment backlog represents the level of investment required to replace all assets that exceed their useful life and to address all rehabilitation activities that are currently past due. Rural and smaller urban needs are estimated using NTD records for vehicle ages and types and records generated for rural smaller urban agency facilities based on counts from NTD. The generated records for rural facilities include estimated facility size, replacement cost, and date built. Each estimated value was substantially revised for this C&P report for two reasons: (1) The replacement costs for facilities used in previous reports were much higher than the costs rural and smaller urban agencies typically face; and (2) Some date-built values were much greater (i.e., the facilities were older) than is typical. For this report, facility size and cost were reassessed based on agency fleet size and facility cost "per vehicle." The age range used to generate date-built values also was tightened to recognize a more realistic distribution of facility ages (based on sample data). These changes significantly reduced the value of these assets and type size of the rural and smaller urban backlogs. Given that the current rate of capital reinvestment is insufficient to address the replacement needs of the existing stock of transit assets, the size of that backlog is projected to increase from the currently estimated level of $89.8 billion to roughly $122.1 billion by 2032.

Exhibit 8-17 Investment Backlog: Sustain 2012 Spending ($10 Billion Annually)

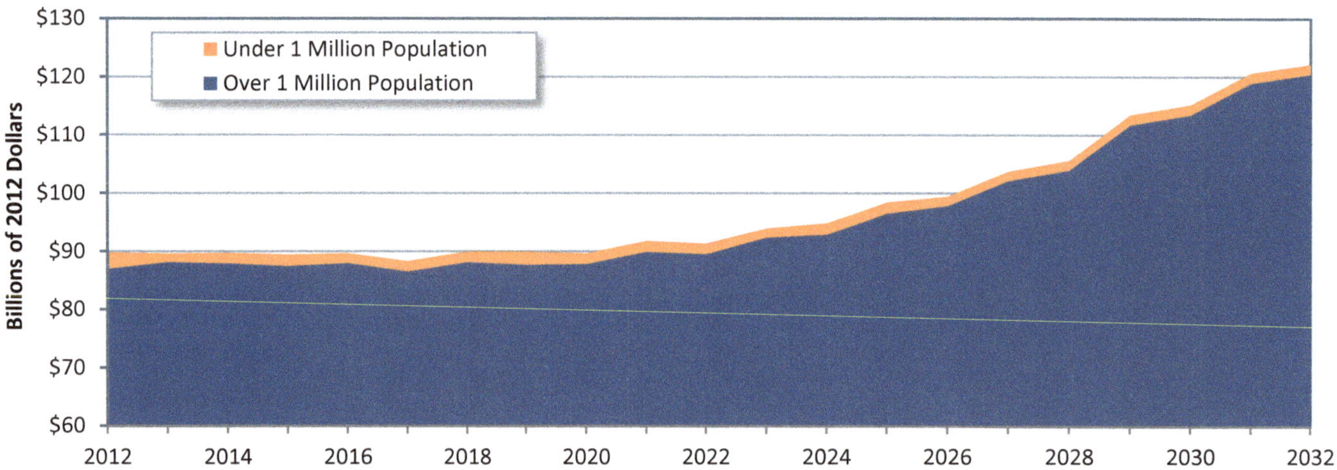

Source: Transit Economic Requirements Model.

The chart in *Exhibit 8-17* also divides the backlog amount according to size of transit service area, with the lower portion showing the backlog for UZAs having populations greater than 1 million and the upper portion showing the backlog for all other UZAs and rural areas combined. This segmentation highlights the significantly higher existing backlog for those UZAs serving the largest number of transit riders. Regardless of the actual allocation, the 2012 expenditure level of $10.0 billion, if sustained, clearly is not sufficient to prevent a further increase in the backlog needs of one or more of these UZA types.

Expansion Investments

In addition to the $10.0 billion spent on preserving transit assets in 2012, transit agencies spent $7.1 billion on expansion investments to support ridership growth and improve transit performance. This section considers the impact of sustaining the 2012 level of expansion investment on future ridership capacity and vehicle utilization rates under the assumptions of both lower and higher growth rates in ridership (i.e., the Low-Growth and High-Growth scenarios). As noted above, recall that the $7.1 billion spent on expansion investments in 2012 was significantly higher than that reported in prior years.

As previously considered in Chapter 7 (see *Exhibit 7-23*), the 2012 rate of investment in transit expansion is not sufficient to expand transit capacity at a rate equal to the rate of growth in travel demand, as projected by the historical trend rate of increase. Under these circumstances, transit capacity utilization (e.g., passengers per vehicle) should be expected to increase, with the level of increase determined by actual growth in demand. Although the impact of this change could be minimal for systems that currently have lower-capacity utilization, service performance on some higher-utilization systems likely would decline as riders experience increased vehicle crowding and service delays. *Exhibit 8-18* illustrates this potential impact. It presents the projected change in vehicle occupancy rates by mode from 2012 through 2032 (reflecting the impacts of spending from 2013 through 2032) under both the Low-Growth and High-Growth scenarios in transit ridership, assuming that transit agencies continue to invest an average of $7.1 billion per year on transit expansion. Under the Low-Growth scenario, capacity utilization—or the average number of riders per transit vehicle—decreases across each of the four modes depicted here, indicating that investment is sufficient or higher than needed to maintain current occupancy levels. For the High-Growth scenario, however, the average number of riders per transit vehicle steadily rises across each mode. Chapter 9 provides more detail on the new methodology for both the Low- and High-Growth scenarios.

Exhibit 8-18 Sustain 2012 Spending Scenario: Capacity Utilization by Mode Forecast, 2012–2032

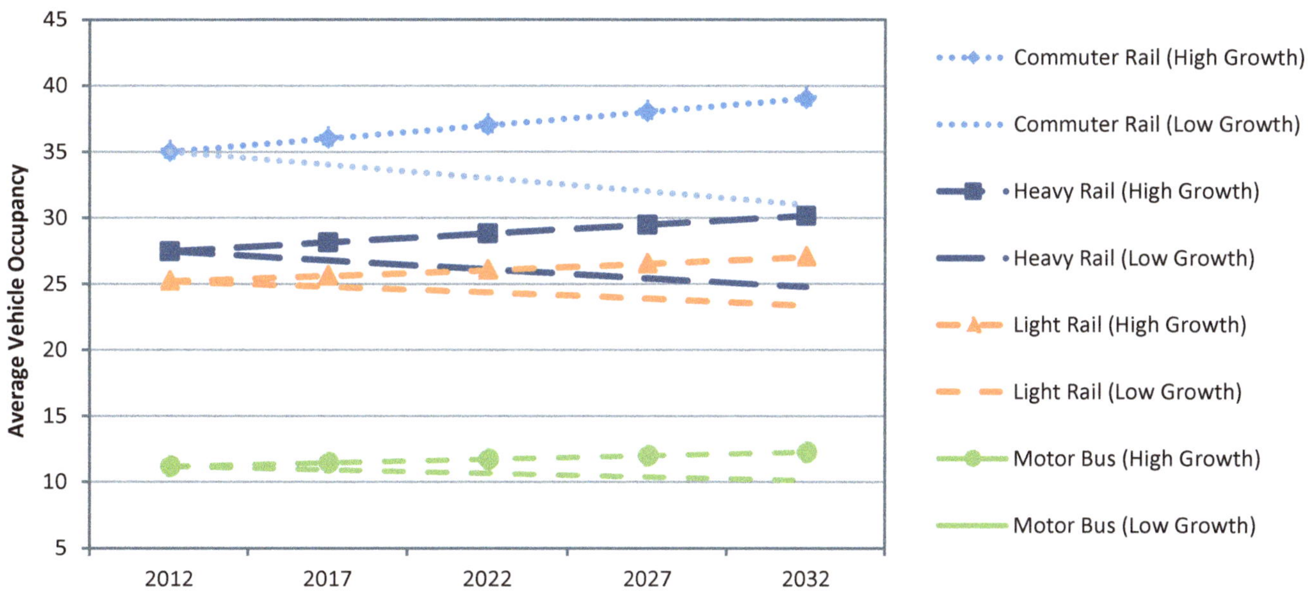

Source: Transit Economic Requirements Model.

Exhibit 8-19 presents the projected growth in transit riders that the 2012 level of investment (keeping vehicle occupancy rates constant) can accommodate as compared with the potential growth in total ridership under both the Low-Growth and High-Growth scenarios. Similar to previous analyses, the $7.1-billion level of investment for expansion can support ridership growth that is similar to the ridership increases projected in the Low-Growth scenario, but is short of that required to support continued ridership under the High-Growth scenario (i.e., without impacting service performance).

Exhibit 8-19 Projected vs. Currently Supported Ridership Growth

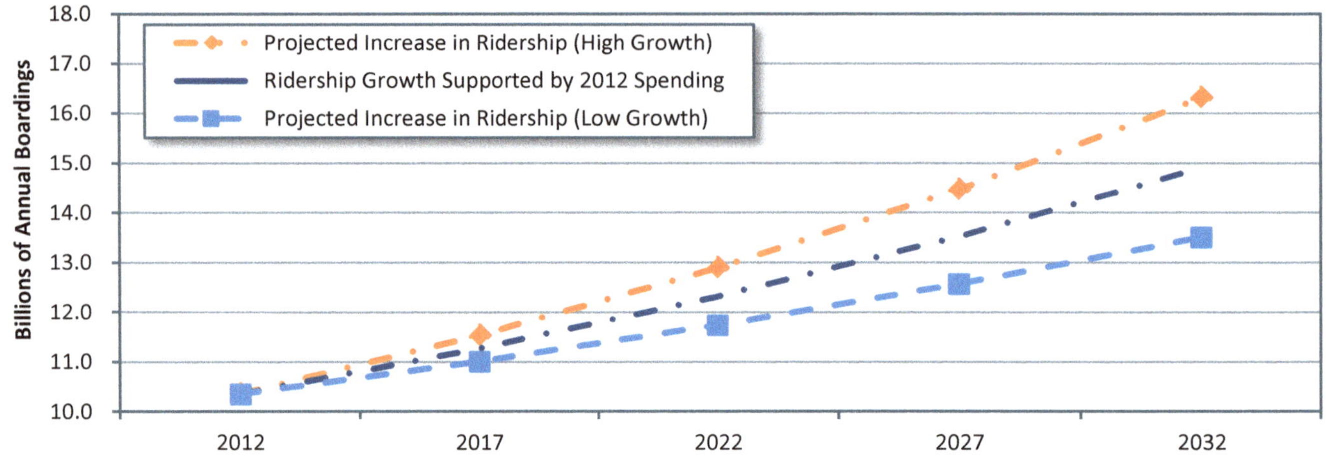

Source: Transit Economic Requirements Model.

State of Good Repair Benchmark

The Sustain 2012 Spending scenario considered the impacts of sustaining transit spending at current levels, which appear to be insufficient to address either deferred investment needs (which are projected to increase) or the projected trends in transit ridership (without a reduction in service performance). In contrast, this section focuses on the level of investment required to eliminate the investment backlog over the next 20 years and to provide for sustainable rehabilitation and replacement needs once the backlog has been addressed. Specifically, the SGR Benchmark estimates the level of annual investment required to replace assets that currently exceed their useful lives, to address all deferred rehabilitation activities (yielding an SGR where the asset has a condition rating of 2.5 or higher), and to address all future rehabilitation and replacement activities as they come due. The SGR Benchmark considered here uses the same methodology as that described in FTA's *National State of Good Repair Assessment*, released June 2012.

> **Q&A What is the definition of state of good repair (SGR)?**
>
> The definition of "state of good repair" used for the SGR Benchmark relies on TERM's assessment of transit asset conditions. Specifically, for this benchmark, TERM considers assets to be in a state of good repair if they are rated at condition 2.5 or higher and if all required rehabilitation activities have been addressed.

Differences from Scenarios: In contrast to the scenarios described in this chapter, the SGR Benchmark does not (1) assess expansion needs or (2) apply TERM's benefit-cost test to investments proposed in TERM. These benchmark characteristics are inconsistent with the SGR concept. First, analyses of expansion investments ultimately focus on capacity improvements and not on the needs of deteriorated assets. Second, application of TERM's benefit-cost test would leave some potential reinvestment improvements unaddressed. The intention of this benchmark is to assess the total magnitude of unaddressed reinvestment needs for all transit assets currently in service, regardless of whether having these assets remain in service would be cost beneficial.

SGR Investment Needs

Annual reinvestment needs under the SGR Benchmark are presented in *Exhibit 8-20*. Under this benchmark, an estimated $ 17.0 billion in annual expenditures would be required over the next 20 years to bring the condition of all existing transit assets to an SGR. Of this amount, roughly $11.7 billion (69 percent) is required to address the SGR needs of rail assets. Note that a large proportion of rail reinvestment needs are associated with guideway elements (primarily aging elevated and tunnel structures) and rail systems (including train control, traction power, and communications systems) that are past their useful lives and potentially are technologically obsolete. Bus-related reinvestment needs are primarily associated with aging vehicle fleets.

Exhibit 8-20 also provides a breakout of capital reinvestment needs by type of UZA. This breakout emphasizes the fact that capital reinvestment needs are most heavily concentrated in the Nation's larger UZAs. Together, these urban areas account for approximately 92 percent of total reinvestment needs (across all mode and asset types), with the rail reinvestment needs of these urban areas accounting for more than one-half the total reinvestment required to bring all assets to an SGR. This high proportion of total needs reflects the high level of investment in older assets found in these urban areas.

Exhibit 8-20 SGR Benchmark: Average Annual Investment by Asset Type, 2013 through 2032

	Urban Area Type		
	Over 1 Million Population	Under 1 Million Population	Total
Asset Type	(Billions of 2012 Dollars)		
Rail			
Guideway Elements	$3.2	$0.1	$3.2
Facilities	$0.7	$0.0	$0.8
Systems	$3.1	$0.0	$3.1
Stations	$3.0	$0.0	$3.0
Vehicles	$1.5	$0.1	$1.6
Subtotal Rail[1]	**$11.5**	**$0.2**	**$11.7**
Nonrail			
Guideway Elements	$0.1	$0.0	$0.1
Facilities	$0.7	$0.0	$0.8
Systems	$0.3	$0.0	$0.3
Stations	$0.1	$0.0	$0.1
Vehicles	$2.9	$1.1	$4.0
Subtotal Nonrail[1]	**$4.1**	**$1.2**	**$5.3**
Total[1]	**$15.7**	**$1.3**	**$17.0**

[1] Note that totals may not sum due to rounding.

Source: Transit Economic Requirements Model.

Impact on the Investment Backlog

A key objective of the SGR Benchmark is to determine the level of investment required to attain and then maintain an SGR across all transit assets over the next 20 years, including elimination of the existing investment backlog. *Exhibit 8-21* shows the estimated impact of the $17.0 billion in annual expenditures under the SGR Benchmark on the existing investment backlog over the 20-year forecast period (compare these data with *Exhibit 8-17*). Given this level of expenditures, the backlog is projected to be eliminated by 2032, with most of this drawdown addressing the reinvestment needs of the UZAs having populations greater than 1 million.

Exhibit 8-21 Investment Backlog: State of Good Repair Benchmark ($16.6 Billion Annually)

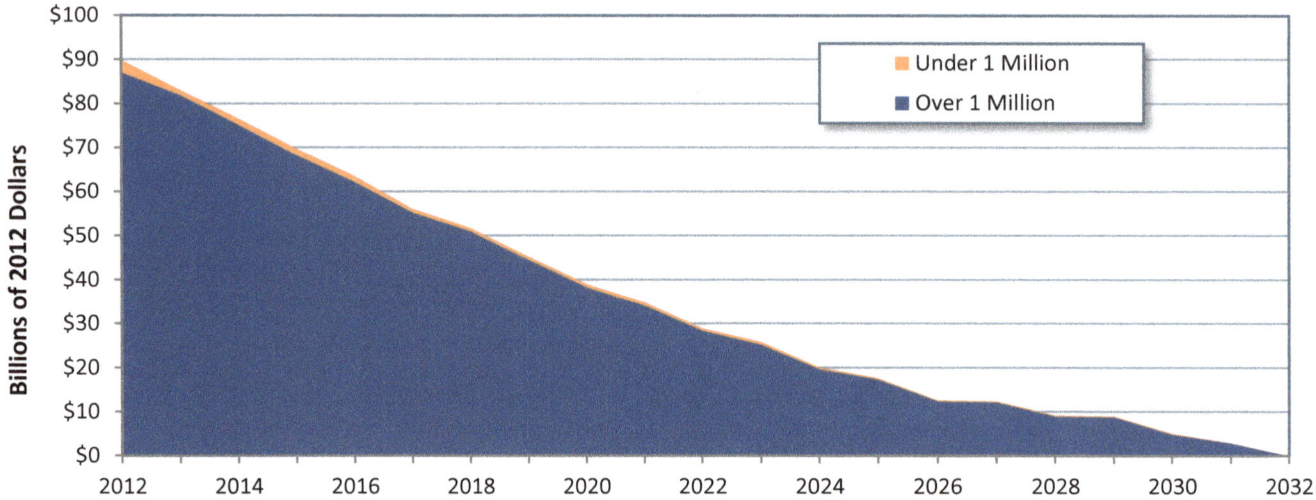

Source: Transit Economic Requirements Model.

Impact on Conditions

In drawing down the investment backlog, the annual capital expenditures of $17.0 billion under the SGR Benchmark also would lead to the replacement of assets with an estimated condition rating of 2.5 or less. Within TERM's condition rating system, these assets would include those in marginal condition having ratings less than 2.5 and all assets in poor condition. *Exhibit 8-22* shows the current distribution of asset conditions for assets estimated to be in a rating condition of 2.5 or less (with assets in poor condition divided into two subgroups). Note that this graphic excludes both tunnel structures and subway stations in tunnel structures because these are considered assets that require ongoing capital rehabilitation expenditures but that are never actually replaced. As with the investment backlog, the proportion of assets at condition rating 2.5 or lower is projected to decrease under the SGR Benchmark from roughly 14 percent of assets in 2012 to less than 1 percent by 2032. Once again, this replacement activity would remove from service those assets with higher occurrences of service failures, technological obsolescence, and lower overall service quality. Important to note is that the assets with condition less than 2.5 presented in *Exhibit 8-22* capture only a subset of assets in the SGR backlog as depicted in *Exhibit 8-21*. Specifically, the total SGR backlog (*Exhibit 8-21*) includes not just those assets in need of replacement (i.e., those at less than condition 2.5), but also those assets in need of rehabilitation or other form of capital reinvestment.

Low-Growth and High-Growth Scenarios

The SGR Benchmark considered the level of investment to bring existing transit assets to an SGR but in doing so did not consider either (1) the economic feasibility of these investments (investments were not required to pass TERM's benefit-cost test) or (2) the level of expansion investment required to support projected ridership growth. The Low-Growth scenario and High-Growth scenario address both these issues. Specifically, these scenarios use the same rules to assess when assets should be rehabilitated or replaced as were applied in the preceding SGR Benchmark (e.g., with assets being replaced at condition 2.5), but also require that these preservation and expansion investments pass TERM's benefit-cost test. In general, some reinvestment activities do not pass this test (i.e., have a benefit-cost ratio less than 1), which can result from low ridership benefits, higher capital or operating costs, or a mix of these factors. Excluding investments that do not pass the benefit-cost test has the effect of reducing total estimated needs.

Exhibit 8-22 Proportion of Transit Assets Not in State of Good Repair (Excluding Tunnel Structures)

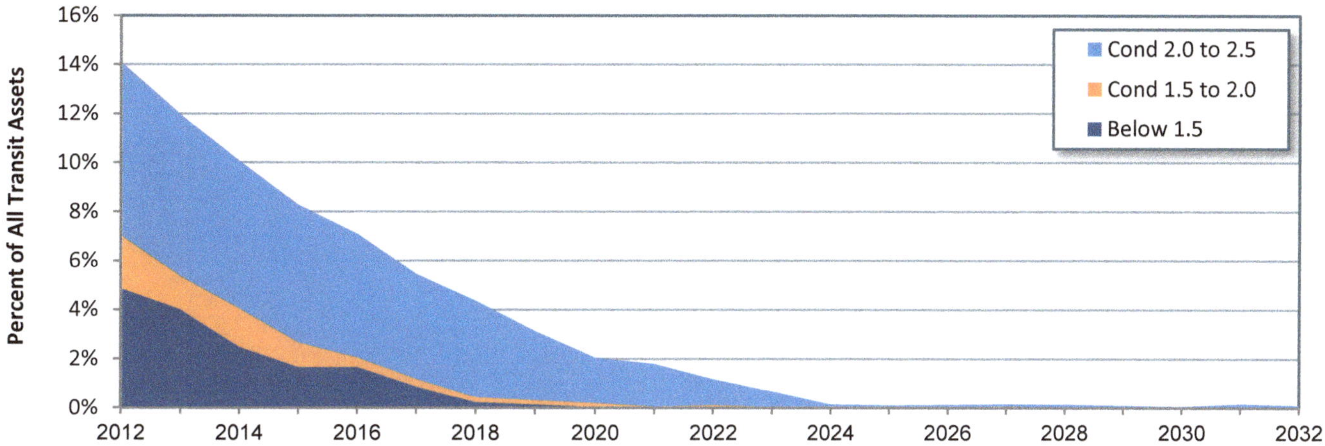

Source: Transit Economic Requirements Model.

In addition, the Low- and High-Growth scenarios also assess transit expansion needs given ridership growth based on the average annual compound rate as experienced over the past 15 years minus 0.5 percent (Low-Growth) or plus 0.5 percent (High-Growth). For the expansion component of this scenario, TERM assesses the level of investment required to maintain current vehicle occupancy rates (at the agency-mode level) subject to the rate of projected growth in transit demand in that UZA and subject to the proposed expansion investment passing TERM's benefit-cost test.

Low- and High-Growth Assumptions

The Low-Growth scenario is intended to provide a lower bound on the level of investment required to maintain current service performance (as measured by transit vehicle capacity utilization) as determined by a relatively low rate of growth in travel demand. In contrast, the High-Growth scenario provides a higher bound on the level of investment required to maintain current service performance as determined by a relatively high rate of growth in travel demand. The methodology for the Low- and High-Growth scenarios has been revised to use a common, consistent approach that better reflects differences in PMT growth by mode. Specifically, these scenarios are now based on the 15-year trend rate of growth in PMT. When calculated across all transit operators and modes, this historical trend rate of growth converts to a national average compound annual growth rate of approximately 1.7 percent during the 20-year period.

Within this new framework, the Low-Growth scenario is defined as the trend rate of growth (by FTA region, population stratum, and mode) less 0.5 percent, while the High-Growth scenario is defined as the trend rate of growth plus 0.5 percent. Hence, the Low-Growth and High-Growth scenarios differ by a full 1.0 percent in annual growth.

Low- and High-Growth Scenario Needs

Exhibit 8-23 presents TERM's projected capital investment needs on an annual average basis under the Low- and High-Growth scenarios, including those for both asset preservation and asset expansion.

Exhibit 8-23 Low-Growth and High-Growth Scenarios: Average Annual Investment by Asset Type, 2013 through 2032

	Lower Growth			Higher Growth		
	Preservation	Expansion		Preservation	Expansion	
Asset Type	(Billions of 2012 Dollars)		Total	(Billions of 2012 Dollars)		Total
Rail						
Guideway Elements	$3.2	$1.2	$4.4	$3.2	$1.7	$4.9
Facilities	$0.7	$0.2	$0.9	$0.8	$0.3	$1.1
Systems	$3.1	$0.3	$3.4	$3.1	$0.3	$3.5
Stations	$3.0	$0.7	$3.7	$3.0	$1.0	$3.9
Vehicles	$1.5	$1.9	$3.4	$1.5	$3.0	$4.5
Other Project Costs	$0.0	$1.3	$1.3	$0.0	$1.7	$1.7
Subtotal Rail[1]	**$11.6**	**$5.5**	**$17.1**	**$11.6**	**$7.9**	**$19.5**
Subtotal UZAs Over 1 Million[1]	**$11.4**	**$5.5**	**$16.9**	**$11.5**	**$7.9**	**$19.3**
Subtotal UZAs Under 1 Million and Rural[1]	**$0.1**	**$0.03**	**$0.2**	**$0.2**	**$0.04**	**$0.2**
Nonrail						
Guideway Elements	$0.1	$0.0	$0.1	$0.1	$0.0	$0.1
Facilities	$0.7	$0.1	$0.8	$0.7	$0.3	$1.0
Systems	$0.2	$0.0	$0.3	$0.3	$0.1	$0.3
Stations	$0.0	$0.0	$0.0	$0.0	$0.0	$0.1
Vehicles	$3.8	$0.7	$4.6	$3.8	$1.5	$5.4
Other Project Costs	$0.0	$0.0	$0.0	$0.0	$0.0	$0.0
Subtotal Nonrail[1]	**$4.9**	**$0.9**	**$5.8**	**$4.9**	**$2.0**	**$6.9**
Subtotal UZAs Over 1 Million[1]	**$3.7**	**$0.4**	**$4.1**	**$3.8**	**$1.1**	**$4.9**
Subtotal UZAs Under 1 Million and Rural[1]	**$1.1**	**$0.5**	**$1.7**	**$1.1**	**$0.9**	**$2.0**
Total Investment[1]	**$16.4**	**$6.4**	**$22.9**	**$16.5**	**$9.9**	**$26.4**
Total UZAs Over 1 Million[1]	**$15.2**	**$5.9**	**$21.1**	**$15.2**	**$9.0**	**$24.2**
Total UZAs Under 1 Million and Rural[1]	**$1.3**	**$0.5**	**$1.8**	**$1.3**	**$0.9**	**$2.2**

[1] Note that totals may not sum due to rounding.

Source: Transit Economic Requirements Model.

Lower-Growth Needs

Assuming the relatively low ridership growth in the Low-Growth scenario, investment needs for system preservation and expansion are estimated to average roughly $22.9 billion each year for the next two decades. Of this amount, roughly 72 percent is for preserving existing assets and approximately $11.6 billion is associated with preserving existing rail infrastructure alone. Note that the $0.6-billion difference between the $17.0 billion in annual preservation needs under the SGR Benchmark and the $16.4 billion in preservation needs under the Low-Growth scenario is entirely due to the application of TERM's benefit-cost test under the Low-Growth scenario. Finally, expansion needs in this scenario total $6.4 billion annually, with 86 percent of that amount associated with rail expansion costs.

Higher-Growth Needs

In contrast, total investment needs under the High-Growth scenario are estimated to be $26.4 billion annually, a 15-percent increase over the total investment needs under the Low-Growth scenario. The High-Growth scenario total includes $16.5 billion for system preservation and an additional $9.9 billion for system expansion. Note that system preservation costs are higher under the High-Growth scenario because the higher growth rate leads to a larger expansion of the asset base as compared to the Low-Growth scenario. Under this scenario, investment in expansion of rail assets is still larger than that for nonrail expansion (80 percent for rail and 20 percent for nonrail). Under the High-Growth scenario, however, rail takes only 80 percent of total expansion investment versus 86 percent of expansion needs under the Low-Growth scenario.

Impact on Conditions and Performance

The impact of the Low- and High-Growth rate preservation investments on transit conditions is essentially the same as that already presented for the SGR Benchmark in *Exhibits 8-21* and *8-22*. As noted above, the Low and High-Growth scenarios use the same rules to assess when assets should be rehabilitated or replaced as were applied in the SGR Benchmark (e.g., with assets being replaced at condition rating 2.5). In terms of asset conditions, the primary difference between the SGR Benchmark and the Low- and High-Growth scenarios relates to (1) TERM's benefit-cost test not applying to the SGR Benchmark (leading to higher SGR preservation needs overall) and (2) the Low- and High-Growth scenarios having some additional needs for replacing expansion assets with short service lives. Together, these impacts tend to work in opposite directions. The result is that the rate of drawdown in the investment backlog and the elimination of assets exceeding their useful lives are roughly comparable between the SGR Benchmark and these scenarios and between the two scenarios.

Similarly, the impact of the Low- and High-Growth rate expansion investments on transit ridership was considered in *Exhibit 8-19*. That analysis demonstrated the significant difference in the level of ridership growth supported by the High-Growth scenario as compared with either the current level of expenditures ($6.6 billion in 2012 for UZAs with populations greater than 1 million) or the rate of growth supported under the Low-Growth scenario.

Scenario Impacts Comparison

Finally, this subsection summarizes and compares many of the investment impacts associated with each of the three analysis scenarios and the SGR Benchmark considered above. Although much of this comparison is based on measures already introduced above, this discussion also considers a few additional investment impact measures. These comparisons are presented in *Exhibit 8-24*. Note that the first column of data in *Exhibit 8-24* presents the current values for each of these measures (as of 2012). The subsequent columns present the estimated future values in 2032, assuming the levels, allocations, and timing of expenditures associated with each of the three investment scenarios and the SGR Benchmark.

Exhibit 8-24 includes the following measures:

- **Average annual expenditures (billions of dollars):** This amount is broken down into preservation and expansion expenditures.
- **Condition of existing assets:** This analysis considers only the impact of investment funds on the condition of those assets currently in service.

  Average physical condition rating: The weighted average condition of all existing assets on TERM's condition scale of 5 (excellent) through 1 (poor).

  Investment backlog: The value of all deferred capital investment, including assets exceeding their useful lives and rehabilitation activities that are past due (this value can approach but never reach zero due to assets continually aging with some exceeding their useful lives). The backlog is presented here both as a total dollar amount and as a percent of the total replacement value of all U.S. transit assets.

  Backlog ratio: The ratio of the current investment backlog to the annual level of investment required to maintain normal annual capital needs once the backlog is eliminated.

- **Performance measures:** The impact of investments on U.S. transit ridership capacity and system reliability.

  New boardings supported by expansion investments: The number of additional riders that transit systems can carry without a loss in performance (given the projected ridership assumptions for each scenario).

  Revenue service disruptions per PMT: Number of disruptions to revenue service per million passenger miles.

  Fleet maintenance cost per vehicle revenue mile: Fleet maintenance costs tend to increase with fleet age (or reduced asset condition). This measure estimates the change in fleet maintenance costs expressed in a per-revenue-vehicle-mile basis.

Exhibit 8-24 Scenario Investment Benefits Scorecard

Measure	Baseline 2012 Actual Spending, Conditions and Performance	Projections for 2032			
		SGR Benchmark	Sustain 2012 Spending Scenario	Low-Growth Scenario	High-Growth Scenario
Average Annual Expenditures (Billions of 2012 Dollars)					
Preservation	$9.9	$17.0	$9.9	$16.4	$16.5
Expansion	$7.1	NA	$7.1	$6.4	$9.9
Total	**$17.0**	**$17.0**	**$17.0**	**$22.8**	**$26.4**
Conditions (Existing Assets)					
Average Physical Condition Rating	3.5	3.2	3.1	3.4	3.6
Investment Backlog (Billions of Dollars)	$89.8	$0.0	$121.7	$0.0	$0.0
Investment Backlog (% of Replacement Costs)	10.6%	0.0%	14.4%	0.0%	0.0%
Backlog Ratio[1]	7.2	0.0	9.7	0.0	0.0
Performance					
Ridership Impacts of Expansion Investments (2012)					
New Boardings Supported by Expansion (Billions)	NA	NA	3.4	3.1	5.8
Total Projected Boardings in 2032 (Billions)	NA	NA	13.7	13.5	16.2
Fleet Performance					
Revenue Service Disruptions per Thousand PMT	8.0	8.2	8.0	8.2	8.2
Fleet Maintenance Cost per Revenue Vehicle Mile	$1.83	$1.85	$1.81	$1.84	$1.85

[1] The backlog ratio is the ratio of the current investment backlog to the annual level of investment to maintain SGR once the backlog is eliminated.

Source: Transit Economic Requirements Model.

Scorecard Comparisons

A review of the scorecard results for each of the three investment scenarios and the SGR Benchmark reveals the impacts discussed below.

Preservation Impacts

Continued reinvestment at the 2012 level is likely to yield a decline in overall asset conditions (from 3.5 in 2012 to 3.1 in 2032) and an increase in the size of the investment backlog (from $89.8 billion in 2012 to $121.7 billion in 2032). Continued reinvestment at the 2012 level, however, likely will cause no change in service disruptions per million passenger miles and a decrease in maintenance costs per vehicle revenue mile. In contrast, with the exception of overall asset conditions, opposite results occur under the SGR Benchmark, the Low-Growth scenario, and the High-Growth scenario. Note that the overall condition rating measures of 3.2, 3.4, and 3.6 under the SGR Benchmark, the Low-Growth scenario, and the High-Growth scenario, respectively, represent sustainable, long-term condition levels for the Nation's existing transit assets over the long term. This is in contrast to the current measure of roughly 3.5, which would be difficult to maintain over the long term without replacing many asset types prior to the conclusion of their expected useful lives.

For this report, expansion assets are included in the overall condition rating measures. This approach is a departure from that in previous reports, in which the goal was to be cognizant of what happens to the SGR of existing assets under alternative scenarios.

Expansion Impacts

Although continued expansion investment at the 2012 level appears sufficient to support a low rate of increase in transit ridership to about 3.4 billion new boardings in 2032, higher rates of growth to nearly 5.8 billion new boardings in 2032 suggest that a significantly higher rate of expansion investment (nearly $3 billion more annually in expansion investment) is required to avoid a decline in overall transit performance (e.g., in the form of increased crowding on high-utilization systems).

chapter 9

Supplemental Scenario Analysis

Highway Supplemental Scenario Analysis

This chapter explores the implications of the highway investment scenarios considered in Chapter 8, starting with a comparison of the scenario investment levels to those presented in previous C&P reports. This section also examines the long-term forecasts of vehicle miles traveled (VMT) presented in earlier C&P reports and compares them to actual outcomes. The highway travel demand forecast by the Highway Performance Monitoring System (HPMS) is also compared with FHWA-modeled forecasts to ensure consistency.

This chapter illustrates the impact of alternative rates of future inflation on the constant-dollar scenario investment levels presented in Chapter 8 and explores alternative assumptions about the timing of investment over the 20-year analysis period. A subsequent section within this chapter provides supplementary analysis regarding the transit investment scenarios.

Comparison of Scenarios with Previous Reports

Each edition of this report presents various projections of travel growth, pavement conditions, and bridge conditions under different performance scenarios. The projections cover 20-year periods, beginning the first year after the data presented on current conditions and performance. Although the scenario names and criteria have varied over time, the C&P report traditionally has included highway investment scenarios corresponding in concept to the Maintain Conditions and Performance scenario and the Improve Conditions and Performance scenario presented in Chapter 8.

Comparison With 2013 C&P Report

One obvious difference between the scenarios presented in this 2015 C&P Report and those in the 2013 C&P Report is that they cover a different 20-year period, 2013 through 2032 rather than 2011 through 2030. The 2013 edition also presented two alternative estimates for each scenario based on different assumptions regarding future travel growth, while the current edition reverts to the traditional approach of reporting only one primary estimate for each scenario, accompanied by sensitivity analyses presented in Chapter 10 that explore the impacts of alternative future travel growth rates.

Aside from these differences, the procedure used to determine the investment levels associated with the Improve Conditions and Performance scenario is the same for both editions. This scenario sets a level of spending sufficient to gradually fund all potential highway and bridge projects that are cost-beneficial over 20 years. Neither the 2013 C&P Report nor this 2015 C&P Report made assumptions about financing mechanisms that would be used to cover the costs of this scenario.

The Maintain Conditions and Performance scenario identifies a level of investment associated with keeping overall conditions and performance in 20 years at base-year levels. As discussed in Chapter 8, the target of the scenario component derived from the National Bridge Investment Analysis System (NBIAS) has been modified to maintain the percentage of total bridge deck area classified as structurally deficient or functionally obsolete, rather than to maintain the average bridge sufficiency rating as in the 2013 C&P Report. The target measures for the component of this scenario derived from the Highway Economic Requirements System (HERS)—average pavement roughness and average delay per VMT—were retained in the current edition but applied in a different way. Rather than following the 2013 C&P Report approach of taking the average of the annual investment levels predicted to be adequate for maintaining each of these two indicators, the scenario in the current edition was based on the higher of the two investment levels. Thus, under the new approach, the HERS component of the scenario reflects the lowest average level of investment at which both pavement roughness and delay either stay the same or improve.

As discussed in Chapter 6, highway construction costs are measured by the Federal Highway Administration's (FHWA's) National Highway Construction Cost Index, which increased by 6.1 percent between 2010 and 2012. Consequently, adjusting the 2013 C&P Report's scenario figures from 2010 dollars to 2012 dollars causes the observed and projected highway construction costs to appear larger. As shown in *Exhibit 9-1*, the 2013 C&P Report estimated the average annual investment level in the current Maintain Conditions and Performance scenario to range from $65.3 to $86.3 billion in 2010 dollars; adjusting for inflation shifts this range to $69.3 to $91.6 billion in 2012 dollars. The comparable amount for the Maintain Conditions and Performance scenario presented in Chapter 8 of this edition is $89.9 billion in 2012 dollars, approximately 1.9 percent lower than the high end of 2013 C&P Report estimate of $91.6 billion dollars.

Exhibit 9-1 Selected Highway Investment Scenario Projections Compared with Comparable Data from the 2013 C&P Report

Highway and Bridge Scenarios—All Roads	2011 Through 2030 Projection (Based on 2010 Data)[1]		2013 Through 2032 Projection (Billions of 2012 Dollars)
	2013 C&P Report (Billions of 2010 Dollars)	Adjusted for Inflation[2] (Billions of 2012 Dollars)	
Maintain Conditions and Performance scenario[3]	**$65.3–$86.3**	**$69.3–$91.6**	**$89.9**
Improve Conditions and Performance scenario	**$123.7–$145.9**	**$131.3–$154.9**	**$142.5**

[1] The 2013 C&P report included two alternative estimates for each scenario based on different assumptions regarding future travel growth.

[2] The investment levels for the highway and bridge scenarios were adjusted for inflation using the FHWA National Highway Construction Cost Index.

[3] In the 2013 C&P Report, the HERS component of this scenario focused on maintaining a composite indicator reflecting average delay and average pavement condition rather than just maintaining average pavement condition; the NBIAS component of the scenario focused on maintaining the average sufficiency rating for bridges rather than the percentage of deck area on deficient bridges.

The average annual investment level in the 2013 C&P Report scenario comparable to the current Improve Conditions and Performance scenario was estimated to be $123.7 to $145.9 billion in 2010 dollars; adjusting for inflation increases this range to $131.3 to $154.9 billion in 2012 dollars. The comparable amount for the current Improve Conditions and Performance scenario presented in Chapter 8 of this edition is $142.5 billion, approximately 8.0 percent lower than the high end of 2013 C&P Report estimate of $154.9 billion dollars.

The changes in the scenario findings in this report relative to the 2013 C&P Report are also partially attributable to changes in the underlying characteristics, conditions, and performance of the bridge system reported in Chapters 2 and 3 and to changes in the analytical methodology and data in HERS and NBIAS. Despite changes in database and model specifications, the estimated investment needs suggest similar patterns for comparable scenarios over different editions of the C&P report.

Comparisons of Implied Funding Gaps

Exhibit 9-2 compares the funding gaps implied by the analysis in the current report with those implied by previous C&P report analyses. The funding gap is measured as the percentage by which the estimated average annual investment needs for a specific scenario exceeds the base-year level

Exhibit 9-2 Comparison of Average Annual Highway and Bridge Investment Scenario Estimates with Base-Year Spending, 1997 to 2015 C&P Reports

Report Year	Relevant Comparison	Percent Above Base-Year Spending	
		Primary "Maintain" Scenario[1]	Primary "Improve" Scenario[1]
1997	Average annual investment scenario estimates for 1996 through 2015 compared with 1995 spending	21.0%	108.9%
1999	Average annual investment scenario estimates for 1998 through 2017 compared with 1997 spending	16.3%	92.9%
2002	Average annual investment scenario estimates for 2001 through 2020 compared with 2000 spending	17.5%	65.3%
2004	Average annual investment scenario estimates for 2003 through 2022 compared with 2002 spending	8.3%	74.3%
2006	Average annual investment scenario estimates for 2005 through 2024 compared with 2004 spending	12.2%	87.4%
2008	Average annual investment scenario estimates for 2007 through 2026 compared with 2006 spending	34.2%	121.9%
2010	Average annual investment scenario estimates for 2009 through 2028 compared with 2008 spending	10.8%	86.6%
2013	Average annual investment scenario estimates for 2011 through 2030 compared with 2010 spending	-13.9%	45.7%
2015	Average annual investment scenario estimates for 2013 through 2032 compared with 2012 spending	-14.6%	35.5%

[1] Amounts shown correspond to the primary investment scenario associated with maintaining or improving the overall highway system in each C&P report; the definitions of these scenarios are not fully consistent among reports. The values shown for this report reflect the Maintain Conditions and Performance and the Improve Conditions and Performance scenarios. Negative numbers signify that the investment scenario estimate was lower than base-year spending.

Sources: Highway Economic Requirements System and National Bridge Investment Analysis System.

of investment. The scenarios examined are this report's Maintain Conditions and Performance scenario and Improve Conditions and Performance scenario, and their counterparts in previous C&P reports.

Prior to the 2013 C&P Report, each C&P report edition showed that actual annual spending in the base year for that report had been below the estimated average investment level required to maintain conditions and performance at base-year levels over 20 years. In both the 2013 C&P Report and this 2015 C&P Report, the trend was reversed and gaps between actual and required amounts for the primary "Maintain" scenario (the higher VMT growth scenario in 2013 C&P Report) became negative. This result dramatically differs from the positive numbers estimated in pre-2013 C&P reports, indicating that base-year spending reported in the 2013 and 2015 C&P Reports was more than the average annual spending levels identified for the Maintain Conditions and Performance scenario. The primary "Improve" scenario follows a similar trend, where the funding gap has dropped steadily since its peak in the 2008 C&P Report.

Changes in actual capital spending by all levels of government combined can substantially alter these spending gaps, as can sudden, large swings in construction costs. The large increase in the gap between base-year spending and the primary Maintain and Improve scenarios presented in the 2008 C&P Report coincided with a large increase of construction costs experienced between 2004 and 2006 (base year for the 2008 C&P Report). On the other hand, the decreases in the gaps presented in recent editions coincided with declines in construction costs since their 2006 peak.

The differences among C&P report editions in the implied gaps reported in *Exhibit 9-2* are not a consistent indicator of change over time in how effectively highway investment needs are addressed. FHWA continues to enhance the methodology used to determine scenario estimates for each edition of the C&P report to provide a more comprehensive and accurate assessment. In some cases, these refinements have increased the level of investment in one or both of the scenarios (the Maintain or Improve scenarios, or their equivalents); other refinements have reduced this level.

Highway Travel Demand Forecasts

For each HPMS sample highway section, States provide the actual traffic volume in the base year and a forecast of traffic volume for a future year, typically 20 years after the base year. The HPMS reporting guidance requires the traffic forecasts States generate to be derived from a technically supportable procedure based on available information concerning the particular section and corridor to which it belongs. Because the HERS model was introduced in the 1995 C&P Report and used through the 2010 C&P Report, the primary highway investment scenarios presented in each of these editions relied on these State-provided forecasts. Without specific information regarding the assumptions built into the State forecasting procedures, an assumption was made that the forecasts reflected the level of future VMT that would occur for each HPMS sample section if average user costs, including costs of travel time, vehicle operation, and crash risk, were unchanged over the 20-year analysis period. Beginning with the 1999 C&P Report, sensitivity analyses also were presented showing the implications of alternative rates of future VMT growth.

The sensitivity analysis was conducted by using the State-reported values for individual sample sections as a starting point. The values then were adjusted upward or downward proportionally, as needed to achieve a particular nationwide level of VMT growth.

The 2013 C&P Report followed the same general approach, except that it presented two alternative sets of scenarios. One scenario was based on the State-provided HPMS forecasts. For the other scenario, these forecasts were proportionally adjusted downward to match the average annual VMT growth rate observed over the preceding 15 years.

The primary advantage of using the State-provided VMT forecasts in the baseline analysis derives from their geographic specificity. Separate forecasts are provided for more than 100,000 HPMS sample sections. States can account for local conditions of a specific section and project long-term travel patterns of the particular routes or corridors accordingly. These section-level forecasts enable more refined projections of future travel demand.

The primary disadvantage of relying solely on the State-provided VMT forecasts is the uncertainty about exactly what they represent. To the extent that some States factor in changes to components of highway user costs in making their predictions, the traditional assumption made in HERS—that these are "constant-price" forecasts that do not reflect such changes—would be incorrect. Thus, the travel demand elasticity procedures in HERS discussed in Chapter 7, which adjust future VMT projections based on changes in "price" of travel as reflected by user costs, could be double counting the effects of some changes. Also, although the forecasts supplied for individual HPMS sample sections might appear reasonable in isolation based on information available at that particular location, when aggregated with all other forecasts for a given State or nationally, they could yield an overall growth rate that appears inconsistent with observed trends.

To address these issues, FHWA has adopted a new national-level VMT forecasting model. The analyses presented in this 2015 C&P Report used State-provided VMT forecasts for highways and bridges in HERS and NBIAS, respectively, as the starting point. Then, these values were proportionally reduced to yield a national-level forecast consistent with the predictions of this new VMT forecasting model. HERS also was modified to account for changes in user costs built into the VMT forecasts when applying its travel demand elasticity procedures to avoid three potential issues: (1) double counting the effects of assumed constant-dollar increases in the value of time, (2) presumed changes in constant-dollar fuel prices, and (3) changes in the share of total travel attributable to single-unit and combination trucks.

New National VMT Forecasting Model

The Volpe National Transportation Systems Center developed the *National Vehicle Miles Traveled Projection* for FHWA. The first projection was released in May 2014. The documentation for the model version used for this forecast is posted at http://www.fhwa.dot.gov/policyinformation/tables/vmt/vmt_model_dev.cfm. The current plan is to release revised forecasts each May; this 2015 C&P Report relies on the 20-year forecasts for the *Baseline Economic Outlook* from the May 2015 release posted at http://www.fhwa.dot.gov/policyinformation/tables/vmt/vmt_forecast_sum.cfm.

The travel forecasting model estimates future changes in passenger and freight VMT based on predicted changes in demographic and economic conditions. Built on economic theory, the national total VMT model establishes a separate but structurally similar econometric model for each of three vehicle categories—light-duty vehicles, single-unit trucks, and combination trucks using time series data beginning in the 1960s. These econometric models include underlying factors that strongly influence user demand to travel, such as demographic characteristics, economic activity, employment, cost of driving, road miles, and transit service availability. The three econometric models are applied to different roadway classes to develop detailed forecasts of future travel demand and VMT growth. A separate model is used for national-level bus VMT.

In addition to the econometric approaches used to construct the aggregate national VMT models, the travel forecasting models include a methodology for forecasting national VMT from a vehicle fleet perspective. This component of the VMT models disaggregates nationwide total VMT by vehicle class, model year or vintage, and vehicle age. The aggregate national-level VMT totals for each vehicle type were estimated and used as control totals for lower-level (functional classification and location) models. The econometric and vehicle fleet approaches are complementary and provide a more accurate forecast of future VMT.

Comparison of FHWA-Modeled Forecasts and HPMS Forecasts

Based on the May 2015 release, the FHWA forecast for the *Baseline Economic Outlook* is for VMT to grow at an average annual rate of 1.04 percent per year. In contrast, aggregating the forecasts for individual HPMS sample sections yields a composite, weighted-average, annual VMT growth rate between the 2012 base year and the forecast year, 2032, of 1.41 percent.

Exhibit 9-3 translates these two average annual VMT growth rates into projected annual VMT for each year from 2012 to 2032 for Federal-aid highways and for all roads combined. Consistent with the approach used in the HERS and NBIAS analyses for this report and other recent editions of the C&P report, future VMT is assumed to grow linearly (so that 1/20th of the additional VMT is added each year), rather than geometrically (growing at a constant annual rate). With linear growth, the annual percentage rate of growth gradually declines over the forecast period. This approach is logically consistent with the FHWA national VMT forecasting model, which projects lower average annual VMT growth rates over 30 years than it does over 20 years.

The VMT on all roads in 2012 was estimated at 2.99 trillion, and by 2032, VMT would reach 3.67 trillion, assuming an average annual VMT growth rate of 1.04 percent per year. The investment analyses presented in Chapters 7 and 8 reflect this assumption. *Exhibit 9-3* also projects that VMT on Federal-aid highways will rise from 2.53 trillion to 3.11 trillion by 2032; these are the values actually modeled in HERS, as it considers only Federal-aid highways.

If future VMT were to rise at an average annual rate of 1.41 percent, consistent with the weighted aggregate VMT growth rate derived from the State-provided HPMS forecasts, total VMT would rise to 3.95 trillion by 2032, with 3.34 trillion VMT occurring on Federal-aid highways. Aggregating the forecasts of future bridge traffic reported in the National Bridge Inventory yields a similar average annual growth rate of 1.46 percent. Chapter 10 includes a sensitivity analysis showing how substituting these State-supplied forecasts would affect the projection of the Maintain Conditions and Performance scenario and the Improve Conditions and Performance scenario presented in Chapter 8.

Exhibit 9-3 Annual Projected Highway VMT Based on HPMS-Derived Forecasts or FHWA VMT Forecast Model

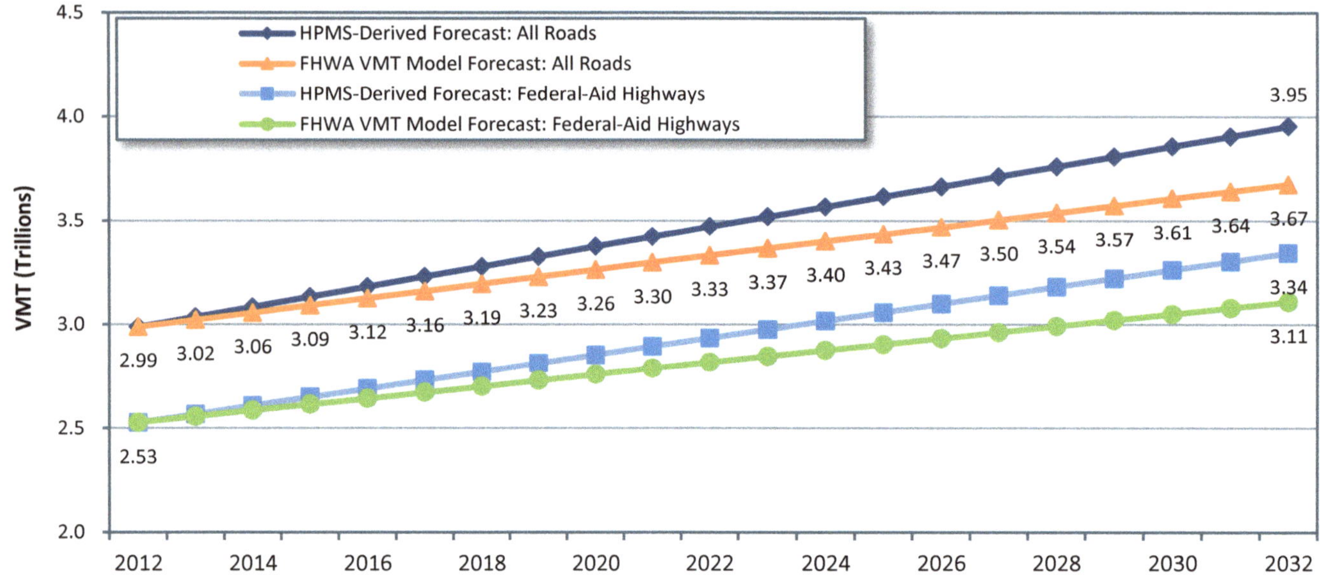

Sources: Highway Performance Monitoring System; FHWA Forecasts of Vehicle Miles Traveled (VMT), May 2015.

Q&A **What has happened with VMT growth since 2012?**

The United States recorded 2.99 trillion vehicle miles traveled (VMT) in 2012. According to preliminary statistics of the Federal Highway Administration, annual VMT is estimated to have increased by 0.6 percent in 2013, 1.7 percent in 2014, and 3.5 percent in 2015 to a level of 3.15 trillion.

A wide variety of factors correlate with the VMT trend, including macroeconomic and demographic factors and the availability of other transportation modes. Key economic variables that contribute to an increase in VMT are a declining unemployment rate and increasing income. Drivers respond to changes in fuel prices, and lower gas prices encourage more discretionary driving. Demographic changes also might explain the development of VMT. For example, a growing population tends to drive more. Other forces also can decrease VMT: The younger generation takes fewer automobile trips, and options to telecommute decrease travel demand.

VMT Forecasts from Previous C&P Reports

Future traffic projection is central to evaluations of capital spending on transportation infrastructure. Forecasting future traffic conditions, however, is extremely difficult because many uncertain circumstances are related to travel behavior. A rich body of literature has examined the accuracy issue of travel demand modeling and found rampant inaccuracy in project-specific traffic forecasts, with most at least 20–30 percent off actual future traffic volumes (see Flyvbjerg, Holm, and Buhl [2005] and Hartgen [2013], for example). This inaccuracy could be attributable to the model's failure to consider influencing factors (e.g., changing demographics and preferences), the effects of certain policies (e.g., pedestrian and bicycle), or treating changes in VMT growth patterns as temporary phenomena instead of long-term trends. These project-level inaccuracies could translate into inaccuracy in national aggregates. Even where the underlying relationships may be correctly modeled, the evolution of key variables (such as expected regional economic growth) could differ significantly from the assumptions made in the VMT forecast.

In light of this uncertainty, that the effective VMT growth rates predicted in the C&P report could be off target is not surprising. *Exhibit 9-4* presents the long-term VMT projections in the 21 C&P reports starting in 1968, including the current report, and compares them to actual highway VMT. The forecasts differed from actual trends in most cases, sometimes underestimating future travel demand and other times overestimating it.

Each of the first five editions of the C&P report (1968 through 1977) underpredicted future total VMT by about 10 to 15 percent. Actual VMT for periods covered by these forecasts grew by more than 3 percent per year; in contrast, the average annual VMT growth rate forecasts among these five editions ranged from 2.2 to 2.7 percent.

The total VMT forecasts in the next five editions of the C&P report (1981 through 1989) were closer to the mark, deviating from actual VMT for the forecast periods by plus or minus 5 percent. The highest annual VMT growth forecast presented in the C&P report series was 2.85 percent per year from 1985 to 2000, presented in the 1987 C&P Report; VMT actually grew slightly faster during this period, increasing at an average annual rate of 2.95 percent. The 1989 C&P Report came the closest to projecting future VMT accurately, as the report's forecast of 3.05 trillion VMT in 2007 was within 0.66 percent of actual 2007 VMT of 3.03 trillion; the average annual VMT growth rate forecast in this edition was 2.34 percent, while VMT actually grew by 2.31 percent over this period.

The next two editions of the C&P report (1991 and 1993) are the last for which the 20-year projection period ended by 2012. Both significantly overpredicted future total VMT, by 16 to 21 percent. Both reports projected annual VMT growth of approximately 2.5 percent per year, but actual VMT growth during their forecast periods had fallen to well below 2 percent per year.

Although the 20-year forecast period for the 1995 C&P Report and later editions had not yet concluded by the end of 2012, each appears to be overpredicting future VMT so far, by about 1 to 2 percent per year. States have gradually reduced their projections of future annual VMT growth, but these reductions have not kept pace with recent declines in the actual rates of VMT growth.

Exhibit 9-4 illustrates that States tended to underpredict future VMT when actual VMT was growing rapidly and to overpredict when actual VMT growth stagnated or declined. This observation suggests that many States have been slow to adjust their models to incorporate emerging socioeconomic trends, as they wait to determine whether new data observations represent one-time phenomena or the start of new long-term trends. The downward shift in the VMT forecasts the States provided as part of the 2012 HPMS submittal was significant, as were the further reductions applied in HERS and NBIAS for this 2015 C&P Report to match the predictions of the new FHWA VMT forecasting model. Had the 1.42-percent-per-year VMT growth rate derived from the 2012 HPMS data been used for this report, it would have been the lowest rate assumed during the C&P report series; the further reduction to the 1.04-percent VMT growth rate

Exhibit 9-4 State-Provided Long-Term VMT Forecasts Compared with Actual VMT, 1965–2032[1]

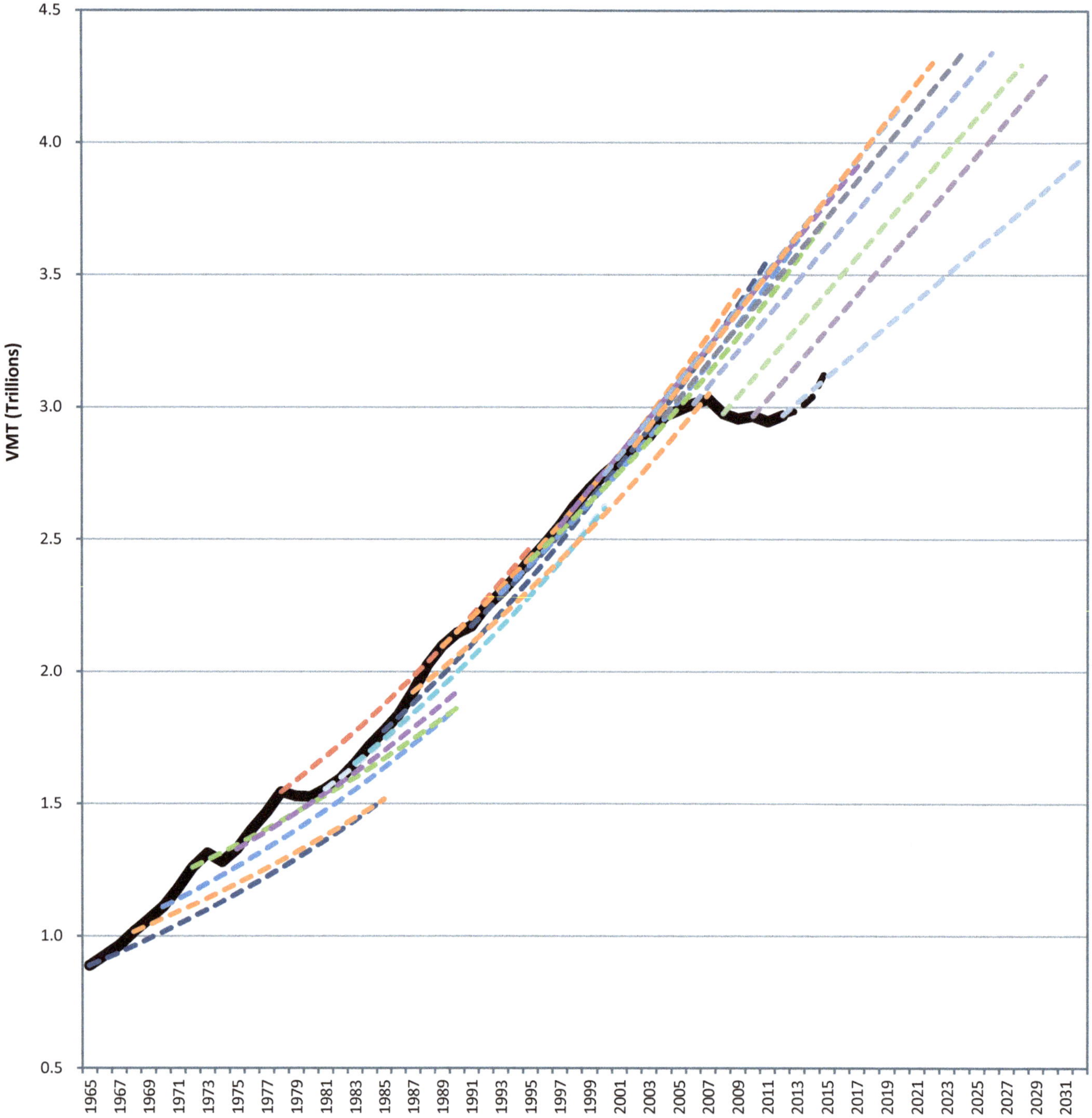

[1] Solid black line represents actual VMT through 2012; dashed black line reflects preliminary estimates through 2015. Other dashed lines represent State-provided long-term VMT forecasts utilized in the C&P report; since 1980, these have been derived from the HPMS.

Source: C&P Report, various years.

assumed in Chapters 7 and 8 represents an even more significant departure from the forecasts in previous reports. Actual VMT growth over the next several years will help inform whether these changes have gone too far or have not gone far enough. The findings presented in *Exhibit 9-4* emphasize that, given the uncertainties involved, considering the implications of the sensitivity analyses presented in Chapter 10 rather than focusing purely on the summary findings presented in Chapter 8 is essential.

Timing of Investment

The investment-performance analyses presented in this report focus mainly on how alternative average annual investment levels over 20 years might impact system performance at the end of this period. Within this period, the timing of investment can significantly influence system performance. The discussion below explores the impacts of three alternative assumptions about the timing of future investment—baseline ramped spending, flat spending, or spending driven by benefit-cost ratio (BCR)—on system performance within the 20-year period analyzed. The average annual investment levels of each scenario analyzed correspond to the baseline HERS analyses for Federal-aid highways and the baseline NBIAS analyses for all bridges presented in Chapter 7.

The baseline ramped spending assumption is consistent with the approach first adopted in the 2008 C&P Report and is discussed in Chapter 7 of the current report. The assumption is that any change from the combined investment level by all levels of government would occur gradually over time and at a constant growth rate. The constant growth rate of the baseline ramped analysis measures future investment in real terms; thus, the distribution of spending among funding periods is driven by the annual growth of spending. To ensure higher overall growth rates for a given amount of total investment, a smaller portion of the 20-year total investment would occur in the earlier years than in the later years.

Some previous editions used different assumptions in the timing of investment. The HERS component in the 2006 C&P Report assumed that combined investment would immediately jump to the average annual level being analyzed, then remain fixed at that level for 20 years. This spending assumption is labeled as flat spending, which is linked directly to the average annual investment levels associated with the baseline analysis. Because spending would stay at the same level in each of the 20 years, the distribution of spending within each 5-year period comprises one-quarter of the total.

The HERS analyses presented in the 2004 C&P Report were tied directly to BCR cutoffs, rather than to particular levels of investment in any given year. This BCR-driven approach resulted in significant front-loading of capital investment in the early years of the analysis, as the existing backlog of potential cost-beneficial investments was first addressed, followed by a sharp decline in later years. This analysis assumed no increase in material and labor costs even though the number of highway construction projects sharply increased.

Alternative Timing of Investment in HERS

This section presents information regarding how the timing of investment would impact the distribution of spending among the four 5-year funding periods considered in HERS, and how these spending patterns could impact performance. Because the timing of investment is varied for any given capital investment level, pavement condition and delay per VMT will change accordingly.

Alternative Investment Patterns

Exhibit 9-5 indicates how alternative assumptions regarding the timing of investment would impact the distribution of spending among the four 5-year funding periods considered in HERS,

Exhibit 9-5 Impact of Investment Timing on HERS Results Reflected in the Improve Conditions and Performance Scenario—Effects on Pavement Roughness and Delay per VMT

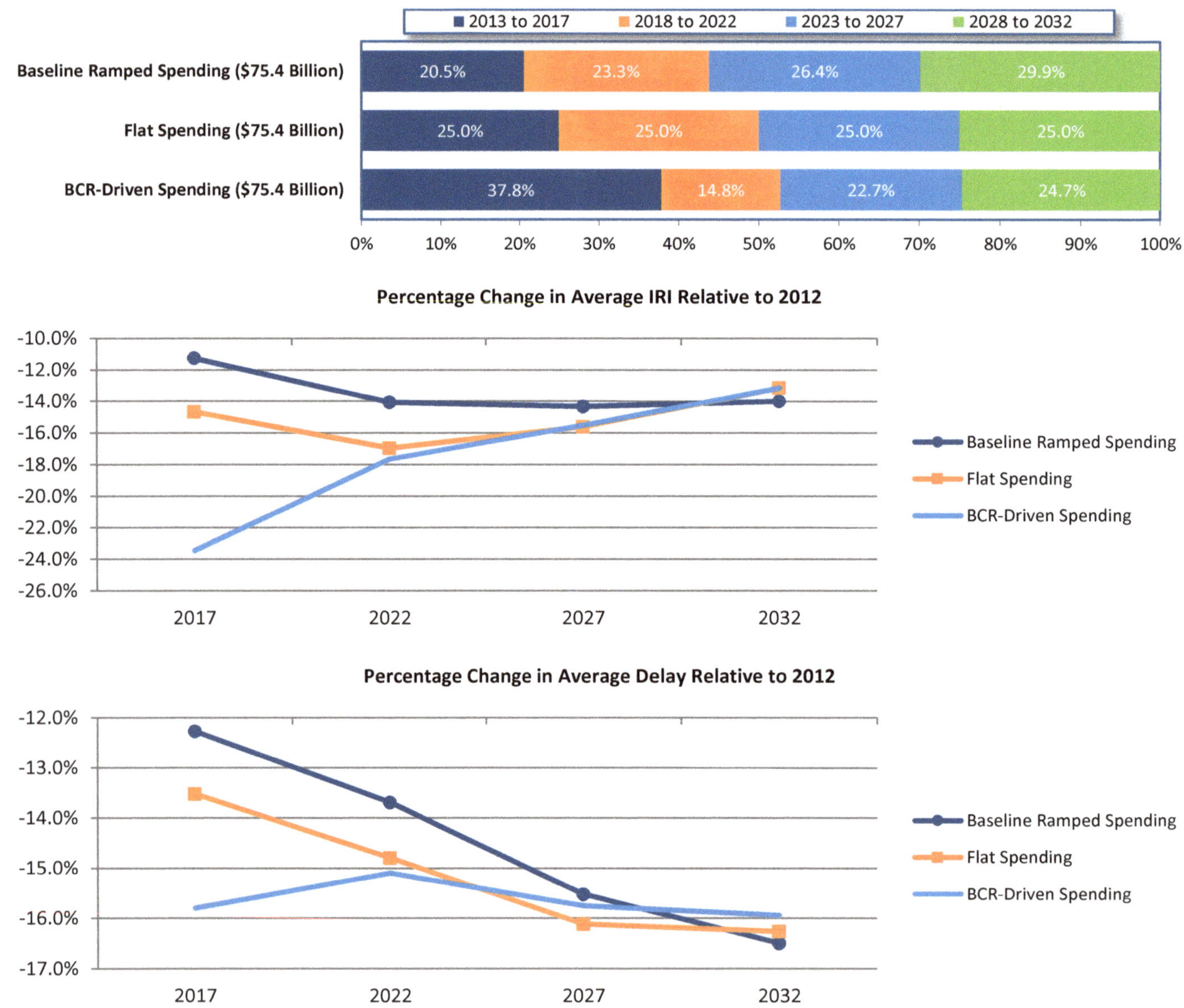

Source: Highway Economic Requirements System.

and how these spending patterns could affect pavement condition (measured using the International Roughness Index [IRI]) and average delay per VMT. The investment levels were selected from the baseline HERS analyses for Federal-aid highways presented in Chapter 7 to compare across the three investment patterns: baseline ramped spending, flat spending, and BCR-driven spending. The average annual investment requirement is kept constant in all three cases to compare the impact of different investment patterns, even when the total amount of spending is identical. The average annual investment requirement is set at $75.4 billion, which is the HERS-derived input to the Improve Conditions and Performance scenario in Chapter 8.

As shown in the top panel of *Exhibit 9-5*, the level of investment grows over time in the baseline ramped spending case assuming a constant growth of real investment. Under this scenario, annual investment would grow by 2.53 percent per year from $57.4 billion in 2013 to $94.6 billion in 2032, which totals $1,508 billion over 20 years or $75.4 billion per year in constant 2012 dollars. Only 20.5 percent of the total 20-year investment occurs in the first 5-year period, 2013 to 2017, while 29.9 percent of total investment occurs in the last 5-year period, 2028 to 2032. Under the flat spending alternative, investment is equally distributed over time so that each 5-year period accounts for exactly one-quarter of the total 20-year investment.

The HERS-modeled and BCR-driven spending alternative displays a different investment pattern. A high proportion of total spending, 37.8 percent of total investment, would occur in the first 5-year period to address the large backlog of cost-beneficial investment the system is facing now (see Backlog discussion in Chapter 8). Under this alternative, investment needs in the second 5-year period would drop to 14.8 percent of the total 20-year need. Investment needs would increase in the last two 5-year periods because many roadways that were rehabilitated in the first 5-year period would need to be resurfaced or reconstructed again.

Impacts of Alternative Investment Patterns

An obvious difference among the three alternative investment patterns is that the higher the level of investment within the first 5-year analysis period, the better the level of performance achieved by 2017.

The middle panel of *Exhibit 9-5* presents percentage changes of average pavement roughness as measured by IRI compared with the 2012 level under the three investment cases. A reduction in average IRI represents improvement in pavement conditions. The graph shows that the BCR-driven spending case yields the greatest improvement in pavement conditions in the first 5-year period, represented by a large drop in average IRI by more than 20 percent from its 2012 level. The improvement under the BCR-driven spending alternative shrinks to about 13 percent by the last 5-year period. Steady pavement improvement over time is achieved in baseline ramped spending and flat spending assumptions. In the first 5 years, average IRI decreases by approximately 15 percent (relative to the 2012 level) under the flat spending case and the descending trend continues across the rest of the analysis periods. The baseline ramped spending assumption leads to an 11-percent drop in average IRI in the first 5-year period and further improvement in pavement afterward, but the improvement is not as pronounced as for the flat spending alternative. The decreases of average IRI are similar by 2032 under all three cases.

The bottom panel of *Exhibit 9-5* illustrates the progress in average delay reduction across three investment cases. The percentage change of average delay, relative to its 2012 level, remains negative, indicating a decrease in average delay of travelers. In the first 5 years, the BCR-driven spending approach results in the largest reduction in average delay per VMT, 16 percent, and the baseline ramped spending the smallest reduction, 12 percent. The percentages of delay reduction grow over time under the baseline ramped and flat spending cases, suggesting sustained benefits through capital investment to improve pavement. The percentage change of average delay is stable under BCR-driven spending. By the end of the 20-year analysis period, the difference between projected average delay and the 2012 delay will be approximately 16 percent under all three alternatives.

These results show that the BCR-driven approach achieves the highest IRI and delay reduction in the medium run (the first 5-year period). The baseline ramped spending approach results in the smallest pavement and delay improvement over the same period. System performance, however, does not differ substantially across investment timing in the long run of 20 years. Based on this analysis, the key advantage to front-loading highway investment is not in reducing 20-year total investment needs; instead, the strength of BCR-driven spending lies in the years of additional benefits that highway users would accrue over time if system conditions and performance were improved earlier in the 20-year analysis period.

Alternative Timing of Investment in NBIAS

Exhibit 9-6 identifies the impacts of alternative investment timing on the share of bridges that are structurally deficient by deck area using the three investment assumptions described above: baseline ramped spending, flat spending, and BCR-driven spending. The average annual investment level of each alternative analyzed ($24.6 billion) corresponds to NBIAS-derived input to the Improve Conditions and Performance scenario presented in Chapter 8.

Similar to the results of pavement investment in HERS presented earlier, investment timing has an impact on structurally deficient bridges. The baseline ramped case for the NBIAS Improve Conditions and Performance scenario assumes constant annual spending growth of 3.72 percent from its 2012 level, with a total 20-year investment of $492.1 billion and an average annual investment of $24.6 billion in constant 2012 dollars. The top panel of *Exhibit 9-6* indicates that more investment occurs in the later years under the baseline ramped case of gradual and constant growth—about 32.2 percent in the last 5-year period. The BCR-driven spending case requires a large portion of the total 20-year investment in the first 5-year period (39.3 percent), then declines to 17.8 percent in the last 5-year period. Spending levels remain constant in the flat spending case.

A different investment pattern produces substantially different outcomes. The middle panel of *Exhibit 9-6* shows that the greatest bridge improvement in the first 5-year period occurs under the BCR-driven spending assumption, as the share of structurally deficient bridges by deck area drops from 7.7 percent in 2012 to 3.9 percent in 2017. During the same period, the share of structurally deficient bridges decreases to 6.4 percent under the flat spending assumption but increases to 9 percent under the baseline ramped spending assumption. In the next 15 years, however, this pattern is reversed. At an average annual investment level of $24.6 billion, NBIAS projects that the lowest share of structurally deficient bridges in 2032 would be achieved under the baseline ramped spending approach with 1.9 percent of bridges that are structurally deficient, compared to 3.4 percent assuming flat spending and 4.1 percent for the BCR-driven spending alternative.

The economic bridge investment backlog also exhibits different trends under the alternative investment timing. The lower panel of *Exhibit 9-6* indicates that, from 2012 to 2017, the average backlog declines sharply under the BCR-driven alternative, with slower declines under the flat spending alternative and baseline ramped spending. The rate of decline is determined by the

Exhibit 9-6 Impact of Investment Timing on NBIAS Results Reflected in the Improve Conditions and Performance Scenario—Effects on Structurally Deficient Bridges and Economic Bridge Investment Backlog

Source: National Bridge Investment Analysis System.

investment timing. High bridge investment in later years under baseline ramped spending leads to the elimination of economic backlog by 2032, while the projected backlog will be $18.5 billion and $25 billion under the flat spending and BCR-driven spending assumptions, respectively.

The indicators suggest that continued baseline ramped spending for bridges would yield the best outcomes by 2032. The share of structurally deficient bridges and investment backlog start to increase in the last 5-year period under the flat spending and BCR-driven spending assumptions, however, highlighting a potential surge of investment needs by the end of the analysis period under these spending assumptions.

Accounting for Inflation

The analysis of potential future investment/performance relationships in the C&P report has traditionally stated future investment levels in constant dollars, with the base year set according to the year of the conditions and performance data supporting the analysis. Throughout Chapters 7 and 8, this edition of the C&P report has stated all investment levels in constant 2012 dollars. For some purposes, however, such as comparing investment spending in a particular scenario with nominal dollar revenue projections, adjusting for inflation to present spending in nominal dollar terms might be desirable. Given an assumption about future inflation, the C&P report's constant-dollar numbers could be converted to nominal dollars or the nominal projected revenues could be converted to constant 2012 dollars for comparison purposes. *Exhibit 9-7* takes the former approach by converting constant-dollar values to nominal dollars.

The average annual increase in highway construction costs over the past 20 years (1992 to 2012) was 2.9 percent. Since the creation of the Federal Highway Trust Fund in 1956, the 20-year period with the smallest increase in construction costs was 1980 to 2000, when costs grew by 2.0 percent per year. (Historic inflation rates were determined using the FHWA Composite Bid Price Index through 2006, and the new FHWA National Highway Construction Cost Index from 2006 to 2012; these indices are discussed in Chapter 6.) *Exhibit 9-7* illustrates how the constant-dollar figures associated with three scenarios for highways and bridges presented in Chapter 8 could be converted to nominal dollars based on two alternative annual inflation rates of 2.0 percent (historically lowest rate) and 2.9 percent (past 20 years' rate).

The systemwide Sustain 2012 Spending scenario presented in Chapter 8 assumes that combined capital spending for highway and bridge improvements would be sustained at its 2012 level in constant-dollar terms for 20 years. Thus, the first column in *Exhibit 9-7* shows $105.2 billion of spending in constant 2012 dollars for each year from 2013 to 2032, for a 20-year total of $2.1 trillion. Applying annual inflation in construction costs at 2.0 percent or 2.9 percent would imply a 20-year total in nominal dollars of $2.6 trillion or $2.9 trillion, respectively.

Exhibit 9-7 Illustration of Potential Impact of Alternative Inflation Rates on Selected Systemwide Investment Scenarios

	Highway Capital Investment (Billions of Dollars)								
	Constant 2012 Dollars[1]			Nominal Dollars (Assuming 2.0 Percent Annual Inflation)			Nominal Dollars (Assuming 2.9 Percent Annual Inflation)		
Year	Sustain 2012 Spending Scenario	Maintain Conditions & Performance Scenario	Improve Conditions & Performance Scenario	Sustain 2012 Spending Scenario	Maintain Conditions & Performance Scenario	Improve Conditions & Performance Scenario	Sustain 2012 Spending Scenario	Maintain Conditions & Performance Scenario	Improve Conditions & Performance Scenario
2012	$105.2	$105.2	$105.2	$105.2	$105.2	$105.2	$105.2	$105.2	$105.2
2013	$105.2	$103.6	$108.2	$107.3	$105.7	$110.3	$108.2	$106.6	$111.3
2014	$105.2	$102.0	$111.2	$109.4	$106.1	$115.7	$111.4	$108.0	$117.7
2015	$105.2	$100.5	$114.3	$111.6	$106.6	$121.3	$114.6	$109.5	$124.6
2016	$105.2	$98.9	$117.5	$113.9	$107.1	$127.2	$117.9	$110.9	$131.8
2017	$105.2	$97.4	$120.8	$116.1	$107.6	$133.4	$121.4	$112.4	$139.4
2018	$105.2	$95.9	$124.2	$118.5	$108.0	$139.9	$124.9	$113.9	$147.5
2019	$105.2	$94.5	$127.7	$120.8	$108.5	$146.7	$128.5	$115.4	$156.0
2020	$105.2	$93.0	$131.3	$123.3	$109.0	$153.9	$132.2	$116.9	$165.1
2021	$105.2	$91.6	$135.0	$125.7	$109.5	$161.3	$136.1	$118.5	$174.6
2022	$105.2	$90.2	$138.8	$128.2	$110.0	$169.2	$140.0	$120.1	$184.7
2023	$105.2	$88.8	$142.7	$130.8	$110.5	$177.4	$144.1	$121.7	$195.4
2024	$105.2	$87.5	$146.7	$133.4	$111.0	$186.1	$148.3	$123.3	$206.8
2025	$105.2	$86.2	$150.8	$136.1	$111.4	$195.1	$152.5	$124.9	$218.7
2026	$105.2	$84.8	$155.1	$138.8	$111.9	$204.6	$157.0	$126.6	$231.4
2027	$105.2	$83.5	$159.4	$141.6	$112.4	$214.6	$161.5	$128.3	$244.8
2028	$105.2	$82.3	$163.9	$144.4	$112.9	$225.0	$166.2	$130.0	$259.0
2029	$105.2	$81.0	$168.5	$147.3	$113.4	$236.0	$171.0	$131.7	$274.0
2030	$105.2	$79.8	$173.3	$150.2	$113.9	$247.5	$176.0	$133.5	$289.9
2031	$105.2	$78.6	$178.1	$153.3	$114.5	$259.5	$181.1	$135.2	$306.6
2032	$105.2	$77.4	$183.1	$156.3	$115.0	$272.1	$186.3	$137.0	$324.4
Total	**$2,104.0**	**$1,797.5**	**$2,850.9**	**$2,607.2**	**$2,205.0**	**$3,597.0**	**$2,879.3**	**$2,424.3**	**$4,003.7**
	0.00%	**-1.52%**	**2.81%**	**Constant-Dollar Growth Rate**					
	$105.2	**$89.9**	**$142.5**	**Average Annual Investment Level in Constant 2012 Dollars**					

[1] Based on average annual investment levels and annual constant-dollar growth rates identified in *Exhibit 8-2*.

Source: FHWA staff analysis.

Chapter 8 indicates that achieving the objectives of the systemwide Maintain Conditions and Performance scenario would require investment averaging $89.9 billion per year in constant 2012 dollars. The investment totals $1.8 billion over 20 years (2013 to 2032) in constant-dollar spending, equivalent to spending that steadily decreases at 1.52 percent per year. *Exhibit 9-7* illustrates the application of this real reduction rate, demonstrating how annual capital investment in constant-dollar terms would decrease from $105.2 billion in 2012 to $77.4 billion in 2032. A 2.0-percent inflation rate applied to these constant-dollar estimates would produce a 20-year total cost of $2.2 trillion in nominal dollars, while a 2.9-percent inflation rate results in a total cost of $2.4 trillion.

Q&A **Why are the investment analyses presented in this report expressed in constant base-year dollars?**

The investment/performance models discussed in this report estimate the future benefits and costs of transportation investments in constant-dollar terms. This practice is standard for this type of economic analysis. Converting the model outputs from constant dollars to nominal dollars is necessary to adjust them to account for projected future inflation.

Traditionally, this type of adjustment has not been made in the C&P report. Because inflation prediction is an inexact science, adjusting the constant-dollar figures to nominal dollars tends to add to the uncertainty of the overall results and make the report more difficult to use if the inflation assumptions are inaccurate. Allowing readers to make their own inflation adjustments based on actual trends observed after publication of the C&P report or based on the most recent projections from other sources is expected to yield a better overall result, particularly given the sharp swings in recent years in material costs for highway construction.

The use of constant-dollar figures also is intended to provide readers with a reasonable frame of reference in terms of an overall cost level that they have recently experienced. When inflation rates are compounded for 20 years, even relatively small growth rates can produce nominal dollar values that appear very large when viewed from the perspective of today's typical costs.

The compounding impacts of inflation are even more evident for the systemwide Improve Conditions and Performance scenario. As described in Chapter 8, this scenario assumes 2.81-percent growth in constant-dollar highway capital spending per year to address all potentially cost-beneficial highway and bridge improvements by 2032. The 20-year total investment level of $2.9 trillion associated with this scenario equates to an average annual investment level of $142.5 billion in constant 2012 dollars. Adjusting this figure to account for inflation of 2.0 percent or 2.9 percent would result in 20-year total nominal dollar costs of $3.6 trillion or $4.0 trillion, respectively.

Transit Supplemental Scenario Analysis

This section provides a more detailed discussion of the assumptions underlying the scenarios presented in Chapters 7 and 8 and of the real-world issues that affect transit operators' ability to address their outstanding capital needs. Specifically, this section discusses the following topics:

- asset condition forecasts under three scenarios: (1) Sustain 2012 Spending, (2) Low-Growth, and (3) High-Growth; in addition, the analysis includes a discussion of the State of Good Repair Benchmark;
- a comparison of recent historic passenger miles traveled (PMT) growth rates with the revised low-growth and high-growth projections;
- an assessment of the impact on the backlog estimate of purchasing hybrid vehicles; and
- the forecast of purchased transit vehicles, route miles, and stations under the Low- and High-Growth scenarios.

Asset Condition Forecasts and Expected Useful Service Life Consumed for All Transit Assets under Three Scenarios and the SGR Benchmark

As in the 2013 edition, this edition of the C&P report uses three condition projection scenarios (i.e., Sustain 2012 Spending, Low-Growth, and High-Growth scenarios) and the State of Good Repair (SGR) Benchmark to understand better which condition outcome is desirable or even sensible. For example, are current asset conditions at an acceptable level or are they too low (or too high) for individual asset types?

To help answer this question, *Exhibit 9-8* presents the condition projections for each of the three scenarios and the SGR Benchmark. Note that these projections predict the condition of all transit assets in service during each year of the 20-year analysis period, including transit assets that exist today and any investments in expansion assets by these scenarios. The Sustain 2012 Spending, Low-Growth, and High-Growth scenarios each make investments in expansion assets while the SGR Benchmark reinvests only in existing assets. Note that the estimated current average condition of the Nation's transit assets is 3.45. As discussed in Chapter 8, expenditures under the financially constrained Sustain 2012 Spending scenario are not sufficient to address replacement needs as they arise, leading to a predicted increase in the investment backlog. This increasing backlog is a key driver in the decline in average condition of transit assets, as shown for this scenario in *Exhibit 9-8*.

Exhibit 9-8 Asset Condition Forecast for All Existing and Expansion Transit Assets

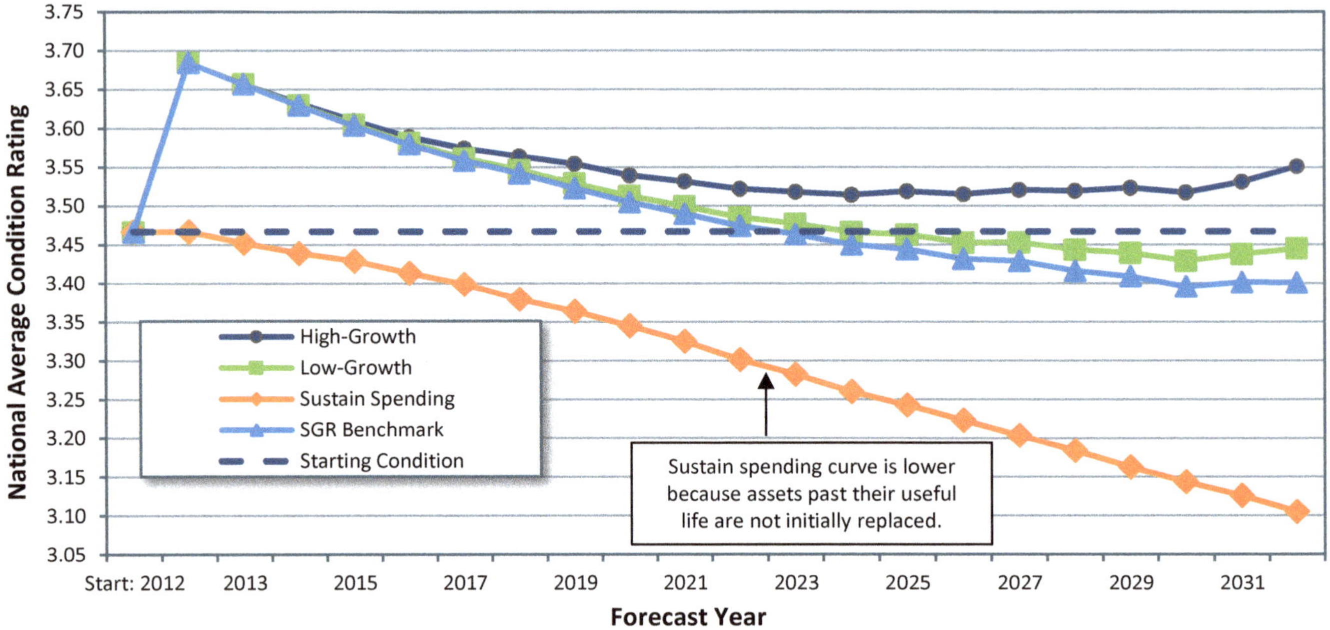

Source: Transit Economic Requirements Model.

If the SGR Benchmark represents a reasonable long-term investment strategy (i.e., replacing assets close to the end of their useful life, which results in a long-term decline in average conditions), investing under the Sustain 2012 Spending scenario implies an investment strategy of replacing assets at later ages, in worse conditions, and potentially after the end of their useful life, as shown in *Exhibit 9-9*. Expenditures on asset reinvestment for the Sustain 2012 Spending scenario are insufficient to address ongoing reinvestment needs, leading to an increase in the size of the backlog. Note that the forecast for 2032 for the Sustain 2012 Spending scenario shown in *Exhibit 9-9* indicates that assets under this scenario will be closer to or beyond the end of their useful lives, when compared with the other scenarios; this difference reflects a larger portion of the national transit assets still in use after the end of their useful lives.

Exhibit 9-9 Sustain 2012 Spending Scenario: Asset Percent of Useful Life Consumed

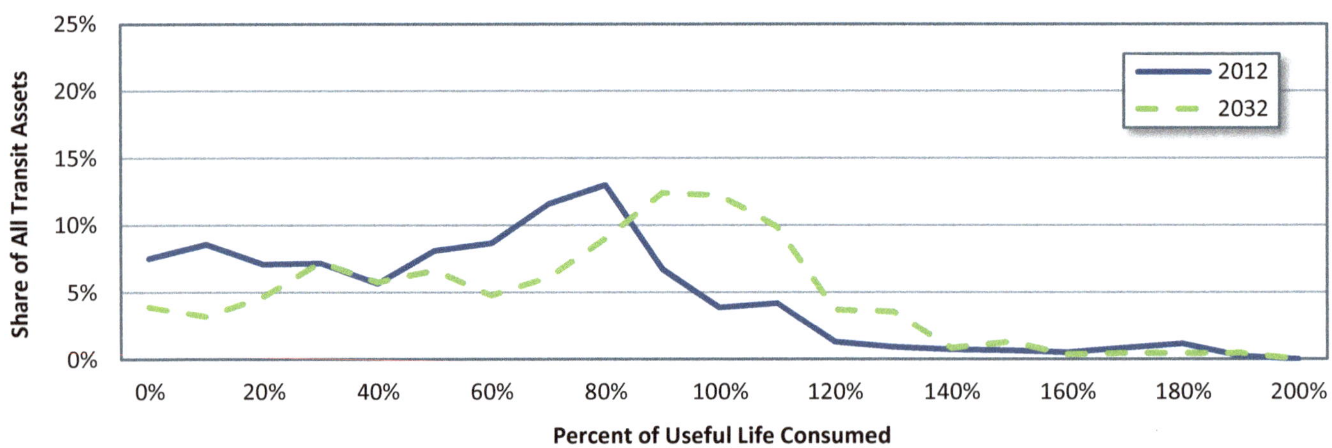

Source: Transit Economic Requirements Model.

In contrast to the Sustain 2012 Spending scenario, the SGR Benchmark is financially (and economically)_unconstrained and considers the level of investment required to both eliminate the current investment backlog and to address all ongoing reinvestment needs as they arise such that all assets remain in an SGR (i.e., a condition of 2.5 or higher). Despite adopting the objective of maintaining all assets in an SGR throughout the forecast period, average conditions under the SGR Benchmark ultimately decline to levels below the current average condition value of 3.45.

This result, although counterintuitive, is explained by a high proportion of long-lived assets (e.g., guideway structures, facilities, and stations) that currently have high average condition ratings and a significant amount of useful life remaining, as shown in *Exhibit 9-10*. The exhibit shows the share of all transit assets (equal to approximately $804 billion in 2012) as a function of useful life consumed. Eliminating the current SGR backlog removes a significant number of over-age assets from service (resulting in an initial jump in asset conditions). The ongoing aging of the longer-lived assets, however, ultimately will draw the average asset conditions down to a long-term condition level that is consistent with the objective of SGR (and hence sustainable) but ultimately measurably below current average aggregate conditions.

Exhibit 9-10 SGR Benchmark: Asset Percent of Useful Life Consumed

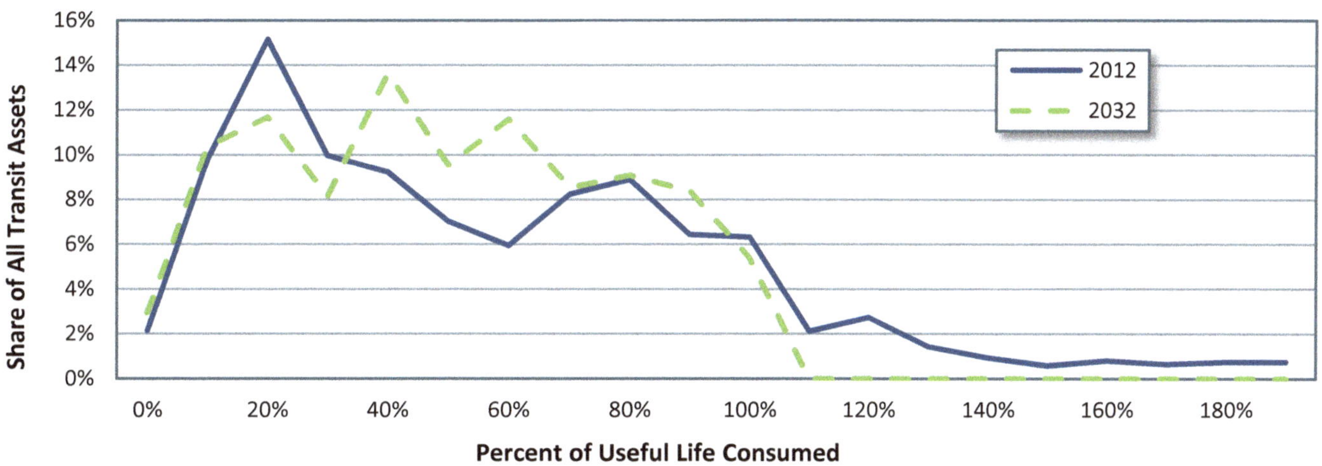

Source: Transit Economic Requirements Model.

To underscore these findings, note that the Low- and High-Growth scenarios include unconstrained investments in both asset replacements and asset expansions. Hence, not only are older assets replaced as needed with an aggressive reinvestment rate, but also new expansion assets are also continually added to support ongoing growth in travel demand. Although initially insufficient to arrest the decline in average conditions completely, the impact of these expansion investments ultimately would reverse the downward decline in average asset conditions in the final years of the 20-year projections. A higher proportion of long-lived assets with more useful life remaining in 2032 than in 2012 also would result, as illustrated in *Exhibit 9-11* and *Exhibit 9-12*, respectively. Furthermore, the High-Growth scenario (*Exhibit 9-12*) adds newer expansion assets at a higher rate than does the Low-Growth scenario (*Exhibit 9-11*), ultimately yielding higher average condition values for that scenario (and average condition values that exceed the current average of 3.45 throughout the entire forecast period).

Exhibit 9-11 Low-Growth Scenario: Asset Percent of Useful Life Consumed

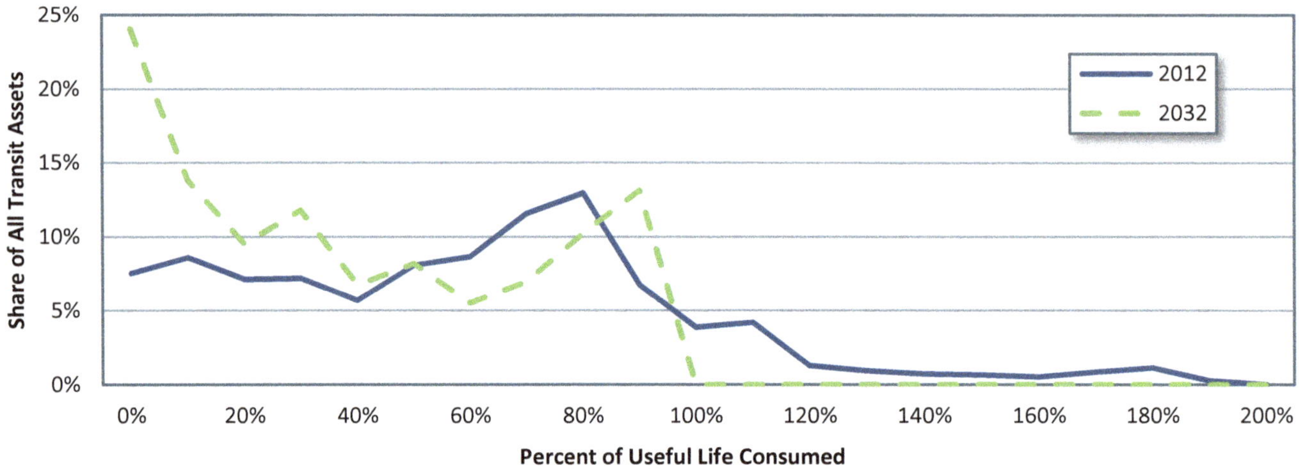

Source: Transit Economic Requirements Model.

Exhibit 9-12 High-Growth Scenario: Asset Percent of Useful Life Consumed

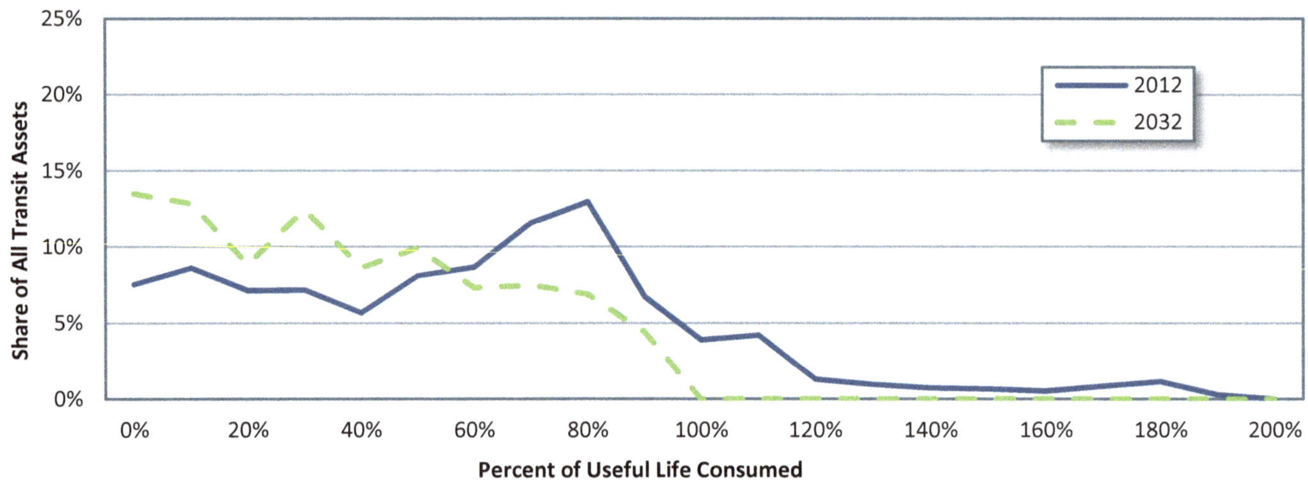

Source: Transit Economic Requirements Model.

Alternative Methodology

When current transit investment practices are considered, the level of investment needed to eliminate the SGR backlog in 1 year is infeasible. Thus, the SGR Benchmark, Low-Growth, and High-Growth scenarios' financially unconstrained assumptions (e.g., spending of unlimited transit investment funds each year) are unrealistic. As indicated in *Exhibit 9-8*, the elimination of the backlog in the first year and the resulting jump in asset conditions in year 1 can be attributed to this unconstrained assumption.

An alternative, more-feasible methodology is to have the Low- and High-Growth scenarios, and the SGR Benchmark, use a financially constrained reinvestment rate to eliminate the SGR backlog by year 20 while maintaining the collective national transit assets at a condition rating of 2.5 or higher. Analysis has determined that investing $16.6 billion annually would eliminate the backlog in 20 years.

Exhibit 9-13 presents the condition projections for the two scenarios and the benchmark using this alternative methodology. The Low- and High-Growth scenarios and SGR Benchmark are financially constrained so the investment strategies result in replacing assets at later ages, in worse conditions, and potentially after the end of their useful lives.

Exhibit 9-13 Asset Condition Forecast for All Existing and Expansion Transit Assets

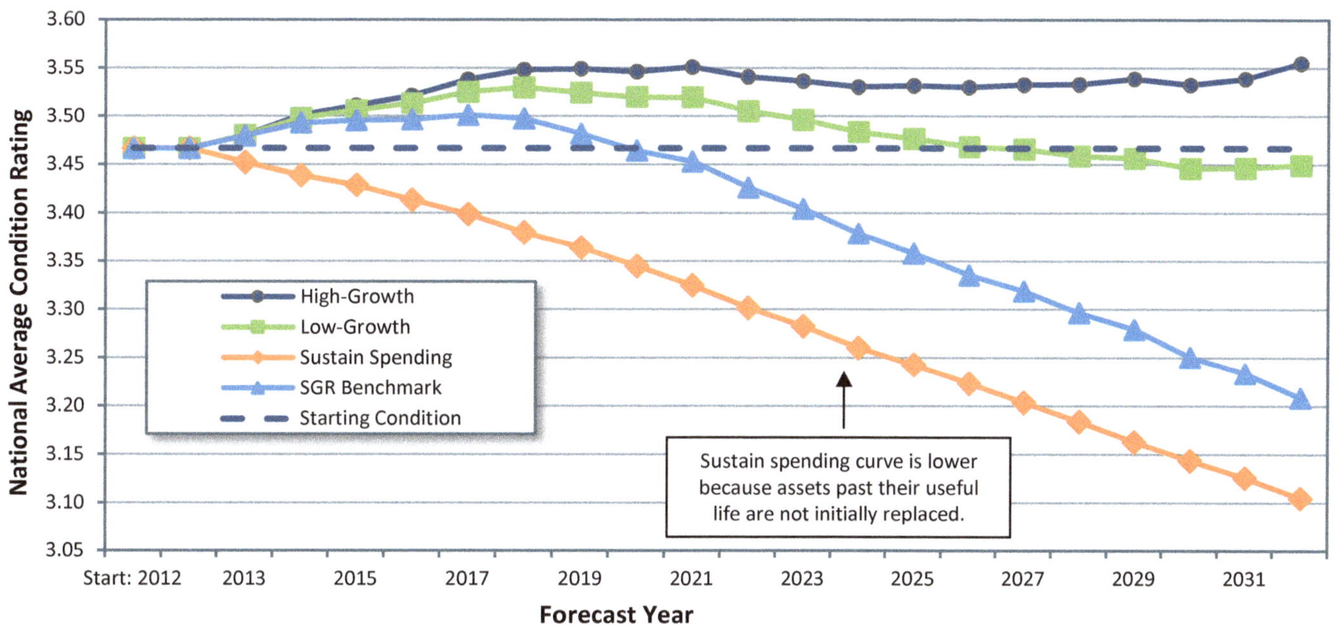

Source: Transit Economic Requirements Model.

Revised Method for Estimating PMT Growth Rates

The Low-and High-Growth scenarios presented in Chapters 7 and 8 estimate the level of investment in expansion assets (e.g., additional passenger vehicles, maintenance facilities, stations, and miles of track) required to support growth in ridership over the upcoming 20-year period. Specifically, these scenarios are designed to expand investment in passenger vehicles and related assets at the same rate of growth as the projected rate of increase in PMT, ensuring that the ratio of riders to transit assets (e.g., riders per vehicle) remains at today's levels throughout the entire forecast period.[i] Here, the Low-Growth scenario provides the level of expansion investment as required to support relatively low PMT growth (lower bound). In contrast, the High-Growth scenario supports an upper-bound estimate of future PMT growth. The actual rate of increase in PMT and related expansion needs are expected to fall somewhere between these bounds.

Change in Methodology: For this report, FTA has applied a new methodology for estimating PMT growth for use in the Low- and High-Growth scenarios. This revised approach is believed to be more accurate and it provides greater consistency between the Low- and High-Growth scenarios compared with the prior approach. Following is an explanation of the methodology used to estimate growth in PMT in previous C&P reports and the new methodology used for this report.

Prior Methodology: Prior to this report, the Low- and High-Growth scenarios used different (and hence inconsistent) approaches to project the rate of increase in PMT (*Exhibit 9-14*). Note that in previous years, PMT growth was modeled at the urbanized area (UZA) level (i.e., the same growth rate would be applied to all agencies and modes within a given UZA).

Low-Growth: For prior-year C&P reports, PMT growth rates for the Low-Growth scenario were obtained from a sample of the Nation's metropolitan planning organizations (MPOs). Specifically, this sample included the MPOs representing all of the Nation's 30 largest UZAs and a sample of 30 or more MPOs from small and mid-sized UZAs. MPOs prepare ridership and PMT projections using detailed ridership models and use the results for urban planning purposes. Note that MPO ridership forecasts are financially constrained and, for this reason, the MPO projections were used for the Low-Growth scenario.

High-Growth: For prior-year C&P reports, PMT growth for the High-Growth scenario was based on the weighted average trend rate of growth for each UZA. These UZA-specific weighted-average growth rates were calculated across all agencies and modes within a given UZA, using the most recent 15 years of historical National Transit Database data. Hence, under this approach, all agencies and modes within a UZA were projected to have the same trend rate of PMT growth.

Revised Methodology (2015 C&P Report): For this report, the PMT growth rates used in the Low- and High-Growth scenarios are calculated using a common approach (*Exhibit 9-14*). Specifically, both scenarios are based on the trend rate of PMT growth for all riders using the same mode, within UZAs of similar size (large, medium, or small) and within the same FTA region. This approach also used 15 years of data from the National Transit Database to establish a trend rate of increase. Finally, the Low-Growth scenario used the trend rate of increase minus 0.5 percent while the High-Growth scenario used the trend rate of increase plus 0.5 percent.

Exhibit 9-14 Prior and Revised Passenger Miles Traveled (PMT) Projection Methodology

<table>
<tr><th></th><th>Level of PMT Projected</th><th>Low-Growth Scenario</th><th>High-Growth Scenario</th></tr>
<tr><td>Prior Reports</td><td>■ Average PMT growth by UZA
■ Same rate applied equally across all agencies and modes within a given UZA</td><td>■ Projected PMT growth by UZA obtained from sample of MPOs; sample includes:
■ 30 largest UZAs
■ 30 or more small and medium UZAs</td><td>■ Weighted average trend PMT growth by UZA
■ Based on NTD data</td></tr>
<tr><td>2013 C&P Report</td><td>■ Average PMT growth stratified by:
■ mode
■ UZA stratum (large, medium, small)
■ FTA region
■ Same rate applied to all agencies of same mode in UZAs of similar size and within same FTA region</td><td colspan="2">■ Low and high growth scenarios both based on same 15-year trend PMT growth segmented by:
■ mode
■ UZA stratum (large, medium, small)
■ FTA region
■ The trend growth rate is adjusted by ± 0.5% to obtain growth rates for the low and high growth scenarios:
■ low-growth rate = trend rate - 0.5%
■ high-growth rate = trend rate + 0.5%</td></tr>
</table>

Impact of Change: A key benefit of this revised approach is the recognition that the PMT growth rates can be and have been significantly different by mode (*Exhibit 9-15*). For example, over the most recent 15-year period, PMT for motor bus has tended to be flat while the rate of increase for heavy rail, demand response, and vanpool has been high. The result is decreased bus expansion needs and increased needs for heavy rail and demand-response expansion as compared to prior-year reports. In addition, the revised approach also recognizes that growth rates can differ significantly by urban area size and geographic region.

Exhibit 9-15 Passenger Miles Traveled: 15-Year Compound Annual Growth Rate by Mode, 1997-2012[1]

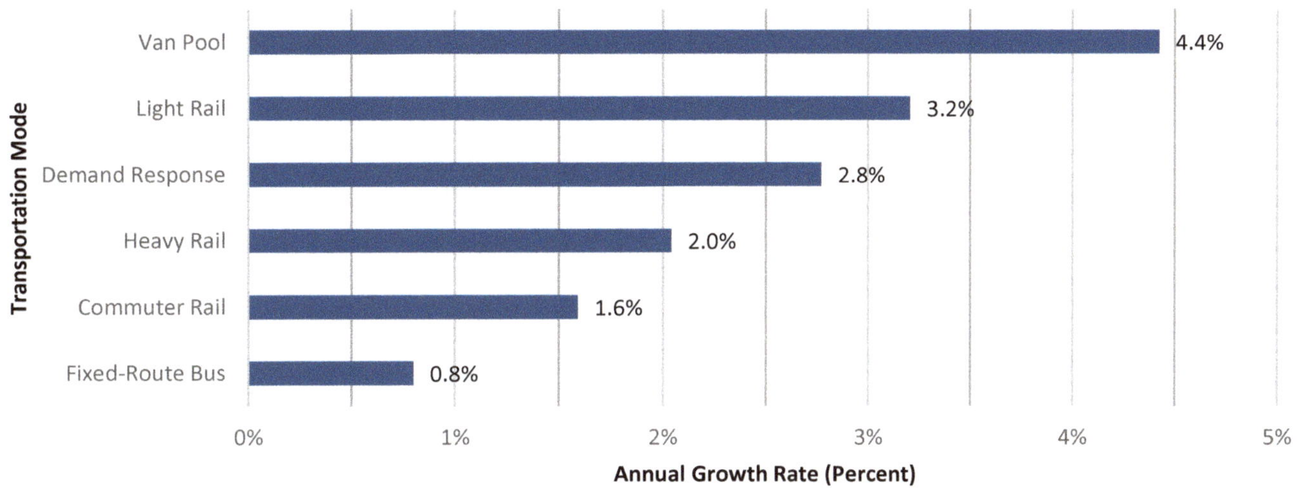

[1] *Adjusted to remove outliers.*

Source: Transit Economic Requirements Model.

Impact of New Technologies on Transit Investment Needs

The investment needs scenarios presented in Chapter 8 implicitly assume that all replacement and expansion assets will use the same technologies as are currently in use today (i.e., all asset replacement and expansion investments are "in kind"). As with most other industries, however, the existing stock of assets used to support transit service is subject to ongoing technological change and improvement, and this change tends to result in increased investment costs (including future replacement needs). Although many improvements are standardized and hence embedded in the asset (i.e., the transit operator has little or no control over this change), numerous instances occur where transit operators have intentionally selected technology options that are significantly more costly than preexisting assets of the same type. A key example is the frequent decision to replace diesel motor buses with compressed natural gas or hybrid buses. Although such options offer clear environmental benefits (and compressed natural gas might decrease operating costs), acquisition costs for these vehicle types are 20 to 60 percent higher than diesel. This increase in cost generally increases current and long-term reinvestment needs and, in a budget-constrained environment, increases the expected future size of the investment backlog. This increase might be offset by lower operating costs from more reliable operation, longer useful lives, and improved fuel efficiency, but this possible offset is not captured in this assessment of capital needs. Again, the effect of technology-driven increases in needs is not included in the needs estimates presented in Chapters 7 and 8 of this report.

In addition to improvements in preexisting asset types, transit operators periodically expand their existing asset stock to introduce new asset types that take advantage of technological innovations. Examples include investments in intelligent transportation system technologies such as real-time passenger information systems and automated dispatch systems—assets and technologies that are common today but were not available 15 to 20 years ago. These improvements typically yield improvements in service quality and efficiency, but they also tend to yield increases in asset acquisition, maintenance, and replacement costs, resulting in an overall increase in reinvestment costs and the expected future size of the SGR backlog.

Impact of Compressed Natural Gas and Hybrid Buses on Future Needs

To provide a better sense of the impact of new technology adoption on long-term needs, the analysis below presents estimates of the long-term cost of the shift from diesel to compressed natural gas and hybrid buses. Important to emphasize is that this analysis is intended to provide only a sense of the significance of this impact on long-term capital needs (including the possible consequences of not capturing this impact in TERM's needs estimates). This assessment is not one of the full range of operational, environmental, or other potential costs and benefits arising from this shift and, hence, it does not evaluate the decision to invest in any specific technology.

Exhibit 9-16 presents historical (2000–2012) and forecast (2013–2030) estimates of the share of transit buses that rely on compressed natural gas and other alternative-fuel vehicles and on hybrid power sources. The forecast estimates assume the current trend rate of increase in alternative and hybrid vehicle shares as observed from 2007 to 2012. Based on this projection, the share of vehicles powered by alternative fuels is estimated to increase from 26 percent in 2012 to 50 percent in 2030. During the same period, the share of hybrid buses is estimated to increase from 6 percent to 21 percent. This results in diesel shares declining from roughly 71 percent today to about 29 percent by 2030.

Exhibit 9-16 Hybrid and Alternative Fuel Vehicles: Share of Total Bus Fleet, 2000–2030

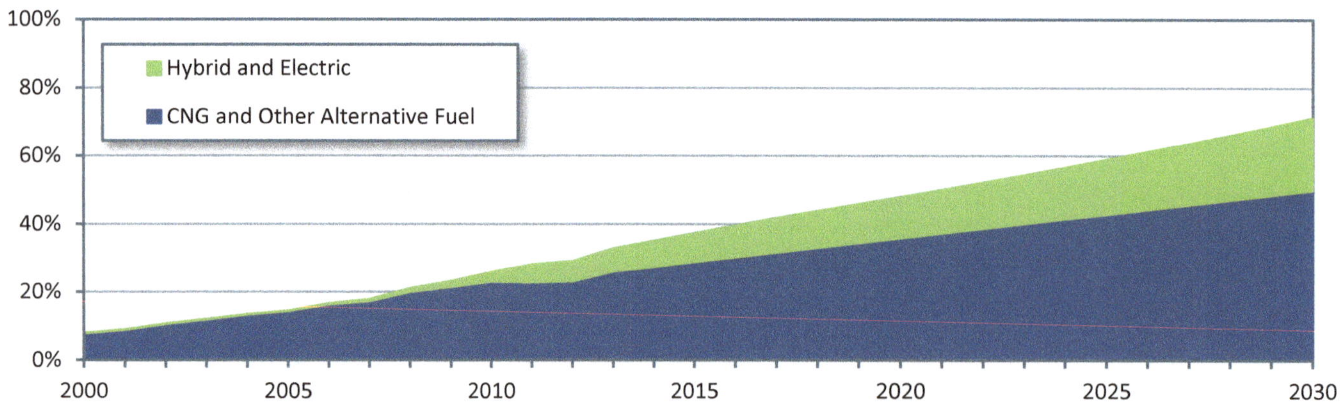

Source: Transit Economic Requirements Model.

Impact on Costs

According to a 2007 report by FTA, *Transit Bus Life Cycle Cost and Year 2007 Emissions Estimation*, the average unit cost of an alternative-fuel bus plus its share of cost for the required fueling station is 15.5 percent higher than that of a standard diesel bus of the same size. Similarly, hybrid buses cost roughly 65.9 percent more than standard diesel buses of the same size. When combined with the current and projected mix of bus vehicle types presented above in *Exhibit 9-16*, these cost assumptions yield an estimated increase in average capital costs for bus vehicles of 14.3 percent from 2012 to 2030 (using the mix of bus types from 2012 as the base of comparison). (Note that this cost increase represents a shift in the mix of bus types purchased and not the impact of underlying inflation, which will affect all vehicle types, including diesel, alternative fuels, and hybrid.) Reductions in operating costs due to the new technology are not shown in this analysis of capital needs but are presumably part of the motivation for agencies that purchase these vehicles.

Impact on Needs

What, then, is the impact of this cost increase on long-term transit capital needs? *Exhibit 9-17* presents the impact of this potential cost increase on annual transit needs as estimated for the Low-Growth scenario presented in Chapter 8. For this scenario, the cost impact is negligible in the

Exhibit 9-17 Impact of Shift to Vehicles Using Hybrid and Alternative Fuels on Investment Needs: Low-Growth Scenario

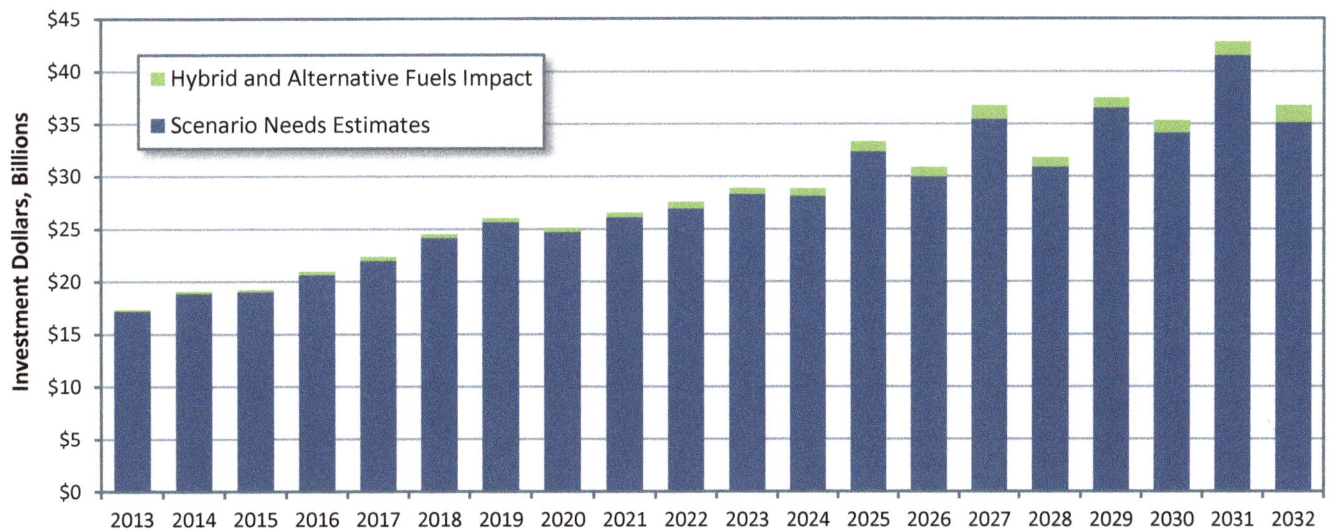

Source: Transit Economic Requirements Model.

early years of the projection period but grows over time as the proportion of buses using alternative fuel and hybrid power increases (note that the investment backlog is not included in this depiction). The impact on total investment needs for Chapter 9 investment scenarios (Low-Growth and High-Growth) and the SGR Benchmark are presented in dollar and percentage terms in *Exhibit 9-18*. Note that the shift to alternative fuels and hybrid buses is estimated to increase average annual replacement needs by $0.5 billion to $0.8 billion, yielding a 2.5- to 3.5-percent increase in investment needs. To provide perspective for these estimated amounts, noting the following is helpful: (1) the shift from diesel to alternative-fuel and hybrid buses is only one of several technology changes that might affect long-term transit reinvestment needs, but (2) reinvestment in transit buses likely represents the largest share of transit needs subject to this type of significant technological change. Hence, the impact of all new technology adoptions (not accounted for in the Chapter 8 scenarios and including new bus propulsion systems) might add 5–10 percent to long-term transit capital needs.

Exhibit 9-18 Impact of Shift from Diesel to Alternative Fuels and Hybrid Vehicles on Average Annual Investment Needs ($B)

Measure	SGR Baseline	Low-Growth	High-Growth
Average Annual Needs	$0.67	$0.69	$0.82
Percent Increase	2.23%	2.12%	2.13%

Source: Transit Economic Requirements Model.

Impact on Backlog

Finally, in addition to affecting unconstrained capital needs, the shift from diesel to hybrid and alternative-fuel vehicles also can affect the size of the future backlog. For example, *Exhibit 9-19* shows the estimated impact of this shift on the SGR backlog as estimated for the Sustain 2012 Spending scenario from Chapter 8. Under this scenario, long-term spending is capped at current levels such that any increase in costs over the analysis period must necessarily be added to the backlog. Moreover, given that the useful lives of buses as estimated by TERM are roughly 7–14 years, all existing and many expansion vehicles will need to be replaced over the 20-year analysis period, meaning that any increase in costs for this asset type will be added to the backlog for the period of analysis.

Exhibit 9-19 Impact of Shift to Vehicles Using Hybrid and Alternative Fuels on Backlog Estimate: Sustain 2012 Spending Scenario

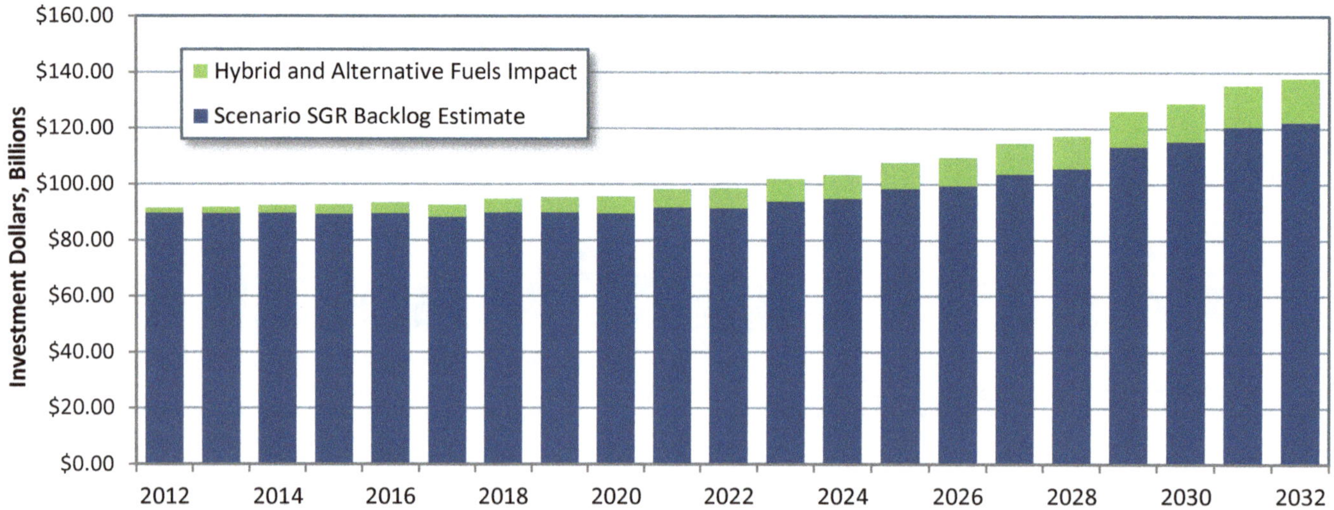

Source: Transit Economic Requirements Model.

As with the analysis above, *Exhibit 9-19* suggests that the initial impact of the shift to hybrid and alternative-fuel vehicles is small but the effect increases over time as the share of the Nation's bus fleet made up by these vehicle types increases. By 2030, this shift is estimated to increase the size of the backlog from $141.7 billion to $151.4 billion, an increase of $9.8 billion or 6.9 percent.

Forecasted Expansion Investment

This section compares key characteristics of the national transit system in 2012 to their forecasted TERM results over the next 20 years for different scenarios. It also includes expansion projections of fleet size, guideway route miles, and stations broken down by scenario to understand better the expansion investments that TERM forecasts.

TERM's projections of fleet size are presented in *Exhibit 9-20*. The projections for the Low- and High-Growth scenarios create upper and lower bounds around the projected Sustain 2012 Spending scenario to preserve existing transit assets at a condition rating of 2.5 or higher and expand transit service capacity to support differing levels of ridership growth while passing TERM's benefit-cost test.

Exhibit 9-20 Projection of Fleet Size by Scenario[1]

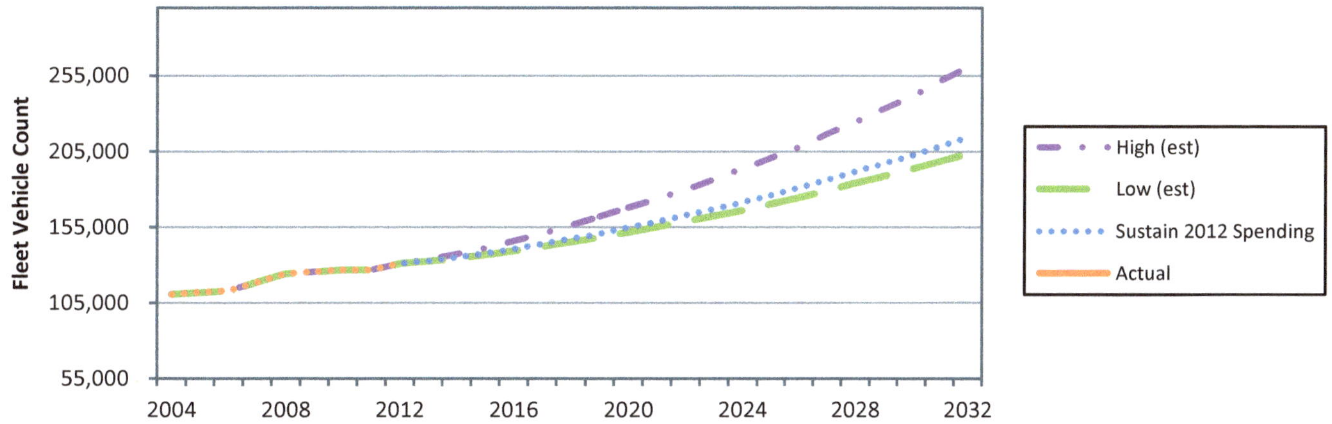

[1] *Data through 2012 are actual; data after 2012 are estimated based on trends.*

Source: Transit Economic Requirements Model.

The projected guideway route miles for the Sustain 2012 Spending scenario are less than for the projected High-Growth scenario, as shown in *Exhibit 9-21*. (Note that TERM's projections of guideway route miles for the Sustain 2012 Spending and Low-Growth scenarios are nearly identical.) Commuter rail has substantially more guideway route miles than heavy and light rail, making accurate projections of total guideway route miles for all rail modes difficult; therefore, the historical trend line is not provided.

TERM's expansion projections of stations by scenario needed to preserve existing transit assets at a condition rating of 2.5 or higher and to expand transit service capacity to support differing levels of ridership growth (while passing TERM's benefit-cost test) are presented *Exhibit 9-22*. TERM's Low-Growth estimates generally are in line with the historical trend, indicating that expansion projections of stations under the Low-Growth scenario could maintain current transit conditions.

Exhibit 9-21 Projection of Guideway Route Miles by Scenario[1]

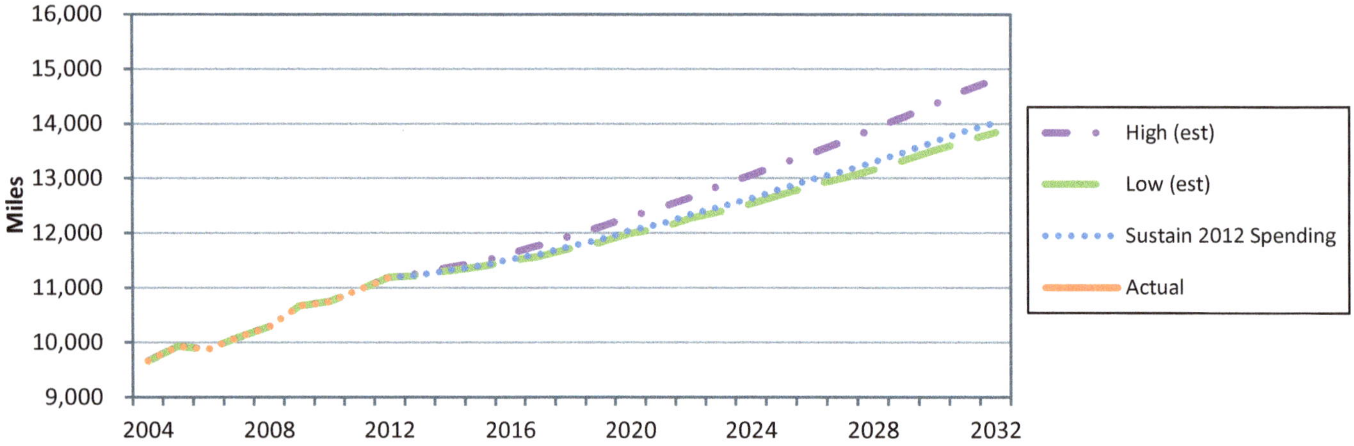

[1] Data through 2012 are actual; data after 2012 are estimated based on trends.

Source: Transit Economic Requirements Model.

Exhibit 9-22 Projection of Stations by Scenario[1]

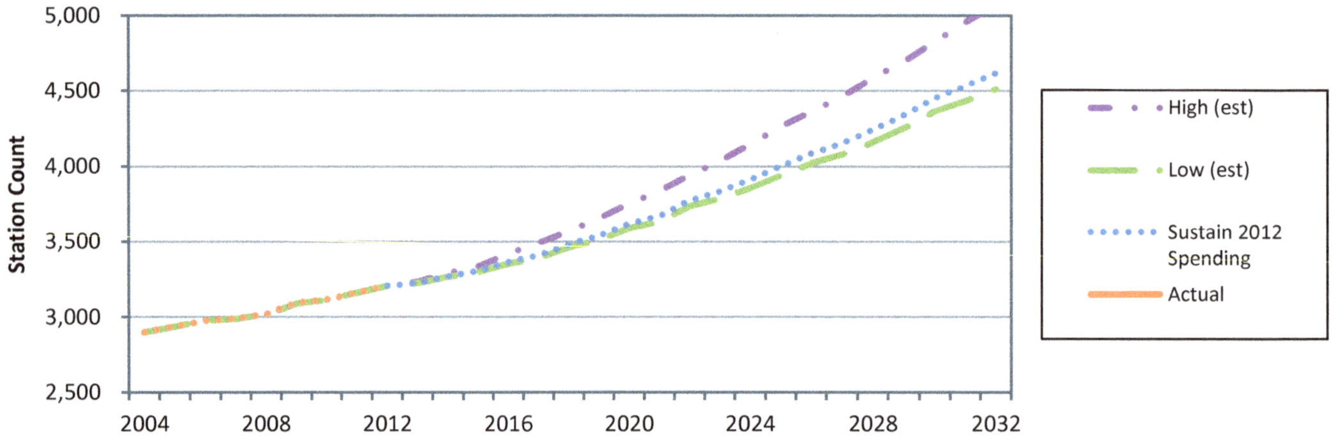

[1] Data through 2012 are actual; data after 2012 are estimated based on trends.

Source: Transit Economic Requirements Model.

For each scenario, TERM estimates future investment in fleet size, guideway route miles, and stations for each of the next 20 years. *Exhibit 9-23* presents TERM's projection for total fixed guideway route miles under a Low-Growth scenario by rail mode. TERM projects different investment needs for each year that are added to the 2012 actual total stock. Heavy rail's share of the projected annual fixed guideway route miles remains relatively constant over the 20-year period, while the amount of fixed guideway route miles increases slightly for light and commuter rail.

Exhibit 9-23 Stock of Fixed Guideway Miles by Year Under Low-Growth Scenario, 2012–2032

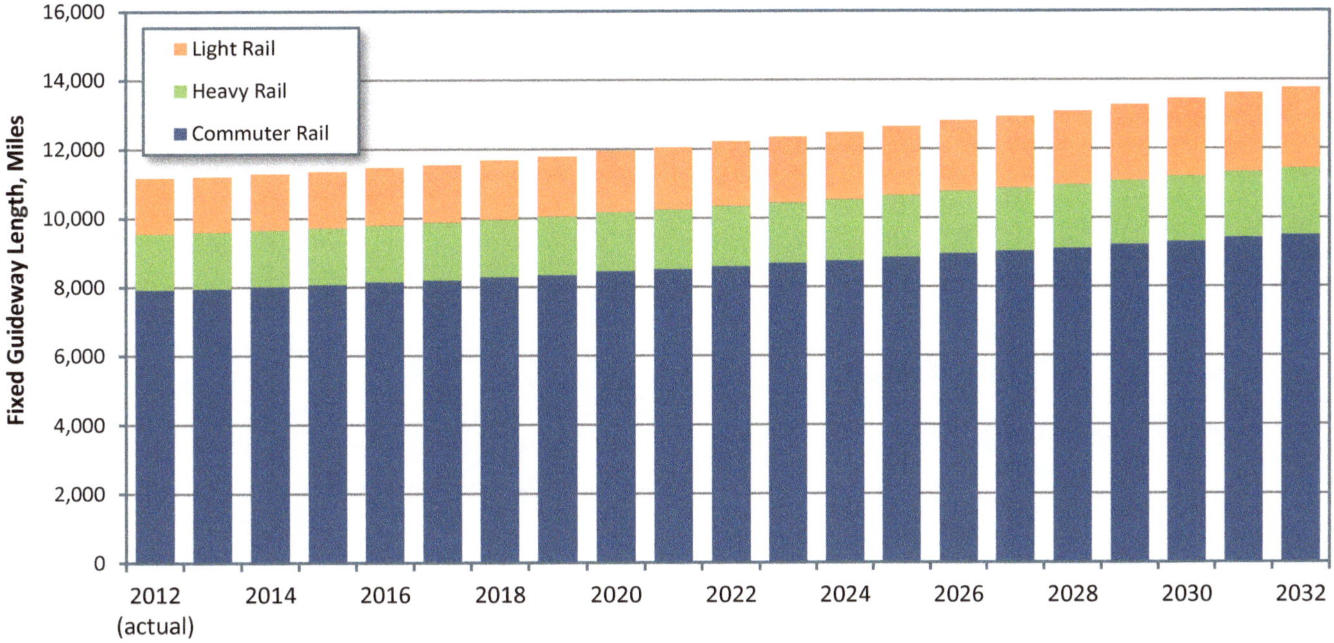

Source: Transit Economic Requirements Model.

---

[i] Subject to some limitations (e.g., agencies must surpass a minimum vehicle occupancy standard before being eligible for expansion investments).

chapter 10

Sensitivity Analysis

Highway Sensitivity Analysis

Sound practice in modeling includes analyzing the sensitivity of key results to changes in assumptions. For the Maintain Conditions and Performance scenario and the Improve Conditions and Performance scenario presented in Chapter 8, this section analyzes how changes in some of the underlying assumptions would affect the estimate of the average annual requirement for highway investment. First to be varied are economic assumptions about the

- Values of traveler time savings and traveler safety,
- Discount rate used to convert future costs and benefits into present-value equivalents,
- Costs of the types of capital improvements modeled,
- Projections for the price of motor fuel, and
- Projected growth in aggregate traffic volumes.

Conducted only within the Highway Economic Requirements System (HERS) are the tests that vary the growth rates assumed for the value of travel time savings and the price of motor fuel; growth in these factors is absent from National Bridge Investment Analysis System (NBIAS) and is presumed to have minor effects on the bridge investment needs within that model's scope—repair, rehabilitation, and functional improvements.

Next varied are the investment strategies assumed in HERS for future deployment of Operations/ Intelligent Transportation System (ITS). A subsequent section within this chapter explores information regarding the assumptions underlying the analyses developed using the Transit Economic Requirements Model (TERM).

An important outcome of the HERS results is that, under both baseline and sensitivity test assumptions, the Maintain Conditions and Performance is equivalent to a scenario in which the metric to be maintained is simply average pavement roughness. As defined, the Maintain Conditions and Performance sets HERS-related spending at the lowest level at which the 2032 projections for each of two measures—the average International Roughness Index (IRI) and average delay per vehicle miles traveled (VMT)—indicate conditions and performance that match or surpass those in the 2012 base year. In each of this report's simulations of this scenario, however, the binding constraint was maintaining average IRI. (The level of HERS-related spending that just sufficed to meet this constraint resulted in a decrease in average delay per VMT below the level in 2012.) For this reason, and because travel time delay depends much more on highway capacity than on pavement condition, any change to HERS assumptions that causes the model to reduce the share of spending for system expansion projects also will decrease the HERS component of spending in the Maintain Conditions and Performance scenario (and conversely).

Alternative Economic Analysis Assumptions

For application in benefit-cost analyses of programs and actions under their purview, the U.S. Department of Transportation (DOT) periodically issues guidance on valuing changes in travel time and traveler safety, and the Office of Management and Budget (OMB) provides guidance on the discount rate. Recognizing the uncertainty regarding these values, the guidance documents include both specific recommended values and ranges of values to be tested. The analyses presented in Chapters 7 and 8 of this report are based on the primary recommendations in DOT and OMB guidance for these economic inputs, whereas the analyses presented in this chapter rely on recommended alternative values to be used for sensitivity testing.

For the HERS analyses presented in Chapters 7 and 8, fuel price projections incorporate the "Reference Case" forecasts from the U.S. Department of Energy's *Annual Energy Outlook* (AEO). The AEO presents a range of potential alternative forecasts. One such alternative assuming lower fuel prices is explored in this section.

Value of Travel Time Savings

The value of travel time savings is a critical component of benefit-cost analysis of transportation investments, often the largest component of the estimated benefits. For HERS and NBIAS, the Federal Highway Administration (FHWA) estimates average values of time savings by vehicle hour traveled by vehicle type. Primarily, these values reflect the benefits from savings in the time travelers spend in vehicles, taking into account that vehicles can have multiple occupants. Time used for travel represents a cost to society and the economy because that time could be used for other more enjoyable or productive purposes. For heavy trucks, the FHWA makes additional allowances for the benefits from freight's arriving at its destination faster and from the opportunities for more intensive vehicle utilization when trips can be accomplished in less time. Even for these types of vehicles, however, the value of travel time savings estimated by FHWA primarily reflects the benefits from the freeing of travelers' time—the time of the truck driver and other vehicle occupants.

For valuation of traveler time, the analysis in this report follows, essentially, DOT's guidance on valuing travel time saved in 2012 (https://www.transportation.gov/office-policy/transportation-policy/guidance-value-time). In the analyses presented in Chapters 7 and 8, traveler time savings are valued per person hour at $12.30 for personal travel and between $27 and $32 for business travel. The value for personal travel is set in the guidance at 50 percent of hourly household income calculated as median annual household income divided by 2,080, the annual work hours of someone working 40 hours every week. The values for business travel are set at the relevant estimate of average hourly labor compensation (wages plus supplements). The variation in these values by vehicle type indicates, for example, that truck drivers typically earn less than business travelers in light-duty vehicles. (For details on the derivation of these values, see Appendix A.)

These values per person hour of travel are estimates subject to considerable uncertainty. Even when personal and business travel purposes are distinguished, estimating an average value of travel time is complicated by substantial variation in the value of travel time among individuals and, even for a given individual, among trips. Contributing to such variation are differences in incomes, employment status and earnings, attitudes, conditions of travel (e.g., the level of traffic congestion), and other factors. Moreover, studies that estimate values of travel time often are difficult to compare because of differences in data and methodology.

In view of the resulting uncertainty, DOT guidance calls for sensitivity tests that set values of travel time lower or higher than for the baseline. For personal travel time, these values are 35 percent and 60 percent of median hourly household income, rather than 50 percent as assumed in the baseline. For business travel time, these values are 80 percent and 120 percent of average hourly labor compensation, rather than the baseline assumption of 100 percent.

Exhibit 10-1 shows the effects of these variations on spending levels in the two scenarios reexamined in this chapter. For the NBIAS-derived component of spending, the effects are very small (well under 1.0 percent), consistent with bridge capacity expansion being outside the model's scope. Except where they would eliminate long detours caused by vehicle weight restrictions on a bridge, the bridge preservation actions evaluated by NBIAS would have minimal effects on travel times.

Exhibit 10-1 Impact of Alternative Value of Time Assumptions on Highway Investment Scenario Average Annual Investment Levels

Alternative Time Valuation Assumptions for Personal and Business Travel as Percentage of Hourly Earnings	Maintain Conditions and Performance Scenario		Improve Conditions and Performance Scenario	
	Billions of 2012 Dollars	Percent Change From Baseline	Billions of 2012 Dollars	Percent Change From Baseline
Baseline[1] (Personal–50%; Business–100%)	**$89.9**		**$142.5**	
HERS-Derived Component	$50.9		$75.4	
NBIAS-Derived Component	$12.2		$24.6	
Other (Nonmodeled) Component	$26.9		$42.6	
Lower (Personal–35%; Business–80%)	**$84.7**	**-5.8%**	**$134.6**	**-5.6%**
HERS-Derived Component	$47.3	-7.0%	$69.8	-7.4%
NBIAS-Derived Component	$12.1	-0.6%	$24.6	0.0%
Other (Nonmodeled) Component	$25.3	-5.8%	$40.2	-5.6%
Higher (Personal–60%; Business–120%)	**$92.7**	**3.1%**	**$147.7**	**3.6%**
HERS-Derived Component	$52.8	3.8%	$78.9	4.7%
NBIAS-Derived Component	$12.2	0.2%	$24.7	0.3%
Other (Nonmodeled) Component	$27.7	3.1%	$44.1	3.6%

[1] The baseline levels shown correspond to the systemwide scenarios presented in Chapter 8. The investment levels shown are average annual values for the period from 2013 through 2032.

Sources: Highway Economic Requirements System and National Bridge Investment Analysis System.

For the HERS-derived component of spending, the percentage reductions with lower values of traveler time are slightly over 7 percent in both scenarios, but the explanations differ. In the Improve Conditions and Performance scenario, the goal is to exploit all opportunities for cost-beneficial investments, which become fewer when the travel time savings are valued less. In the Maintain Conditions and Performance scenario, valuing travel time savings less increases the share of spending that HERS allocates to capacity expansion, making funds available for the system preservation improvements that reduce pavement roughness. For this reason, and because the binding constraint in this scenario is maintaining average pavement roughness, the required level of HERS-related spending decreases. Conversely, that spending increases when the higher values of time are assumed, by 3–4 percent in both scenarios.

Nonmodeled Highway Investments

The HERS-derived component of each scenario represents spending on pavement rehabilitation and capacity expansion on Federal-aid highways. The NBIAS-derived component represents rehabilitation spending on all bridges, including those off the Federal-aid highways. The nonmodeled component corresponds to system enhancement spending, plus pavement rehabilitation and capacity expansion on roads not classified as Federal-aid highways.

In the Sustain 2012 Spending scenario presented in Chapter 8, the values for these HERS and NBIAS components sum to $73.8 billion. In 2012, nonmodeled spending accounted for 29.9 percent of total investment ($31.4 billion of $105.2 billion) and is assumed to form the same share in all scenarios presented in Chapter 8.

Likewise, for the sensitivity analysis for the Maintain Condition and Performance and the Improve Condition and Performance scenarios presented in this section, the nonmodeled component is set at 29.9 percent of the total investment level. As the combined levels of the HERS-derived and NBIAS-derived scenario components increase or decrease, the nonmodeled component changes proportionally. Consequently, the percentage change in the nonmodeled component of each alternative scenario relative to the baseline always matches the percent change in the total investment level for that scenario.

Growth in the Value of Time

The opportunity cost of time spent traveling generally increases when real earnings or real incomes increase. Higher hourly pay usually reflects an increase in the value that an hour of labor contributes to production, and hence in the value of an hour of travel time saved on the job. On higher incomes, people are more able, and hence more willing, to pay for savings in personal travel time.

In addition, the long-term trend in U.S. economic history is for real growth over time in both average household incomes and average hourly earnings. In factoring this trend into its guidance on the value of travel time savings in benefit-cost analysis, DOT assumes that such growth will occur in the future at 1.2 percent per year (based on Congressional Budget Office projections for real median household income) and that the average value of travel time savings will increase at the same rate. In this report, these assumptions are built into the baseline analyses with HERS that are presented in Chapters 7 and 8.

Exhibit 10-2 shows the results of sensitivity tests with HERS that assume zero future growth in the value of travel time and, alternatively, 2.4 percent growth. Qualitatively, the results are the same as in the sensitivity test that changed the base-year value of travel time, and the explanations are also the same. Quantitatively, relative to the baseline assumption of 1.2 percent growth, assuming zero future growth in the value of time reduces the scenario investment levels by about 4 percent, and assuming higher values of time increases the scenario investment levels by about 3 percent.

The modeled changes in future economic growth also could shift future demand for highway travel, which in turn would affect the investment levels in this report's scenarios, but these shifts are not reflected in the present analysis. In theory, the direction of these shifts is ambiguous. Although affluence tends to generate demand for travel, higher wage levels increase the opportunity cost of time spent traveling rather than at the workplace, which could dampen the demand for highway travel. Similarly, higher household incomes could generate demands for uses of time that compete with personal travel—for example, with the additional money, someone might purchase video games that incline them to spend more time at home rather than engaging in outside pursuits that require travel. Although a preliminary literature review that FHWA has undertaken suggests that increasing affluence will increase demand for highway travel overall, further investigation is needed to confirm and quantify this effect; this research is among the priorities for the HERS program.

Exhibit 10-2 Impact of Alternative Assumptions About Growth in the Real Value of Time on Highway Investment Scenario Average Annual Investment Levels

	Maintain Conditions and Performance Scenario		Improve Conditions and Performance Scenario	
Alternative Value of Time Growth Assumptions	Billions of 2012 Dollars	Percent Change From Baseline	Billions of 2012 Dollars	Percent Change From Baseline
Baseline[1] (1.2%-increase per year)	$89.9		$142.5	
HERS-Derived Component	$50.9		$75.4	
Lower (0.0%-increase per year)	$86.5	-3.8%	$136.3	-4.4%
HERS-Derived Component	$48.5	-4.7%	$71.0	-5.8%
Higher (2.4%-increase per year)	$92.6	3.1%	$146.3	2.6%
HERS-Derived Component	$52.8	3.8%	$78.0	3.5%

[1] The baseline levels shown correspond to the systemwide scenarios presented in Chapter 8. The investment levels shown are average annual values for the period from 2013 through 2032.

Sources: Highway Economic Requirements System and National Bridge Investment Analysis System.

Value of Traveler Safety

One of the most challenging questions in benefit-cost analysis is what monetary cost to place on injuries of various severities. Few people would consider any amount of money to be adequate compensation for a person's being seriously injured, much less killed. On the other hand, people can attach a value to changes in their risk of suffering an injury, and indeed such valuations are implicit in their everyday choices. For example, a traveler may face a choice between two travel options that are equivalent except that one carries a lower risk of fatal injury but costs more. If the additional cost is $1, a traveler who selects the safer option is manifestly willing to pay at least $1 for the added safety—what economists call "revealed preference." Moreover, if the difference in risk is, say, one in a million, then a million travelers who select the safer option are collectively willing to pay at least $1 million for a risk reduction that statistically can be expected to save one of their lives. In this sense, the "value of a statistical life" among this population is at least $1 million.

Based on the results of various studies of individual choices involving money versus safety tradeoffs, some government agencies estimate an average value of a statistical life for use in their regulatory and investment analyses. Although agencies generally base their estimates on a synthesis of evidence from various studies, the decision as to which value is most representative is never clear-cut, thus warranting sensitivity analysis. DOT issued guidance in 2013 recommending a value of $9.1 million for analyses with a base year of 2012, as is the case in this C&P report (https://www.transportation.gov/office-policy/transportation-policy/guidance-treatment-economic-value-statistical-life). The guidance also required that regulatory and investment analyses include sensitivity tests using alternative values of $5.2 million as the lower bound and $12.9 million for the upper bound. For nonfatal injuries, the guidance sets values per statistical injury as percentages of the value of a statistical life; these vary according to the level of severity, from 0.3 percent for a "minor" injury to 59.3 percent for a "critical" injury. (The injury levels are from the Abbreviated Injury Scale.)

Impact of Alternatives on HERS Results

HERS contains equations for each highway functional class to predict crash rates per VMT and parameters to determine the number of fatalities and nonfatal injuries per crash. The model assigns to crashes involving fatalities and other injuries an average cost consistent with DOT guidance, including the use of alternative values for sensitivity tests. As shown in *Exhibit 10-3*, the sensitivity tests reveal only minor impacts on the average annual requirement for HERS-related investment; relative to a baseline in which the value of a statistical life is set at $9.1 million, increasing or decreasing that value by about $3.8 million alters the estimated investment requirement by well under 1 percent in each case. One reason for this insensitivity is that crash costs are estimated in HERS to form a small share of total highway user costs (14.0 percent in 2012). In addition, as Chapter 7 revealed, the crash costs are much less sensitive than travel time and vehicle operating costs to changes in the level of total investment within the scope of HERS. (Data limitations preclude that scope from including highway improvements that primarily address safety issues.)

Exhibit 10-3 Impact of Alternative Value of Life Assumptions on Highway Investment Scenario Average Annual Investment Levels

	Maintain Conditions and Performance Scenario		Improve Conditions and Performance Scenario	
Alternative Value of Statistical Life Assumptions (2012 Dollars)	Billions of 2012 Dollars	Percent Change From Baseline	Billions of 2012 Dollars	Percent Change From Baseline
Baseline[1] ($9.1 Million)	**$89.9**		**$142.5**	
HERS-Derived Component	$50.9		$75.4	
NBIAS-Derived Component	$12.2		$24.6	
Other (Nonmodeled) Component	$26.9		$42.6	
Lower ($5.2 Million)	**$89.0**	**-1.0%**	**$138.3**	**-3.0%**
HERS-Derived Component	$50.6	-0.5%	$74.9	-0.6%
NBIAS-Derived Component	$11.8	-3.1%	$22.1	-10.2%
Other (Nonmodeled) Component	$26.6	-1.0%	$41.3	-3.0%
Higher ($12.9 Million)	**$90.6**	**0.8%**	**$144.2**	**1.2%**
HERS-Derived Component	$51.2	0.7%	$75.8	0.6%
NBIAS-Derived Component	$12.3	1.2%	$25.3	3.0%
Other (Nonmodeled) Component	$27.1	0.8%	$43.1	1.2%

[1] The baseline levels shown correspond to the systemwide scenarios presented in Chapter 8. The investment levels shown are average annual values for the period from 2013 through 2032.

Sources: Highway Economic Requirements System and National Bridge Investment Analysis System.

Impact of Alternatives on NBIAS Results

Changes in the valuation of traveler safety affect the NBIAS-derived component of the scenario investment levels more significantly. In *Exhibit 10-3*, reducing the assumed value of a statistical life from the baseline of $9.1 million to $5.2 million decreases the estimate of investment needed for bridge repair, rehabilitation, and functional improvement by 3.1 percent in the Maintain Conditions and Performance scenario and by 10.9 percent in the Improve Conditions and Performance scenario. In comparison with these decreases, the estimated percentage increases in NBIAS-related investment when the assumed value of a statistical life increases by about the same amount above the baseline (from $9.1 million to $12.9 million) are less than half as large.

Discount Rate

Benefit-cost analyses use a discount rate that weighs benefits and costs expected to arise farther in the future less than those that would arise sooner. Thus far, in this report's applications of HERS, NBIAS, and TERM have set the discount rate at 7 percent; this means that deferring a benefit or cost for a year reduces its real value by approximately 6.5 percent (1/1.07). This choice of real discount rate conforms to the "default position" in the 1992 OMB guidance on discount rates, in Circular A-94, for benefit-cost analyses of Federal programs or policies. Subsequently, in 2003, OMB's Circular A-4 recommended that regulatory analyses use both 3 percent and 7 percent as alternative discount rates (http://www.whitehouse.gov/sites/default/files/omb/assets/omb/circulars/a004/a-4.pdf). The justifications for these recommendations apply equally to benefit-cost analyses of public investments, so the sensitivity tests in this section include the use of the 3-percent discount rate as an alternative to the 7-percent rate used in the baseline simulations.

Q&A **Could the discount rate be higher than 7 percent?**

The 2003 OMB guidance calls for using a discount rate higher than 7 percent as a further sensitivity test in some instances. In the context of public investment, this recommendation applies when the likelihood is that (1) the investment's opportunity cost will consist largely of displaced private investment, and (2) the displaced investment would have generated an average real rate of return exceeding 7 percent annually. Although the first of these conditions could be valid for some public investments in highways and transit systems, that displaced private investments will average rates of return above 7 percent annually could be difficult to justify. In 2003, OMB referred to its own recent estimate that the average real rate of return on private investment remained near the 7 percent level that OMB estimated in 1992. Although OMB noted that the average real rate of return on corporate capital in the United States was approximately 10 percent in the 1990s, whether the current economic outlook could justify the expectation of a rate of return averaging above 7 percent during this report's analysis period is by no means clear.

For infrastructure improvements, including those that HERS and NBIAS consider, the normal sequence is for an initial period in which net benefits are negative, reflecting the costs of construction, followed by many years of positive net benefits, reflecting the benefits of improved infrastructure in place. Because the positive net benefits materialize farther in the future than the costs of construction, a reduction in the discount rate increases the weight attached to the positive net benefits relative to the construction costs, resulting in a higher benefit-cost ratio. Moreover, with all potential projects now having a higher benefit-cost ratio, when the investment objective is to exhaust all opportunities for implementing cost-beneficial projects, the indicated amount of investment will increase. Accordingly, *Exhibit 10-4* shows that in the Improve Conditions and Performance scenario, a reduction in the assumed annual discount rate from 7 percent to 3 percent increases the total level of investment by 20.3 percent, and the HERS and NBIAS components by a similar percentage.

Exhibit 10-4 Impact of Alternative Discount Rate Assumption on Highway Investment Scenario Average Annual Investment Levels

	Maintain Conditions and Performance Scenario		Improve Conditions and Performance Scenario	
Alternative Assumptions About Discount Rate	Billions of 2012 Dollars	Percent Change From Baseline	Billions of 2012 Dollars	Percent Change From Baseline
Baseline[1] (7% discount rate)	**$89.9**		**$142.5**	
HERS-Derived Component	$50.9		$75.4	
NBIAS-Derived Component	$12.2		$24.6	
Other (Nonmodeled) Component	$26.9		$42.6	
Alternative (3% discount rate)	**$88.0**	**-2.1%**	**$171.5**	**20.3%**
HERS-Derived Component	$50.4	-0.9%	$90.8	20.5%
NBIAS-Derived Component	$11.3	-7.0%	$29.5	19.8%
Other (Nonmodeled) Component	$26.3	-2.1%	$51.2	20.3%

[1] The baseline levels shown correspond to the systemwide scenarios presented in Chapter 8. The investment levels shown are average annual values for the period from 2013 through 2032.

Sources: Highway Economic Requirements System and National Bridge Investment Analysis System.

For the Maintain Conditions and Performance scenario, the reduction in the discount rate has more complex effects within the models. At any given level of HERS-related spending, the model determines that allocating a slightly higher share to system preservation projects would be cost-beneficial; this is because, in HERS, benefits arising relatively late in the project life cycle tend to be more important for system rehabilitation than for system expansion projects. Because the preservation share of spending increases, the $50.9 billion of spending from the baseline (7-percent discount rate) would more than suffice to maintain IRI at the base-year level. Thus, a reduction in the discount rate leads the model to slightly reduce spending in the Maintain Conditions and Performance scenario.

The NBIAS-derived component of spending in the Maintain Conditions and Performance scenario is much more sensitive to the discount rate. Reducing the discount rate from 7 percent to 3 percent causes this component to decrease by 7.0 percent.

Costs of Capital Improvements

The HERS database includes a cost matrix that indicates typical cost for each type of modeled improvement. For example, the current matrix indicates that in 2012, reconstructing and widening a lane of rural Interstate highway typically cost $3,180 per lane mile. The matrix is periodically updated—the most recent full update, which obtained cost data from a survey of projects, produced estimates for 2002. These estimates have since been updated by simply using a general highway construction cost index. Applying the same general index to all types of improvements ignores changes that might have occurred in the relative costs of different types of projects.

Even for updating the overall level of improvement costs, the indexing approach has been problematic because of challenges in splicing together the FHWA Composite Bid Price Index and its successor, the National Highway Construction Cost Index. During the period in which they overlapped, 2002–2006, quality and coverage of the supporting data were deteriorating for the Bid Price Index and improving for the Construction Cost Index. To splice these series together for the C&P reports, FHWA chose 2006 as the year to switch to the Construction Index. This choice is arguable, however, and the selection of a different year could have made a material difference, given the significant divergences in movements of the two indices during the overlap years. The period under consideration was one of marked volatility in highway construction costs, so the divergences could owe in part to challenges in measuring costs when they are fluctuating sharply. (For further discussion of this issue, see Chapter 10 of the 2010 C&P Report.) FHWA is currently conducting a study to update the HERS improvement cost matrix using project-level data.

Furthermore, even without the complications from switchover between indices, simple inflation adjustments are inadequate to reflect many factors that have changed since 2002. These factors include changes in the construction materials typically used, greater reliance of off-peak or night work (with resulting higher labor costs), and changes in the nature of typical reconstruction projects. (In particular, as the system ages, reconstruction projects more frequently require replacement through the sub-base).

The uncertainty that such problems introduce into the base-year estimates of improvement warrants sensitivity testing. This is also true of the base-year improvement costs in NBIAS, which could be too low. *Exhibit 10-5* shows the sensitivity of the HERS and NBIAS results to increasing all

Exhibit 10-5 Impact of an Increase in Capital Costs on Highway Investment Scenario Average Annual Investment Levels

	Maintain Conditions and Performance Scenario		Improve Conditions and Performance Scenario	
Alternative Assumptions Regarding Unit Costs for Capital Improvements	**Billions of 2012 Dollars**	**Percent Change From Baseline**	**Billions of 2012 Dollars**	**Percent Change From Baseline**
Baseline[1]	**$89.9**		**$142.5**	
HERS-Derived Component	$50.9		$75.4	
NBIAS-Derived Component	$12.2		$24.6	
Other (Nonmodeled) Component	$26.9		$42.6	
Alternative (25% above baseline)	**$108.0**	**20.2%**	**$145.9**	**2.3%**
HERS-Derived Component	$62.8	23.5%	$77.6	3.0%
NBIAS-Derived Component	$13.0	6.6%	$24.7	0.3%
Other (Nonmodeled) Component	$32.3	20.2%	$43.6	2.3%

[1] The baseline levels shown correspond to the systemwide scenarios presented in Chapter 8. The investment levels shown are average annual values for the period from 2013 through 2032.

Sources: Highway Economic Requirements System and National Bridge Investment Analysis System.

base-year capital improvement costs by 25 percent. In the Improve Conditions and Performance scenario, the increase in the estimate of required spending is 3.0 percent for the HERS component and 0.3 percent for the NBIAS component, far smaller than the assumed 25-percent increase in unit improvement costs. This is because the target of the Improvement Conditions and Performance scenario is to implement all cost-beneficial improvements, and an increase in improvement costs reduces the pool of projects that pass a benefit-cost test. The reduction in this pool nearly offsets the direct effect of the cost increase on the scenario's requirement for investment spending.

In the Maintain Conditions and Performance scenario, the sensitivity test yields notably different results between HERS and NBIAS. The required level of HERS-related spending, 23.5 percent, nearly matches the assumed 25-percent increase in unit improvement costs, as expected: Meeting the scenario's goal of keeping average pavement roughness unchanged requires a similar scale of improvement, regardless of cost. In contrast, the estimate of the NBIAS-derived component of spending increases by only 6.6 percent because the model shifts a large amount of spending on bridge replacement to more cost-beneficial bridge maintenance projects. This substitution occurs because, when the costs of improving bridges rise, benefit-cost analysis more strongly influences the model's project selection decisions, leading to the rejection of some aggressive bridge replacement projects that have low benefit-cost ratios. The sharper focus on benefit-cost ratio reduces the estimate of the total spending required to maintain conditions and performance, and this substantially offsets the direct effect of the modeled 25-percent increase in improvement costs.

Motor Fuel Prices

The projections of motor fuel prices in this report's baseline analysis conform to those in the Reference case of the 2014 AEO, released by the U.S. Energy Information Administration. The 2014 release was the most current available when the data inputs to the modeling in this report were being prepared. AEO projections for prices of motor fuel and other energy products are constant-dollar, or "real," measures that show changes after adjusting for general inflation. For this report's analysis period, 2013–2032, the Reference case projections indicated average retail prices of gasoline significantly below the 2012 level for the first decade and then substantially recovering; the projections for 2022 and 2032 are 14.1 percent and 4.9 percent higher than the 2012 level.

In addition to the Reference case, the AEO includes alternative cases that explore important areas of uncertainty for markets, technologies, and policies in the U.S. energy economy. For the Low Oil Price case, the 2014 AEO projects low oil prices resulting from a combination of low demand for petroleum and other liquids in developing economies and higher global supply. Gasoline prices projected for 2022 and 2032 are each about 30 percent below the 2012 level.

The Low Oil Price case has projected motor fuel prices more accurately than the Reference case to date: For 2015, the average real price per gallon of gasoline (2012 dollars, Consumer Price Index-deflated) was $2.35, much closer to the $2.63 projected in the Low Oil Price case than the $3.12 projected in the Reference case. Past experience has shown, however, that motor fuel and other energy prices are volatile and hard to predict, so this result does not indicate future relative performance of the fuel price projections over the entire two decades for which this report projects highway conditions and performance.

Broader Sensitivity Test of Economic Assumptions Related to Oil Prices

The sensitivity tests presented here for motor fuel prices omit various indirect impacts. Lower fuel prices reduce the importance consumers attach to fuel economy in vehicle purchase decisions, which gradually reduces average fuel economy by changing the composition of the vehicle fleet. More immediately, travelers will adjust to lower fuel prices in other ways that reduce fuel economy. In particular, those with more than one vehicle at their disposal (as in multi-vehicle households) will tend to use the less fuel-efficient vehicles more intensively. These responses will also affect vehicle miles traveled, but in ways more complex than the HERS model can adequately represent at present.

In addition to these responses, the sensitivity tests for motor fuel prices omit consideration of the differences between the two AEO cases in macroeconomic outcomes. For real GDP, the average annual growth rate projected over 2012–2040 is 0.5 percent higher in the Low Oil Price case than in the Reference case (2.4 percent versus 1.9 percent), and higher growth will increase vehicle miles traveled as well as the value of travel time savings. Although other sensitivity tests presented in the chapter treat uncertainty in both vehicle miles traveled and the value of travel time, future C&P reports could examine the overall effects of incorporating into HERS the AEO projections for the High or Low Oil Price alternatives to the Reference case.

Replacing the Reference case projections for gasoline and diesel fuel prices with those from the Low Oil Price case increases the HERS-derived component of spending by 1.4 percent in the Maintain Conditions and Performance scenario and 3.8 percent in the Improve Conditions and Performance scenario (*Exhibit 10-6*). This increase reflects partly that lower fuel prices stimulate travel demand. In the Maintain Conditions and Performance scenario, VMT in the final year of the analysis period, 2032, are projected to be 2.0 percent greater under the low fuel price assumptions than in the baseline.

Exhibit 10-6 Impact of Alternative Future Fuel Price Assumption on Highway Investment Scenario Average Annual Investment Levels

	Maintain Conditions and Performance Scenario		Improve Conditions and Performance Scenario	
Alternative Assumptions About Future Fuel Prices	Billions of 2012 Dollars	Percent Change From Baseline	Billions of 2012 Dollars	Percent Change From Baseline
Baseline[1] (AEO Reference Case)	**$89.9**		**$142.5**	
HERS-Derived Component	$50.9		$75.4	
Alternative (AEO Low Oil Price Case)	**$90.9**	**1.2%**	**$146.6**	**2.8%**
HERS-Derived Component	$51.6	1.4%	$78.2	3.8%

[1] The baseline levels shown correspond to the systemwide scenarios presented in Chapter 8. The investment levels shown are average annual values for the period from 2013 through 2032.

Sources: Highway Economic Requirements System and National Bridge Investment Analysis System.

In the Improve Conditions and Performance scenario, where the corresponding difference in VMT in 2032 is 2.1 percent, the increase in spending also reflects that in HERS, additional highway spending produces negative savings in fuel consumption per mile traveled. This is because the improvements funded out of the additional spending, particularly expansions of highway capacity, lead to increases in average travel speed, which in HERS, degrades fuel economy. Lower prices for motor fuel prices reduce the cost attached to this disbenefit, resulting in more improvements passing the benefit-cost test, and hence more spending in a scenario where all such improvements are funded. That said, the HERS equations for fuel and other vehicle operating costs are dated and are not altogether accurate in representing the effects of highway improvements on fuel economy; in particular, they make no allowance for the extra fuel consumption due to speed variability from congestion and incident delay. The FHWA is currently conducting a project to update and revamp these equations.

Traffic Growth Projections

In this report's baseline analyses, projections for traffic growth rates by vehicle class were taken from an econometric model for forecasting VMT. Chapter 7 described this model and its application in HERS and NBIAS. For 2013–2032, total VMT were projected to increase at an average annual rate of 1.04 percent in the baseline, and are projected to increase at lower and higher rates in the sensitivity tests (*Exhibit 10-7*). The lower rate, 0.74 percent, is the growth rate in the official projections for the U.S. resident population. The higher rates, 1.41 percent for the HERS simulations and 1.48 percent for the NBIAS simulations, derive from the projections of traffic volumes by highway section in the Highway Performance Monitoring System and by bridge in the National Bridge Inventory. The low and high growth rates for heavy trucks are based on the "pessimistic" and "optimistic" assumptions in the econometric forecasting model. As in the baseline projections, they exceed the growth rate projected for traffic overall, which means that light-duty vehicle traffic would grow at a rate below that for traffic overall.

Exhibit 10-7 Projected Average Percent Growth per Year in Vehicle Miles Traveled by Vehicle Class, 2013–2032

Vehicle Class	Baseline Growth Rate	Basis	Low-Growth Growth Rate	Basis	High-Growth Growth Rate	Basis
All Vehicles	1.04%	Econometric Model Forecast	0.74%	Equals Projected Population Growth Rate (U.S. Census)	1.41%	HPMS Section-level Traffic Projections, Aggregated
Single-Unit Trucks	2.15%	Econometric Model Forecast	1.26%	Econometric Model Forecast (Pessimistic Assumptions)	2.94%	Econometric Model Forecast (Optimistic Assumptions)
Combination Trucks	2.12%	Econometric Model Forecast	1.57%	Econometric Model Forecast (Pessimistic Assumptions)	2.67%	Econometric Model Forecast (Optimistic Assumptions)

Sources: FHWA National Vehicle Miles Traveled Projection; Highway Performance Monitoring System; U.S. Bureau of the Census

In both scenarios, assuming the lower traffic growth rates reduces the HERS-derived component of spending by about 12 percent, while assuming the higher traffic growth rates increases it about 15 percent (*Exhibit 10-8*). On the other hand, the NBIAS-derived component responds minimally to these changes in assumptions. This difference in sensitivity of results partly reflects a difference in benefit composition between the types of investment evaluated in HERS and NBIAS. In general, the benefits from the bridge improvements that NBIAS evaluates are predominantly savings in agency maintenance costs; unlike in HERS, savings in the user costs of travel are a small component. Also, the performance of many types of bridge elements is primarily influenced by age and environmental conditions rather than traffic volume.

Exhibit 10-8 Impact of Alternative Travel Growth Forecasts on Highway Investment Scenario Average Annual Investment Levels

Alternative Assumptions About Future Annual VMT Growth[1]	Maintain Conditions and Performance Scenario		Improve Conditions and Performance Scenario	
	Billions of 2012 Dollars	Percent Change From Baseline	Billions of 2012 Dollars	Percent Change From Baseline
Baseline[2] (1.04% per year)	**$89.9**		**$142.5**	
HERS-Derived Component	$50.9		$75.4	
NBIAS-Derived Component	$12.2		$24.6	
Other (Nonmodeled) Component	$26.9		$42.6	
Lower (Tied to Projected Rate of Population Growth—0.74% per year)	**$81.3**	**-9.6%**	**$129.0**	**-9.5%**
HERS-Derived Component	$44.9	-11.7%	$66.0	-12.4%
NBIAS-Derived Component	$12.1	-0.5%	$24.5	-0.5%
Other (Nonmodeled) Component	$24.3	-9.6%	$38.6	-9.5%
Higher (Tied to State Forecasts—HPMS at 1.41% per year; NBI at 1.48% per year)	**$101.1**	**12.5%**	**$159.8**	**12.1%**
HERS-Derived Component	$58.6	15.2%	$87.2	15.7%
NBIAS-Derived Component	$12.3	1.3%	$24.9	1.1%
Other (Nonmodeled) Component	$30.2	12.5%	$47.8	12.1%

[1] The VMT growth rates identified represent the forecasts entered into the HERS and NBIAS models. The travel demand elasticity features in HERS modify these forecasts in response to changes in highway user costs resulting from future highway investment.

[2] The baseline levels shown correspond to the systemwide scenarios presented in Chapter 8. The investment levels shown are average annual values for the period from 2013 through 2032.

Sources: Highway Economic Requirements System and National Bridge Investment Analysis System.

Alternative Strategies

Sensitivity tests can be conducted with HERS and NBIAS not only for alternative technical assumptions, but also for selected policy alternatives. One such alternative pertains to accelerating the future rate of deployment of Operations/ITS strategies modeled in HERS.

Accelerating Operations/ITS Deployments

As described in Chapter 7, the HERS model considers the impacts on highway conditions and performance of various types of ITS and other operational enhancements to highways. Appendix A describes the types of strategies considered (including arterial management, freeway management, incident management, and traveler information systems) and three scenarios for future deployment. Although it incorporates assumptions about future deployment, HERS does not subject operational enhancements to benefit-cost analysis or to other economic evaluation; thus, the preceding chapters in this report referred to spending on these and other system enhancements as "nonmodeled." The only spending that HERS models in this sense is on highway pavement rehabilitation and capacity expansion, although spending on operational enhancements is represented.

In the Maintain Conditions and Performance scenario, annual spending on HERS-modeled improvements averaged $50.9 billion under the baseline assumptions about future deployment of operational improvements. If HERS-modeled spending were held at that level while future deployment of operational improvements were assumed to be more aggressive, overall conditions and performance in 2030 relative to 2010 would be improved rather than maintained. To attain the scenario goal, HERS-modeled spending must therefore be lower when the alternative deployment assumptions replace the baseline, which assumes continuation of existing deployment trends. The "aggressive" alternative adjusts the various triggers for deployment—for example, how congested a freeway has to be for ramp metering to be introduced—such that the rates of deployment are 30-60 percent higher than in the baseline. The other alternative considered would deploy all the operational improvements selected in the aggressive alternative "immediately" – i.e. in the first five years of the 20-year analysis period.

For the "aggressive" deployment alternative, *Exhibit 10-9* shows the HERS-modeled capital spending to average $49.3 billion per year and spending on operational enhancements (including capital, and operations and maintenance costs) to be $0.6 billion per year more than in the baseline. The sum of these figures, $49.9 billion, indicates a $1.0-billion decrease in total spending relative to the baseline value of $50.9 billion to achieve the objectives of the Maintain Conditions and Performance scenario. For the "full immediate deployment alternative," total spending is $49.8 billion, slightly lower than for the aggressive deployment alternative.

Exhibit 10-9 Impact of Alternative Operations Strategies Deployment Rate Assumptions on Selected Performance Indicators and Highway Investment Scenarios

Operations/ITS Deployments Assumption[1]	Average Annual Highway Investment, 2013 through 2032 (Billions of 2012 Dollars)			
	HERS-Derived Component			
	HERS Modeled Spending	Additional Deployment Spending[2]	Total HERS	Total
Maintain Conditions and Performance Scenario				
Baseline (continue existing trends)	**$50.9**	**N/A**	**$50.9**	**$89.9**
Aggressive deployments alternative	$49.3	$0.6	$49.9	$88.5
Full immediate deployments alternative	$39.7	$10.1	$49.8	$88.4
Improve Conditions and Performance Scenario				
Baseline (continue existing trends)	**$75.4**	**N/A**	**$75.4**	**$142.5**
Aggressive deployments alternative	$74.0	$0.6	$74.6	$141.5
Full immediate deployments alternative	$64.2	$10.1	$74.3	$141.0

[1] The analyses presented in this table assume one of the following: (1) existing trends in ITS deployments will continue for 20 years; (2) an aggressive pattern of deployment will occur over the next 20 years; or (3) all of the aggressive deployments will occur immediately rather than being spread out over 20 years. The costs associated with the more aggressive deployments were deducted from the budget available in HERS for pavement and widening investments.

[2] Amounts reflect additional capital and operation and maintenance costs associated with the alternative Operations/ITS deployment strategies relative to the baseline.

Source: Highway Economic Requirements System.

In the Improve Conditions and Performance scenario, more aggressive deployment of operational enhancements marginally reduces the amount of highway rehabilitation and capacity investment that HERS finds to be cost-beneficial. HERS-modeled rehabilitation and capacity investment decreases from $75.4 billion per year assuming baseline deployment to $74.6 billion per year assuming the aggressive deployment alternatives and to $74.3 billion assuming the full immediate deployment alternative. Notwithstanding the offsetting increases in spending for operational improvements, total average annual spending represented in HERS decreases by $0.8 billion if the aggressive deployment alternative replaces the baseline and by another $0.3 billion if the alternative changes from aggressive to full immediate deployment.

Transit Sensitivity Analysis

This section examines the sensitivity to key inputs of the estimates of transit investment needs that the Transit Economic Requirements Model (TERM) produces. The sensitivity of the estimates is evaluated in response to variations in the values of these key inputs:

- asset replacement timing (condition threshold),
- capital costs,
- value of time, and
- discount rate.

The alternative projections presented in this chapter assess how the estimates of baseline investment needs for the State of Good Repair (SGR) Benchmark and the Low-Growth and High-Growth scenarios discussed in Chapter 8 vary in response to changes in the assumed values of the input variables, above. Note that, by definition, funding under the Sustain 2012 Spending scenario does not vary with changes in any input variable, and thus this scenario is not considered in this sensitivity analysis.

Changes in Asset Replacement Timing (Condition Threshold)

Each of the four investment scenarios examined in Chapter 8 assumes that assets are replaced at condition rating 2.50 as determined by TERM's asset condition decay curves (in this context, 2.50 is referred to as the "replacement condition threshold"). Recall that TERM's condition rating scale runs from 5.0 for assets in "excellent" condition through 1.0 for assets in "poor" condition. In practice, this assumption implies replacement of assets within a short period (e.g., roughly 1 to 5 years, depending on asset type) of their having attained their expected useful lives. Replacement at condition 2.50 can therefore be thought of as providing a replacement schedule that is both realistic and potentially conservative. This replacement schedule is realistic because, in practice, few assets are replaced exactly at their expected useful life value due to many factors, including the time to plan, fund, and procure an asset replacement. It is a potentially conservative schedule because the needs estimates would be higher if all assets were to be replaced at precisely the end of their expected useful lives.

Exhibit 10-10 shows the effect of varying the replacement condition threshold by increments of 0.25 on TERM's projected asset preservation needs for the SGR Benchmark and the Low-Growth and High-Growth scenarios. Note that selection of a higher replacement condition threshold results in assets being replaced at a higher condition (i.e., at an earlier age). This, in turn, reduces the length of each asset's service life, thus increasing the number of replacements over any given period of analysis and driving up scenario costs. Reducing the replacement condition threshold would have the opposite effect. As shown in *Exhibit 10-10*, each of these three scenarios shows significant changes to total estimated preservation needs from quarter-point changes in the replacement condition threshold. Relatively small changes in the replacement condition threshold frequently translate into significant changes in the expected useful life of some asset types; hence, small changes can also drive significant changes in replacement timing and replacement costs.

Exhibit 10-10 Impact of Alternative Replacement Condition Thresholds on Transit Preservation Investment Needs by Scenario (Excludes Expansion Impacts)

	SGR Benchmark		Low-Growth Scenario		High-Growth Scenario	
Replacement Condition Thresholds	Billions of 2012 Dollars	Percent Change From Baseline	Billions of 2012 Dollars	Percent Change From Baseline	Billions of 2012 Dollars	Percent Change From Baseline
Very late asset replacement (2.00)	$16.27	-4.3%	$15.73	-4.4%	$15.84	-4.2%
Replace assets later (2.25)	$16.57	-2.6%	$16.01	-2.6%	$16.10	-2.7%
Baseline (2.50)	$17.01		$16.44		$16.54	
Replace assets earlier (2.75)	$17.61	3.5%	$16.99	3.3%	$17.13	3.6%
Very early asset replacement (3.00)	$18.02	5.9%	$17.35	5.5%	$17.53	6.0%

Source: Transit Economic Requirements Model.

Changes in Capital Costs

The asset costs used in TERM are based on actual prices paid by agencies for capital purchases as reported to Federal Transit Administration (FTA) in the Transit Electronic Award Management (TEAM) System and in special surveys. Asset prices in the current version of TERM have been converted from the dollar-year replacement costs in which assets were reported to FTA by local agencies (which vary by agency and asset) to 2012 dollars using the RSMeans© construction cost index. Given the uncertain nature of capital costs, a sensitivity analysis has been performed to examine the effect that higher capital costs would have on the dollar value of TERM's baseline projected transit investment.

As *Exhibit 10-11* shows, TERM projects that a 25-percent increase in capital costs (i.e., beyond the 2012 level used for this C&P report) would be fully reflected in the SGR Benchmark, but only partially realized under the Low-Growth or High-Growth scenarios. This difference in sensitivity results is driven by the fact that investments are not subject to TERM's benefit-cost ratio in computing the SGR Benchmark (i.e., increasing costs has no consequences), whereas the two cost-constrained scenarios do employ this test. Hence, for the Low-Growth or High-Growth scenarios, any increase in capital costs (without a similar increase in the value of transit benefits) results in lower benefit-cost ratios and the failure of some investments to pass this test. Therefore, for these latter two scenarios, a 25-percent increase in capital costs would yield a roughly 13- to 15-percent increase in needs that pass TERM's benefit-cost test.

Exhibit 10-11 Impact of Increase in Capital Costs on Transit Investment Estimates by Scenario

	SGR Benchmark		Low-Growth Scenario		High-Growth Scenario	
Capital Cost Increases	Billions of 2012 Dollars	Percent Change From Baseline	Billions of 2012 Dollars	Percent Change From Baseline	Billions of 2012 Dollars	Percent Change From Baseline
Baseline (no change)	$17.01		$22.88		$26.42	
Increase Costs 25%	$21.27	25.0%	$26.48	15.7%	$29.90	13.2%

Source: Transit Economic Requirements Model.

Changes in the Value of Time

The most significant source of transit investment benefits, as assessed by TERM's benefit-cost analysis, is the net cost savings to users of transit services, a key component of which is the value of travel time savings. Therefore, the per-hour value of travel time for transit riders is a key model input and a key driver of total investment benefits for those scenarios that use TERM's benefit-cost test. Readers interested in learning more about the measurement and use of the value of time for the benefit-cost analyses that TERM, the Highway Economic Requirements System, and the National Bridge Investment Analysis System perform should refer to the related discussion presented earlier in the highway section of this chapter.

For this C&P report, the Low-Growth and High-Growth scenarios are the only scenarios with investment needs estimates that are sensitive to changes in the benefit-cost ratio. (Note that the Sustain 2012 Spending scenario uses TERM's estimated benefit-cost ratios to allocate fixed levels of funding to preferred investments, while the computation of the SGR Benchmark does not.)

Exhibit 10-12 shows the effect of varying the value of time on the needs estimates of the Low-Growth and High-Growth scenarios. The baseline value of time for transit users is currently $12.50 per hour, based on Department of Transportation guidance. TERM applies this amount to all in-vehicle travel, but then doubles it to $25.00 per hour when accounting for out-of-vehicle travel time, including time spent waiting at transit stops and stations.

Given that value of time is a key driver of total investment benefits, changes in this variable lead to changes in investment ranging from an increase of roughly 7 percent to a decrease of 13 percent. The resulting different magnitudes of percent changes is because the absolute value of the changes from the baseline differ ($6.25 is a 50-percent change from baseline and $25 is a 100-percent change from baseline). In addition to this issue, we observe that the High-Growth scenario appears to be more sensitive to the value of time than the Low-Growth scenario. This is because the High-Growth scenario is associated with higher investment levels than is the Low-Growth scenario; therefore, any changes in the value of time will be magnified accordingly.

Exhibit 10-12 Impact of Alternative Value of Time Rates on Transit Investment Estimates by Scenario

	Low-Growth Scenario		High-Growth Scenario	
Changes in Value of Time	Billions of 2012 Dollars	Percent Change From Baseline	Billions of 2012 Dollars	Percent Change From Baseline
Reduce 50% ($6.25)	$20.21	-11.7%	$22.99	-13.0%
Baseline ($12.50)	$22.88		$26.42	
Increase 100% ($25.00)	$23.84	4.2%	$28.14	6.5%

Source: Transit Economic Requirements Model.

Changes to the Discount Rate

Finally, TERM's benefit-cost module uses a discount rate of 7 percent in accordance with guidance provided by the White House Office of Management and Budget. Readers interested in learning more about the selection and use of discount rates for the benefit-cost analyses that TERM, the Highway Economic Requirements System, and the National Bridge Investment Analysis System perform should refer to the related discussion presented earlier in the highway section of this chapter. For this sensitivity analysis and for consistency with the discussion above on Highway Economic Requirements System and National Bridge Investment Analysis System discount rate sensitivity, TERM's needs estimates for the Low-Growth and High-Growth scenarios were reestimated using a 3-percent discount rate. The results of this analysis are presented in *Exhibit 10-13*. These results show that this approximately 57-percent reduction in the discount rate leads to a range in total investment needs (or changes in the proportion of needs passing TERM's benefit-cost test) of a greater than 17-percent increase to a less than 1-percent decrease.

Under this sensitivity test, investment needs are usually higher for the lower (3 percent) discount rate as compared to the higher base rate (7 percent). This means that use of the lower rate allows more investments to pass TERM's benefit cost test. This situation is primarily the result of differences in the timing of the flows of benefits vs costs for the underlying scenario. Specifically, this test has based off of a fully (financially) unconstrained scenario that completely eliminates the large investment backlog at the start of the period of analysis and then invests incrementally as needed at a much lower rate to maintain this "perfect state of good repair" for the remaining 20 years of analysis. In contrast, investment benefits tend to be more evenly distributed throughout the 20-year period of analysis. So, with a high proportion of costs concentrated very early in the period of analysis and evenly distributed benefits, the ratio of discounted benefits to discounted costs tends to decline as the discount rate increases.

Exhibit 10-13 Impact of Alternative Discount Rates on Transit Investment Estimates by Scenario

	Low-Growth Scenario		High-Growth Scenario	
Discount Rates	Billions of 2012 Dollars	Percent Change From Baseline	Billions of 2012 Dollars	Percent Change From Baseline
7% (Baseline)	$22.88		$26.42	
3%	$22.85	-0.2%	$30.95	17.2%

Source: Transit Economic Requirements Model.

part III

Special Topics

Introduction

Chapters 11 and 12 provide additional insights into topics touched on elsewhere in this report and highlight related issues. Chapter 11 presents information on pedestrian and bicycle transportation, nonmotorized modes of travel that are essential components of personal mobility. Chapter 12 provides information on transportation serving Federal and Tribal lands, a subset of the transportation system that is not explored in depth in the analyses presented in Chapters 1 through 10.

Chapter 11, **Pedestrian and Bicycle Transportation**, describes national policies and plans over the past 25 years to promote bicycle and pedestrian use. It also discusses Federal investment over time, activity levels, and safety trends. It concludes with a discussion of ongoing initiatives and research projects aimed at increasing pedestrian and bicycle mode share and improving safety for these modes.

Chapter 12, **Transportation Serving Federal and Tribal Lands**, examines the transportation systems serving Federal and Tribal lands, including resources and types of lands served, and the role of these systems. It also discusses the condition, sources of funding, and expenditures. The chapter concludes with a discussion of the future of the Federal and Tribal transportation systems.

chapter 11

Pedestrian and Bicycle Transportation

Pedestrian and Bicycle Transportation

Improving pedestrian and bicycle safety is a top priority at the U.S. Department of Transportation (DOT). The agency is committed to making walking and bicycling safer and more comfortable transportation options for everyone. Providing multimodal transportation options such as walking and biking improves access and mobility, fosters Ladders of Opportunity,[1] and contributes to a range of policy goals related to equity, health, economic development, and the environment.

This chapter outlines policies and plans that frame and provide context for ongoing activities to advance pedestrian and bicycle transportation in the United States. It summarizes trends in funding and walking and bicycling activity, while also highlighting selected current projects and initiatives. Information on pedestrian and bicycle safety, including data on fatality statistics and trends, is provided in Chapter 4.

Background and Context for Pedestrian and Bicycle Network Development, Safety, and Usage Trends

The following summary of trends, discussed in more detail throughout this chapter, highlights progress made by DOT; partner agencies; advocacy organizations; and local, metropolitan planning organization (MPO), and State stakeholders over the past three decades toward advancing safe, comfortable, and well-utilized pedestrian and bicycle transportation networks.

- The Federal goal set in the *1994 National Bicycling and Walking Study* to reduce pedestrian and bicycle injuries and fatalities by 10 percent has been exceeded. Injuries have decreased 17 percent for pedestrians and 20 percent for bicyclists, while fatalities have dropped 16 percent among pedestrians and 13 percent among bicyclists. Nevertheless, the rate of injuries increased between 2009 and 2013, after steadily dropping between 1994 and 2008.
- Progress also has been made toward the 1994 goal to double the share of nationwide trips made by pedestrians and bicyclists from 7.9 percent to 15.8 percent; it has risen to 11.5 percent, almost halfway to the target.
- Federal funding for pedestrian and bicycle transportation has increased significantly, from $113 million in 1994 to a peak level of $1.2 billion in 2009; funding for 2014 was $820 million.
- Federal policies and guidance supporting the inclusion of pedestrian and bicycle transportation into routine transportation planning, design, and construction have advanced multimodal planning and project development at all levels. Hundreds of communities, MPOs, and State Departments of Transportation (DOTs) have adopted Complete Streets policies that require the formal consideration of all modes of travel throughout the project planning and

development process. States and communities now routinely accommodate people with disabilities when developing pedestrian facilities and pedestrian access routes.

- Context Sensitive Solutions (CSS), a collaborative, interdisciplinary, and holistic approach to the development of transportation projects, has become increasingly accepted by a broad range of stakeholders in all phases of program delivery, including long-range planning, programming, environmental studies, design, construction, operations, and maintenance.
- Livability, or the linkage between the quality and location of transportation facilities and broader opportunities such as access to good jobs, affordable housing, quality schools, and safer streets and roads, also has become increasingly commonplace in transportation planning and design at all levels. For more information on livability and FHWA's Livability Initiative, see Chapter 5.
- The field of pedestrian and bicycle transportation engineering and planning has evolved, enabling practitioners at all levels to become more effective in improving safety and mobility for pedestrians and bicyclists. Professional organizations such as the Association of Pedestrian and Bicycle Professionals and pedestrian and bicycle advocacy organizations have played a key role in this process.
- Information-sharing resources such as the Pedestrian and Bicycle Information Center have been established, and professional training programs, guidebooks, and other educational resources have been developed.
- Improvements have been made to pedestrian and bicycle data collection methods and analysis tools, and research activities have increased.

Trends in Pedestrian and Bicycle System Usage

In 1994, FHWA and NHTSA submitted the final report of the National Bicycling and Walking Study to Congress. The study set two overall goals:

- Double the percentage of trips made by bicycling and walking in the United States from 7.9 percent to 15.8 percent of all travel trips. Although the percentage of trips has increased to nearly 12 percent, the goal to double the mode share has not yet been reached.
- Reduce the number of bicyclists and pedestrians killed or injured in traffic crashes by 10 percent. This goal has been met, although recent trends indicate slowdowns and reversals of the progress achieved between 1994 and 2009. For more information, see Chapter 4.

Activity Level Trends

According to the most recent National Household Travel Survey (NHTS),[2] 11.5 percent of all trips were made by bicycling or walking in 2009, compared with 7.9 percent in 1994 (note that the NHTS is not performed annually). This change represents an increase of 45 percent, which demonstrates progress but falls short of the goal to double the share. Most of the increase is attributed to more walking: The percentage of all trips pedestrians made increased from 7.2 to 10.5 percent, while the share of trips made by bicycles increased from 0.7 to 1 percent (see *Exhibit 11-1*).

According to the 2014 American Community Survey,[3] an ongoing survey that provides information about travel behavior and trends, the number of American workers who commute by bicycle increased from 532,364 in 2005 (0.4 percent of all commute trips) to 857,774 in 2013 (0.6 percent of all commute trips). Although still modest when compared to all commute trips, this 60-percent increase among bicyclists was the largest change among the types of commuter modes reported. The number of people walking to work increased from 3,327,276 (2.5 percent of all commute trips) to 4,002,946 (2.8 percent of all commute trips) over the same period, representing a 20-percent increase. By comparison, the number of people driving alone increased by only 4 percent, from 89,875,050 in 2005 to 93,713,554 in 2013, and this group's representation among all commuters dropped from 77 to 76 percent (see *Exhibit 11-2*).

Exhibit 11-1 Bicycle and Pedestrian Travel Trends as Percentage of All Trips, 1994 and 2009

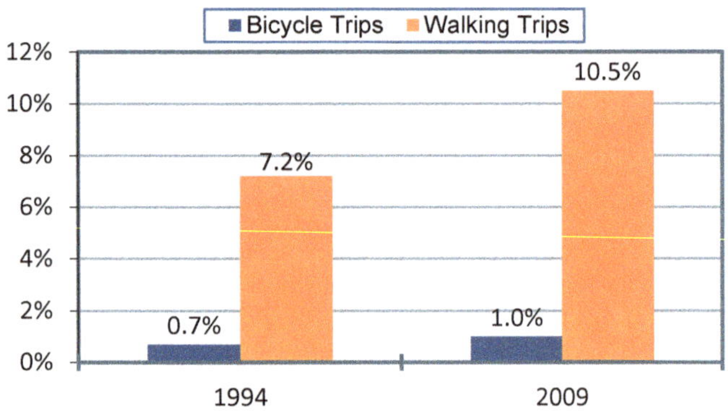

Source: National Household Travel Survey.

Exhibit 11-2 Commuting Trends as Percent Increase, 2005–2013

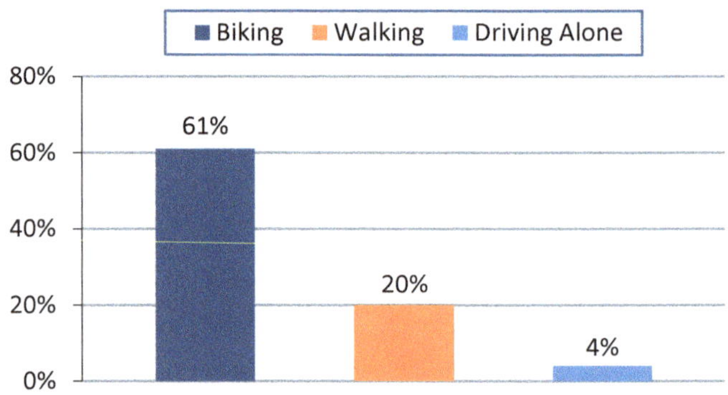

Source: American Community Survey.

In conjunction with the Centers for Disease Control and Prevention's Healthy Community Design Initiative, the Alliance for Biking & Walking publishes a biennial Benchmarking Report on bicycle and walking travel and behavior within the 50 United States.[4] *Exhibit 11-3* lists the "top ten" States and major cities where commuter bicycling and walking levels are higher than the national average. *Exhibit 11-4*, also drawn from the 2014 Benchmarking Report, illustrates the percentage of pedestrian and bicycle commuters in large cities, while also illustrating the recently growing interest in nonmotorized traffic counting programs among American cities.

Exhibit 11-3 Bicycle and Pedestrian Commuter Mode Share in States and Large Cities, 2009–2011

State Commuter Mode Share (Highest to Lowest)						Large City Commuter Mode Share (Highest to Lowest)					
Bicycling to Work			Walking to Work			Bicycling to Work			Walking to Work		
1	Oregon	2.3%	1	Alaska	7.9%	1	Portland, OR	6.1%	1	Boston	15.0%
2	Montana	1.4%	2	New York	6.4%	2	Minneapolis	3.6%	2	Washington, DC	11.8%
3	Colorado	1.3%	3	Vermont	5.8%	3	Seattle	3.4%	3	New York City	10.3%
4	Idaho	1.1%	4	Hawaii	4.8%	4	San Francisco	3.3%	4	San Francisco	9.9%
5	Alaska	1.0%	5	Montana	4.8%	5	Washington, DC	2.9%	5	Honolulu	9.7%
6	California	1.0%	6	Massachusetts	4.7%	6	Tucson	2.5%	6	Philadelphia	8.8%
7	Arizona	1.0%	7	South Dakota	4.3%	7	Oakland	2.5%	7	Seattle	8.6%
8	Hawaii	0.9%	8	Oregon	3.9%	8	New Orleans	2.3%	8	Baltimore	6.8%
9	Wyoming	0.9%	9	Pennsylvania	3.9%	9	Sacramento	2.3%	9	Minneapolis	6.3%
10	Washington	0.9%	10	Maine	3.8%	10	Denver	2.2%	10	Chicago	6.3%

Source: National Alliance for Bicycling and Walking 2014 Benchmarking Report.

Exhibit 11-4 Pedestrian and Bicycle Mode Share Trends, Outcome Benchmark Changes, 2005–2012

Mode Share	Years of Benchmarking				Data Source
	2005/2006	2007/2008	2009/2010	2011/2012	
% of commuters who walk: national average	2.5%	2.8%	2.9%	2.8%	ACS 1 year est.
% of commuters who walk: large city average	4.5%	4.8%	4.9%	5.0%	ACS 3 year est.
% of commuters who bicycle: national average	0.4%	0.5%	0.6%	0.6%	ACS 1 year est.
% of commuters who bicycle: large city average	0.7%	0.8%	0.9%	1.0%	ACS 3 year est.
# of cities counting bicyclist trips	-	-	36/51	43/52	City survey
# of cities counting pedestrian trips	-	-	26/51	37/52	City survey
# of States counting bicyclist trips	-	-	24	38	State survey
# of States counting pedestrian trips	-	-	24	36	State survey

Source: National Alliance for Bicycling and Walking 2014 Benchmarking Report.

Safety Trends

NHTSA collects and distributes information regarding traffic crashes, injuries, and fatalities. Fatal injuries are tracked via the Fatality Analysis Reporting System, and injuries are tracked via the National Automotive Sampling System – General Estimates System. The following points are based on NHTSA data from 1994 to 2013:[5]

- The numbers of pedestrian injuries and fatalities have dropped by 16 percent and 17 percent, respectively, since 1994, while bicyclist injuries and fatalities have dropped by 20 percent and 13 percent, respectively. The goal to improve safety by 10 percent has been exceeded. Gains made toward reducing the rates of injuries and fatalities among pedestrians and bicyclists during the first 15 years of the period, however, were undercut by increases during the most recent 4 years.
- About 1.5 million pedestrians have been injured over the past 20 years. Another 1 million bicyclists were injured between 1994 and 2013. Meanwhile, more than 50 million people were injured while riding in cars and light-duty trucks.

- Approximately 100,000 pedestrians died during the past 20 years, compared to fewer than 75,000 motorcyclists and 15,000 bicyclists, along with about 580,000 people in cars and light trucks.
- In 1994, pedestrians and bicyclists suffered more than twice as many injuries as motorcyclists and three times as many fatalities. By 2013, the numbers of injuries and fatalities among pedestrians and bicyclists had dropped moderately, while the number of motorcycle injuries and fatalities rose dramatically. Today, the numbers of deaths among motorcyclists and pedestrians are virtually the same; about 18,000 people in each group died in 2010 through 2013. The rates of injury and death among motorcyclists have begun to drop in recent years, while the rates for pedestrians and bicyclists have been rising.

National Policies, Programs, and Initiatives

Federal policies and investments to promote pedestrian and bicycle transportation have evolved steadily over the past 25 years. The overall tone and content of Federal, State, and local policy statements regarding nonmotorized transportation have shifted to reflect a significant increase in its perceived value—from a forgotten mode in the 1980s to routine consideration, proactive support, and leadership today.

Federal Funding for Pedestrian and Bicycle Transportation

In 1990, the year before the Intermodal Surface Transportation Efficiency Act of 1991 (ISTEA) was enacted, Federal-aid obligations for pedestrian and bicycle improvements amounted to about $6 million. The new emphasis on multimodal transportation under ISTEA, particularly the introduction of the Transportation Enhancement (TE) activities, resulted in a rapid expansion of funds allocated to pedestrian and bicycle transportation. The 1992 obligation of almost $23 million was nearly four times the 1990 obligation.

By 1997, the obligation for pedestrian and bicycle projects totaled $238 million, about three-fourths of which came from TE activities. Federal obligations for these modes under the Transportation Equity Act for the 21st Century (TEA-21) increased from $204 million in 1999 to more than $427 million in 2004, with about two-thirds from TE activities. Funding levels remained around the $500-million mark until 2009, when they increased again to the current annual range of $800 million to $1 billion (with a spike in 2009 and 2010 related to the American Recovery and Reinvestment Act) (see *Exhibit 11-5*).

Today, pedestrian and bicycle projects are broadly eligible for funding throughout the Federal-aid and Federal Lands programs. Funds from the National Highway Performance Program (NHPP), Surface Transportation Block Grant (STBG) Program, Congestion Mitigation and Air Quality Improvement Program (CMAQ), Highway Safety Improvement Program, Transportation Alternatives (TA) Set-Aside from STBG (including the Recreational Trails Program set-aside and Safe Routes to School projects), Tribal Transportation Program, Federal Lands Transportation Program, and Federal Lands Access Program may be used for bicycle transportation and pedestrian walkways. Pedestrian and bicycle projects also are eligible under some Federal Transit Administration programs. FHWA has a table of Pedestrian and Bicycle Funding Opportunities describing available transit, highway, and safety funds.

Exhibit 11-5 Federal Obligations for Bicycle and Pedestrian Projects, 1992–2015

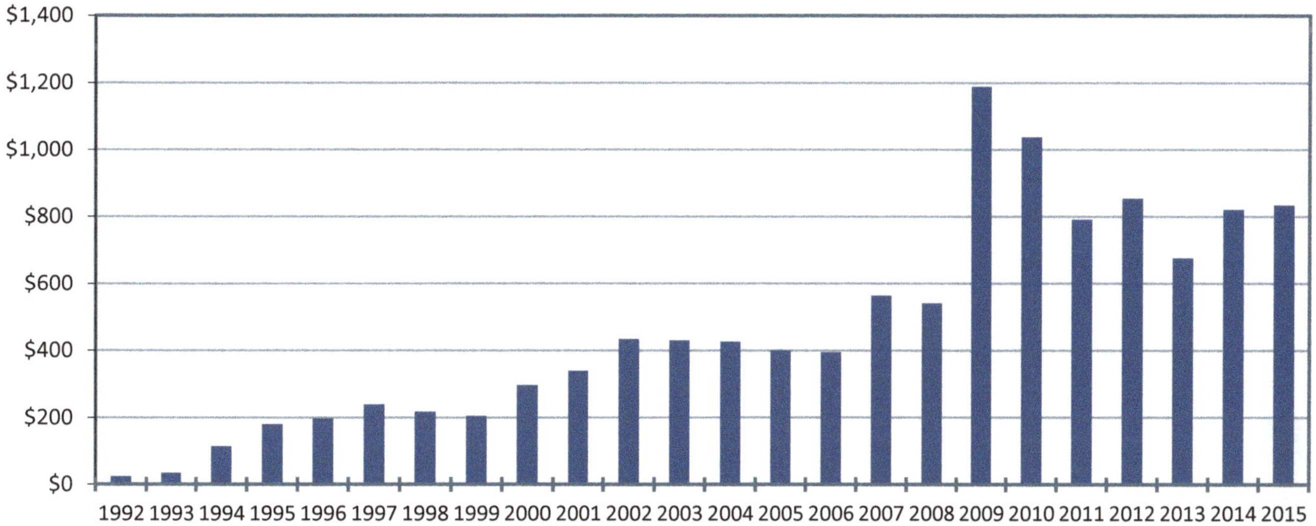

Source: FHWA Fiscal Management Information System.

STBG and CMAQ funds may be used to construct pedestrian walkways and bicycle transportation facilities and to carry out nonconstruction projects related to safe bicycle use. NHPP funds may be used to construct pedestrian walkways and bicycle transportation facilities on land adjacent to any highway on the NHS. Funds from the Federal Lands Transportation Program and Federal Lands Access Program authorized for forest highways, forest development roads and trails, public lands development roads and trails, park roads, parkways, Indian reservation roads, and public lands highways may be used to construct bicycle transportation facilities and pedestrian walkways.

Federal Strategic Plans, Policies, and Guidance: 1994–2016

In addition to legislative directives and increased funding, Federal support for pedestrian and bicycle transportation has grown significantly over the past 20 years through policy and regulatory documents. Selected plans and policies are highlighted below.

The FHWA ***1994 National Bicycling and Walking Study: Transportation Choices for Changing America*** represented the first comprehensive examination of the state of nonmotorized transportation in the United States. The 1994 study also included a 9-point Federal Action Plan, supported by approximately 60 strategies. Status reports on progress toward the 1994 goals and strategies were developed in 1999, 2004, and 2010. These reports documented progress toward the original commitments DOT made to establish pedestrian and bicycle travel as a meaningful element of a safe, convenient transportation system.

Federal Transportation Legislation Supporting Pedestrian and Bicycle Transportation

The Intermodal Surface Transportation Efficiency Act of 1991 (ISTEA) represented the first major shift away from the traditional national focus on high-speed roadway networks. ISTEA provided a framework for creating an intermodal transportation system that links every point to all other points by at least one mode, enabling the movement of all people and goods at reasonable speeds and economic costs.[6] The legislation authorized general eligibility for pedestrian and bicycle projects under the Federal-aid highway program; special funding categories for small-scale projects, such as Transportation Enhancement (TE) activities; and funding of pedestrian and bicycle projects under the Surface Transportation Program (STP), Congestion Mitigation and Air Quality Improvement Program (CMAQ), National Highway System (NHS), and Federal Lands Highway Program. Each State was required to consider and provide safe bicycle accommodations when Federal-aid funds were used to replace or rehabilitate bridge decks, except on full-access controlled highways. Pedestrian walkways and bicycle transportation facilities could be designated as highway projects with a Federal share for construction costs of 80 percent. ISTEA required metropolitan planning organizations (MPOs) and States to give due consideration to pedestrian walkways and bicycle transportation facilities in long-range plans.

The National Highway System (NHS) Designation Act of 1995 included several clauses that advanced Federal support for bicycle and transportation investments, allowing the Federal share for pedestrian and bicycle projects to be the same as that for Federal-aid projects in general, including use of sliding-scale approaches. The Act also provided for an advance payment option and allowed categorical exclusions from environmental impact assessments. The Transportation Equity Act for the 21st Century (TEA-21) of 1998 advanced pedestrian and bicycle transportation by allowing the use of NHS funds for pedestrian walkways and previously eligible bicycle facilities on any route of the NHS. TEA-21 also lifted restrictions to accommodate bicycle access on bridges where access was fully controlled, added pedestrian and bicyclist needs as considerations in the development of comprehensive transportation plans, required FHWA to issue design guidance for accommodating pedestrian and bicycle travel, replaced the National Recreational Trails Fund Act with the Recreational Trails Program (RTP), and established funding for a National Bicycle and Pedestrian Clearinghouse.

Enacted in 2005, the Safe, Accountable, Flexible, Efficient, Transportation Equity Act: a Legacy for Users (SAFETEA-LU) continued the RTP with increases in funding each year through Fiscal Year 2012, established the national Safe Routes to School (SRTS) program, continued funding for a National Bicycle and Pedestrian Clearinghouse, and established the Nonmotorized Transportation Pilot Program to study the impact of concentrated investment in pedestrian and bicycle facilities.

The Moving Ahead for Progress in the 21st Century (MAP-21) Act in 2012 created a new Transportation Alternatives Program (TAP) to replace the Transportation Enhancements, Recreational Trails, and SRTS programs. TAP provides competitive grant funds, administered through the States and MPOs. MAP-21 turned the RTP into a set-aside from TAP and made SRTS activities and projects eligible for TAP, subject to TAP requirements.

The Fixing America's Surface Transportation (FAST) Act of 2015 modified Federal law to require federally funded projects on the NHS to consider access for other modes of transportation. It broadened design guidelines for pedestrian and bicycle facilities, providing greater design flexibility. It renamed the STP as the Surface Transportation Block Grant (STBG) Program, and replaced TAP with the Transportation Alternatives (TA) Set-Aside under the STBG Program, maintaining funding levels and eligibility. It also established a NHTSA safety fund to reduce bicycle and pedestrian fatalities, and broadened design guidelines for pedestrian and bicycle facilities.

The FHWA 2000 publication, ***Accommodating Bicycle and Pedestrian Travel: A Recommended Approach***, focuses on the design and inclusion of pedestrian and bicycle facilities funded by FHWA and the Federal-aid highway program. FHWA offices worked directly with State DOTs to implement the 2000 policy. The 2000 policy also outlined legitimate exceptions to the expectation that bicycling and walking facilities be added to Federal-aid projects.

DOT issued its ***Policy Statement on Bicycle and Pedestrian Accommodations, Regulations, and Recommendations*** in March 2010. The policy states:

> "The DOT policy is to incorporate safe and convenient walking and bicycling facilities into transportation projects. Every transportation agency, including DOT, has the responsibility to improve conditions and opportunities for walking and bicycling and to integrate walking and bicycling into their transportation systems. Because of the numerous individual and community benefits that walking and bicycling provide—including health, safety, environmental, transportation, and quality of life—transportation agencies are encouraged to go beyond minimum standards to provide safe and convenient facilities for these modes."

On August 20, 2013, FHWA issued a **memorandum to support flexibility in pedestrian and bicycle facility design.** The memorandum recognizes the American Association of State Highway and Transportation Officials (AASHTO) *Guide for the Planning, Design, and Operation of Pedestrian Facilities*; AASHTO *Guide for the Development of Bicycle Facilities*; Institute of Transportation Engineers' *Designing Walkable Urban Thoroughfares* document; and the National Association of City Transportation Officials' *Urban Bikeway Design Guide* as resources to inform the design of safe, comfortable, and context-sensitive pedestrian and bicycle facilities. In a subsequent communication, FHWA also noted that the *Urban Street Design Guide* can be used, in conjunction with other design resources, to inform the planning and design process.

The **Design Resource Index** identifies the specific location of information in key national design manuals for various pedestrian and bicycle design treatments. The Design Resource Index, developed by the Pedestrian and Bicycle Information Center and FHWA, helps practitioners quickly access the right resources and will reduce the amount of time needed to search through multiple design guides to find the information. The Design Resource Index consists of three separate matrices: On-Street Bicycle Facilities, Shared Use Paths, and Pedestrian Facilities. The Design Resource Index incorporates national resource manuals and guidelines published by FHWA, Institute of Transportation Engineers (ITE), AASHTO, National Association of City Transportation Officials, and the U.S. Access Board.

FHWA released the ***Separated Bike Lane Planning and Design Guide*** in May 2015. It outlines planning considerations for separated bike lanes (also called "cycle tracks" and "protected bike lanes") and provides a menu of design options covering typical one- and two-way scenarios. It highlights options for providing separation, while also documenting midblock design considerations for driveways, transit stops, accessible parking, and loading zones. The guide provides detailed intersection design information covering topics such as turning movement operations, signalization, signage, and on-road markings. Case studies highlight best practices and lessons learned throughout the document. The guide identifies potential future research, highlights the importance of ongoing peer exchange and capacity building, and emphasizes the need to create holistic ways to evaluate the performance of a separated bike lane.

In September 2015, FHWA provided **updated guidance** on Bicycle and Pedestrian Provisions of Federal Transportation Legislation. The update included policy and legislative references and provided guidance on funding eligibility, planning, project delivery procedures, project selection, and design references. This guidance is consistent with FHWA initiatives related to performance-based practical design and design flexibility, accelerated project delivery, proven safety countermeasures, and Every Day Counts. It describes the range of opportunities to improve conditions for bicycling and walking. FHWA expects to further update this guidance to incorporate the FAST Act by the end of 2016.

The ***Statewide Pedestrian and Bicycle Planning Handbook*** helps State DOTs develop or update State pedestrian and bicycle plans. Based on research that includes interviews with 9 State DOTs and critical evaluations of documents from 15 States, this handbook covers statewide planning from plan inception and scoping to engaging stakeholders and the public; developing goals, objectives, and strategies; collecting and analyzing data; linking to the larger statewide transportation planning process; and implementation. For each stage of the planning process, this handbook uses recent experiences and noteworthy practices from DOTs around the country, helping inform a new generation of statewide nonmotorized planning and implementation.

The **Achieving Multimodal Networks: Applying Design Flexibility and Reducing Conflicts** report helps practitioners address topics such as intersection design, road diets, pedestrian crossing treatments, transit and school access, freight, and accessibility. It highlights ways to apply design flexibility, while focusing on reducing multimodal conflicts and achieving connected networks.

The **Guidebook for Developing Pedestrian and Bicycle Performance Measures** helps communities develop performance measures that can fully integrate pedestrian and bicycle planning in ongoing performance management activities.

The **Incorporating On-Road Bicycle Networks into Resurfacing Projects** report helps communities integrate on-road bicycle facilities as part of their routine roadway resurfacing process. This is an efficient and cost-effective way for communities to create connected networks of bicycle facilities.

Other resources released in 2016 include Pursuing Equity in Pedestrian and Bicycle Planning, the Bike Network Mapping Idea Book, and Bicycle Network Planning and Facility Design Approaches in the Netherlands and the United States.

Current Federal Initiatives

DOT is currently engaged in a range of planning, design, promotion, and project development initiatives to advance pedestrian and bicycle transportation safety, accessibility, and connectivity. The following section describes several key activities currently underway.

As part of a **Safer People, Safer Streets: Pedestrian and Bicycle Safety Initiative,** DOT division and field office staff convened and led pedestrian and bicycle safety assessments in every State in 2014 and 2015. The initiative included the *Mayors' Challenge for Safer People, Safer Streets*, which challenged mayors and local elected officials to take significant action to improve safety for bicyclists and pedestrians of all ages and abilities. DOT is working with university transportation centers and other stakeholders to identify and remove barriers to improving nonmotorized safety.

A newly formed Pedestrian and Bicycle Safety Action Team of the DOT Safety Council is implementing the initiative.

The **Ladders of Opportunity** initiative involves a range of activities to enhance access to economic opportunities for all Americans by investing in transportation projects that better connect communities to essential services. Its aim is to promote prosperity and improve quality of life for all individuals, with a focus on low-income, underserved, vulnerable, and disadvantaged populations. Key objectives of the initiative are to build and restore physical connections, develop workforce capacity, and catalyze neighborhood revitalization. Toward this end, DOT launched the Ladders of Opportunity Transportation Empowerment Pilot in seven U.S. cities in April 2015. Entitled LadderSTEP, the program provides technical assistance and support to help communities attract public and private resources to game-changing transportation projects. As part of DOT's Every Place Counts Design Challenge, in July 2016 the Department convened two-day design sessions with communities in Nashville, TN, Philadelphia, PA, Spokane, WA, and St. Paul-Minneapolis, MN to provide on-site technical assistance in visioning and identifying innovative community design solutions that bridge the infrastructure divide and reconnect people to opportunity (See: https://www.transportation.gov/opportunity/challenge).

The **EDC-4/Community Connections** initiative will promote the use of innovative transportation planning and project delivery strategies to lead to community-focused transportation projects that support community revitalization. Webinars and summits throughout the country will focus on various transportation components to enhance the transportation process and improve connectivity between disadvantaged populations and essential services.

Focus States and Cities: Since 2004, FHWA has focused extra resources on the States and cities with the highest pedestrian fatalities. Beginning in 2015, the list of States and cities was revised to include cyclist fatalities. Under this effort, FHWA concentrates its technical assistance on evaluating, planning, and resolving pedestrian and cyclist safety issues in States with the highest pedestrian and cyclist fatalities. For example, FHWA provides free technical assistance and courses to each of these States and cities, and free bimonthly webinars on subjects of interest.

Certain processes, infrastructure design techniques, and highway features have been highly effective in improving safety. FHWA actively encourages practitioners to consider these **proven safety countermeasures** in projects. Road diets, or a roadway reconfiguration that enhances safety, mobility, and access for all, are one of the proven safety countermeasures. Road diets are promoted through the **Every Day Counts** initiative, which is intended to identify and rapidly deploy proven but underutilized innovations to shorten the project delivery process, enhance roadway safety, reduce congestion, and improve environmental sustainability.

Performance-based practical design is an approach grounded in a performance management framework. The approach encourages cost savings by using the flexibility that exists in current design guidance and regulations. These cost savings will enable cities, MPOs, and States to deliver more projects (for example, projects that will create or significantly improve connected pedestrian and bicycle networks). The planning and design process should consider both short- and long-term project and system goals and should focus on scoping projects to stay within the core purpose and need.

FHWA's **PedSafe** and **BikeSafe** countermeasure selection systems provide State and local transportation officials with information on countermeasures and other treatments that can be installed to help improve pedestrian and cyclist safety. Most recently updated in 2013 and 2014 respectively, the PedSafe and BikeSafe selection tools enable users to input a specific location, select the goals of the treatment (i.e., reduce traffic volumes or mitigate crashes), and describe the location (in terms of the roadway's speed limit, traffic volume, etc.). The tool then provides the user with a list of recommended treatments, describes those treatments and factors to consider prior to installation, and provides case studies of where the treatment has been implemented.

The FHWA focus on **Connected Pedestrian and Bicycle Networks** builds on the agency's long-standing support of pedestrian and bicycle transportation through policies, planning, and funding. To advance this work, FHWA is increasingly focusing on the documentation and promotion of safe and accessible pedestrian and bicycle *networks*, which are interconnected pedestrian and bicycle transportation facilities that enable people of all ages and abilities to travel where they want to go, safely and conveniently.

FHWA's **Bicycle-Pedestrian Count Technology Pilot** funds the purchase of a limited number of portable automatic counters to collect count data at various locations within 10 MPOs. The pilot research project requires counts to be collected over 1 year, and the data and experiences will be shared with FHWA. Participants will have access to a series of internal webinars and other technical assistance opportunities.

An FHWA-funded project will modify the **Traffic Monitoring Analysis System (TMAS)** to receive and report on pedestrian and bicycle counts based on the *Traffic Monitoring Guide* data format. FHWA maintains TMAS to support statistical analysis of travel trends. Using TMAS, FHWA computes basic reports from data generated from automatic collection programs for motorized vehicles, vehicle classification counts, and weigh-in-motion counters. The pedestrian and bicycle data enhancements will be included in the next version of TMAS (Version 3.0), scheduled for testing in 2016. FHWA is providing resources and information to enable communities to collect data in the format recommended in the *Traffic Monitoring Guide* so they can be incorporated into TMAS when that functionality becomes available. FHWA is also leading a related effort to develop a regional pedestrian and bicycle count database. The project will include assessing the feasibility of moving counts from regional collection centers to the TMAS database.

FHWA is addressing professional **Capacity Building**. For example, on August 20, 2015, FHWA issued questions and answers related to *Bicycle and Pedestrian Funding, Design, and Environmental Review: Addressing Common Misconceptions*. This document answered several questions that FHWA had received from the public.

FHWA is partnering with ITE in creating a Practitioner's Guide to the ***Designing Walkable Urban Thoroughfares: A Context Sensitive Approach, an ITE Recommended Practice*** **(ITE CSS RP)** report. The new *Practitioner's Guide for Walkable Urban Thoroughfare Design* will enhance the practices and principles published in the ITE CSS RP by providing an attractive and easy-to-use resource that clearly communicates the principles, techniques, and design solutions portrayed in the ITE CSS RP. This new resource will serve as a catalyst for increased State, regional, and local implementation of multimodal principles in the design of urban thoroughfares.

Moving Forward

The **Strategic Agenda for Pedestrian and Bicycle Transportation** will inform FHWA's pedestrian and bicycle activities in the next 3 to 5 years and is organized around four goals: (1) Networks, (2) Safety, (3) Equity, and (4) Trips. Each goal includes actions relating to (a) Capacity Building, (b) Policy, (c) Data, and (d) Research. It emphasizes collaboration and partnerships, building capacity around existing resources, implementing existing policies, and building on USDOT's Policy Statement on Bicycle and Pedestrian Accommodations. The volume and type of activities described demonstrates FHWA's ongoing national leadership on multimodal transportation and represents the agency's commitment to institutionalize and mainstream multimodal issues.

The Strategic Agenda establishes the following National goals that will inform FHWA's pedestrian and bicycle activities in the coming years:

- Achieve an 80-percent reduction in pedestrian and bicycle fatalities and serious injuries in 15 years and zero pedestrian and bicycle fatalities and serious injuries in the next 20 to 30 years.
- Increase the percentage of short trips represented by bicycling and walking to 30 percent by 2025. This will indicate a 50-percent increase over the 2009 value of 20 percent. Short trips are defined as trips 5 miles or less for bicyclists and 1 mile or less for pedestrians.

Implementation of the Strategic Agenda will involve coordinating policies, leveraging investments, promoting partnerships, and enhancing access to opportunity in communities and neighborhoods throughout the United States. These challenges include:

- Improve network connections and multimodal **connectivity** among bicycle, pedestrian, and transit routes; and measure change in networks over time. This will help meet the goals established in the Strategic Agenda.
- Improve and coordinate bicycle, pedestrian, and transit **safety**, including infrastructure, promotional campaigns, and educational resources. Improving safety is particularly critical, given the recent reversal of progress toward reducing pedestrian and bicycle injuries and fatalities.
- Strengthen **collaboration** among Federal agencies, State DOTs, MPOs, and transit agencies on strategies to integrate pedestrian and bicycle programs within their existing activities. With increased interest in the impact of transportation on public health, continuing to address the relationship between public health and investments in pedestrian and bicycle infrastructure will be key.

- Foster **professional development** among pedestrian and bicycle planners and engineers through training and networking opportunities.
- Improve and promote **pedestrian and bicycle data** collection methods and analysis tools for local, regional, State, and national agencies.
- Increase submittal of pedestrian and bicycle data to **TMAS** by local, regional, State, and Federal agencies so that they can become the basis for valuable research, evaluation, and project prioritization.
- Strengthen coordination and breadth of pedestrian and bicycle **research** programs, addressing topics that directly address the needs of pedestrian and bicycle practitioners at all levels.
- Advance **technology transfer** and information-sharing methods to apply research findings and promote best practices.

---

[1] Ladders of Opportunity is a Federal initiative to promote prosperity and improve quality of life for all individuals, with a focus on low-income, underserved, vulnerable, and disadvantaged populations.
[2] http://nhts.ornl.gov/introduction.shtml.
[3] American Community Survey, 2014: http://www.census.gov/acs/www/.
[4] Alliance for Bicycling and Walking, Bicycling and Walking in the United States: 2014 Benchmarking Report, 2014: http://www.bikewalkalliance.org/resources/benchmarking.
[5] Charts developed by study team based on data provided by NHTSA.
[6] https://www.fhwa.dot.gov/publications/publicroads/94fall/p94au1.cfm.

chapter 12

Transportation Serving Federal and Tribal Lands

Transportation Serving Federal and Tribal Lands

This chapter documents transportation that serves Federal and Tribal lands, a subset of the transportation system that the analyses presented in Chapters 1 through 10 do not explore in depth. Included are discussions of the types of lands, access to Tribal communities, resources served, role of transportation in the use of Federal and Tribal lands, role of Federal lands in the U.S. economy, condition of the transportation system, sources of funding, expenditures of funds for construction and maintenance of transportation infrastructure, and the future of transportation on Federal and Tribal lands.

Types of Federal and Tribal Lands

The Federal government has title to about 650 million acres,[1] or about 30 percent of the country's total area of 2.3 billion acres.[2] Additionally, the Federal government holds in trust approximately 55 million acres of land on behalf of Tribal governments, located primarily in the West. Various Federal land management agencies (FLMAs) manage Federal lands, primarily within the Departments of the Interior (DOI), Agriculture (USDA), and Defense (DOD). DOI's Bureau of Indian Affairs primarily holds Tribal lands in trust, but many Tribes own land in addition to these trust lands. *Exhibit 12-1* illustrates the major Federal and Tribal lands (note that this exhibit shows only the large units; many smaller units are not shown due to the scale of the image). *Exhibit 12-2* highlights resources that eight FLMAs manage.

Accessing Tribal Communities

An Indian reservation is land reserved for a Tribe when the Tribe relinquished its other land areas to the United States through treaties. More recently, congressional acts, Executive Orders, and administrative acts have officially recognized additional Tribes and their lands. Tribal communities exist across the country. Some are located in the cities or suburbs, but most are located in rural America. The 229 Alaska Native Villages are found at their historical locations throughout Alaska.

Access to basic community services for the 566 federally recognized sovereign Tribal governments is primarily by road, but in remote Alaskan villages also can be by ice roads, trails for snow machines and all-terrain vehicles, airfields, and waterways. Some Tribes operate transit service within their communities. This transportation infrastructure (roads, bridges, trails, or transit systems) can be owned by the Bureau of Indian Affairs, Tribes, States, or counties and other local governments.

Many roads accessing Tribal lands can be characterized as substandard native surface roadways, accessible only during periods of good weather. Access to many critical community services, jobs, stores, schools, hospitals, emergency services, or intercommunity commerce can be compromised by a common rain event or a thaw of an Alaskan river or permafrost. More than 8 billion vehicle miles are traveled annually on the Tribal Transportation Program road system, even though it is among the most rudimentary of any transportation network in the United States, with more than 60 percent of the system unpaved.

Exhibit 12-1 Major Federal Lands

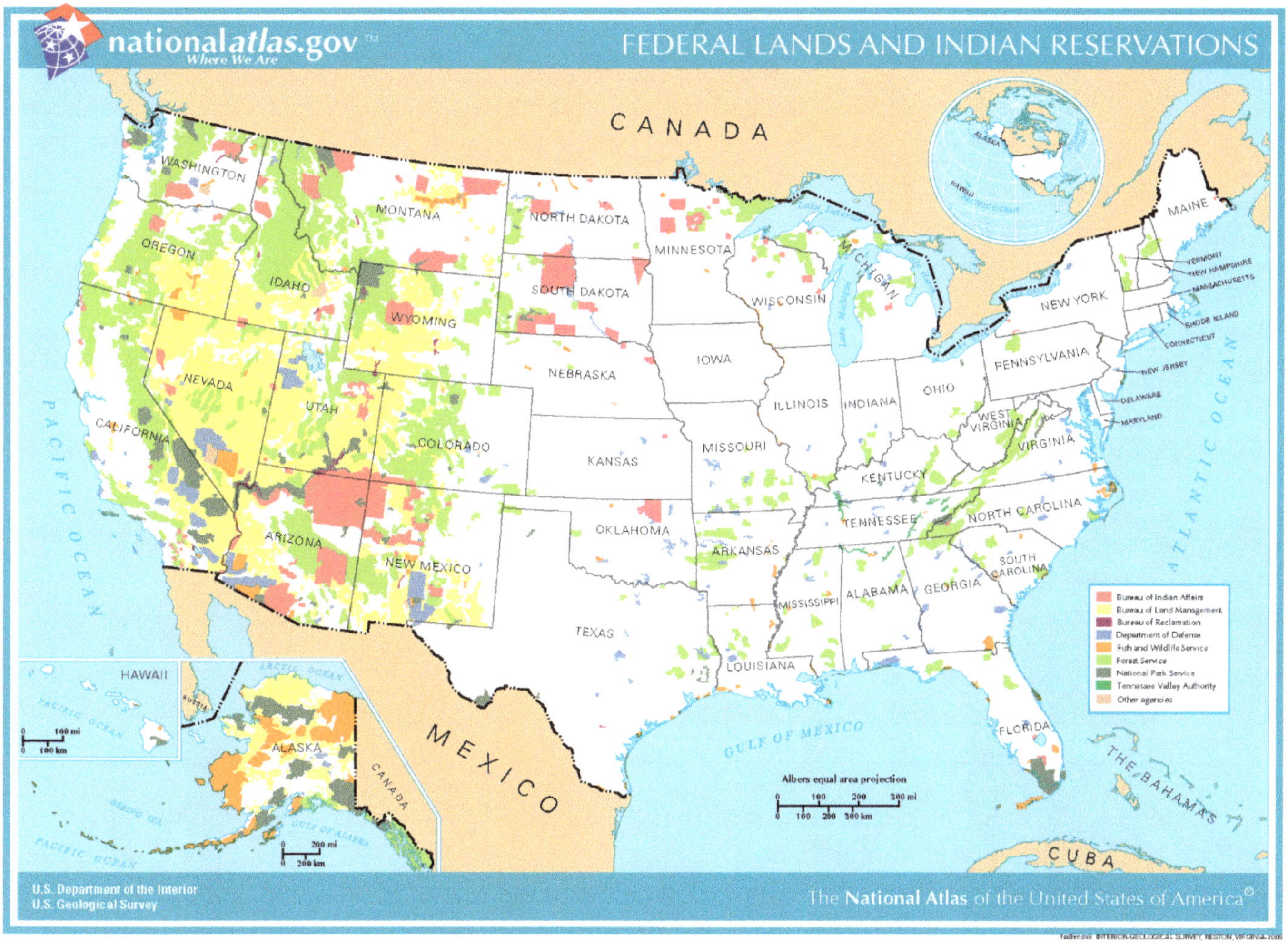

Source: The National Atlas of the United States of America.

Resources Served within Federal Lands

The natural and cultural resources of Federal and Tribal lands are among the Nation's greatest assets. The unique mission of each site shapes how the FLMAs manage the resources and provide access to and around those resources for the public and the citizens living on those lands to enjoy. Most FLMAs are charged with managing the use of resources for the benefit of present and future generations. These Federal lands provide some of the richest resources and most breathtaking scenery in the Nation, clean air and drinking water for millions of Americans, and contributing to hundreds of thousands of jobs for the broader economy. Resource management includes preserving and protecting natural, cultural, historic, and wildlife areas. Many sites have multiple uses, while others have very limited, specific uses. Approximately one-half the Federal lands are managed under multiple use and sustained yield policies that rely on transportation. The remaining lands have protected use management policies, but even so, transportation systems are essential to their resource management, development, recreational use, and protection.

Exhibit 12-2 Types of Lands Managed by Federal Land Management Agencies

Federal Agency	Federal Lands Served
Department of Agriculture	
Forest Service	193 million acres of public lands; 155 National Forests, 20 National Grasslands, and 9 National Monuments; 9100 miles of Scenic Byways; 5,000 miles of Wild and Scenic Rivers; 4,300 Campgrounds; 27 million annual visits to 122 Ski Areas; and 12,000 miles of National Historic and Scenic Trails
Department of the Interior	
National Park Service	412 National Park System units[1]
Fish and Wildlife Service	556 Wildlife Refuges, 38 Wetland Management Districts, 70 Fish Hatcheries, and 43 administrative sites
Bureau of Land Management	247.5 million acres of public lands; 700 million acres of subsurface mineral estate; 2,500 recreation sites; 700 administrative sites; BLM's National Conservation System includes: 16 National Conservation Areas, 17 National Monuments, 221 Wilderness Areas, 2,400 miles of Wild and Scenic Rivers, 545 Wilderness Study Areas, and 5,343 miles of National Historic and Scenic Trails.
Bureau of Indian Affairs	566 federally recognized Indian Tribes
Bureau of Reclamation	476 dams, 338 reservoirs, 187 recreation areas, and 53 power plants
Department of Defense	
Military Installations	4,169 DOD sites, 28.8 million acres
U.S. Army Corps of Engineers - Civil Works Facilities	420 Water Resource Projects

[1] http://www.nps.gov/faqs.htm

Source: FLMAs.

Federal lands have many uses, including national defense facilitation, recreation, education, livestock grazing, timber and minerals extraction, energy generation and transmission, watershed management, fish and wildlife management, and wilderness. These lands are also managed to protect natural, scenic, scientific, and cultural values. In recent years, mineral extraction and timber cutting have been significantly reduced. At the same time, recreation use has significantly increased. *Exhibit 12-3* summarizes annual recreation use and visits on Federal lands. Recreation on Federal lands is measured in recreation visitor days, equivalent to one 12-hour visit.

Role of Transportation in Use of Federal and Tribal Lands

Tribal communities, national defense, recreation, travel and tourism, and resource extraction depend on quality transportation infrastructure. Transportation plays a key role in how people access and enjoy Federal and Tribal lands, and in providing access to jobs and resources. Visiting our Federal lands without the hundreds of thousands of miles of Federal and Tribal roads, trails, and transit systems providing access to and within these lands is inconceivable. This transportation infrastructure provides opportunities for employment, recreational travel and tourism, protection and enhancement of resources, sustained economic development in rural and urban areas, access to educational and health benefits, and national and international access to our Nation's most pristine natural, cultural, and historic resources.

Exhibit 12-3 Summary of Annual Recreation Use and Visits[1]

Federal Agency	Recreation Visits (Millions)	Recreation Visitor Days (Millions)	# of Sites
Department of Agriculture			
Forest Service	173	288	175
Department of the Interior			
National Park Service	307	110	412
Fish and Wildlife Service[2]	47	46	464
Bureau of Land Management	58	58	2,800
Bureau of Indian Affairs	N/A	N/A	N/A
Bureau of Reclamation	25.5	25.5	187
Department of Defense			
Military Installations	N/A	N/A	N/A
U.S. Army Corps of Engineers - Civil Works Facilities	360	N/A	2,857
Total	**970.5**	**527.5**	**6,895**

[1] Data shown are not for a consistent year, but instead reflect the latest available information as of late 2014 when these data were obtained from the FLMAs. Recreation data are based on definitions of the FLMAs, which might not be fully consistent.

[2] 464 of 594 sites open to public use.

Source: FLMAs.

Federal agencies, Tribes, and States have designated and manage numerous roadways as Scenic Byways, many of which are Federal and Tribal roads. Based on archeological, cultural, historic, natural, recreational, and scenic qualities, 150 of these roadways in 46 States are designated as National Scenic Byways and All-American Roads. The USDA Forest Service began designating National Forest Scenic Byways in 1988; as of 2012, more than 130 routes have been designated, totaling 9,000 miles in 36 States. The National Park Service has 80 National Park Service Units designated or affiliated with a National Scenic Byway or an All-American Road. The U.S. Fish and Wildlife Service (FWS) manages 83 National Wildlife Refuge Systems or National Fish Hatcheries along 83 National Scenic Byways or All-American Roads. In 1989, the Bureau of Land Management (BLM) began designating Back Country Byways; more than 54 routes have been designated to date, totaling 3,100 miles in 10 States. BLM has another 60 routes, totaling 5,300 miles in 7 States that are classified as scenic, historic, or other road type. Although just as important, these 60 routes have not been classified as designated Back Country Byways.

DID YOU KNOW?

The Bureau of Land Management has more than 5,700 miles of national scenic and historic trails.

Public roads comprise significant portions of the transportation systems serving these Federal and Tribal lands. In many areas—both urban and rural—transit, bicycle, and pedestrian use supplement this road network, although most agencies do not track this usage. In many remote areas, motorized and nonmotorized trails, waterways, and air transports serve as the primary mode of transportation. The broad range of needs that depends on transportation access to Federal lands is summarized in *Exhibit 12-4*.

Exhibit 12-4 Federal Land Use

Federal Agency	Recreation	Wildlife	Minerals, Oil, & Gas	Grazing & Farming	Water Resources	Timber	Industry	Energy	Housing	National Defense
Department of Agriculture										
Forest Service	X	X	X	X	X	X	X	X	X	X
Department of the Interior										
National Park Service	X	X			X				X	
Fish and Wildlife Service	X	X	X	X	X					
Bureau of Land Management	X	X	X	X	X	X	X	X		
Bureau of Indian Affairs	X	X	X	X	X	X	X	X	X	
Bureau of Reclamation	X	X	X	X	X			X		
Department of Defense										
Military Installations	X	X		X	X		X		X	X
U.S. Army Corps of Engineers - Civil Works Facilities	X	X	X	X	X	X		X		

Source: FLMAs.

Condition and Performance of Roads Serving Federal and Tribal Lands

Although the primary focus of this C&P report is on the Nation's highways, bridges, and transit systems as a whole, the Federal government has a special interest and responsibility for public roads and transportation that provide access to and within federally and tribally owned lands. The transportation systems serving various Federal and Tribal lands are discussed below. Roads serving these lands are summarized in *Exhibit 12-5*.

Forest Service

The Forest Service has jurisdiction over the National Forest System that contains 155 national forests and 20 grasslands spanning approximately 193 million acres in 40 States plus Puerto Rico and the Virgin Islands. The system comprises 30 percent of federally owned lands or approximately 8 percent of the land in the United States.

Approximately 372,000 miles of National Forest System Roads are under the jurisdiction of the Forest Service. About 102,000 miles are reserved for future use, and are not open to or maintained for traffic. Of miles that are maintained for traffic, 205,000 are managed for high-clearance vehicles. These roads are generally native surface roads that are impassable by passenger cars. Approximately 65,000 miles are maintained for passenger car use. Some 9,500 miles of these roads are paved, and the rest are gravel or native surface. Of these, 137 (9,126 miles) are designated byways in the National Forest Scenic Byways Program.

Exhibit 12-5 Roads Serving Federal Lands[1]

Federal Agency	Public Paved Road Miles	Paved Road Condition[2]			Public Unpaved Road Miles	Public Bridges		Backlog of Deferred Maintenance (Transportation Only)
		Good	Fair	Poor		Total	Structurally Deficient[3]	
Forest Service	9,500	42%	55%	3%	362,500	4,200	11%	$2.9 billion[4]
National Park Service	5,500	59%	29%	12%	4,100	1,442	3%	$6 billion
Bureau of Land Management	500	65%	20%	15%	600	835	3%	$350 million
Fish and Wildlife Service	400	60%	25%	15%	5,200	281	7%	$1 billion
Bureau of Reclamation	762	65%	25%	10%	1,253	331	12%	N/A
Bureau of Indian Affairs	8,800	N/A	N/A	N/A	20,400	929	15%	N/A
Tribal Governments	3,300	N/A	N/A	N/A	10,200	N/A	N/A	N/A
Military Installations	27,900	N/A	N/A	N/A	N/A	1,418	26%	N/A
U.S. Army Corps of Engineers	5,247	56%	30%	14%	2,549	416	6.20%	N/A

[1] Data shown are not for a consistent year, but instead reflect the latest available information as of late 2014 when these data were obtained from the FLMAs.

[2] Road condition categories are based on definitions of the FLMAs, which are not fully consistent.

[3] Structural deficiencies are classified using a uniform definition consistent with that presented in the Trends in Bridge Structural Deficiencies section in Chapter 3.

[4] Deferred maintenance figure for the Forest Service is for Passenger Car roads only.

Source: FLMAs.

Timber harvest volumes have declined by 80 percent since the 1980s. The loss of road maintenance support from the timber sale program, reduced work related to resource projects, and increased recreation use have resulted in significant deterioration of the entire road system. The agency currently has a $2.9-billion backlog of deferred maintenance on the 65,000 miles of roads maintained for passenger cars. As shown in *Exhibit 12-6*, of these roads, 42 percent are in good condition, 55 percent are in fair condition, and 3 percent are in poor condition.

Exhibit 12-6 Forest Service Pavement Conditions (Passenger Car Roads Only)

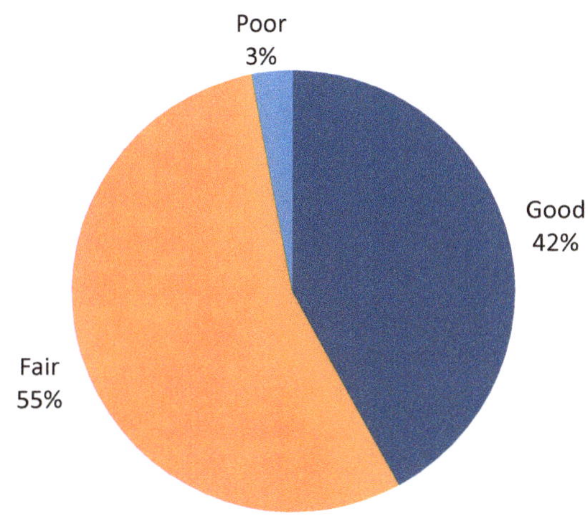

Source

Public National Forest System Roads have approximately 4,200 bridges, 11 percent of which are structurally deficient. Nonpublic National Forest System Roads have approximately 1,000 bridges, 20 percent of which are structurally deficient.

The Forest Service manages approximately 158,000 miles of trails and approximately 6,500 trail bridges. Trails can be motorized trails or nonmotorized and vary in surface and length. The primary distinction between a road and a trail is width: A right-of-way less than 50 inches wide is generally considered a trail. About 32,000 miles of trails are inside wilderness areas and just over 12,000 miles are designated national scenic and historic trails, such as the Appalachian National Scenic Trail, Continental Divide National Scenic Trail, Florida National Scenic Trail, Pacific Crest National Scenic Trail, Nez Perce National Historic Trail, Arizona National Scenic Trail, and Pacific Northwest National Scenic Trail.

National Park Service

The NPS system includes more than 84 million acres[3] across 412 national park units. These units include national parks, national parkways, national monuments, national historic sites, national military parks, national battlefields, national memorials, national recreation areas, national scenic waterways, and national seashores.

Roads continue to be the primary method of access to and within the NPS system. The NPS transportation network is composed of both motorized and nonmotorized facilities that accommodate surface, marine, and aviation modes, enabling high-quality access to park units. With few exceptions, travel by private vehicle or tour buses is the only means of getting to and moving within the system. As a result, some of the most conspicuous problems in units of the NPS system with high visitation levels stem from difficulty accommodating increasing volumes of traffic, larger vehicles, and the growing demand for visitor parking.

Exhibit 12-7 National Park Service Pavement Conditions (Paved Roads Only)

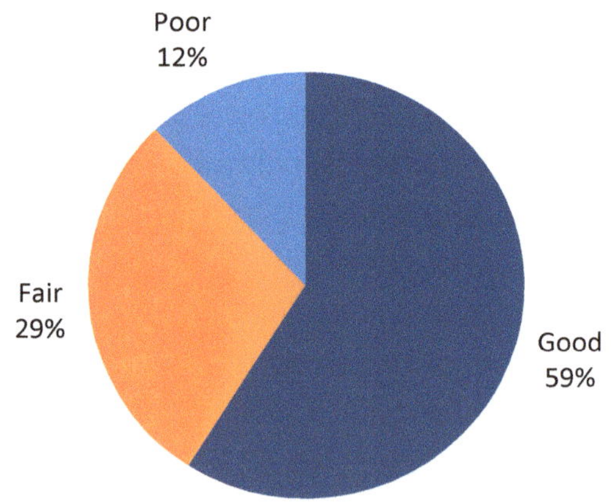

Source. ... .

About 9,600 miles of publicly accessible park roads and parkways exist, approximately 5,500 miles of which are paved. As shown in *Exhibit 12-7*, the condition rankings of paved roads are 59 percent good, 29 percent fair, and 12 percent poor. The NPS network includes approximately 1,442 publicly accessible bridges and 63 publicly accessible tunnels. About 3 percent of the bridges are structurally deficient resulting from deterioration. An additional 23 percent of the bridges are functionally obsolete, and are labeled as such, based on current design standards. NPS owns several historic bridges, which also are often functionally obsolete. The number of fatalities in the NPS system due to crashes varies between 40 and 60 per year, with an annual average of 47.

NPS reports the backlog of improvement needs for paved roads and bridges approaches $6 billion. To prioritize annual funding allocations, NPS implements a performance-based investment strategy, using analytical tools to maximize investment decisions in terms of pavement, bridge, congestion, and safety metrics, as well as mechanisms that ensure preventive maintenance for those assets. In addition to this backlog, investments are needed to address vehicle congestion, transportation safety concerns, and alternative transportation solutions.

Approximately 500 miles of paved and unpaved roads are intended for nonpublic use (i.e., roads restricted to official use), which are not funded from the Federal Lands Transportation Program (FLTP), but are funded from DOI appropriations. NPS also uses NPS Fee Program dollars and various other funding avenues (both public and private) to cover the cost to build, operate, and maintain all the different aspects of the NPS transportation system.

NPS manages 121 discrete transit systems in 63 of the 412 NPS units.[4] These transit systems accommodate 36.5 million passenger boardings annually. Shuttle, bus, van, and tram systems make up the largest share of all system types (50 percent), followed by boat and ferry systems (29 percent), planes (11 percent), snow coaches (7 percent), and trains and trolleys (3 percent). NPS owns and operates 18 of these systems directly and 12 operate under service contracts; together, they account for 35 percent of all passenger boardings. An additional 78 systems operate under concession contracts and represent the majority (49 percent) of all passenger boardings. The final 13 systems operate under a cooperative agreement and represent 16 percent of passenger boardings. Of these systems, 41 provide the sole access to an NPS unit because of resource or management needs and geographic constraints. Twelve systems are operated by a local transit agency under a specific agreement with NPS. In total, these transit systems include 982 vehicles, including 274 vehicles NPS owns or leases. Of NPS-owned vehicles, 61 percent operate on alternative fuel, while 13 percent of non-NPS-owned vehicles operate on alternative fuel.

Bicycle and pedestrian usage in the national parks is integral in the visitor's experience and serves a critical nonmotorized transportation function providing multimodal access to and within the park, as well as to areas unreachable by motorized travel. Bicycling, hiking, and walking are effective and pleasurable alternatives to motor vehicle travel. NPS is exploring the use of these and other transportation alternatives to offer additional visitor access and experiences. Bicycle and pedestrian access can accommodate more park visitors while alleviating congestion, protecting park resources, and improving the visitor experience. All park trails are open to pedestrians, and 28 percent are paved and also used by bicyclists. Bicycle and pedestrian access provides an interface between different transportation modes (i.e., park shuttle and public transportation systems) and many times serves as the primary transportation facility linking visitors (including disabled visitors) with the resources they want to see and experience. The NPS trails inventory includes 17,872 miles of trails, of which 5,012 miles (28 percent) consist solely of front-country paved trails. The total replacement value of these trails is approximately $2.5 billion. The approximate deferred maintenance value exceeds $259 million. Approximately 21 percent of front-country paved trails (1,070 miles) are in fair, poor, or serious condition.

NPS generally does not track usage of bicycle or pedestrian trails. Some NPS units track bicycle or pedestrian usage, however, in multimodal contexts. For example, the Cuyahoga Valley Scenic Railroad has served an average of 21,000[5] "Bike Aboard!" passengers each year since its inception in 2008. Cuyahoga Valley National Park in Ohio partnered with the Cuyahoga Valley Scenic Railroad to offer "Bike Aboard!" so that bicyclists can ride the Towpath Trail and use the railroad to return to their starting location. This program offers visitors the flexibility to pedal as far as they want with an option to return by train. It also provides the opportunity to view the park from two different perspectives. Another example is the 45-mile historic Carriage Path network in Acadia National Park in Maine, a crushed-stone aggregate system of paths providing access to pedestrians and nonmotorized equipment users (e.g., bicycles, skis) to park resources directly from surrounding towns without the need for a vehicle. In conjunction with the Carriage Path network, the Acadia Island Explorer public transportation system, which was inaugurated in 1999, carried more than 500,000 visitors in 2014.[6] Each bus has the capacity to transport bicycles to points throughout the park to access the Carriage Path network. A dedicated Bicycle Express route carried more than 21,000 riders in 2014 alone. Ridership of this transportation system has increased 74 percent since 2001.

Fish and Wildlife Service

FWS manages the National Wildlife Refuge System, which consists of 563 national wildlife refuges and 38 wetland management districts encompassing 150 million acres of lands and waters. It receives nearly 48.5 million recreation visits per year and has a variety of roads, trails, boat ramps, access points, bicycle trails, and viewing areas. FWS also operates 72 National Fish Hatcheries and one historic hatchery open to the public for visits and tours and owns more than 15,500 miles of roads, including 5,300 miles of public roads across all Service lands.

Of those 5,300 miles of public roads, approximately 400 miles are paved; the remaining 4,900 miles consist of gravel and native surfaced roads open to the public. The condition of the public-use roads during the 2008–2012 condition assessments were 60 percent excellent to good, 25 percent fair, and 15 percent poor to failed, as shown in *Exhibit 12-8*. About 300 bridges and 5,150 parking lots are associated with the public road system. Approximately 7 percent of the bridges are structurally deficient.

Exhibit 12-8 Fish and Wildlife Service Pavement Condition

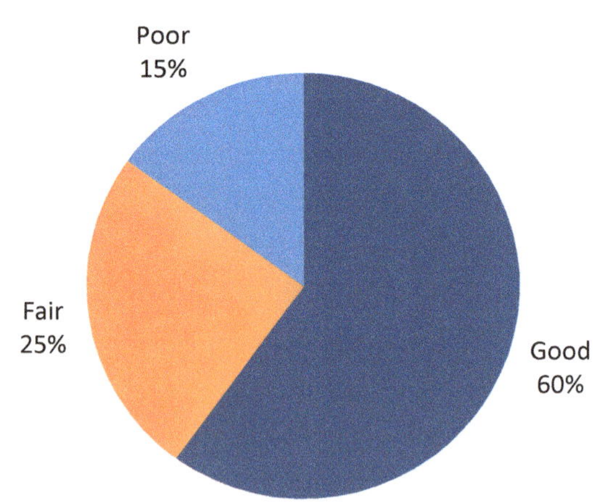

Source

The 2008–2012 inventory and condition assessment identified a maintenance backlog that approaches $1 billion for all public roads and bridges. Using estimated life cycles of 10 years for gravel roads and 20 years for paved roads, prorated annual infrastructure replacement costs amount to approximately $100 million a year to maintain the existing system.

FWS owns and operates 16 permanent transit systems, with temporary service expanded to other units during special events, such as the 3-day Festival of the Cranes at Bosque Del Apache National Wildlife Refuge in New Mexico. Further, at least eight urban transit systems currently serve FWS units. Local agencies and refuges have transit access information on their websites, and are adding transit stops on or immediately adjacent to refuge access points. Additionally, the 2013 FWS Urban Refuge Program implementation strategy included, as a "standard of excellence," the increase of equitable access to urban refuges by all modes, with an emphasis on transit and trails for refuges within 25 miles of urban areas having populations greater than 250,000, by working with local, regional, and State partners.

FWS recorded more than 11 million uses of the designated automobile tour routes on National Wildlife Refuges in FY 2014. Pedestrian and bicycle use continue to be important ways for visitors to experience FWS lands. FWS logs nearly 1 million visits on bicycles on FWS lands and more than 15 million uses of FWS footpaths annually. In FY 2016 and 2017 bicycle and pedestrian counters will be installed at 30 refuges nationally and the data will be incorporated into the national bicycle/pedestrian database being developed by the Federal Highway Administration. FWS maintains 2,187 miles of trails, 95 percent of which are in excellent to good condition. Approximately 32 percent of the miles are paved or boardwalk, and the remainder are gravel, native surface, chipped wood, or mowed vegetation. These trails have a current replacement value of $186 million, with a deferred maintenance backlog of $1.3 million, which yields a trails facility condition index of 0.007 (trails facility condition index is the ratio of deferred maintenance to current replacement value).

Bureau of Land Management

BLM manages 16 percent of the surface area of the United States and is the largest manager (40 percent) of Federal lands. BLM maintains a transportation system that serves as one of the primary means of connectivity to more than one-eighth of the United States, providing access to 247.5 million acres of BLM-administered public lands concentrated primarily in the 11 western States and Alaska. These lands comprise 20–80 percent of the individual States or their political subdivisions. These lands play a significant role in the environmental and socioeconomic fabric of the Nation, the West, Alaska, and local governments. BLM also manages 700 million acres of subsurface mineral estate throughout the United States.

As the national parks and national forest have become increasingly overcrowded, more people have begun using facilities on BLM-managed lands. Visits to BLM lands and facilities have significantly increased, due to an increase in outdoor recreational activities and the number of resources available throughout all FLMAs. Outdoor recreation has increased approximately 5 percent annually in the past several years despite the economic recession. Comprehensive transportation planning is a major priority for BLM. BLM established its Travel and Transportation Management Program to identify and classify all roads and trails, including well-maintained FLTP-eligible roads, temporary access routes for commercial uses, high-clearance primitive roads, and various types of recreational trails through a formal decision-making process. Classification of roads and trails is necessary for proper management of access to and impacts on vegetation; sensitive species and their habitats; soils, air and water quality; and cultural and visual resources. BLM is moving toward a multimodal travel and transportation network that addresses the access and recreational needs of multiple user groups, including both motorized and nonmotorized forms of travel, on a designated system of routes. Completing management travel plans by inventorying and evaluating roads and areas, and deciding how roads or areas will be designated, is an enormous task. Travel plans on more than 212 million acres (83 percent) of overall inventory remain to be completed.

BLM owns approximately 86,000 miles of public lands development roads and trails (PLDR&T), which is the primary road system on BLM lands. The roads in the PLDR&T are not considered public roads. About 1,100 miles of BLM roads, however, are being proposed for inclusion in the FLTP system under Moving Ahead for Progress in the 21st Century (MAP-21, Public Law 112-141). Many of the roads have public uses and special purposes, such as those that serve recreational development areas. The PLDR&T system evolved from a user-established system dating to initial settlement of the West. BLM has completed its 10-year effort to inventory and assess the condition of its road system. This effort identified deferred maintenance and capital replacement costs and gathered basic inventory and geospatial data over what is currently considered the agency's road system (approximately 42,000 miles). Additionally, BLM has an inventory of approximately 29,000 miles of primitive roads, which comprise another set of assets in BLM's formal transportation system. Primitive roads, or high-clearance roads, do not normally meet any BLM road design standards.

BLM owns approximately 500 paved miles and 600 miles of unpaved public roadways and about 835 public bridges and major culverts. As shown in *Exhibit 12-9*, the condition of paved and surfaced roads is 65 percent good, 20 percent fair, and 15 percent poor. Approximately 3 percent of the public bridges are structurally deficient. BLM reports the backlog of improvement needs is $350 million.

Exhibit 12-9 Bureau of Land Management Pavement Conditions (Paved Roads Only)

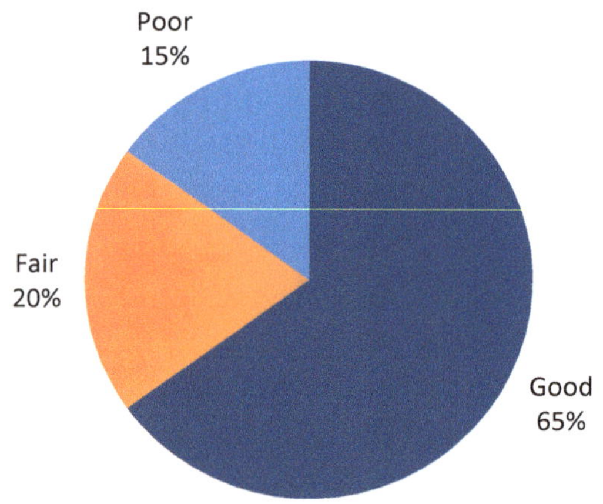

Source: BLM.

Bureau of Reclamation

The Bureau of Reclamation (Reclamation) administers 476 dams and 348 reservoirs in the 17 western States, and manages 187 recreation areas in partnership with other non-Federal recreation partners such as State, county, and city governments. One of the most notable reservoirs, created by Hoover Dam, is Lake Mead, which NPS administers. Reclamation is the ninth largest electric utility and second largest producer of hydropower in the United States, with 58 power plants producing an average 40 billion kilowatt-hours annually. Reclamation is also the Nation's largest wholesale water supplier, delivering 10 trillion gallons of water to more than 31 million people each year and providing 1 of 5 western farmers with irrigation water.

Reclamation owns approximately 2,015 miles of roads open to the public, 762 miles of which are paved. As shown in *Exhibit 12-10*, the condition of Reclamation roads is 65 percent good, 25 percent fair, and 10 percent poor. Reclamation also owns 331 public bridges, approximately 12 percent of which are structurally deficient. In addition, Reclamation owns an estimated 8,000 miles of administrative roads and operations and maintenance roads, which are not open to the public.

Exhibit 12-10 Bureau of Reclamation Pavement Conditions (Paved Roads Only)

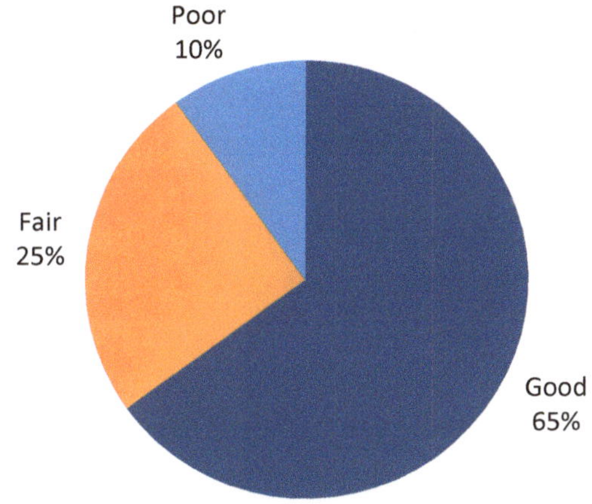

Source

Bureau of Indian Affairs

The United States has a unique legal and political relationship with Indian Tribes and Alaska Native entities as provided for by the Constitution of the United States, treaties, court decisions, and Federal statutes. Within the government-to-government relationship, the Bureau of Indian Affairs (BIA) provides services directly or through contracts, grants, or compacts to 567 federally recognized Tribes with a service population of about 1.9 million American Indian and Alaska Natives. BIA offers an extensive scope of programs that covers the entire range of Federal, State, and local government services. Programs administered through BIA include social services, natural resources management on trust lands (55 million surface acres and 57 million acres of subsurface mineral estates), economic development programs in some of the most isolated and economically depressed areas of the United States, law enforcement and detention services, administration of Tribal courts, implementation of land and water claim settlements, housing improvement, disaster relief, replacement and repair of schools, repair and maintenance of roads and bridges, and repair of structural deficiencies on high-hazard dams. BIA operates a series of irrigation systems and provides electricity to rural parts of Arizona.

BIA is responsible for approximately 29,200 miles of roads that are open to the public, 8,800 miles of which are paved. Tribal governments own an additional 13,500 miles of public-use roads, including 3,300 miles that are paved. Neither number includes any mileage for future or proposed roads that are in the inventory. Approximately 17 percent of total BIA and tribally owned roads are in acceptable condition. Additionally, BIA owns 929 public bridges, approximately 15 percent of which are structurally deficient and 68 percent of which are in acceptable condition. The number and condition of tribally owned bridges is currently unknown, given they first required inspection in 2013 with the passage of the MAP-21.

Department of Defense

The mission of the Department of Defense (DOD) is to provide the military forces needed to deter war and to protect the security of our country. DOD owns or manages more than 28.8 million acres of land within the continental United States. More than 4,100 military sites are within the U.S. installations in the United States. A site is any site larger than 10 acres with a replacement value greater than $10 million. DOD has many smaller sites that also have traffic needs. The economic benefit DOD provides to the country as a whole has not been precisely calculated, but many States and communities have publicly declared the economic benefits their installations provide. For instance, South Carolina reported in 2012 that their military bases generated more than 138,000 jobs and contributed over $13 billion dollars in revenue to their communities.[7] Many examples in other States also support the economic benefits of DOD.

People assume that DOD installations generally are not open to the public due to the overriding military mission of those specific areas. Many installation roads, however, are open to use by military family members, visitors, and other members of the public, even though stopping at a gate area might be required. Roads on military installations serve housing offices, commissaries, base exchanges, recreation facilities, unrestricted training facilities, hospitals, and traffic crossing the installation. Many installations allow the public access for hunting and fishing. Given this mixed use, the public street system must reflect the street system in surrounding areas. In practice, DOD makes every effort to comply with the same traffic and transportation engineering guidance of the civilian road systems. Military installations also offer pristine habitat for our Nation's threatened and endangered species. Other Federal agencies collaborate with DOD to manage ecological endeavors.

The DOD *Base Structure Report – Fiscal Year 2014 Baseline* indicates that DOD, excluding the U.S. Air Force, owns or manages almost 28,000 miles of roadways deemed open for public travel. Travelers on installation roads consist of military personnel and their dependents, civilian work force on military installations, contractors performing work for the military, civilian personnel operating businesses, and visitors (including nonmilitary-associated sportsmen). DOD has 1,418 public bridges, 25 percent of which are classified as structurally deficient, as defined in Chapter 3. Addressing bridge investment needs presents a challenge given recent trends in reduced DOD operating funds. The DOD maintenance and construction of roadways are prioritized at the local installation level. Further, there is no central DOD "pot of funds" for roadway work. Roads must compete with buildings, structures, parking lots, runways, and any other infrastructure for a limited amount of funding. In addition, just as each installation's mission might vary greatly from another's, the infrastructure needs from one installation to another could vary greatly. Therefore, DOD does not track roadway condition for all installations in any one central repository nor does it centrally track the amount spent on roadway improvements. That tracking is completed at the local level, where it will remain. DOD does, however, record and document to the Federal Highway Administration the condition and performance of all bridge structures. Despite the differences at the installation level, DOD does strive to comply with accepted traffic and transportation engineering guidance, which seeks to ensure consistency in all aspects from geometrics to sign standards. DOD's policy is to adhere, whenever possible, to the same standards non-DOD public roadways are held. For instance, DOD policy is that all DOD roadways are subject to the Manual of Uniform Traffic Control Devices and should be operated in conformance to these standards. Therefore, even though DOD does not centrally control and manage roadways and bridges, activities are undertaken to ensure consistency across all installations.

DOD contributes to the highway trust funds at the national and State level. The Hayden Cartwright Act of 1934 directed the collection of fuel taxes on military installations. Many military installations operate fuel service stations that sell fuel to military members, dependents, retirees, and civilians. Each year, those sales include more than 450 million gallons of gasoline and more than 5 million gallons of diesel fuel. (Note these sales do not include fuel used by military or government vehicles.) Those figures contribute more than $100 million to the Federal Highway Trust Fund alone. These sales were to DOD civilians and military personnel who in general live and shop outside military installations.

DOD is one of the largest FLMAs and relies primarily on its own resources for transportation management. It strives to operate and manage a transportation system that is functional, multimodal, and considerate of the environment.

United States Army Corps of Engineers

The USACE Civil Works Program helps build and maintain America's infrastructure through its three main mission areas: commercial navigation, flood risk management, and aquatic ecosystem restoration. USACE also protects the Federal lands and waters and provide recreational opportunities at these water resource projects. USACE currently manages nearly 12 million acres of land and water at 420 lakes and waterways throughout the United States.

USACE owns about 6,500 miles of the more than 7,700 miles of public roads that serve USACE lakes and waterways. More than 5,200 miles are paved. Most of these roadways are found within recreation areas distributed around the water resource project. The system of recreation areas and the road network are established and not expected to grow. Although this road network consists of relatively short segments, they are heavily used by the more than 335 million visitors annually (the highest visitation of any FLMA). USACE also owns and maintains 416 public bridges, of which 6 percent are structurally deficient. As shown in *Exhibit 12-11*, the condition of USACE roads is 56 percent good, 30 percent fair, and 14 percent poor.

Exhibit 12-11 U.S. Army Corps of Engineers Pavement Conditions

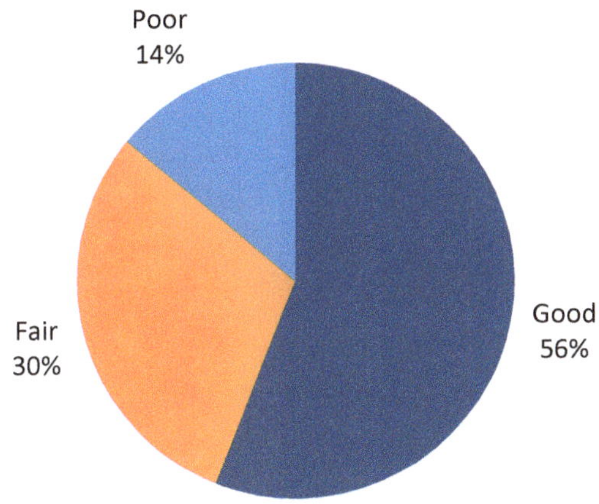

Source

Transportation Funding for Federal and Tribal Lands

Providing access within Federal and Tribal lands generally is not a State or local responsibility, but that of the Federal government. Before the 1980s, all road improvements depended on the unpredictability of the various annual Federal agency appropriations competing with non-transportation needs, which resulted in many road systems on Federal and Tribal lands falling into disrepair. The Surface Transportation Assistance Act of 1982 established the Federal Lands Highway Program. This program brought together, for the first time, a consolidated and coordinated long-range program funded under the Highway Trust Fund.

Under the Safe, Accountable, Flexible, Efficient Transportation Equity Act: A Legacy for Users (Public Law 109-59 SAFETEA-LU), the Federal Lands Highway Program has provided funding for NPS Park Roads and Parkways, BIA Indian Reservation Roads, FWS Refuge Roads, and two components of the Public Lands Highway Program—Forest Highways and a discretionary component called the Public Lands Highway Discretionary Program. The funding categories and annual authorizations are shown for FY 1983 through FY 2012 in *Exhibit 12-12*.

Exhibit 12-12 Federal Lands Highway Program Annual Authorizations, 1983–2012

	Annual Authorizations by Program (Millions of Dollars)						
Authorization	FY	FH	PLHD	IRR	PRP	RR	Total
STAA	1983	50	50	75	75	0	**250**
	1984	50	50	100	100	0	**300**
	1985	50	50	100	100	0	**300**
	1986	50	50	100	100	0	**300**
STURAA	1987	55	40	80	60	0	**235**
	1988	55	40	80	60	0	**235**
	1989	55	40	80	60	0	**235**
	1990	55	40	80	60	0	**235**
	1991	55	40	80	60	0	**235**
ISTEA	1992	94	49	159	69	0	**371**
	1993	113	58	191	83	0	**445**
	1994	113	58	191	83	0	**445**
	1995	113	58	191	83	0	**445**
	1996	114	58	191	84	0	**447**
	1997	114	58	191	84	0	**447**
TEA-21	1998	129	67	225	115	20	**556**
	1999	162	84	275	165	20	**706**
	2000	162	84	275	165	20	**706**
	2001	162	84	275	165	20	**706**
	2002	162	84	275	165	20	**706**
	2003	162	84	275	165	20	**706**
TEA-21 Extension	2004	162	84	275	165	20	**706**
SAFETEA-LU	2005	172	88	314	180	29	**783**
	2006	185	95	344	195	29	**848**
	2007	185	95	384	210	29	**903**
	2008	191	99	424	225	29	**968**
	2009	198	102	464	240	29	**1,033**
SAFETEA-LU	2010	198	102	464	240	29	**1,033**
Extension	2011	198	102	464	240	29	**1,033**
	2012	198	102	464	240	29	**1,033**
	Total	**3,762**	**2,095**	**7,086**	**4,036**	**372**	**17,351**

Source: FLHP.

On July 6, 2012, President Obama signed MAP-21 into law. This transformative law realigned and expanded the component programs of the Federal Lands Highway Program into three more comprehensive Federal Lands and Tribal Transportation Programs, funded at $1 billion annually for FY 2013 and FY 2014 (see *Exhibit 12-13*). The Tribal Transportation Program, funded at $450 million annually for FY 2013 and FY 2014, replaces the Indian Reservation Roads program. The FLTP, funded at $300 million annually for FY 2013 and FY 2014, merges the Park Roads and Parkways and Refuge Roads programs and expands the program to include transportation facilities owned by BLM, USACE, and the Forest Service to address improvements to transportation facilities owned by the largest Federal recreation providers. The Federal Lands Access Program is funded at $250 million annually for FY 2013 and FY 2014 and takes attributes from the Forest Highways program and the Public Lands Highway Discretionary Program to address transportation needs comprehensively on non-Federal roads, which provide access to all types of Federal lands.

Exhibit 12-13 Federal Lands and Tribal Transportation Program Authorizations, 2013–2014

Program	2013 ($ Millions)	2014 ($ Millions)
Federal Lands Transportation Program	300	300
Federal Lands Access Program	250	250
Tribal Transportation Program	450	450
Total	**1,000**	**1,000**

Source: FLTTP

Funds in the Federal Lands Highway Program and Federal Lands and Tribal Transportation Programs may be used for transportation planning, research engineering, and construction of roadways. They may also be used to fund transit facilities that provide access to or within Federal and Tribal lands. Maintenance, rehabilitation, and reconstruction of transportation facilities also may be funded through various other FLMA appropriations.

Increasing Walking, Biking, and Transit Use on Federal and Tribal Lands

Growth in public use of Federal and Tribal lands has created a need for additional investment in transportation facilities for transit, bicycle, and pedestrian uses on Federal and Tribal lands. High visitation levels, to both large and small sites, are causing problems due to the growing volumes of traffic and demands for visitor parking. In many areas, the problem is not too many people but too many motor vehicles and too many visits concentrated in certain periods. Specific examples of successful park investments in transit are shuttle bus systems in Denali National Park and Preserve, Acadia National Park, Cape Cod

Transit in the Parks

The National Park Service currently has 121 alternative transportation systems in 63 park units nationwide.

Exhibit 12-14 Alternative Transportation Systems in National Parks, Percentages by Mode

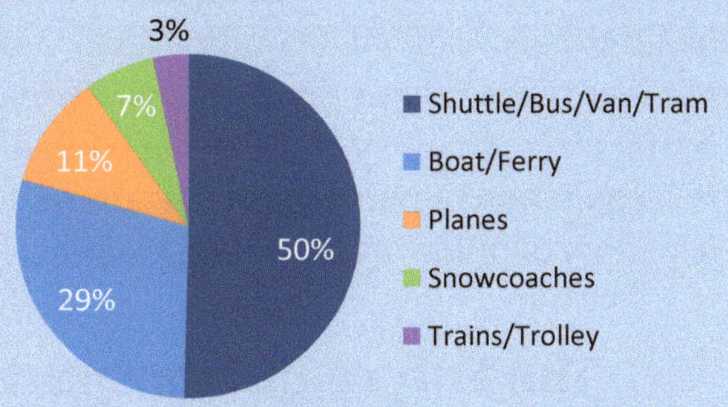

Source: 2014 NPS National Transit Inventory.

National Seashore, Zion National Park, and Grand Canyon National Park; the train system serving Cuyahoga National Park; and the ferry system serving Fire Island National Seashore. In addition, FLMAs actively continue to pursue partnerships to meet visitor demand. An example of a successful partnership is the San Antonio Missions National Historical Park B-Cycle Bike Share Expansion Project. The project enhances alternative transportation options for visitors at the park and provides connections to the surrounding community by offering bike share stations along the San Antonio River and at each of the four NPS missions.

DID YOU KNOW?

The U.S. Fish and Wildlife Service has 16 alternative transportation systems.

A 2004 U.S. Department of Transportation study estimated transit needs on Forest Service lands. This study identified 30 sites that would benefit from new or supplemental transit investments and estimated that approximately $698 million in 2003 dollars ($714 million in 2004 dollars or $60 million per year) would be needed in these areas between 2003 and 2022. A joint 2001 Transit Authority/Federal Highway Administration study estimated transit investment needs on NPS, BLM, and FWS lands—all part of DOI. Total DOI needs for 2002 to 2020 were estimated to be $1.71 billion in 1999 dollars ($2.16 billion in 2004 dollars or $180 million per year). An estimated 91 percent of these needs were for NPS, 7 percent for FWS, and 2 percent for BLM.

In 2005, the Paul S. Sarbanes Transit in the Parks (TRIP) Program was established under SAFETEA-LU and provided approximately $26 million of Federal funding annually. The TRIP Program was established to help develop new alternatives for enjoying our parks and public lands while protecting resources. The program funded transportation in the parks and public lands; helped conserve natural, historic, and cultural resources; reduced congestion and pollution; improved visitor mobility and accessibility; enhanced the visitors' experience; and helped ensure access to all, including persons with disabilities. The TRIP Program was repealed under the most recent surface transportation authorization, MAP-21.

Also in 2005, SAFETEA-LU created the Tribal Transit Program. SAFETEA-LU authorized funding for this program beginning in FY 2006 at $8 million, increasing to $10 million in FY 2007, to $12 million in FY 2008, and to $15 million annually in FY 2009 through FY 2012. MAP-21 increased the funding to $30 million in FY 2013 and 2014. Federally recognized Tribes may use the funding for capital, operating, planning, and administrative expenses for public transit projects that meet the growing needs of rural Tribal communities. Examples of eligible activities include capital projects; operating costs of equipment and facilities for use in public transportation; and the acquisition of public transportation services, including service agreements with private providers of public transportation services.

The Future of Transportation on Federal and Tribal Lands

In examining future transportation needs on Federal and Tribal lands, FLMAs need to address challenges in identifying and involving all stakeholders and in gaining a better understanding of the complex relationship among these entities. Along with this, the following significant issues continually need to be addressed.

- **New Technology and Innovative Transportation Solutions**: As population increases, the demand for access to Federal and Tribal lands will grow. Providing access will require fully considering and implementing new technology and innovative transportation solutions, including efficient intermodal transfers among the available modes of transportation (walking, bicycles, cars, buses, recreational vehicles, transit, ferries, or aircraft), adaptive signal control, and connected/automated vehicles. Intelligent transportation systems will continue to play an increasingly important role as a way to communicate congestion and provide information on alternative routes and times to visit Federal and Tribal lands.
- **Urban Growth**: In many instances, urban growth is encroaching on Federal and Tribal lands. As these lands become part of urban areas, FLMAs and Tribes face challenges with issues affecting urban transportation officials. In close cooperation with metropolitan, local, and other transportation officials, these agencies need to undertake and implement effective land use and urban transportation planning. FLMAs and Tribes are focusing on intermodal solutions to challenges of increasing demands for access and balancing those desires with impacts on natural, cultural, and historic resources and the environment, including air and water quality.
- **Transportation Funding**: As public transportation funding is likely to continue to be constrained, ensuring more effective coordination among Federal agencies, Tribal governments, and State/local transportation agencies becomes paramount. A critical need is the effective development and implementation of transportation investment practices that fully use products of transportation planning and management systems for bridge, safety, pavement, and congestion.
- **Aging Population**: The average age of drivers on Federal and Tribal lands will continue to increase, requiring continued improvements in signs and information systems and accommodation for visitors with disabilities. This need will be especially important in urban areas where effective destination guidance is a challenge to implement.

Deferred Maintenance

Due to a lack of funding for routine maintenance, the deferred maintenance backlog is over $10 billion.

Exhibit 12-15 Deferred Maintenance Backlog

Federal Agency	Backlog of Deferred Maintenance ($ Million)
Forest Service	$2,900
National Park Service	$6,000
Fish and Wildlife Service	$1,000
Bureau of Land Management	$350
Bureau of Reclamation	N/A
Bureau of Indian Affairs	N/A
Tribal Governments	N/A
Military Installations	N/A
U.S. Army Corps of Engineers	N/A
Total	**$10,250**

Source: FLMAs.

- **VMT Reductions and Generational Differences**: Nationally, VMT last peaked in 2007. Millennials (aged 18 through 34 years in 2015) are the Nation's largest population, but their transportation choices noticeably differ from those of prior generations. The Boomer Generation (51 through 69 years old in 2015) is defined by the motor vehicle. Millennials are less inclined to use the automobile, however, tending to use transit and other modes of transportation, like biking and walking. In addition, they are often characterized as being detached from nature and uninterested in outdoor-based experiences. As VMT growth remains uncertain and per capita attendance at parks and recreation areas declines, developing strategies is increasingly important for connecting with and attracting Millennials to the Nation's recreation areas and providing alternative transportation solutions to ensure access to Federal lands.

---

[1] http://nationalmap.gov/small_scale/printable/printableViewer.htm?imgF=images/preview/fedlands/fedlands3.gif&imgW=588&imgH=450

[2] "Public Land Statistics 2011," Bureau of Land Management, Department of Interior, May 2012. http://www.blm.gov/public_land_statistics/pls11/pls2011.pdf

[3] http://www.nps.gov/aboutus/index.htm

[4] "NPS National Transit Inventory, 2014," http://ntl.bts.gov/lib/55000/55500/55568/NPS_WASO_2014_National_Transit_Inventory.pdf

[5] Cuyahoga Valley National Park visitor use statistics.

[6] Acadia National Park visitor use statistics.

[7] "The Economic Impact of the Military Community in South Carolina," November 2012, Research Division, South Carolina Department of Commerce.

part IV

Recommendations for HPMS Changes

Recommendations for HPMS Changes

Section 52003 of the Moving Ahead for Progress in the 21st Century Act (MAP-21) added a requirement that this report include recommendations for changing the Highway Performance Monitoring System (HPMS).

The changes are to address:

> "(i) improvements to the quality and standardization of data collection on all functional classifications of Federal-aid highways for accurate system length, lane length, and vehicle-mile of travel; and
>
> (ii) changes to the reporting requirements authorized under section 315 to reflect recommendations under this paragraph for collection, storage, analysis, reporting, and display of data for Federal-aid highways and, to the maximum extent practical, all public roads."

Part IV of this 2015 C&P Report fulfills this requirement. Future report editions will present updates on progress in implementing improvements to HPMS and other potential changes as they are identified.

HPMS is a major data source for the analyses presented in Chapters 2, 3, and 7–10 of this report. Appendices A and D include material that also derives from HPMS.

Background

Each year, HPMS collects information on the extent, condition, performance, use, and operating characteristics of the Nation's highways. Jointly developed by the Federal Highway Administration (FHWA) and the States in 1978, the system replaced the States' numerous uncoordinated annual data reports and special studies. HPMS includes key data on all public roads, detailed data for a sample of the arterial and collector functional systems, full-extent coverage data for the Interstate and other principal arterials, and other statewide summary data.

HPMS provides essential information for apportioning Federal funds to the States and for assessing highway system performance under FHWA's strategic planning process. Data on pavement condition, congestion, and traffic are used to measure progress in meeting the objectives embodied in the FHWA's Performance Plan and other strategic goals. HPMS also supports this biennial report to Congress.

In addition, HPMS provides data to the States, metropolitan planning organizations, and local governments for assessing highway condition, performance, air quality trends, and future investment requirements. The system provides much of the information that FHWA includes in its annual *Highway Statistics Series* and other publications.

HPMS is a collaborative effort between FHWA and the States. The States collect and report the data, and FHWA reviews the data for quality and consistency, provides guidance on data collection, and offers technical support on improving data quality. States strive to use common practices to promote consistency, such as adhering to the standards set by the American Association of State Highway and Transportation Officials and the American Society for Testing and Materials. An ongoing National Cooperative Highway Research Program study (20-24[82]), "Increasing Consistency in the Highway Performance Monitoring System for Pavement Reporting," is identifying and prioritizing measures to help reduce inconsistencies in pavement performance information.

Periodically, HPMS is reassessed to ensure it is maintaining its role as the repository for national highway performance data and to recommend changes for its improvement. The most recent reassessment began in 2006 and led to the elimination of data items users no longer need and the addition of data items users now require. The assessment resulted in the introduction of a new geospatial data model to improve data processing efficiency and geospatial analysis. After a series of intensive outreach workshops and webinars, FHWA published the final report, *HPMS Reassessment 2010+*, in September 2008.

The HPMS requirements identified in *HPMS Reassessment 2010+* became effective with the submittal of data collected in 2009. A new database management system was developed that incorporates geospatial data from State DOT linear referencing systems and integrates these data into a national dataset. HPMS data are now associated with a State's highway map, which enables mapping and spatial analysis of the HPMS data.

Additional Changes to HPMS

MAP-21 indirectly made two changes to HPMS by expanding the National Highway System (NHS) to include all principal arterials. This expansion has led to increased data collection for truck travel data in HPMS, as States must report such data for the entire NHS. For non-NHS routes, truck travel data will continue to be required only for a representative sample of highway sections. This expansion also has led to increased data collection of International Roughness Index (IRI) data, as States must collect such data annually for NHS; for non-NHS routes, data collection continues to be required only biennially.

HPMS will serve as the foundation for linking FHWA data systems through spatial relationships. This linkage will enable analyses that are more comprehensive by combining the financial and bridge data with the highway information in HPMS. On August 7, 2012, FHWA notified the States that—starting with the 2014 data submittal—they need to provide geospatial information for their road networks on all public roads. The requirement is referred to as ARNOLD or the All Road Network of Linear Referenced Data. FHWA then will build a national basemap for an integrated system of highway attributes for analyzing safety, bridge, freight, and planning data. Included is a requirement for States to represent divided facilities as dual carriageways, enabling the States to provide FHWA with data on highway attributes by roadway direction, increasing data accuracy. Many States support this requirement as it more closely mimics their internal systems.

Q&A **What is a dual carriageway?**

The geospatial networks, or maps, that States currently submit to HPMS are considered a single centerline network, which means that the networks use only single lines to represent all roads, regardless of whether the roads are two-lane collectors or divided Interstate. In contrast, the dual carriageway has two lines for all divided highways, one for each directional roadway. Using a dual carriageway allows for a more accurate spatial representation of divided highways, improves data quality for these roads, and enhances analysis capabilities.

FHWA is considering a possible change to the reporting requirements of the IRI data used for measuring performance of pavement condition. This change would standardize the section length required for reporting IRI so that comparisons are consistent. Currently, States use various IRI section lengths.

part V

Appendices

Introduction

Appendices A, B, and C describe the modeling techniques used to generate the investment/performance analyses and capital investment scenarios highlighted in Chapters 7 through 10. Appendix D discusses an ongoing initiative, "Reimagining the C&P Report in a Performance Management-Based World."

Appendix A describes selected technical aspects of the **Highway Economic Requirements System** (HERS), which is used to analyze potential future investments for highway resurfacing and reconstruction and highway and bridge capacity expansion.

Appendix B details the **National Bridge Investment Analysis System** (NBIAS), which is used to examine potential future bridge rehabilitation and replacement investments.

Appendix C presents technical information on the **Transit Economic Requirements Model** (TERM), which is used to explore potential future transit investments in urbanized areas. TERM includes modules that estimate the future funding required to replace and rehabilitate transit vehicles and other assets and to invest in new assets to accommodate future growth in transit ridership.

Appendix D discusses two ongoing FHWA-sponsored research efforts aimed at identifying opportunities to enhance the analytical approaches used for assessing future investment needs and to improve the communication of information in the print and Web versions of the C&P report.

appendix A

Highway Investment Analysis Methodology

Highway Investment Analysis Methodology

Investments in highway resurfacing and reconstruction and in highway and bridge capacity expansion are modeled using the Highway Economic Requirements System (HERS), which has been used since the publication of the 1995 C&P Report. This appendix describes the basic HERS methodology and approach in more detail than is presented in Part II, including the model features that have changed significantly from those used for the 2013 C&P Report: the valuation of travel time and the representation of pavement quality.

Highway Economic Requirements System

HERS begins the investment analysis process by evaluating the current state of the highway system using information on pavements, geometry, traffic volumes, vehicle mix, and other characteristics from the Highway Performance Monitoring System (HPMS) sample dataset. Using section-specific traffic growth projections, HERS forecasts future conditions and performance across several funding periods. As used in this report, the future analysis covers four consecutive 5-year periods. At the end of each period, the model checks for deficiencies in eight highway section characteristics: pavement condition, surface type, volume/service flow (V/SF) ratio (a measure of congestion), lane width, right shoulder width, shoulder type, horizontal alignment (curves), and vertical alignment (grades).

After HERS determines that a section's pavement or capacity is deficient, it identifies potential improvements to correct some or all of the section's deficient characteristics. The HERS model evaluates seven kinds of improvements: resurfacing, resurfacing with shoulder improvements, resurfacing with widened lanes (i.e., minor widening), resurfacing with added lanes (i.e., major widening), reconstruction, reconstruction with widened lanes, and reconstruction with added lanes. For reconstruction projects, the model allows for upgrades of low-grade surface types when warranted by sufficient traffic volumes. For improvements that add travel lanes, HERS further distinguishes between two capacity additions: those that can be made at "normal cost" and those on sections where obstacles to widening are present, making capacity additions feasible only at "high cost." HERS might also evaluate alignment adjustments to improve curves, grades, or both.

When evaluating which potential improvement, if any, should be implemented on a particular highway section, HERS employs incremental benefit-cost analysis. Such an analysis compares the benefits and costs of a candidate improvement to those of a less aggressive alternative—for example, reconstructing and adding lanes to a section could be compared with reconstruction alone. HERS defines benefits as reductions in direct highway user costs, agency costs, and societal costs. Highway user benefits include reductions in travel time costs, crash costs, and vehicle operating costs (e.g., fuel, oil, and maintenance costs); agency benefits include reduced routine maintenance costs (plus the residual value of projects with longer expected service lives than the alternative); and societal benefits include reduced vehicle emissions. Increases in any of these costs resulting from a highway improvement (such as higher emissions rates at high speeds or the increased delay associated with a work zone) would be factored into the analysis as a negative benefit ("disbenefit").

Dividing these improvement benefits by the capital costs associated with implementing the improvement results in a benefit-cost ratio (BCR) that is used to rank potential projects on different highway sections. HERS implements improvements in order of BCR, with the improvement having the highest BCR implemented first. Thus, as each additional project is implemented, the marginal BCR declines, resulting in a decline in the average BCR for all implemented projects. Until the point at which the marginal BCR falls below 1.0 (i.e., costs exceed benefits), however, total net benefits continue to increase as additional projects are implemented. Investment beyond this point is not economically justified because a decline in total net benefits would result.

Because HERS analyzes each highway section independently rather than the entire transportation system, it cannot fully evaluate the network effects of individual highway improvements. Although efforts have been made to account indirectly for some network effects, HERS is fundamentally reliant on its primary data source—the national sample of independent highway sections contained in HPMS. Fully recognizing all network effects would require developing significant new data sources and analytical techniques.

HERS Improvement Costs

For the 2004 C&P Report, significant changes were made to the structure of the HERS improvement cost matrix, the assumed unit costs in that matrix, and the manner in which those values were applied. The improved cost updates reflected in the 2004 C&P Report were based on highway project data from six States. The 2004 update disaggregated the improvement cost values in urban areas by functional class and by urbanized area size. Three population groupings were used: small urban (populations of 5,000 to 49,999), small urbanized (populations of 50,000 to 200,000), and large urbanized (populations of more than 200,000).

For the 2006 C&P Report, additional project cost data were collected for large urbanized areas, rural mountainous regions, and high-cost capacity improvements. These data were used to update the HERS improvement cost matrix, which was also modified to include a new category for major urbanized areas with populations of more than 1 million. The HERS improvement cost matrix was adjusted further for the 2008 C&P Report based on additional analysis of the data previously collected. For this 2015 C&P Report, the only change made to the cost matrix was an adjustment for the change in the National Highway Construction Cost Index between 2006 and 2012.

Exhibit A-1 identifies the costs per lane mile assumed by HERS for different types of capital improvements. For rural areas, separate cost values are applied by terrain type and functional class, while costs are broken down for urban areas by population area size and type of highway. These costs are intended to reflect the typical values for these types of projects in 2012, and thus do not reflect the large variation in cost among projects of the same type, even in a given year.

Such variation is evident in the project-level data on which these typical values are based and are attributable to several location-specific factors. For example, the costs assumed for highway widening projects are predicated on each section's having several bridges typical for the section's length, but in reality some sections will have more bridges than other sections of equal length, which adds to costs. Among other factors that could make costs unusually high are complicated interchanges, major environmental issues, and other extreme engineering issues.

Exhibit A-1 Typical Costs per Lane Mile Assumed in HERS by Type of Improvement

	Typical Costs (Thousands of 2012 Dollars per Lane Mile)								
Category	Reconstruct and Widen Lane	Reconstruct Existing Lane	Resurface and Widen Lane	Resurface Existing Lane	Improve Shoulder	Add Lane, Normal Cost	Add Lane, Equivalent High Cost	New Alignment, Normal	New Alignment, High
Rural									
Interstate									
Flat	$1,496	$977	$847	$347	$65	$1,923	$2,666	$2,666	$2,666
Rolling	$1,677	$1,003	$975	$370	$106	$2,085	$3,374	$3,374	$3,374
Mountainous	$3,180	$2,195	$1,615	$547	$223	$6,492	$7,600	$7,600	$7,600
Other Principal Arterial									
Flat	$1,169	$782	$706	$279	$43	$1,541	$2,205	$2,205	$2,205
Rolling	$1,319	$804	$803	$310	$72	$1,650	$2,662	$2,662	$2,662
Mountainous	$2,562	$1,810	$1,556	$438	$95	$5,824	$6,706	$6,706	$6,706
Minor Arterial									
Flat	$1,069	$687	$658	$247	$41	$1,400	$1,966	$1,966	$1,966
Rolling	$1,290	$761	$819	$266	$75	$1,605	$2,532	$2,532	$2,532
Mountainous	$2,143	$1,405	$1,556	$365	$168	$4,916	$5,900	$5,900	$5,900
Major Collector									
Flat	$1,125	$728	$680	$252	$52	$1,455	$1,965	$1,965	$1,965
Rolling	$1,232	$739	$765	$267	$70	$1,486	$2,418	$2,418	$2,418
Mountainous	$1,869	$1,157	$1,113	$365	$108	$3,147	$4,111	$4,111	$4,111
Urban									
Freeway/Expressway/Interstate									
Small Urban	$2,440	$1,690	$1,923	$410	$75	$3,061	$10,022	$4,126	$14,085
Small Urbanized	$2,623	$1,704	$1,989	$485	$99	$3,345	$10,991	$5,562	$18,986
Large Urbanized	$4,184	$2,790	$3,081	$651	$376	$5,598	$18,777	$8,158	$27,849
Major Urbanized	$8,368	$5,580	$5,979	$1,078	$752	$11,197	$46,691	$16,315	$62,414
Other Principal Arterial									
Small Urban	$2,127	$1,436	$1,760	$344	$76	$2,602	$8,500	$3,253	$11,102
Small Urbanized	$2,275	$1,453	$1,840	$406	$102	$2,819	$9,244	$4,013	$13,698
Large Urbanized	$3,251	$2,129	$2,692	$511	$328	$4,126	$13,786	$5,509	$18,804
Major Urbanized	$6,501	$4,259	$5,384	$825	$656	$8,252	$31,988	$11,018	$47,693
Minor Arterial/Collector									
Small Urban	$1,567	$1,084	$1,331	$252	$55	$1,922	$6,225	$2,347	$8,011
Small Urbanized	$1,642	$1,097	$1,343	$286	$68	$2,025	$6,580	$2,880	$9,830
Large Urbanized	$2,210	$1,466	$1,837	$351	$184	$2,807	$9,321	$3,748	$12,792
Major Urbanized	$4,421	$2,932	$2,779	$585	$368	$5,614	$31,988	$7,496	$39,585

Source: Highway Economic Requirements System.

The values shown in *Exhibit A-1* for adding a lane at "Normal Cost" reflect costs of projects for which sufficient right-of-way is available or readily obtained to accommodate additional lanes. The values for adding lane equivalents at "High Cost" are intended to reflect situations in which conventional widening is infeasible and alternative approaches are required to add capacity to a given corridor. Such alternatives include the construction of parallel facilities, double decking, tunneling, or the purchase of extremely expensive right-of-way. HERS models these lane equivalents as though they are part of existing highways, but some of this capacity could be from new highways or other modes of transportation.

Allocating HERS Results Among Improvement Types

Highway capital expenditures can be divided among three types of improvements: system rehabilitation, system expansion, and system enhancements (see Chapters 6 and 7 for definitions and discussion). Improvements selected by HERS that do not add lanes to a facility are classified as part of system rehabilitation. Highway projects that add lanes to a facility normally include resurfacing or reconstructing existing lanes; HERS therefore splits the costs of such projects between system rehabilitation and system expansion.

Pavement Condition Modeling

HERS incorporates information on pavement condition when evaluating deficiencies, determining speed, calculating vehicle operating costs, estimating agency maintenance costs, and forecasting pavement deterioration. Building from a multiyear research effort beginning in 2004, this C&P report reflects a new set of HERS pavement performance equations.

The new HERS procedures are based on a series of equations from a Mechanistic-Empirical Pavement Design Guide (MEPDG) formula sponsored by the American Association of State Highway and Transportation Officials (AASHTO) and the Federal Highway Administration (FHWA) through the National Cooperative Highway Research Program. Previous editions of the C&P report relied on HERS equations derived from prediction models of pavement performance AASHTO developed in 1993, which are no longer considered representative of current pavement design practices.

The new HERS equations for pavement performance rely heavily on a set of new data items related to pavement characteristics and distresses that were added to the HPMS data reporting requirements in 2009 (*Exhibit A-2*). The equations also incorporate numerous default values for the multiple variables in the MEPDG algorithms; FHWA adopted this approach to reduce the number of new data elements that States would need to report. State Departments of Transportation (DOTs) provide default parameters for many of the variables. Additional data from the Long-Term Pavement Performance (LTPP) program and MEPDG supplement the HPMS sample data with required detail not available within the HPMS sample:

- LTPP climate and geographical data offer a finer level of detail used to supplement the broader climate zone values from the HPMS sample.
- LTPP and MEPDG soil data supplement the soil type indicators from the HPMS sample.
- LTPP and MEPDG data supplement HPMS sample data on base type.

Exhibit A-2 HERS Pavement Performance Equations Input Data by Type and Source

General Characteristics		
Data from HPMS	Location: State and county	
	Section length	
	Urban/rural status	
	Functional class	
	Facility type (e.g., one-way, two-way)	
	Number of lanes	
	Lane width	
	Speed limit	
	State ownership status	
Data from multiple sources (in parentheses)	Climate	Climate zone (HPMS)
		Rainfall (LTPP)
		Freezing index (LTPP)
		Air freeze-thaw cycles (LTPP)
		Mean monthly temperature (LTPP)
		Depth of ground water table (LTPP)
Travel Demand		
Data from HPMS	Annual average daily traffic (AADT)	
	Forecast AADT: forecast value and year of forecast	
Surface, Base, and Soil Characteristics		
Data from HPMS	Surface type	
	Year of last improvement	
	Year of last reconstruction	
	Thickness of rigid pavement: measured thickness and design/construction thickness	
	Thickness of flexible pavement: measured thickness and design/construction thickness	
	Previous overlay thickness (or typical design/construction thickness)	
	IRI (International Roughness Index)	
	PSR (Pavement Serviceability Rating)	
	Extent of fatigue cracking	
	Extent of transverse cracking	
	Average rutting	
	Average faulting (vertical displacement between adjacent panels)	
	Binder type	
	Typical joint spacing	
	Shoulder: type and tied shoulder status	
Data from multiple sources (in parentheses)	Dowel bar characteristics	Dowel bar status (HPMS)
		Dowel bar diameter (default)
	Soil	Soil type (HPMS)
		Sand fraction, by State and HPMS soil type (LTPP)
		Silt fraction, by State and HPMS soil type (LTPP)
		Clay fraction, by State and HPMS soil type (LTPP)
		Plasticity index, by State and HPMS soil type (LTPP)
		Soil resilient modulus, by HPMS surface type code and soil type code (MEPDG)
	Base characteristics	Base type (HPMS)
		Measured thickness (HPMS)
		Design/construction thickness (HPMS)
		Base material modulus (LTPP and MEPDG)
	Portland cement concrete properties	Portland cement concrete/base interface loss of friction age [debonding] (HPMS)
		Modulus, by State (LTPP)
		Compressive strength, by State (LTPP)
		Air content, by State (LTPP)
		Water-to-cement ratio, by State (LTPP)
	Asphalt layer	Gradation, by State (MEPDG and LTPP)
		Air voids, by State (MEPDG and LTPP)
		Binder content by volume, by State (MEPDG and LTPP)
		Air content, by State (LTPP)
		Asphalt PG (performance grade) model parameters, by high and low temperature grade (AASHTO)
		Asphalt viscosity grade model parameters, by asphalt binder grade (AASHTO)
Other data	Sealant type, by State (default)	

- For Portland cement concrete surfaces identified in the HPMS sample, data from the LTPP and Jointed Plain Concrete Pavement programs offer a range of attributes representing Portland cement concrete strength.

- For asphalt surfaces identified in the HPMS sample, MEPDG and LTPP data offer a range of attributes representing asphalt strength, while AASHTO model parameters governing asphalt performance grade and viscosity seed the pavement model equations directly.

The new HERS pavement performance equations project the level or severity of three distresses for concrete pavements—roughness, as measured by the International Roughness Index (IRI), faulting, and cracking, and three distresses for flexible composite pavements—roughness, rutting, and cracking. Among the pavement distresses HERS predicts, pavement roughness has the most direct influence on the model's calculations because the impact of pavement roughness on highway user costs is taken into account when HERS computes the benefit-cost ratio for potential improvements. Roughness, faulting, and cracking values do not directly influence HERS projections of user costs or agency maintenance costs. These distresses are taken into account in the computations indirectly, however, because they influence predicted pavement roughness.

The projected IRI values for rigid pavements are computed as a function of fatigue cracking, spalling, and faulting. Future IRI for flexible pavements is computed as a function of fatigue cracking, rutting, and transverse (low-temperature) cracking. The IRI for composite pavement is projected as a function of rutting, fatigue cracking, and reflective cracking. Each pavement type also has an age-based component and a climate-based component that are independent of individual distresses.

For each HPMS sample section analyzed, HERS first compares the State-reported IRI with a predicted IRI based solely on the characteristics of the pavement and the date the section was last constructed, reconstructed, or resurfaced as reported by the State. If the predicted IRI for the base year differs from the actual base-year IRI value, an adjustment factor is applied when predicting future IRI for that section. This procedure accounts for the impact of variables beyond those considered directly by HERS that might influence current pavement performance and is similar in concept to an adjustment procedure HERS uses when a State reports a different capacity for a sample section in HPMS than the software computes. This approach, however, makes the model highly dependent on the accuracy of the State-reported dates when each highway sample section was last improved.

Valuation of Travel Time Savings

HERS uses estimates of the value of travel time per vehicle hour for different vehicle types. FHWA recently conducted extensive research to expand and update to 2012 the estimated hourly values of travel time used in HERS. The primary objectives of this effort were to (1) identify reliable and recent sources of information on the major components of the values of travel time and (2) develop more comprehensive estimates of the value and amount of work-related business travel in light-duty passenger vehicles (automobiles and light trucks). This second objective required expanding previous estimates of business travel in HERS, which included only work-related trips using household vehicles (i.e., vehicles owned by households rather than by organizations), to include work-related travel using corporate and government fleet vehicles, rental cars, emergency vehicles, and taxi service. This update also sought to distinguish between hourly values of travel time for buses and those for three- or four-axle single-unit trucks, which were previously combined into a single vehicle class in HERS. Finally, this update ensured that the values of travel time for vehicle occupants used in HERS were consistent with DOT's most recent official guidance.

The vehicle types (VT) considered in HERS are:

- VT1: Small Auto
- VT2: Medium Auto
- VT3: Four-Tire Truck
- VT4: Six-Tire Truck
- VT5a: Three- or Four-Axle Truck
- VT5b: Bus
- VT6: Four-Axle Combination Truck
- VT7: Five- or More Axle Combination Truck

Several factors were considered in computing the value of time per vehicle hour for these VTs, including the value of vehicle occupants' travel time, vehicle occupancy, travel purpose (business or personal), capital costs for vehicles used for business, and inventory value of cargo carried by trucks.

The estimates of vehicle occupants' values of travel time, average vehicle occupancy, and the distribution of vehicle use between work-related and personal travel were constructed for individual vehicle classes as they are commonly identified in government and commercial information sources, surveys of vehicle ownership and use, and published research. These classes include passenger automobiles; light-duty trucks with different chassis and body configurations, such as minivans, SUVs (sport-utility vehicles), and pickup trucks; and various types of medium- and heavy-duty trucks usually identified by axle configuration, purpose, or body type.

For this update, values of vehicle occupants' travel time, average occupancy, and the amounts of work-related and personal travel were first developed for different categories of users for each vehicle class. The user categories used in this update include:

- households;
- businesses (those not primarily engaged in providing transportation services);
- rental agencies;
- government agencies;

- emergency service providers (such as police and fire departments);
- taxi operators;
- transit authorities;
- operators of intercity, charter, and tour bus services;
- school bus operators;
- freight carriers (including both for-hire carriers and private or in-house subsidiaries of other businesses); and
- a broad category of "commercial truck operators."

This approach was taken to enable ready use of data reported in government and industry publications, which typically report data for a single user category (such as rental companies, government agencies, or school bus operators), or obtained from surveys of vehicle use including the National Household Travel Survey (NHTS) and Vehicle Inventory and Use Survey (VIUS), which apply to clearly identifiable user categories. It also enabled incorporation of significant differences in the values of travelers' time, vehicle occupancy patterns, and the distribution of travel by purpose among different users of the same class of vehicles into the updated estimates.

Several of the HERS vehicle types correspond directly to a single class of vehicles for which such vehicle class data are commonly available, so the values of occupants' travel time, average vehicle occupancy, and the fractions of personal and business use for that vehicle class could be "mapped" readily to a single HERS vehicle type. These included VT1 (small automobiles), VT4 (six-tire trucks), VT5a (three- or four-axle single-unit trucks), VT5b (buses, including transit, intercity, charter and tour, and school buses), VT6 (four-axle combination trucks), and VT7 (combination trucks with five or more axles). This mapping was more complex for VT2 (medium automobiles) and VT3 (four-tire trucks), as each includes multiple classes of vehicles: VT2 includes automobiles and other light-duty vehicles used primarily to carry passengers such as minivans and SUVs; VT3 consists of large light-duty vehicles used to carry both passengers and cargo, including large passenger vans, cargo vans, and pickup trucks. Work-related and personal travel by all users of each vehicle class were totaled and used to determine the overall shares of these travel purposes for VT2 and VT3. The shares of work-related and personal travel for each vehicle class then were used to construct weighted-average values of occupant's values of time and average occupancy for VT2 and VT3.

Values of Vehicle Occupants' Travel Time

DOT's *Guidance on the Value of Travel Time in Economic Analysis*[1] outlines procedures for quantifying the value of time for vehicle occupants. For personal travel, this guidance recommends valuing vehicle occupants' time in local and intercity personal travel at different fractions of their hourly earnings rates. Because the data required to apportion nationwide travel between local and intercity trips are unavailable, however, this update adheres to the previous HERS practice of valuing personal travel time using the DOT recommendation for local travel. The value from DOT guidance was assumed to apply to drivers and to other occupants of all light-duty vehicles used for personal travel, including household vehicles, corporate fleet vehicles that employees are permitted to use for personal travel, rental vehicles, and taxis.[2] When multiple trip purposes were recorded, the distribution of person hours for the vehicle trip was applied to allocate vehicle hours recorded for the trip across reported trip purposes. The distribution also was used to value travel by bus passengers, including users of public transit, intercity, charter and tour, and school buses.

For business travel in light-duty vehicles (HERS VT1 through VT3) used by households, corporate fleets, and rental agencies, this update follows DOT guidance by valuing both drivers' and passengers' travel time at 100 percent of the average pre-tax hourly wage rate for all U.S. workers, including an allowance for the dollar value of fringe benefits. Wage rate and fringe benefit data for all U.S. workers and for the more detailed occupational categories used in the calculations described below were obtained for 2012 from U.S. Bureau of Labor Statistics (BLS) publications.[3] Travel in light-duty vehicles by government employees, which was assumed exclusively work-related, is valued at 100 percent of mean hourly earnings and fringe benefits for State and local government administrative workers. Similarly, travel in light-duty vehicles by police officers and fire department employees was assumed exclusively work-related and was valued at 100 percent of the mean hourly wage plus fringe benefits for police patrol officers. Taxi drivers' time is valued at 100 percent of the mean wage rate (plus the usual allowance for fringe benefits) as reported by BLS, while travel time for taxi passengers was valued identically to personal and business travel by household members in light-duty vehicles.

Travel time for drivers and other occupants of commercial vehicles (HERS VT4 through VT7) was valued at 100 percent of the average wage rate for corresponding occupational categories reported in BLS publications, including allowances for the hourly value of fringe benefits.[4] Drivers of vehicles included in HERS VT4 were assumed to have hourly wage rates corresponding to the BLS occupational category "light truck or delivery service drivers." Drivers of vehicles included in HERS VT5 through VT7 were assumed to earn wage rates reported by BLS for "heavy and tractor-trailer truck drivers." Other occupants of commercial trucks belonging to HERS VT4 were assumed to be paid work crewmembers engaged in work-related travel, whose travel time was valued using wage and fringe benefit rates for the BLS occupational category "general laborers." Travel time for bus drivers (HERS VT5) was valued using an estimate of the average hourly earnings rate (including wages plus the hourly value of fringe benefits) for public transit, intercity, and school bus drivers developed from BLS sources.[5]

Exhibit A-3 summarizes the resulting estimates of the average hourly values of individual vehicle occupants' time for business and personal travel using various classes of vehicles (in 2012 constant dollars). Although the entries in *Exhibit A-3* apply to each person hour of travel, they are average values that reflect the distribution of travel using vehicles in that class among various user categories (e.g., households, businesses, government agencies, commercial freight carriers), as well as their typical occupancies when used for business and personal travel by each category of user. For example, *Exhibit A-3* indicates that the average hourly value of travel time for business travelers representing all users of automobiles—including households, businesses, government agencies, and other user categories—is $31.65, while the corresponding average for all users of automobiles engaged in personal travel is $12.30. The entries for the "Bus" vehicle class in *Exhibit*

A-3 indicate that bus drivers' travel time, which represents business travel, is valued at an hourly rate of $23.79, while the travel time of bus passengers—who are traveling for personal reasons—is valued at $12.30 per hour.

Exhibit A-3 Average Hourly Value of Vehicle Occupants' Travel Time by User Category, Vehicle Class, and Travel Purpose (Values in Constant 2012 Dollars)

Vehicle Class	HERS Vehicle Type	Travel Purpose		
		Business	Personal	Total
Auto (small and medium)	VT1, VT2	$31.65	$12.30	$14.01
Minivan	VT2	$22.74	$12.30	$12.76
Large passenger van	VT2	$27.76	$12.30	$15.92
Cargo van	VT3	$27.52	$12.30	$27.43
SUV	VT3	$30.50	$12.30	$13.17
Pickup	VT3	$30.58	$12.30	$15.20
6-tire truck	VT4	$29.49	$12.30	$29.32
3- or 4-axle truck	VT5a	$30.29	$12.30	$30.28
Bus	VT5b	$23.79	$12.30	$13.18
4-axle combination truck	VT6	$28.00	$12.30	$27.92
5+-axle combination truck	VT7	$28.00	$12.30	$27.99

Source: U.S. DOT Revised Guidance on the Value of Travel Time in Economic Analysis (Revision 2 - 2015 Update), Bureau of Labor Statistics.

Vehicle Occupancy

For this update, average vehicle occupancy values were estimated by combining values obtained from the 2009 NHTS for household vehicles with detailed estimates of average vehicle occupancy for other vehicle types calculated from the National Highway Traffic Safety Administration's Fatality Analysis Reporting System (FARS) for 2010 to 2012.[6] In cases for which these estimates are not directly comparable, they appear to agree closely, suggesting that relying on occupancy counts from fatal accidents is unlikely to produce biased estimates of overall average vehicle occupancy.

These data were supplemented with estimates of the average number of qualified drivers carried by trucks and buses in long-haul service, where team or replacement drivers are occasionally used. These estimates were tabulated from approximately 3.5 million roadside inspection records for 2010 and 2011 obtained from the Federal Motor Carrier Safety Administration's Motor Carrier Management Information System.[7]

Corporate fleet and rental vehicles were assumed to have average occupancy identical to that of household vehicles of the same types (automobiles, mini vans, SUVs, and pickups) when used for the same travel purpose. Thus, for example, corporate fleet and rental automobiles were assumed to carry an average of 1.24 persons—the figure derived from the 2009 NHTS for work-related business travel using household automobiles—when used for work-related trips. No published estimates of occupancy were available for government and emergency service vehicles, which were assumed to be used exclusively for work-related travel; for this update, government fleet and emergency service vehicles were assumed to carry an average of one passenger in addition to the driver.

The number of fatal accidents involving buses reported in FARS was judged insufficient to yield reliable estimates of their average occupancy. Instead, average passenger occupancy estimates for intercity, charter, and tour buses were obtained from trade association publications.[8] Similarly, passenger occupancy estimates for transit buses were calculated from information reported by the American Public Transit Association.[9] No published estimates of occupancy were available for school buses, so these were assumed to carry 10 passengers on average. School and transit buses were assumed to carry a single driver, while roadside inspection records from the Federal Motor Carrier Safety Administration suggested that approximately 1 percent of intercity coaches carry a second driver.

Estimates of average occupancy for HERS VT4 (six-tire trucks) and VT5 (three- or four-axle single-unit trucks) obtained from FARS records suggest that these vehicles frequently carry occupants other than drivers. As indicated previously in the description of values of travel time, these additional occupants were assumed paid work crewmembers. Although FARS records also suggest that some combination trucks (HERS VT6 and VT7) carry occupants in addition to drivers, these additional occupants were assumed primarily companions to drivers or other passengers traveling for personal rather than work-related reasons. Inspection records from the Federal Motor Carrier Safety Administration indicated that approximately 2 percent of combination trucks carried a second qualified driver.

Exhibit A-4 summarizes the resulting estimates of average occupancy—including drivers and any passengers—for each vehicle class. It shows, for example, that automobiles operated by all categories of users (e.g., households, businesses, government agencies) carry an average of 1.24 occupants when used for business travel and 1.57 occupants when used for personal travel. These averages reflect variation in the typical occupancies of vehicles within each class when they are operated by different users, such as households, corporate fleets, government agencies, vocational operators (such as suppliers of construction materials and services), and freight carriers. The entries for the Bus vehicle class in *Exhibit A-4* indicate that buses are occupied by a single driver engaged in work-related travel, plus an average of 12.10 occupants engaged in personal travel.

Exhibit A-4 Average Vehicle Occupancy (Persons) by User Category, Vehicle Class, and Travel Purpose

		Travel Purpose		
Vehicle Class	HERS Vehicle Type	Business	Personal	Total
Auto (small and medium)	VT1, VT2	1.24	1.57	1.53
Minivan	VT2	1.39	2.27	2.21
Large passenger van	VT2	1.39	2.38	2.04
Cargo van	VT3	1.52	1.52	1.52
SUV	VT3	1.20	1.92	1.87
Pickup	VT3	1.29	1.50	1.47
6-tire truck	VT4	1.38	1.38	1.38
3- or 4-axle truck	VT5a	1.14	1.14	1.14
Bus	VT5b	1.00	12.10	13.10
4-axle combination truck	VT6	1.02	1.02	1.02
5+-axle combination truck	VT7	1.02	1.02	1.02

Sources: National Household Travel Survey, Fatality Analysis Reporting System, and Federal Motor Carrier Safety Administration.

Distribution of Vehicle Use by Purpose

The fractions of total vehicle use (represented in terms of vehicle hours traveled) that represent business (or work-related) and personal travel were used as weights to combine the separate hourly values of time for those two travel purposes into a single average hourly value for each HERS vehicle type. Ideally, the fractions of business and personal travel using vehicles assigned to each HERS vehicle type would be calibrated with respect to the number of vehicle hours traveled for these purposes, to improve their consistency with HERS' use of vehicle hours as a basis for estimating values of travel time. This calibration was possible only for household vehicles, for which measures of travel in vehicle hours were available from NHTS. This update used estimates of the number of vehicle hours that household automobiles, vans, SUVs, and pickups were used for work-related and personal travel tabulated from the 2009 NHTS to estimate the shares of use of these vehicles for each purpose.

For non-household vehicles, vehicle hours of work-related and personal travel were calibrated with respect to vehicle miles traveled, assumptions about the distribution of vehicle miles traveled between personal and business purposes, and estimates of average travel speed.

- Total vehicle miles driven by corporate fleet automobiles were estimated from the number of vehicles and their average monthly use for 2012 reported in trade association sources. Total vehicle miles were allocated between work-related and personal travel using the distribution of travel between those purposes by minivans and SUVs—the classes of vehicles for which use patterns appear to be most closely comparable to those of automobiles—reported in VIUS to be based at non-household locations.[10] Estimated vehicle miles were converted to vehicle hours by assuming that corporate automobiles used for business and personal travel achieved the same average speeds as household automobiles used for those same purposes.
- Vehicle miles of work-related and personal travel using all other light-duty vehicles—including minivans, large passenger vans, cargo vans, SUVs, and pickups—operated by corporate fleets were estimated from VIUS for vehicles reported to be based at non-household locations. Vehicle miles were converted to vehicle hours by assuming that these vehicles operate at the same average speeds as household vehicles of the same types when used for the same purpose.
- Total vehicle miles traveled using each class of light-duty vehicles—automobiles, minivans, large passenger vans, cargo vans, SUVs, and pickups—operated by Federal, State, and local government agencies and emergency service providers (police and fire) were estimated from the total number of such vehicles and their average annual use reported in published sources.[11] Average annual use of each vehicle type by State and local government agencies and by emergency service providers were assumed equal to those for Federal government fleet vehicles. All use of these vehicles was assumed to be for work-related purposes. Estimates of vehicle miles were again converted to vehicle hours by assuming that these vehicles operate at the same average speeds as household vehicles of the same types used for work-related travel.
- Total vehicle miles traveled using rented light-duty vehicles were estimated from the number of automobiles and light-duty trucks operated by rental agencies during 2012.[12] The total number of light-duty trucks owned by rental agencies was allocated among minivans, large

passenger vans, cargo vans, SUVs, and pickups using the distribution of all light trucks operated by corporate fleets among these same vehicle classes. Average annual utilization of rental vehicles was calculated from estimates of their average age and odometer reading at the time of their resale by rental agencies, and was assumed identical for automobiles and all types of light-duty trucks operated by rental agencies.[13] The distribution of rental vehicle use between business and personal travel was estimated from the reported distribution of rental transactions among the purposes of business, leisure (assumed to correspond to personal travel), insurance replacement, and service/maintenance reported in published sources.[14] Use of each type of vehicle (e.g., automobile, minivan) rented for insurance replacement and service/maintenance purposes was assumed distributed between work-related and personal travel in the same proportions as use of household vehicles of these same types because vehicles rented for these purposes are presumably temporary replacements for mainly household vehicles. Vehicle miles of personal and work-related travel using each type of rental vehicle were converted to vehicle hours by assuming that they operate at the same average speeds as household vehicles of the same type when used for the same purpose.

- Total vehicle miles of use by taxi operators were calculated from published estimates of the total number of vehicles in the U.S. taxi fleet and their average annual utilization.[15] The distribution of total vehicle hours of taxi use between business and personal travel was estimated by combining the estimates of annual taxi trips for each purpose, their average duration, and the average number of persons traveling together on work-related and personal taxi trips reported in the 2009 NHTS.
- Total use of each class of light-duty vehicles for work-related and personal travel was calculated as the sum of its use for each of those purposes by households, corporate fleets, government agencies, emergency responders, rental agencies, and taxi operators. *Exhibit A-5* shows the resulting shares of business and personal travel using different vehicle classes. For example, the exhibit indicates that 10.9 percent of travel by all users of automobiles represents work-related business travel, while the remaining 89.1 percent represents personal travel. The entry in *Exhibit A-5* for the Bus vehicle class indicates that 7.6 percent of total person hours of bus travel represents the time of bus drivers, who are assumed engaged in business travel, while the remaining 92.4 percent represents bus passengers' travel time, which is assumed exclusively personal travel.

In previous updates of HERS' travel time values, all use of medium and heavy trucks—those included in VT4 through VT7—was assumed work-related travel. As part of this update, this assumption was tested using estimates of personal and work-related use of medium and heavy trucks from VIUS, which asked survey respondents to report the percentage of each vehicle's use for personal transportation.[16] Almost no personal use of the classes of trucks included in HERS VT5 through VT7—single-unit trucks with three or four axles and combination trucks—was reported in the 2002 VIUS, so retaining the assumption used previously in HERS that use of single-unit trucks with three or four axles and combination trucks is exclusively work-related appears justified.

Exhibit A-5 Shares of Business and Personal Vehicle Use by User Category, Vehicle Class, and Travel Purpose

Vehicle Class	HERS Vehicle Type	Travel Purpose		
		Business	Personal	Total
Auto (small and medium)	VT1, VT2	7.8%	92.2%	100.0%
Minivan	VT2	4.8%	95.2%	100.0%
Large passenger van	VT2	5.7%	94.3%	100.0%
Cargo van	VT3	21.4%	78.6%	100.0%
SUV	VT3	99.1%	0.9%	100.0%
Pickup	VT3	11.6%	88.4%	100.0%
6-tire truck	VT4	100.0%	0.0%	100.0%
3- or 4-axle truck	VT5a	100.0%	0.0%	100.0%
Bus	VT5b	7.6%	92.4%	100.0%
4-axle combination truck	VT6	99.5%	0.5%	100.0%
5+-axle combination truck	VT7	99.9%	0.1%	100.0%

Sources: NHTS and VIUS.

The VIUS data indicated that some six-tire trucks were based at households and that their owners made some use of these vehicles for personal travel. The fraction of personal use for six-tire trucks vehicles was so small, however, that retaining the assumption that they are used exclusively for work-related travel appears justified. Government agencies also operate some six-tire trucks, but these are presumably used exclusively for work-related purposes.

Vehicle Capital Costs

Like other capital assets, vehicles depreciate over their lifetimes because of use and aging, which occurs independently of accumulating use. In addition, vehicle owners incur opportunity costs on the investment represented by vehicles' remaining value, and these costs continue throughout vehicles' useful lifetimes. The HERS procedure for estimating vehicle operating costs captures depreciation that occurs as a consequence of their use, but does not include depreciation related simply to their aging or the opportunity cost of the capital investment they represent. Although use-related depreciation occurs because of the number of miles or hours vehicles are operated, the decline in their value with the passage of time and the opportunity cost on their remaining capital value are more closely related to their original value. Thus, the magnitude of these latter two costs for a fleet of vehicles depends on both the number of vehicles it includes and their original purchase prices.

Many of the potential highway improvements evaluated by HERS would increase average travel speeds. For vehicle fleets with sizes determined primarily by the daily or weekly number of scheduled vehicle trips and their expected duration, improved travel speeds would shorten the time required by some trips and allow for a reduction in the required number of vehicles. Allowing some fleet operators to reduce the number of vehicles they employ would in turn lower both the time-related component of vehicle depreciation costs and the value of their investment in vehicles—and with the latter, the opportunity cost on that investment—although the use-related component of vehicle depreciation would not necessarily be reduced.

To capture this effect, the values of travel time for medium and heavy trucks (HERS VT4 through VT7), incorporate the hourly equivalent value of time-related vehicle depreciation and opportunity costs on capital investment in vehicles. Thus, any reduction in the number of hours that these vehicles are operated because of increased travel speeds will be reflected in savings in these components of hourly costs, in addition to the hourly value of their occupants' travel time. The following procedures and assumptions were used:

- Average purchase prices of new vehicles were obtained from published sources and converted to equivalent annual capital costs using estimates of expected vehicle lifetimes and the 7-percent real annual opportunity cost of private capital estimated by the White House Office of Management and Budget.[17] These estimates were constructed using annual capital recovery factors that incorporate each vehicle type's expected lifetime and the 7-percent opportunity cost of capital, and thus include both use- and time-related depreciation and opportunity costs on invested capital.
- Average annual miles of use were estimated from the 2002 VIUS and converted to average annual vehicle hours of use by dividing by average travel speeds for VT4 through VT7 derived from sample HERS outputs.
- Annual capital costs for each vehicle type were divided by its estimated average annual vehicle hours of use to determine total capital costs per hour of vehicle use.
- HERS estimates of use-related depreciation costs per vehicle mile for each vehicle type were converted to an hourly basis by multiplying by average speed. These results were subtracted from total hourly capital costs to determine time-related capital costs, including time-related depreciation plus opportunity costs on invested capital.

The results of these computations are shown in *Exhibit A-6.*

Exhibit A-6 Hourly Capital Costs by HERS Vehicle Type

		Costs in 2012 Dollars				
Vehicle Class	HERS Vehicle Type	Annual Capital Cost	Annual Hours of Use	Hourly Capital Cost	Use-Related Depreciation ($/hour)	Time-Related Capital Cost ($/hour)
6-tire truck	VT4	$4,883	284	$17.21	$4.76	$12.44
3+-axle single-unit truck	VT5a	$10,908	341	$31.98	$12.18	$19.81
All buses	VT5b	$14,497	1,164	$12.45	$4.61	$7.84
Transit bus	VT5b	$44,066	3,260	$13.52	$5.15	$8.37
Motorcoach	VT5b	$35,594	1,156	$30.78	$11.72	$19.06
School bus	VT5b	$9,439	1,200	$7.87	$2.99	$4.87
4-axle combination truck	VT6	$10,893	533	$20.43	$4.73	$15.70
5+-axle combination truck	VT7	$21,597	1,199	$18.01	$5.00	$13.01

Sources: VIUS and published vehicle purchase price data.

This procedure was not applied to VT1 through VT3, as the size of commercial vehicle fleets was assumed determined by considerations other than the number and duration of scheduled vehicle trips. The size of corporate automobile fleets, for example, seems more likely determined by peak demand for work-related travel during the typical workday and by corporate policies on employees' eligibility to use company-owned vehicles. Similarly, the size of rental vehicle fleets also seems likely affected by geographic and temporal (i.e., by time of day or day of week) variation in vehicle demand.

Value of Cargo Carried by Freight Vehicles

The final component included in HERS hourly vehicle costs is the inventory value of cargo carried by freight trucks, which is included only for VT6 and VT7. An important limitation of this approach is that the inventory value of cargo does not capture potential costs associated with delays of freight shipments outside of the direct accounting cost of holding cargo (e.g., disruptions of production schedules, spoilage). To estimate this value, large combination trucks of various sizes and configurations were first assigned to HERS VT6 and VT7. Detailed FHWA data on the distribution of vehicle miles by operating weight for each individual truck size and configuration were aggregated to produce distributions of total vehicle miles for HERS VT6 and VT7, and these distributions were used to compute the mileage-weighted average operating weight of trucks included in VT6 and VT7.[18] The empty weight of typical trucks included in each vehicle type was then subtracted from their average operating weight to yield an estimate of the average weight of cargo carried; these estimates were 22,900 pounds for VT6 and 36,800 pounds for VT7.

Data from the 2012 Commodity Flow Survey were used to estimate the average dollar value per pound for commodities shipped by truck, including those operated by for-hire freight carriers and trucks used for private or in-house freight carriage. These per-pound values were applied to the estimated shipment weights derived previously to calculate the total value of the typical cargo loads carried by combination trucks with three or four axles (HERS VT6) and those with five or more axles (VT7). Finally, the hourly values of cargo carried by trucks included in VT6 and VT7 were calculated by converting the 7-percent annual opportunity cost of capital used previously to its hourly equivalent and applying the result to the total value of cargo carried by trucks in VT6 and VT7. The resulting estimates were $0.11 per hour for VT6 and $0.17 per hour for VT7.

Estimated 2012 Values of Travel Time by Vehicle Type

Exhibit A-7 shows components of the hourly value of travel time for each HERS vehicle type, reports the overall average values of time per vehicle hour in 2012 dollars, and compares these to (1) the 2010 values used in the 2013 C&P Report and (2) the 2008 values used in the 2010 C&P Report. The estimated values of business travel time per vehicle hour presented in this report are higher than in the 2013 C&P Report for three key reasons. First, the estimated values of business travel time in this report are considerably higher than the 2010 values used in the 2013 C&P Report for all vehicle types except bus (which reflects only a small increase in the value of time for bus drivers). For all vehicle types except bus, the estimated value of business travel time per person hour is approximately $30 (ranging from $28.00 to $31.65), compared to a range of $22.98 to $23.98 in the 2013 C&P Report.

Additionally, the estimated average vehicle occupancy for business travel is also considerably higher for four-tire (VT3) vehicles and six-tire (VT4) trucks in this report. In the 2013 C&P Report, the estimates of vehicle occupancy for business travel were universally close to zero (ranging from

1.01 to 1.04); in this report, estimates of vehicle occupancy for business travel range from 1.24 (for small [VT1] and medium [VT2] automobiles) to 1.38 (for six-tire trucks). The estimated value of travel time per vehicle hour for business travel is the product of the estimated per-person hour value of time and vehicle occupancy. Both sets of estimates are higher for four-tire vehicles and six-tire trucks in this report than in the 2013 C&P Report, and the corresponding estimates of the value of time per vehicle hour are much larger, with increases ranging from around 34 percent for four-tire trucks to 74 percent for six-tire trucks.

Exhibit A-7 Estimated 2012 Values of Travel Time by Vehicle Type

2012 Travel Time Cost Element	VT1 Small Auto	VT2 Medium Auto	VT3 4-Tire Truck	VT4 6-Tire Truck	VT5a 3–4 Axle Truck	VT5b Bus	VT6 4-Axle Combination	VT7 5+-Axle Combination
Business Travel								
Value of time per person hour	$31.65	$30.70	$29.79	$29.49	$30.29	$23.79	$28.00	$28.00
Average vehicle occupancy	1.24	1.24	1.32	1.38	1.14	1.00	1.02	1.02
Vehicle capital cost per vehicle	N/A	N/A	N/A	$12.44	$19.81	$7.84	$15.70	$13.01
Inventory value of cargo	N/A	N/A	N/A	N/A	N/A	N/A	$0.11	$0.17
Value of time per vehicle hour	**$39.21**	**$38.15**	**$39.35**	**$53.15**	**$54.34**	**$31.66**	**$44.37**	**$41.75**
Personal Travel								
Value of time per person hour	$12.30	$12.30	$12.30	N/A	N/A	$12.30	N/A	N/A
Average vehicle occupancy	1.57	1.74	1.62	N/A	N/A	12.10	N/A	N/A
Value of time per vehicle hour	**$19.25**	**$21.40**	**$19.90**	**N/A**	**N/A**	**$148.85**	**N/A**	**N/A**
Share of vehicle use for personal travel	89.1%	90.1%	75.9%	0.0%	0.0%	92.4%	0.0%	0.0%
Average Values per Vehicle Hour								
2012	**$21.43**	**$23.06**	**$24.58**	**$53.15**	**$54.34**	**$180.51**	**$44.37**	**$41.75**
2010 (from 2013 C&P)	**$16.89**	**$16.92**	**$19.75**	**$30.47**	**$58.80**	**$58.80**	**$32.17**	**$31.44**
2008 (from 2010 C&P)	**$20.96**	**$21.00**	**$24.51**	**$29.88**	**$34.35**	**$34.35**	**$38.32**	**$38.00**

Source: U.S. DOT Revised Guidance on the Value of Travel Time in Economic Analysis (Revision 2 – 2015 Update) and internal DOT estimates.

Finally, the estimated vehicle capital cost per vehicle hour is approximately twice as large in this report as in the 2013 C&P Report. Vehicle capital cost is a component of the estimated value of time per vehicle hour for six-tire trucks, three- and four-axle trucks (VT5a), buses (VT5b), and all combination trucks (VT6 and VT7). Thus, in this report, the combined effects of increased value of time per person hour and vehicle capital cost per vehicle hour increase the estimated value of time per vehicle hour for business travel across all vehicle types, including vehicle types that do not have considerably larger estimated vehicle occupancy (three- and four-axle trucks, buses, and all combination trucks).

Exercising caution is essential when comparing the estimated values of time per vehicle hour for business travel for three- and four-axle trucks and buses across C&P reports. In the 2013 C&P

Report, three- and four-axle trucks and buses were reported in an aggregate category (i.e., buses were classified as three- and four-axle trucks). In particular, aggregating vehicle types obscures the impacts of changes to estimates of the value of time per person hour and vehicle capital cost for three- and four-axle trucks.

The estimated values of personal travel time per vehicle hour are also higher for small and medium automobiles in this report than in the 2013 C&P Report. The estimated increases are driven primarily by increases in estimated vehicle occupancy for personal travel (increasing from 1.38 to 1.57 for small automobiles and from 1.38 to 1.74 for medium automobiles). The estimated values of time per person hour for personal travel increased only slightly, from $11.89 per person hour to $12.30 per person hour. As with comparisons of the estimated changes to values of time per vehicle hour for business travel, comparisons of the estimated values of time per vehicle hour for personal travel for three- and four-axle trucks and buses across C&P reports also should be made cautiously. The estimates of average travel time cost per vehicle hour for each vehicle class in *Exhibit A-7* were specified as weighted averages of values of time per vehicle hour for business and personal travel, calibrated with respect to estimated shares of vehicle use for business and personal travel. For four vehicle classes—six-tire trucks, three- and four-axle trucks, four-axle combination trucks, and five- or more axle combination trucks—all travel was specified as business travel; for these vehicle classes, the estimated travel time cost per vehicle hour is equal to the estimated value of time per vehicle hour for business travel. For the other four vehicle classes, the estimated shares of vehicle use for personal travel range from 75.9 percent (for four-tire trucks) to 92.4 percent (for buses).

The estimated average travel time cost for small and medium automobiles increased from around $17 per vehicle hour in the 2013 C&P Report to $21.43 for small automobiles and $23.06 for medium automobiles in this report, as shown in *Exhibit A-7*. The larger increase for medium automobiles was driven chiefly by the larger increase in the estimate of average vehicle occupancy for personal travel in medium automobiles relative to small automobiles.

The relative increase in the current estimated average travel time cost for four-tire trucks ($24.58 per vehicle hour versus $19.75 in the 2013 C&P Report) is approximately equal to the corresponding increase for small automobiles. This result was driven by offsetting relative changes in estimated vehicle occupancy for business travel (a higher increase for four-tire trucks) and personal travel (positive for small automobiles). For four-tire trucks and small automobiles, the estimated average travel time costs in *Exhibit A-7* are similar to the corresponding estimates in the 2010 C&P Report; for medium automobiles, the current estimated average travel time cost is higher than in the previous two reports.

In the 2013 C&P Report, the estimated average travel time costs for six-tire trucks and for all combination trucks were similar ($30.47 per vehicle hour for six-tire trucks, $32.17 per vehicle hour for four-axle combination trucks, and $31.44 per vehicle hour for combination trucks with five or more axles). In *Exhibit A-7*, the estimated average travel time cost for six-tire trucks ($53.15 per vehicle hour) is between around $7 and $9 higher than the corresponding estimates for combination trucks ($44.37 and $41.75 for four-axle and five-axle combination trucks, respectively); this result was driven by a strong upward revision to estimated vehicle occupancy for business travel in six-tire trucks. For six-tire trucks and all combination trucks, the estimated average travel time costs in *Exhibit A-7* are higher than the corresponding estimates in the 2010 C&P Report.

Disaggregating buses and three- and four-axle trucks results in distinct estimated average travel time costs for these vehicles, relative to each other and to the aggregated estimates in previous reports. The estimated average travel time cost per vehicle hour for three- and four-axle trucks is more than $4 lower than the 2010 values from the 2013 C&P report ($54.34 versus $58.80), and much lower than the corresponding estimate for buses ($180.51). The estimate for three- and four-axle trucks represents only business travel with relatively few occupants, while the estimate for buses represents predominantly personal travel with many occupants. Although the estimated cost of business travel per vehicle hour is higher for three- and four-axle trucks, the large number of estimated bus occupants traveling for personal purposes on buses yields a much larger estimated average value of travel time for buses.

The estimated average travel time costs presented in *Exhibit A-7* represent the values of travel time HERS applies to base-year travel. DOT guidance directs FHWA to assume that values of travel time will grow at a rate of 1.2 percent per year when forecasting travel time impacts, to project the effects of real wage growth on travel time costs. Thus, the values of time specified in HERS for a given year, *t*, are equal to the base-year values in *Exhibit A-7*, multiplied by 1.012^{t-2012}.

Costs of Air Pollutant Emissions

Greenhouse Gas Emissions

Road traffic generates an appreciable share of anthropogenic emissions of greenhouse gases (GHG). In the United States, passenger vehicles alone account for roughly 20 percent of emissions of carbon dioxide, and CO_2 emissions account for about 95 percent of the total global warming potential from all U.S. emissions of GHGs. In line with CO_2 emissions as the dominant concern relating to global warming, HERS has the capability to quantify and cost these emissions starting with the version used for the 2010 C&P Report.

The quantification of CO_2 emissions from motor vehicle traffic is based on the amounts of gasoline and diesel fuel consumed (alternative fuels have yet to be incorporated into the model). Emissions directly from vehicles amount to 8,852 grams of CO_2 per gallon of gasoline consumed, and 10,239 grams per gallon of diesel fuel.[19] These emissions are often referred to as tailpipe emissions, because they result from the fuel combustion process in motor vehicles' engines. In addition to these direct emissions, the fuel production and distribution processes produce CO_2 emissions, which are often referred to as upstream emissions. For this report, the HERS analysis added upstream emissions, which quantitatively are more uncertain, to estimates of direct or tailpipe CO_2 emissions. The HERS estimates of upstream emissions are 2,072 grams CO_2 per gallon of gasoline consumed and 2,105 grams CO_2 per gallon of diesel fuel consumed.

HERS uses these estimates of CO_2 emissions per gallon of fuel consumed to convert consumption rates of vehicle fuel to CO_2 emissions per vehicle mile. The resulting estimates of CO_2 emissions per vehicle mile are then converted to dollar costs using estimates of climate-related economic damages caused by CO_2 emissions. A recent study by a Federal interagency working group (Interagency Working Group on Social Cost of Carbon 2010) estimated the costs to society from future climate-related economic damages caused by incremental CO_2 emissions. The group's estimates of this social cost of carbon were intended to include, at a minimum, the monetized impacts of emissions-induced climate change on net agricultural productivity, human health, property damages from increased flood risk, and the value of ecosystem services. Low, medium, high, and very high estimates of the social cost per metric ton of carbon were developed for each year from 2010 through 2050 using alternative discount rates.

The analyses presented in this report use the medium estimates, updated to 2012 dollars using the gross domestic product price deflator (as was done in a recent analysis of corporate average fuel economy standards conducted by the National Highway Traffic Safety Administration). The adjusted values of CO_2 damage costs increase annually from $37 per metric ton in 2012 to $57 by 2032, the final year for which this report projects highway conditions and performance. For use as HERS inputs, the values were averaged to produce estimates of CO_2 damage costs for each 5-year HERS funding period; a 3-percent discount rate was applied to all estimated impacts on CO_2 emissions when calculating benefits associated with improvements in HERS.

Emissions of Criteria Air Pollutants

For the 2013 C&P Report, FHWA conducted new research to enhance and update HERS procedures for estimating economic damage costs from motor vehicle emissions of criteria air pollutants or their chemical precursors: carbon monoxide, volatile organic compounds, nitrogen oxides, sulfur dioxide, and fine particulate matter.

HERS estimates of economic damages from vehicle emissions of air pollutants were updated by first estimating new emission rates—measured in mass per vehicle mile traveled—for criteria pollutants and their precursors. These updated estimates were developed using the U.S. Environmental Protection Agency's (EPA's) Motor Vehicle Emission Simulator (MOVES) model. Average emissions per vehicle mile traveled of each pollutant vary among the roadway functional classes used in HERS because the typical mix of vehicles operating on each functional class varies and different types of vehicles emit these pollutants at different rates per vehicle mile traveled. The MOVES emission rates also vary with travel speed and other driving conditions that affect vehicles' power output.

Repeated runs of the MOVES model were conducted to develop a schedule of average emissions per vehicle mile traveled of each pollutant by travel speed for each roadway functional class during the midpoint year of each 5-year funding period used by HERS. Because MOVES uses different roadway classes than HERS uses, the most appropriate MOVES roadway class was used to represent each HERS functional class.

HERS combines these schedules of average emissions per vehicle mile traveled for different pollutants with estimates of the average dollar cost of health damages caused per unit mass of each pollutant to calculate damage costs per vehicle mile traveled for each pollutant. The dollar costs per unit of each pollutant used in HERS were updated using estimates for 2015, 2020, 2030, and 2040, supplied by EPA; these were interpolated to produce estimates for the midpoint of each 5-year funding period.[20] HERS then sums the estimates of damage costs for individual pollutants to calculate total air-pollution-related costs per vehicle mile traveled at different speeds. This process resulted in updated schedules of the average dollar cost of air-pollution-related damages per vehicle mile traveled by speed for each HERS functional class and funding period.

Motor vehicles emission rates for each criteria pollutant are projected to decline significantly in the future as new vehicles that meet more stringent emissions standards gradually replace older models in the vehicle fleet. At the same time, however, EPA projects that economic damage costs per unit of each criteria air pollutant (except carbon monoxide) will increase rapidly over time; projections of unit damage costs for nitrogen oxides, sulfur dioxide, and fine particulate matter are all projected to increase around 24 to 30 percent from 2015 to 2030.

Effects on HERS Results

Potential improvement projects evaluated by HERS can affect air pollution and CO_2 damage costs by increasing the volume of travel on a section during future funding periods and by increasing the average speed of travel on that section. Higher travel volumes invariably increase emissions and damage costs, but emission and fuel consumption rates are more complex functions of travel speeds, so increasing travel speed on a sample section can cause air pollution and CO_2 damage costs to either increase or decrease. Because the speed-mediated effect is often to reduce emissions, the overall effect of an improvement project on air pollution or CO_2 damage costs could be either an increase or a decrease. Net reductions in air pollution costs represent one component of the benefits from a potential improvement to a HERS sample section, while net increases represent one component of the costs (disbenefits).

Highway Operational Strategies

One of the key modifications to HERS featured in previous reports was the ability to consider the impact of highway management and operational strategies, including Intelligent Transportation Systems (ITSs), on highway system performance. This feature is continued in this report with only minor modifications. Current and future investments in operations are modeled outside of HERS, but the impacts of these deployments affect the model's internal calculations and, thus, also affect the capital improvements considered and implemented in HERS.

Among the many operational strategies available to highway agencies, HERS considers only certain types based on the availability of suitable data and empirical impact relationships. Grouped by category, these are:

- Arterial Management
    - Adaptive Traffic Signal Control
    - Electronic Roadway Monitoring (considered a supporting deployment necessary to other operations strategies)
    - Variable Message Signs (VMS)
- Freeway Management
    - Adaptive Ramp Metering
    - Electronic Roadway Monitoring (considered a supporting deployment necessary to other operations strategies)
    - VMS
    - Integrated Corridor Management, with and without comprehensive deployment of Vehicle Infrastructure Integration (VII) technologies.[21] Integrated Corridor Management coordinates the operation of the infrastructure elements within a corridor—for example, the timing of traffic signals near freeway interchanges with freeway incident management and ramp metering.
    - Active Traffic Management, which includes lane controls (dynamic junction control, dynamic lane reversal/contraflow lane reversal, dynamic lane use control, dynamic merge control), queue warning systems, dynamic shoulder lanes, queue warning, and Variable Speed Limits (VSL), also known as "speed harmonization"
- Incident Management (freeways only)
    - Incident Detection (free cell phone call number and detection algorithms)
    - Incident Verification (surveillance cameras)
    - Incident Response (on-call service patrols)
- Traveler Information
    - 511 Systems
    - Advanced In-vehicle Navigation Systems with real-time traveler information (enabled by VII deployment)
    - Incident Response (on-call service patrols).

Creating the operations improvements input files for use in HERS involved four steps: (1) determining current operations deployment, (2) determining additional operations deployments for the HERS funding periods, (3) determining the cost of future operations investments, and (4) determining the impacts of operations deployments. Different levels and types of deployments can be selected for an individual scenario.

Current Operations Deployments

To determine current operations deployments on the HPMS sample sections, data from the ITS Deployment Tracking Survey (http://www.itsdeployment.its.dot.gov/) were merged with 2012 HPMS sample panel section data. The ITS data were assigned to HPMS sample sections for each urbanized area using existing congestion and traffic levels on those sections as criteria.

Future Operations Deployments

For future ITS and operational deployments, projections were developed based on two alternatives. For the "Continuation of Existing Deployment Trends" alternative, existing deployments in urban areas were correlated with the congestion level and area population to predict, based on these factors, where future deployments will occur. This alternative is reflected in the analyses presented in Chapters 7 and 8.

The "Aggressive Deployment" alternative is reflected in sensitivity analysis presented in Chapter 10. This alternative assumes that deployment accelerates above existing trends and expands to more advanced strategies. Under this alternative, advanced in-vehicle navigation systems that provide real-time traveler information would supersede the current 511 systems. *Exhibit A-8* identifies the strategies employed in each alternative.

Operations Investment Costs

The unit costs for each deployment item were taken from the DOT ITS Benefits Database and Unit Costs Database and supplemented with costs based on the ITS Deployment Analysis System model. Costs were broken down into initial capital costs and annual operating and maintenance costs. Additionally, costs were determined for building the basic infrastructure to support the equipment and for the incremental costs per piece of equipment deployed.

Exhibit A-8 Types of Operations Strategies Included in Each Scenario

Operations Strategy	Scenario: Continue Existing Trends	Scenario: Aggressive Deployment
Arterial Management		
Signal control	•	•
Emergency vehicle signal preemption	•	•
Variable message signs		•
Advanced traveler information		•
Freeway Management		
Ramp metering	•	•
Variable message signs	•	•
511 traveler information	•	
Advanced traveler information		•
Integrated corridor management		•
Active traffic management		•
Incident Management (Freeways Only)		
Detection	•	•
Verification	•	•
Response	•	•

Source: Highway Economic Requirements System.

Impacts of Operations Deployments

Exhibit A-9 shows the estimated impacts of the different operations strategies considered in HERS. These effects include:

- Incident Management: Incident duration and the number of crash fatalities are reduced. Incident duration is used as a predictor variable in estimating incident delay in the HERS model.

Exhibit A-9 Impacts of Operations Strategies in HERS

Operations Strategy	Impact Category	Impact Details
Arterial Management		
Signal control	Congestion/delay	Signal Density Factor = n(nx + 2)/(n + 2), where n = no. of signals per mile x = 1 for fixed time control 2/3 for traffic actuated control 1/3 for closed loop control 0 for real-time adaptive control/Split Cycle Offset Optimization Technique (SCOOT)/Sydney Coordinated Adaptive Traffic System (SCATS)® Signal Density Factor used to compute zero-volume delay due to traffic signals
Electronic roadway monitoring	Congestion/delay	Supporting deployment for corridor signal control (two highest levels) and traveler information
Emergency vehicle signal preemption	None	Reflected in costs but no impact currently simulated
Variable message signs	Congestion/delay	-0.5% incident delay
Freeway Management		
Ramp metering		
Preset	Congestion/delay	New delay = ((1 - 0.13)(original delay)) + 0.16 hrs. per 1000 VMT
Traffic actuated	Congestion/delay	New delay = ((1 - 0.13)(original delay)) + 0.16 hrs. per 1000 VMT
	Safety	-3% number of injuries and property damage only accidents
Electronic roadway monitoring	Congestion/delay	Supporting deployment for ramp metering and traveler information
Variable message signs	Congestion/delay	-0.5% incident delay
Integrated corridor management	Congestion/delay	-7.5% total delay without VII, 12.5% total delay with VII
Active traffic management	Congestion/delay	-7.5% total delay
	Safety	-5% fatalities
Incident Management (Freeways Only)		
Detection algorithm/free cell	Incident characteristics	-4.5% incident duration
	Safety	-5% fatalities
Surveillance cameras	Incident characteristics	-4.5% incident duration
	Safety	-5% fatalities
On-call service patrols		
Typical	Incident characteristics	-25% incident duration
	Safety	-10% fatalities
Aggressive	Incident characteristics	-35% incident duration
	Safety	-10% fatalities
All combined	Incident characteristics	Multiplicative reduction
	Safety	-10% fatalities
Traveler Information		
511 only	Congestion/delay	-1.5% total delay, rural only
Advanced traveler information (VII-enabled)	Congestion/delay	-3% total delay, all highways

Source: Highway Economic Requirements System.

- Signal Control: The effects of the different levels of signal control are directly considered in the HERS delay equations.
- Ramp Meters, VMS, ATM, Integrated Corridor Management, VSL, and Traveler Information: Delay adjustments are applied to the basic delay equations in HERS. VSL also is assumed to have a small impact on fatalities.

Based on the current and future deployments and the impact relationships, an operations improvements input file was created for each deployment scenario. Each file contains section identifiers, plus current and future values (for each of the four funding periods in the HERS analysis) for the following five fields:

- Incident Duration Factor
- Delay Reduction Factor
- Fatality Reduction Factor
- Signal Type Override
- Ramp Metering.

Future HERS Enhancements Currently Underway

As part of an ongoing program of model revisions and improvements, the matrix of typical costs per mile for the various types of highway capital improvements modeled in HERS as reflected in *Exhibit A-1* are currently being updated. As part of this effort, the matrix will be expanded to capture differences in costs associated with "typical reconstruction" versus "total reconstruction," which would involve complete reconstruction of the roadway starting at the subgrade. The current distinction between "normal-cost" capacity expansion and "high-cost" capacity expansion will be broadened to consider the impact on expansion costs resulting from different types of obstacles to widening that are now coded by the States in HPMS. Other aspects of this research effort include developing procedures for adjusting the cost matrix to remove costs associated with culverts and bridge replacements in conjunction with highway widening projects in anticipation that future enhancements to the National Bridge Investment Analysis System will allow it to compute such needs more accurately than HERS can. Procedures also will be developed to facilitate analysis of the variable costs associated with different overlay depths.

Work is also underway to refine and update the new pavement performance equations recently introduced into HERS. These equations were based on an early version of the AASHTO Mechanistic-Empirical Pavement Design Guide algorithms, some of which have subsequently been revised. This research is also intended to address certain anomalies encountered in translating the simplified mechanistic-empirical equations into the HERS framework.

FHWA has initiated a major effort to update the equations for predicting vehicle fuel economy and other vehicle operating costs currently included in HERS and in several other public and private-sector tools for highway benefit-cost analysis. The current HERS procedures are based on a 1982 study and are not considered adequately reflective of current vehicle technology and driving patterns. The new study is building from the Strategic Highway Research Program 2 Naturalistic Driving Study and the Road Information Database to develop driving cycles that will be used to model the relationship between vehicle speed and fuel consumption. The impacts of road curvature and pavement roughness on fuel consumption also will be explored. This project includes modeling the relationships among pavement roughness, speed, roadway characteristics, and vehicle operating costs such as repair and maintenance, tire wear, mileage-related vehicle depreciation, and oil consumption.

Another research effort currently underway will update the costs and benefit associated with the types of operations strategies currently incorporated into the HERS operations preprocessor as referenced in *Exhibits A-8* and *A-9*. This effort includes an evaluation of the potential for simulating the impacts of connected vehicles and the potential for modeling the impacts of managed lanes.

FHWA is sponsoring research targeted at improving the specification of business travel time costs in HERS, including both refinements to the content and use of existing data sources and methodological improvements. A key data-centered effort involves identifying approaches for capturing and applying data on business travel from NHTS. The set of methodological improvements under investigation includes an effort to incorporate travel time reliability into the measurement of benefits associated with travel time improvements. HERS currently uses a proxy for reliability-based benefits, by adding a premium to account for lost time under unexpected delay due to traffic incidents. The premium for incident delay time also features in the ITS Deployment Analysis System model, which FHWA developed as a tool for benefit-cost analysis of ITS deployments.

FHWA sponsors research to develop and implement an updated plan for valuing personal travel time in HERS. Focal areas of the updated plan include the potential to differentiate values of travel time savings along dimensions such as trip length, the level of congestion, and trip purpose (e.g., commute travel versus discretionary travel). Consistent with the research on business travel time discussed above, this research includes efforts to incorporate travel time reliability into the measurement of benefits associated with travel time improvements.

A related research effort FHWA is sponsoring uses HERS outputs as inputs to a national economic model to capture the impact of highway investment on macroeconomic performance. After analyzing the capabilities of various macroeconomic models (econometric, input-output, and computable general equilibrium), the United States Applied General Equilibrium (USAGE) model was selected for further development and scenario analysis. USAGE is a 500-industry, dynamic computable general equilibrium model of the U.S. economy developed at Monash University (now housed at Victoria University) in collaboration with the U.S. International Trade Commission. USAGE was the only model among the candidates reviewed that satisfied all the following criteria considered important for estimating the economic effects of transportation investments:

- The freight-carrying transportation modes are represented as separate industries.
- Substitution between freight transportation modes can be represented.

- The model can represent changes in productivity in freight modes through changes in technical parameters defining the industry.
- Changes in prices of freight service influence demand for freight services, consistent with economic theory.
- Prices and demand can adjust in response to changes in fiscal and monetary policy (e.g., through changes in budget deficits, income taxes, and fuel taxes).
- Short-term Keynesian effects of government spending under the presence of slack resources (i.e., stimulus effects) can be represented.

The first phase of the research centered on the customization of USAGE to map outputs from HERS to impacts within the national economy. The customized version of the model, USAGE-Hwy, uses key outputs from HERS as model inputs, including levels of highway investment and impacts on travel time (specified separately for light-duty vehicles and heavy-duty trucks), operating costs (specified separately for light-duty vehicles and heavy-duty trucks), fuel consumption, vehicle miles traveled, and highway fatalities. FHWA anticipates including analyses based on USAGE-Hwy in future C&P reports to investigate the sensitivity of macroeconomic outcomes to changes in highway spending levels and to associated changes in highway travel costs and vehicle miles traveled.

---

[1] See http://www.dot.gov/administrations/office-policy/2015-value-travel-time-guidance.

[2] Median household income data for 2012 were obtained from U.S. Bureau of the Census, Historical Income Tables – Households, Table H-6 (http://www.census.gov/hhes/www/income/data/historical/household/), and were converted to their hourly equivalent assuming 2,080 paid working hours per year.

[3] Hourly wage rates for All Occupations during 2012 are reported in Bureau of Labor Statistics, *Occupational Employment and Wages – May 2012*, March 29, 2013, Table 1 (http://www.bls.gov/news.release/ocwage.t01.htm). Hourly values of fringe benefits during 2012 were estimated from fractions of Total Compensation for Civilian Workers, reported in Bureau of Labor Statistics, *Employer Costs for Employee Compensation – June 2012*, September 11, 2012, Table 2 (http://www.bls.gov/news.release/archives/ecec_09112012.pdf).

[4] Hourly wage rates for 2012 were obtained from Bureau of Labor Statistics, *Occupational Employment and Wages – May 2012*, March 29, 2013, Table 1 (http://www.bls.gov/news.release/ocwage.t01.htm) for the occupational categories of light truck or delivery service drivers and heavy and tractor-trailer truck drivers. Hourly values of fringe benefits were estimated from fractions of Total Compensation for the "Transportation and material moving" occupational group, reported in Bureau of Labor Statistics, *Employer Costs for Employee Compensation – June 2012*, September 11, 2012, Table 2 (http://www.bls.gov/news.release/archives/ecec_09112012.pdf).

[5] Mean wage rate for bus drivers during 2012 was estimated using a weighted average of mean wage rates for BLS occupational categories Bus drivers – transit and intercity and Bus drivers – school or special client, reported in Bureau of Labor Statistics, *Occupational Employment and Wages – May 2012*, March 29, 2013, Table 1 (http://www.bls.gov/news.release/ocwage.t01.htm). Weights used in calculating this average are the product of employment in each category (reported in same source as wage rates) and estimates of average number of hours worked per week for school and all other bus drivers during May 2012, constructed using data tabulated from BLS Current Employment Survey (http://www.bls.gov/ces/data.htm).

[6] See http://www.nhtsa.gov/FARS.

[7] https://portal.fmcsa.dot.gov/login.

[8] ABA Foundation, Motorcoach Census 2013, February 27, 2014, Table 4-1, p. 19 (http://www.buses.org/assets/images/uploads/general/Report%20-%20Census2013data.pdf) suggests an average passenger occupancy of 36.5 persons.

[9] http://www.apta.com/resources/statistics/Documents/FactBook/2013-APTA-Fact-Book.pdf.

[10] These include vehicles reported to be based at a company office/headquarters, terminal, manufacturing plant, or distribution center; see definition of variable HB_TYPE, *Vehicle Inventory and Use Survey 2002, Microdata Data Dictionary*, p. 27. (https://www.census.gov/svsd/www/vius/datadictionary2002.pdf).

[11] Federal Fleet Report, Tables 2-5, 2-6, and 4-2 (http://www.gsa.gov/portal/category/102859), *Government Fleet Fact Book 2012*, *Fleet Size by Unit Type*, p. 28, and *State, County, and Municipal Vehicle Totals*, p. 30. (http://www.government-fleet.com/fileviewer/1556.aspx), and Automotive Fleet, U.S. Fleet Statistics by Industry Segment (http://www.automotive-fleet.com/statistics/statsviewer.aspx?file=http%3a%2f%2fwww.automotive-fleet.com%2ffc_resources%2fstats%2faffb12-9-fleetstats.pdf&channel).

[12] Automotive Fleet, U.S. Fleet Statistics by Industry Segment (http://www.automotive-fleet.com/statistics/statsviewer.aspx?file=http%3a%2f%2fwww.automotive-fleet.com%2ffc_resources%2fstats%2faffb13fleetstats.pdf&channel).

[13] Reported in http://online.wsj.com/news/articles/SB10001424127887324463604579040870991145200.

[14] *Auto Rental News*, Fact Book, various issues 2000–2003, http://www.autorentalnews.com/content/research-statistics.aspx.

[15] Automotive Fleet, U.S. Fleet Statistics by Industry Segment (http://www.automotive-fleet.com/statistics/statsviewer.aspx?file=http%3a%2f%2fwww.automotive-fleet.com%2ffc_resources%2fstats%2faffb13fleetstats.pdf&channel).

[16] See definition of variable OPCLASS _PSL, *Vehicle Inventory and Use Survey 2002, Microdata Data Dictionary*, p. 33 (https://www.census.gov/svsd/www/vius/datadictionary2002.pdf).

[17] See White House Office of Management and Budget, Office of Information and Regulatory Affairs, Circular A-4, *Regulatory Analysis: A Primer*, August 15, 2011 (https://www.whitehouse.gov/sites/default/files/omb/inforeg/regpol/circular-a-4_regulatory-impact-analysis-a-primer.pdf, p. 11. Purchase prices for trucks were obtained from IHS Automotive, Truck Pricing: GWV Class 3-8, 2013. Estimate for transit buses is total for Bus, Trolley Bus, Commuter Bus, and Bus Rapid Transit from American Public Transit Association, *Transit Fact Book 2013* Table 6, p.12 (data for 2011) (http://www.apta.com/resources/statistics/Documents/FactBook/2013-APTA-Fact-Book.pdf). Motor coach estimate from ABA Foundation, Motorcoach Census 2013, February 27, 2014 (http://www.buses.org/assets/images/uploads/general/Report%20-%20Census2013data.pdf), Table 4-1, p. 19 (data for 2012). School bus estimate from *School Bus Fleet 2015 Factbook*, Volume 60, No. 11, (http://digital.schoolbusfleet.com/2015FB/Default/3/0/2414989#&pageSet=0), *School Transportation Statistics: 2012–13 School Year*, pp. 29–30.

[18] Federal Highway Administration, *1997 Highway Cost Allocation Study*, Chapter II, Table II-8 (http://ntl.bts.gov/lib/5000/5900/5940/final.pdf).

[19] The chemical properties of fuels were obtained from Wang, M.Q., *GREET 1.5 — Transportation Fuel-Cycle Model: Volume 1, Methodology, Use, and Results*, ANL/ESD-39, Vol.1, Center for Transportation Research, Argonne National Laboratory, Argonne, Ill., August 1999, Table 3.3, p. 25 (available at http://greet.es.anl.gov/index.php?content=publications&by=date&order=up#Technical_Publications).

[20] For a description of these estimated damage costs, see U.S. EPA and National Highway Traffic Safety Administration, Joint Technical Support Document, Final Rulemaking for 2017–2025 Light-Duty Vehicle Greenhouse Gas Emission Standards and Corporate Average Fuel Economy Standards, August 2012, pp. 4-42 to 4-48 (available at http://www.nhtsa.gov/fuel-economy).

[21] The VII program at DOT has evolved into the Connected Vehicle Program: http://www.its.dot.gov/landing/cv.htm. As of this writing, for HERS, the strategy enabled by VII technologies is advanced traveler information. Additional strategies covered under the Connected Vehicle program have not been incorporated.

appendix B

Bridge Investment Analysis Methodology

Bridge Investment Analysis Methodology

The National Bridge Investment Analysis System (NBIAS) was developed to assess national bridge investment needs and the tradeoff between funding and performance. NBIAS, first introduced in the 1999 C&P Report, is used to model investments in bridge repair, rehabilitation, and functional improvements. Over time, the system has been used increasingly as an essential decision-support tool for analyzing policy and providing information to the U.S. Congress.

NBIAS is based on an analytical framework similar to that used in the Pontis bridge management system developed by the Federal Highway Administration (FHWA) in 1992 and subsequently adopted by the American Association of State Highway and Transportation Officials (AASHTO). The system incorporates economic forecasting tools to project the multiyear funding needs required to meet user-selected performance objectives over the length of a user-specified performance period. NBIAS differs from Pontis in that it works with bridge condition data as reported by the States, Federal agencies, and Tribal governments for the National Bridge Inventory (NBI) in addition to the element/condition state inspection regime used in Pontis. NBIAS combines statistical models with engineering principles and heuristic rules to synthesize representative elements so they can be defined and manipulated using the same structure of condition states, actions, deterioration, costs, and effectiveness probabilities used in Pontis, which makes them compatible with the predictive models and analytical routines in Pontis. NBIAS extends the Pontis element model by introducing the climate zone dimension into the stratification scheme and adding user cost components to the cost model. Effective in version 4.0 (2011), NBIAS also features an enhanced element optimization model that integrates selected maintenance policies.

General Methodology

Using linear programming optimization, NBIAS generates a set of prototype maintenance policies for defined subsets of the NBI. Models of element deterioration, feasible actions, and the cost and effectiveness of those actions are incorporated as major inputs for each subset of the inventory. For functional deficiencies and improvements, NBIAS uses a model similar to the bridge level-of-service standards and user cost models of Pontis, augmented by a bridge improvement model the Florida Department of Transportation developed.

With a set of synthesized projects developed from the maintenance and functional improvement models, NBIAS calculates a tradeoff structure showing the effect of hypothetical funding levels on each of more than 200 performance measures. For this analysis, the system uses an adaptation of an incremental benefit-cost model with graphical output showing the tradeoff between funding and performance. To estimate functional improvement needs, NBIAS applies a set of improvement standards and costs, which the user can modify, to each bridge in the NBI. The system uses the available NBI data to predict detailed structural element data for each bridge. It measures repair and rehabilitation needs at the bridge-element level using a Markov decision model and then applies the obtained maintenance strategy, along with the improvement model, to each bridge.

Replacement costs for structures are determined based on State-reported values FHWA gathers. Improvement costs are consistent with those in Pontis and are adjusted to account for inflation. In evaluating functional improvement needs and repair and rehabilitation needs, the system uses a set of unit costs for various improvement and preservation actions. State-specific cost-adjustment factors are applied to the unit costs.

The NBIAS user can specify hypothetical budget constraints in several ways, by setting (1) a range of constant budgets, which directs the software to find the performance levels achievable with each budget level within the range; (2) a range of budget growth rates; or (3) a minimum benefit/cost ratio, in which case, the software determines the funding level corresponding to that benefit/cost ratio. All of these options have applications in the preparation of the C&P report and could be useful for specific owner agencies that might want to use NBIAS to analyze the funding vs. performance tradeoff for their transportation asset management plans or other planning purposes.

Determining Functional Improvement Needs

The standards for functional improvement address lane width, shoulder width, load rating, and clearances (vertical and horizontal). NBIAS includes a set of standards by functional class, additional standards derived from sufficiency rating calculations, and those standards the Florida DOT models prescribe.

The standards used in NBIAS initially were set to be the same as the default standards specified in Pontis, which were established as an early effort to define level-of-service standards for AASHTO. The standards used in the previous editions of the C&P report were reviewed and compared with design standards in the AASHTO Green Book, and adjustments were made where warranted. A revised set of standards has subsequently been added that triggers consideration of a functional improvement whenever a deduction in sufficiency rating occurs due to road width, load rating, or clearances. Adopting the Florida improvement model enabled further fine-tuning of the analysis logic of functional needs.

NBIAS estimates needs for the following types of bridge functional improvements: widening existing bridge lanes, raising bridges to increase vertical clearances, and strengthening bridges to increase load-carrying capacity. Functional improvement needs are determined by applying user-specified standards to the existing bridge inventory, subject to benefit-cost considerations. For example, a need to raise a bridge will be identified if the vertical clearance under the bridge fails to meet the specified standard and if the stream of discounted excess cost of diverting commercial vehicles around the bridge exceeds the cost of improving the bridge.

If functional improvement is infeasible due to the bridge design or impractical because of deteriorated structural condition, a replacement need is generated. Replacement need might also be identified if a user-specified replacement rule is triggered. For example, one or more replacement rules can be introduced in NBIAS based on the threshold values for age, sufficiency rating, and health index.

Because the benefit predicted for a functional improvement increases proportionally with the amount of traffic, whether a functional improvement is justified, and how much benefit is derived from that improvement, greatly depends on predicted traffic. In the current version of NBIAS, traffic predictions are made for each year in an analysis period based on NBI data. NBIAS allows the user to apply either linear or exponential traffic growth projections. Linear growth was selected for this edition of the C&P report, consistent with the assumption used in the Highway Economic Requirements System. When NBIAS selects a structure for replacement, the cost of the replacement is based on the number of lanes on the existing bridge. The cost of adding lanes to satisfy increased capacity demands is not included in the cost to construct the replacement structure. Additional costs for expanding bridges to meet increased capacity demands are included in the cost to construct a lane mile of highway used in the Highway Economic Requirements System.

Determining Repair and Rehabilitation Needs

To determine repair and rehabilitation needs, NBIAS estimates the type, quantity, and condition of elements that exist for each bridge in the NBI by statistical means and applies a set of deterioration and cost models to the estimated elements. This allows NBIAS to determine the optimal preservation actions for maintaining the bridge inventory in a state of good repair while minimizing user and agency costs.

Predicting Bridge Element Composition

The NBIAS analytical approach relies on structural element data not available in the NBI. To develop such data, NBIAS uses a set of Synthesis, Quantity, and Condition (SQC) models to predict the elements that exist on each bridge in the NBI and the condition of those elements.

The synthesis part of the SQC model is implemented as a decision tree, in which the choice of the elements for a bridge is dictated by its design, material, and several other characteristics available in the NBI. Element quantities are estimated based on the geometric dimensions of the bridge, its design, and material. The condition of the synthesized elements is modeled in the form of a percentage-based distribution of element quantities across condition states. Such distributions are evaluated based on the structural ratings (superstructure, substructure, and deck) of the bridge to which statistically tabulated lookup data and Monte Carlo simulation are applied.

The current version of NBIAS can accept the direct import of structural element data when such data are available, but this capability was not used for the development of this report. Although most States now routinely collect such data on State-owned bridges as part of their bridge inspection process, these data are not currently part of the NBI data set.

Moving Ahead for Progress in the 21st Century (MAP-21) requires the use of element-level data to analyze the performance of the bridges on the National Highway System (NHS). All other bridges have the minimum data recorded and require element-level data to be generated. Therefore, bridges on the NHS with detailed element data are combined with non-NHS bridges with generated element data.

Calculating Deterioration Rates

NBIAS takes a probabilistic approach to modeling bridge deterioration based on techniques first developed for Pontis. In the system, deterioration rates are specified for each bridge element through a set of transition probabilities that specify the likelihood of progression from one condition state to another over time. For each element, deterioration probability rates vary across nine climate zones.

Forming of the Optimal Preservation Policy

The policy of maintenance, repair, and rehabilitation (MR&R) in NBIAS is generated with the help of two optimization models: long-term and short-term. The long-term model is formulated as a linear program with the objective of keeping the element population in a steady-state condition that requires the minimum cost to maintain. The short-term model, not being concerned with the steady state, seeks to find a policy of remedial actions that minimize the cost of moving the inventory to conditions the long-term solution recommends. The short-term MR&R model is implemented as the Markov decision model solved as a linear programming problem.

In the earlier versions of NBIAS, only one MR&R strategy was available. In the course of developing NBIAS version 4.0, a study was conducted to develop alternative MR&R models. The result was three additional MR&R strategies reflecting approaches for maintaining a bridge network that are more diverse, as discussed in the following sections.

Minimize MR&R Costs

This strategy involves identifying and implementing a pattern of MR&R improvements that minimizes long-term MR&R spending. This model was adopted from Pontis and used for the NBIAS analyses presented in the 2010 C&P Report and all previous editions. This strategy is intended to prevent a catastrophic decrease in bridge network performance rather than to maintain or improve the overall condition of the bridge network. Some Pontis users and participants on expert peer-review panels for NBIAS had raised concerns that this strategy was not consistent with typical bridge management strategies, and that following such a strategy could call for a bridge to be replaced sooner than might be the case if a more aggressive MR&R approach were used.

One consequence of having initially developed this strategy as the only MR&R option in NBIAS was that most measures of bridge performance (such as the health index or percentage of deficient bridges) would always worsen over the 20-year analysis period, even if all the potential bridge improvements identified in NBIAS as cost-beneficial were implemented. The exception was the estimated backlog of bridge needs, which is why this report has focused on that metric in the past. The MR&R strategy influences the estimated backlog; assuming a less aggressive MR&R strategy reduces the estimated MR&R backlog but also increases the estimated bridge replacement backlog, generally resulting in a higher combined backlog estimate.

Maximize Average Returns

This strategy seeks to maximize the degree of bridge system performance improved per dollar of MR&R expenditure. Following this strategy results in more MR&R spending than under the Minimize MR&R strategy, but still generally results in an increase in the number of deficient bridges over time.

Sustain Steady State

This strategy was used for the analyses presented in the 2013 C&P Report. It involves identifying and implementing a pattern of MR&R improvements that would achieve an improved steady state in terms of overall bridge system conditions, without frontloading MR&R investment. Following this strategy results in more MR&R spending than under the Maximize Average Returns strategy, but still generally results in increases in deficient bridges over time.

State of Good Repair

This strategy seeks to bring all bridges to a good condition that can be sustained via ongoing investment. MR&R investment is frontloaded under this strategy, as large MR&R investments would be required in the early years of the forecast period to improve bridge conditions, while smaller MR&R investments would be needed in the later years to sustain bridge conditions.

The selection of MR&R policy can significantly influence the results of an NBIAS analysis. Based on the results of the comparison of life-cycle costs for MR&R relative to replacement, the system might simulate more or fewer bridge replacements. Given the MR&R and replacement costs developed for this C&P report, the State of Good Repair strategy, although the most aggressive, generates results more consistent with agency practices and recent trends in bridge condition than the other three strategies evaluated. It has been adopted for use in the baseline analyses presented in Chapters 7 and 8 of this report. (Please note that, despite the similarity in names, the correspondence is not one to one between the NBIAS State of Good Repair strategy and the State of Good Repair benchmark presented in Chapter 8. The State of Good Repair benchmark includes all investments identified as cost-beneficial by NBIAS and includes both MR&R investments and functional improvements.)

Applying the Preservation Policy

Using transition probability data, and information on preservation action costs and user costs for operating on deteriorated bridge decks, NBIAS applies the Markov decision model to determine the optimal set of repair and rehabilitation actions for each bridge element based on the element's condition. During the simulation process, the preservation policy is applied to each bridge in the NBI to determine bridge preservation work needed to minimize user and agency costs over time.

In analyzing potential improvement options, NBIAS compares the cost of performing preservation work with the cost of completely replacing a bridge, to identify situations in which replacement would be more cost effective. If the physical condition of the bridge has deteriorated to the point of being considered unsafe (the system user specifies the threshold for such a determination), the system might consider bridge replacement to be the only feasible alternative.

Future NBIAS Enhancements Currently Underway

As part of an ongoing program of model revisions and improvements, NBIAS is being enhanced to enable the user to assign individual budgets for specific work categories, such as maintenance, rehabilitation, and replacement of structurally deficient bridges, instead of providing a single budget for all actions. This capability will enable the user to consider a broader array of potential alternative future investment strategies. NBIAS also will be modified to improve its ability to determine budget levels required to meet user-defined performance measures. This feature will enable the user quickly to determine the annual level of funding required over a specified period to change the current value of a performance measure to a user-specified target value.

The standard element definitions for Commonly Recognized elements have been superseded by the new National Bridge Element (NBE) standard. A provision in the MAP-21 transportation legislation requires States to report element-level data to FHWA for all bridges on the NHS. NBIAS will be updated to use data reported according to the NBE standard, allowing for better incorporation of available State data (which are now collected using the NBE) and support future use of the system. At the same time, the NBIAS element performance algorithms will be recalibrated to improve the model's prediction of various bridge condition measures. These algorithms, which were last fully recalibrated in 2006, are no longer fully consistent with current bridge management practices.

Currently, data for approximately 125,000 culverts are included in NBI. The NBIAS model does not contain the algorithms needed to conduct a full analysis of culverts because, unlike typical bridges, culverts do not have a deck, superstructure, or substructure. Instead, they are self-contained units located under roadway fill and typically are constructed of concrete or corrugated steel pipes. When multiple pipes or box culverts placed side by side below a public roadway span a total length greater than 6.1 meters, they are considered structures and are subject to the NBI reporting requirements. A feature will be added to the NBIAS model that will enable analysis of culvert deterioration, projection of future overall culvert conditions, and estimation of the costs of culvert maintenance and replacement.

appendix C

Transit Investment Analysis Methodology

Transit Investment Analysis Methodology

The Transit Economic Requirements Model (TERM), an analytical tool the Federal Transit Authority (FTA) developed, forecasts transit capital investment needs over a 20-year period. Using a broad array of transit-related data and research results, including data on transit capital assets, current service levels and performance, projections of future travel demand, and a set of transit asset-specific condition decay relationships, the model generates the forecasts that appear in the biennial C&P report.

This appendix provides a brief technical overview of TERM and describes the various methodologies used to generate the estimates for the 2015 C&P Report.

Transit Economic Requirements Model

TERM forecasts the level of annual capital expenditures required to attain specific physical condition and performance targets within a 20-year period. These annual expenditure estimates cover the following types of investment needs: (1) asset preservation (rehabilitation and replacement); and (2) asset expansion to support projected ridership growth.

TERM Database

The capital needs that TERM forecasts rely on a broad range of input data and user-defined parameters. Gathered from local transit agencies and the National Transit Database (NTD), the input data are the foundation of the model's investment needs analysis and include information on the quantity and value of the Nation's transit capital stock. The input data in TERM are used to draw an overall picture of the Nation's transit landscape; the most salient data tables that form the backbone of the TERM database are described below.

Asset Inventory Data Table

The asset inventory data table documents the asset holdings of the Nation's transit operators. Specifically, these records contain information on each asset's type, transit mode, age, and expected replacement cost. Because the FTA does not directly measure the condition of transit assets, asset condition data are not maintained in this table. Instead, TERM uses asset-decay relationships to estimate the current and future physical conditions, as required for each model run. These condition forecasts then are used to determine when each type of asset identified in the asset inventory table is due for either rehabilitation or replacement. The decay relationships are statistical equations that relate asset condition to asset age, maintenance, and utilization. The decay relationships and the way TERM estimates asset conditions are further explained later in this appendix.

The asset inventory data are derived from a variety of sources, including NTD, responses by local transit agencies to FTA data requests, and special FTA studies. The asset inventory data table is the primary data source for the information used in TERM's forecast of preservation needs. Note that FTA does not currently require agencies to report on all asset types (with the exception of data for revenue vehicles, these data are provided only when requested). Furthermore, the transit industry has no standards for collecting or recording such data. Because of this, TERM analyses must rely on asset inventory data in the format and level of detail as provided by those agencies that respond to FTA's asset data requests. On July 2012, Congress passed the new Moving Ahead for Progress in the 21st Century (MAP-21) reauthorization act, which is the current law. The Act requires transit agencies to develop a capital asset report that inventories their capital assets and to evaluate the conditions of those assets. These data significantly enhance the consistency and availability of the Nation's asset base, resulting in greater accuracy of TERM's estimates of capital investment needs.

Urban Area Demographics Data Table

This data table stores demographic information on nearly 500 large, medium-sized, and small urbanized areas and for 10 regional groupings of rural operators. TERM uses fundamental demographic data, such as current and anticipated population, in addition to more transit-oriented information, such as current levels of vehicle miles traveled and transit passenger miles, to predict future transit asset expansion needs.

Agency-Mode Statistics Data Table

The agency-mode statistics table contains operations and maintenance data on each mode operated by approximately 725 urbanized transit agencies and more than 1,700 rural operators. Specifically, TERM uses the agency-mode data on annual ridership, passenger miles, operating and maintenance costs, mode speed, and average fare data to help assess current transit performance, future expansion needs, and the expected benefits from future capital investments in each agency-mode (both for preservation and expansion). All data in this portion of the TERM database come from the most recently published NTD reporting year. When reported separately, directly operated services and contracted services are merged into a single agency-mode within this table.

Asset Type Data Table

The asset type data table identifies approximately 500 different asset types the Nation's public transit systems use in support of transit service delivery (either directly or indirectly). Each record in this table documents each asset's type, unit replacement cost, and expected timing and cost of all life-cycle rehabilitation events. Some of the asset-decay relationships used to estimate asset conditions are also included in this data table. The decay relationships—statistically estimated equations relating asset condition to asset age, maintenance, and utilization—are discussed more in the next section of this appendix.

Benefit-Cost Parameters Data Table

The benefit-cost parameters data table contains values used to evaluate the merit of different types of transit investments TERM forecasts. Measures in the data table include transit rider values (e.g., value of time and links per trip); auto costs per vehicle miles traveled (e.g., congestion delay, emissions costs, and roadway wear); and auto user costs (e.g., automobile depreciation, insurance, fuel, maintenance, and daily parking costs).

Mode Types Data Table

The mode types data table provides generic data on all mode types used to support U.S. transit operations—including their average speed, average headway, and average fare—and estimates of transit riders' responsiveness to changes in fare levels. Similar data are included for nontransit modes, such as private automobile and taxi costs. The data in this table are used to support TERM's benefit-cost analysis.

The input tables described above form the foundation of TERM but are not the sole source of information used when modeling investment forecasts. In combination with the input data, which are static—meaning that the model user does not manipulate them from one model run to the next—TERM contains user-defined parameters to facilitate its capital expenditure forecasts.

Investment Policy Parameters

As part of its investment needs analysis, TERM predicts the current and expected future physical condition of U.S. transit assets over 20 years. These condition forecasts are then used to determine when each individual asset identified in the asset inventory table is due for rehabilitation or replacement. The investment policy parameters data table enables the model user to set the physical condition ratings at which rehabilitation or replacement investments are scheduled to occur (although the actual timing of rehabilitation and replacement events might be deferred if the analysis is budget constrained). Unique replacement condition thresholds can be chosen for the following asset categories: guideway elements, facilities, systems, stations, and vehicles. For the 2015 C&P Report, all of TERM's replacement condition thresholds have been set to trigger asset replacement at condition 2.5 (under the Sustain 2012 Spending scenario, many of these replacements would be deferred due to insufficient funding capacity).

In addition to varying the replacement condition, users can also vary other key input assumptions intended to reflect better the circumstances under which existing assets are replaced and the varying cost impacts of those circumstances. For example, users can assume that existing assets are replaced under full service, partial service, or a service shut down. Users can also assume assets are replaced either by agency (force-account) or by contracted labor. Each affects the cost of asset replacement for rail assets.

Financial Parameters

TERM also includes two key financial parameters. First, the model enables the user to establish the rate of inflation used to escalate the cost of asset replacements for TERM's needs forecasts.

Note that this feature is not used for the C&P report, which reports all needs in current dollars. Second, users can adjust the discount rate used for TERM's benefit-cost analysis.

Investment Categories

The data tables described above enable TERM to estimate different types of capital investments, including rehabilitation and replacement expenditures, expansion investments, and capital projects aimed at performance improvements. These three different investment categories are described below.

Asset Rehabilitation and Replacement Investments

TERM's asset rehabilitation and replacement forecasts are designed to estimate annual funding needs for the ongoing rehabilitation and replacement of the Nation's existing transit assets. Specifically, these needs include the normal replacement of assets reaching the end of their useful lives, mid-life rehabilitations, and annual "capital expenditures" to cover the cost of smaller capital reinvestment amounts not included as part of asset replacement or rehabilitation activities.

To estimate continuing replacement and rehabilitation investments, TERM estimates the current and expected future physical condition of each transit asset identified in TERM's asset inventory for each year of the 20-year forecast. These projected condition values then are used to determine when individual assets will require rehabilitation or replacement. TERM also maintains an output record of this condition forecast to assess the impacts of alternative levels of capital reinvestment on asset conditions (both for individual assets and in aggregate). In TERM, the physical conditions of all assets are measured using a numeric scale of 5 through 1; see *Exhibit C-1* for a description of the scale.

TERM currently allows an asset to be rehabilitated up to five times throughout its life cycle before being replaced. During a lifecycle simulation, TERM records the cost and timing of each reinvestment event as a model output and adds it to the tally of national investment needs (provided they pass a benefit-cost test, if applied).

Exhibit C-1 Definitions of Transit Asset Conditions

Rating	Condition	Description
Excellent	4.8–5.0	No visible defects, near new condition.
Good	4.0–4.7	Some slightly defective or deteriorated components.
Adequate	3.0–3.9	Moderately defective or deteriorated components.
Marginal	2.0–2.9	Defective or deteriorated components in need of replacement.
Poor	1.0–1.9	Seriously damaged components in need of immediate repair.

Source: Transit Economic Requirements Model.

TERM's process of estimating rehabilitation and replacement needs is represented conceptually for a generic asset in *Exhibit C-2*. In this theoretical example, asset age is shown on the horizontal axis, the cost of life-cycle capital investments is shown on the left vertical axis (as a percentage of acquisition cost), and asset conditions are shown on the right vertical axis. At the acquisition date, each asset is assigned an initial condition rating of 5, or "excellent," and the asset's initial purchase cost is represented by the tall vertical bar at the left of the chart. Over time, the asset's condition begins to decline in response to age and use, represented by the dotted line, requiring periodic life-cycle improvements, including annual capital maintenance and periodic rehabilitation projects. Finally, the asset reaches the end of its useful life, defined in this example as a physical condition rating of 2.5, at which point the asset is retired and replaced.

Exhibit C-2 Scale for Determining Asset Condition Over Time, From Acquisition to Replacement

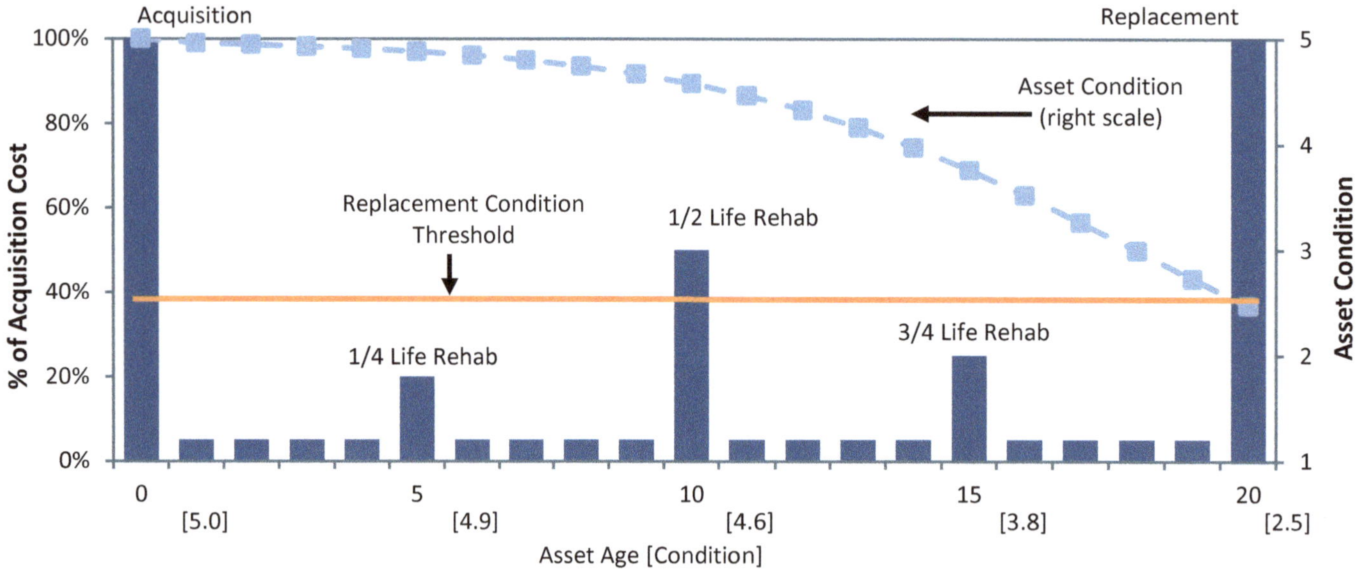

Asset Expansion Investments

In addition to devoting capital to the preservation of existing assets, most transit agencies invest in expansion assets to support ongoing growth in transit ridership. To simulate these expansion needs, TERM continually invests in new transit fleet capacity as required to maintain at current levels the ratio of peak vehicles to transit passenger miles. The rate of expansion is projected individually for each of the Nation's roughly 500 urbanized areas (UZAs) (e.g., based on the UZA's specific growth-rate projections or historic rates of transit passenger mile growth), while the expansion needs are determined at the individual agency-mode level. TERM will not invest in expansion assets for agency-modes with current ridership per peak vehicle levels that are well below the national average (these agency modes can become eligible for expansion during a 20-year model run if projected growth in ridership is sufficient for them to rise above the expansion investment threshold).

In addition to forecasting fleet expansion requirements to support the projected ridership increases, the model also forecasts expansion investments in other assets needed to support that fleet expansion. This includes investment in maintenance facilities and, in the case of rail systems, additional guideway miles, including guideway structure, track work, stations, train control, and traction power systems. Like other investments forecast by the model, TERM can subject all asset expansion investments to a benefit-cost analysis. Finally, as TERM adds the cost of newly acquired vehicles and supporting infrastructure to its tally of investment needs, it also ensures that the cost of rehabilitating and replacing the new assets is accounted for during the 20-year period of analysis.

TERM's estimates for capital expansion needs in the Low- and High-Growth scenarios are driven by the projected growth in passenger miles traveled (PMT). For this report, FTA has applied a new methodology for estimating growth in PMT, which is believed to be more accurate and provides greater consistency between the Low- and High-Growth scenarios.

In prior years, the Low-Growth scenario was driven by PMT projections obtained from metropolitan planning organizations (MPOs). Specially, UZA-level PMT growth projections were obtained from MPOs representing the Nation's 30 largest UZAs along with a sample of projections for MPOs representing smaller UZAs (less than 1 million population). These projections then were used to estimate transit capital expansion needs for the Low-Growth scenario (UZA growth rates for smaller UZAs not included in the sample were based on an average for UZAs of comparable size and region of the country). In contrast, the High-Growth scenario was driven by the historical (compound average annual) trend rate of growth, also at the UZA level, based on NTD data for the most recent 15 years.

For this report, the Low- and High-Growth scenarios now use a common, consistent approach that better reflects differences in PMT growth by mode. Specifically, these scenarios are now based on the trend rate of growth in PMT, now calculated as the compound average annual PMT growth by FTA region, UZA stratum, and mode over the most recent 15 years (hence, all bus operators located in the same FTA region in UZAs of the same population stratum are assigned the same growth rate). Use of the 10 FTA regions captures regional differences in PMT growth, while use of population strata (more than 1 million population; 1 million to 500,000; 500,000 to 250,000; and less than 250,000) capture differences in urban area size. Perhaps more importantly, the revised approach now recognizes differences in PMT growth trends by mode. Over the past decade, the rate of PMT growth has differed significantly across transit modes, being highest for heavy rail, vanpool, and demand response, and low to flat for motor bus. These differences are now recognized in the expansion needs projections for the Low- and High-Growth scenarios.

Asset-Decay Curves

Asset-decay curves were developed expressly for use within TERM and are comparable to asset-decay curves used in other modes of transportation, bridge, and pavement deterioration models. Although collecting asset condition data is not uncommon within the transit industry, TERM asset-decay curves are believed to be the only such curves developed at a national level for transit assets. Most of the TERM key decay curves were developed using data FTA collected at multiple U.S. transit properties specifically for this purpose.

TERM decay curves serve two primary functions: (1) to estimate the physical conditions of groups of transit assets and (2) to determine the timing of rehabilitation and replacement reinvestment.

Estimating Physical Conditions

One use of the decay curves is to estimate the current and future physical conditions of transit asset groups. The groups can reflect all national transit assets or specific subsets, such as all assets for a specific mode. For example, *Exhibit C-3* presents a TERM analysis of the distribution of transit asset conditions at the national level as of 2012.

Exhibit C-3 shows the proportion and replacement value of assets in each condition category (excellent, good, etc.), segmented by asset category. TERM produced this analysis by first using the decay curves to estimate the condition of individual assets identified in the inventory of the national transit assets and grouping these individual asset condition results by asset type.

Exhibit C-3 Distribution of Asset Physical Condition by Asset Type for All Modes

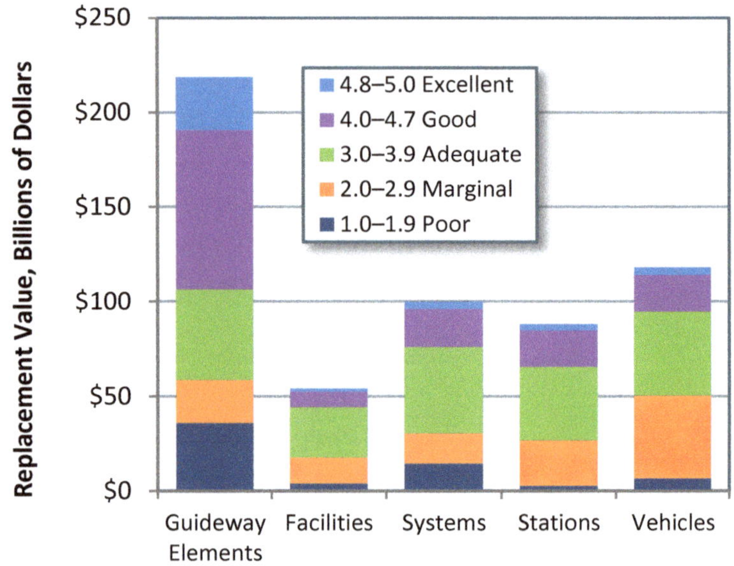

Source: Transit Economics Requirements Model.

TERM also uses the decay curves to predict expected future asset conditions under differing capital reinvestment funding scenarios. An example of this type of analysis is presented in *Exhibits C-4* and *C-5*, which present TERM forecasts of the future condition of the national transit assets, assuming the national level of reinvestment remains unchanged. *Exhibit C-4* shows the future condition values estimated for each asset identified in the asset inventory (weighted by replacement value) to generate annual point estimates of average future conditions at the national level by asset category. *Exhibit C-5* presents a forecast of the proportion of assets in either marginal or poor condition, assuming limited reinvestment funding for a subset of the national transit assets.

Exhibit C-4 Weighted Average by Asset Category, 2010–2029

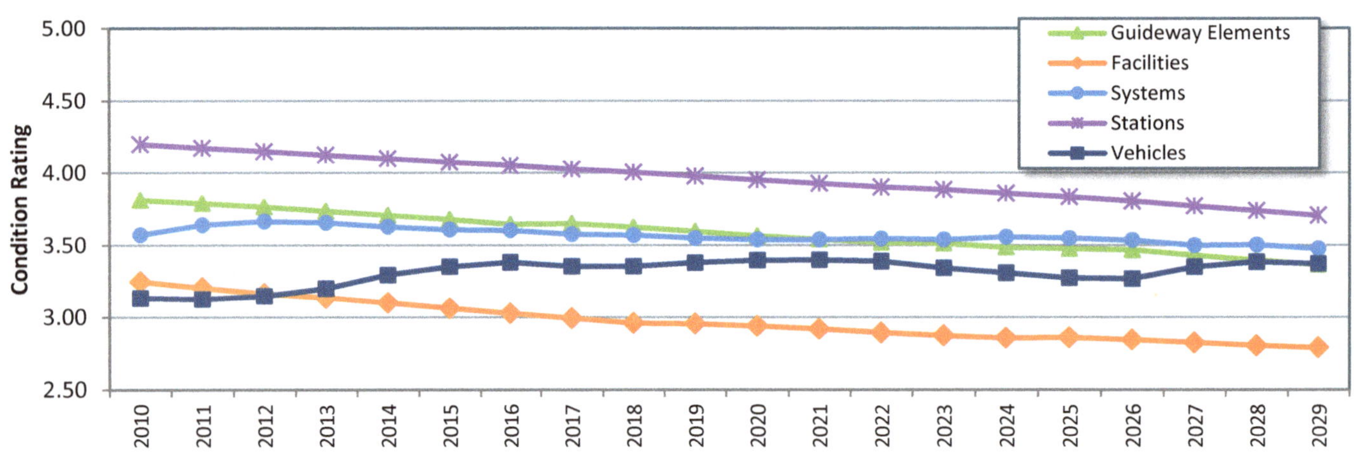

Source: TERM, Sustain 2010 Spending.

Exhibit C-5 Assets in Marginal or Poor Condition, 2010–2029

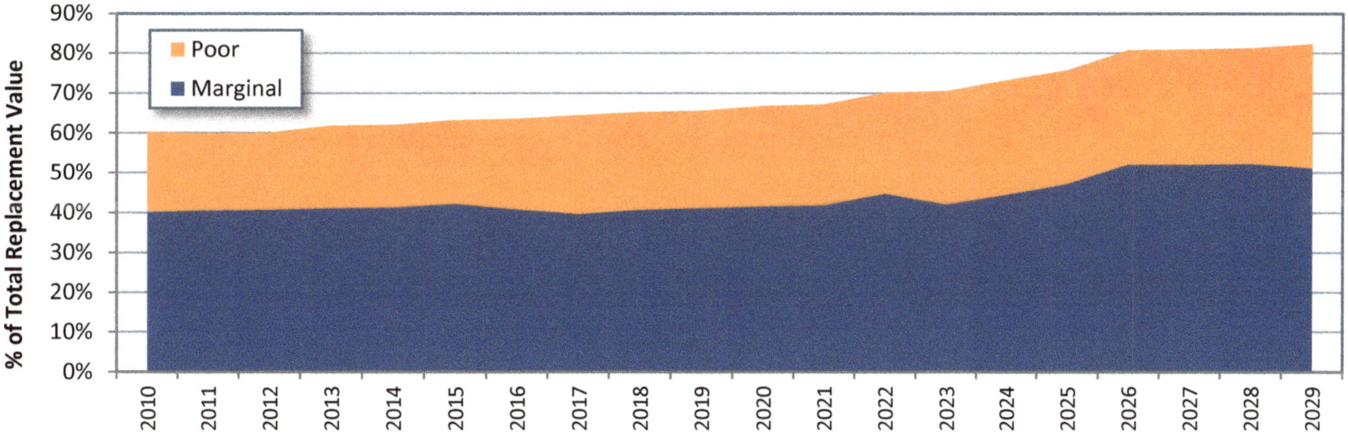

Source: TERM, Sustain 2010 Spending (Excludes Unreplaceable Assets).

Determine the Timing of Reinvestment

Another key use of the TERM asset-decay curves is to determine when the individual assets identified in the asset inventory will require rehabilitation or replacement, with the ultimate objective of estimating replacement needs and the size of the state of good repair backlog. Over the 20-year period covered by a typical TERM simulation, the model uses the decay curves to monitor the declining condition of individual transit assets continually as they age. As an asset's estimated condition value falls below predefined threshold levels (known as "rehabilitation condition threshold" and "replacement condition threshold"), TERM will seek to rehabilitate or replace that asset accordingly. If sufficient funding is available to address the need, TERM will record this investment action as a need for the specific period in which it occurs. If insufficient funding remains to address a need, that need will be added to the state of good repair backlog. These rehabilitation and replacement condition thresholds are controlled by asset type and can be changed by the user. Some asset types, such as maintenance facilities, undergo periodic rehabilitation, while others, such as radios, do not.

Development of Asset-Decay Curves

Asset-decay curves are statistically estimated mathematical formulas that rate the physical condition of transit assets on a numeric scale of 5 (excellent) to 1 (poor).

Most TERM decay curves are based on empirical condition data obtained from a broad sample of U.S. transit operators; hence, they are considered representative of transit asset-decay processes at the national level. An example decay curve showing bus asset condition as a function of age and preventive maintenance based on observations of roughly 900 buses at 43 different transit operators is presented in *Exhibit C-6* below.

Exhibit C-6 TERM Asset Decay Curve for 40-Foot Buses

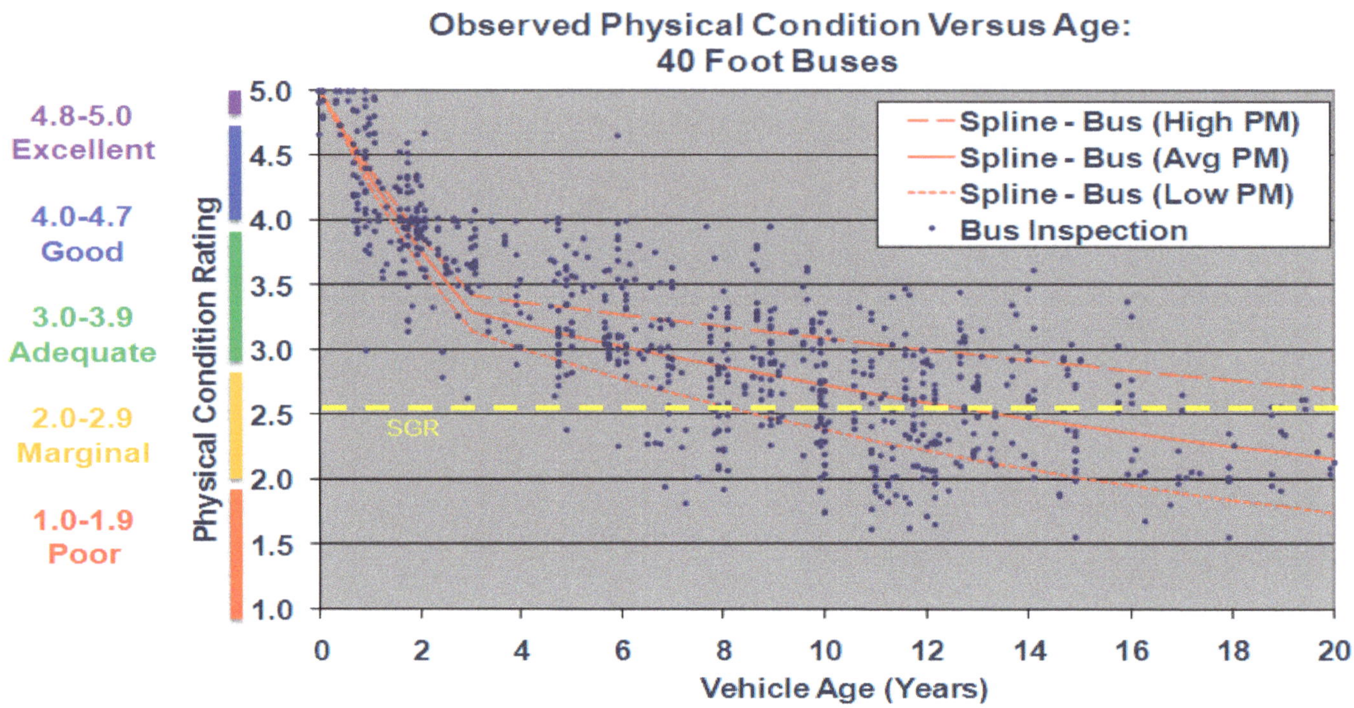

Source: FTA; empirical condition data obtained from a broad sample of U.S. transit operators.

Benefit-Cost Calculations

TERM uses a benefit-cost (B/C) module to assess which of a scenario's capital investments are cost effective and which are not. The purpose of this module is to identify and filter investments that are not cost effective from the tally of national transit capital needs. Specifically, TERM can filter all investments where the present value of investment costs exceeds investment benefits (B/C < 1).

The TERM B/C module is a business case assessment of each agency-mode (e.g., "Metroville Bus" or "Urban City Rail") identified in NTD. Rather than assessing B/C for each investment need for each agency-mode (e.g., replacing a worn segment of track for Urban City Rail), the module compares the stream of future benefits arising from continued future operation for an entire agency-mode against all capital (rehab-replace and expansion) and operating costs required to keep that agency-mode in service. If the discounted stream of benefits exceeds the costs, TERM includes that agency-mode's capital needs in the tally of national investment needs. If the net present value of that agency-mode investment is less than 1 (B/C < 1), TERM scales back these agency-mode needs until the benefits are equal to costs, as discussed below.

In effect, the TERM B/C module conducts a system-wide business case analysis to determine if the value generated by an existing agency-mode is sufficient to warrant the projected cost to operate, maintain, and potentially expand that agency-mode. If an agency-mode does not pass this system-wide business case assessment, TERM will not include some or all of that agency-mode's identified reinvestment needs in the tally of national investment needs. The benefits assessed in this analysis include user, agency, and social benefits of continued agency operations.

The specific calculations used by the TERM B/C module—comparing the stream of investment benefits for agency-mode "j" against the stream of ongoing costs calculated over the TERM 20-year analysis horizon—is presented below in Equation (1).

$$Benefit - Cost\ Ratio_{investments\ for\ agency-mode\ j} =$$

$$\left\{ \frac{\sum_{t=1}^{20} \left\{ \left(\left(User,\ Social\ Benefits_{j,t=0} \right) * \left(1 + TPM\ Growth_j \right)^t \right) / (1+i)^t \right\}}{\sum_{t=1}^{20} \left\{ \left(Replacement\ Needs_{j,t} + Expansion_{j,t} + \left(O\&M\ Costs_{j,t} * \left(1 + TPM\ Growth_j \right)^t \right) \right) / (1+i)^t \right\}} \right\}$$

The Benefit-Cost Ratio equation above has the following parameters:

- j: Agency Mode (where costs and benefits are assessed by agency mode and summed across all agency modes reported to National Transit Database; e.g., NYCT Heavy Rail is an "Agency Mode")
- t: time measured in years (year 1 to year 20)
- i: discount rate
- User Social Benefits: combination of user benefits and social benefits for users in Agency Mode j. User benefits consists of travel time saving and reduced auto costs. Social benefits is associated with reduced vehicle miles traveled, which results in reductions in air and noise emissions, roadway wear, and accidents.
- TPM Growth: recent growth in Total Passenger Miles for Agency Mode j (15 year)
- Replacement Needs: TERM's projected reinvestment needs for Agency Mode j in year t
- Expansion: TERM's projected expansion needs for Agency Mode j in year t
- O&M Costs: TERM's projected operating and maintenance costs needs for Agency Mode j in year t

Why Use a System-Wide Business Case Approach?

TERM considers the cost-benefit of the entire agency rail investment versus simply considering the replacement of a single rail car. Costs and benefits are grouped into an aggregated investment evaluation and are not analyzed at the level of individual asset investment actions (e.g., replacement of a segment of track) for two primary reasons: (1) lack of empirical benefits data and (2) transit asset interrelationships.

Lack of empirical benefits data: The marginal benefits of transit asset reinvestment are very poorly understood for some asset types (e.g., vehicles) to nonexistent for others. Consider this example: Replacement of an aging motor bus will generate benefits in the form of reduced maintenance costs, improved reliability (fewer in-service failures and delays) and improved rider comfort, and potentially increased ridership in response to these benefits. The magnitude of each benefit will depend on the age of the vehicle retired (with benefits increasing with increasing age of the vehicle being replaced). However, what is the dollar value of these benefits? Despite the fact that transit buses are the most numerous of all transit assets and a primary component of most transit operations, the relationship between bus vehicle age and operations and maintenance cost, reliability, and the value of rider comfort is poorly understood (no industry standard metrics tie bus age to reliability and related agency costs). The availability of reinvestment benefits for other transit asset types is even more limited (perhaps with the exception of rail cars, where the understanding is comparable to that of buses).

Transit asset interrelationships: The absence of empirical data on the benefits of transit asset replacement is compounded further by both the large number of transit assets that must work together to support transit service and the high level of interrelatedness between many of these assets. Consider the example of a (1) rail car operating on (2) track work equipped with (3) train control circuits and (4) power supply (running through the track), all supported by (5) a central train control system and located on (6) a foundation, such as elevated structure, subway, retained embankment, etc. This situation represents a system that depends on the ongoing operation of multiple assets, each with differing costs, life cycles, and reinvestment needs—and yet completely interdependent. Now consider the benefits of replacing a segment of track that has failed. The cost of replacement (thousands of dollars) is insignificant compared to the benefits derived from all the riders that depend on that rail line for transit service in maintaining system operations. The fallacy in making this comparison is that the rail line benefits depend on ongoing reinvestment in all components of that rail line (track, structures, control systems, electrification, vehicles, and stations) and not just from reinvestment in specific components.

Incremental Benefit-Cost Assessment

TERM's B/C module is designed to assess the benefits of incremental levels of reinvestment in each agency-mode in a three-step approach:

- Step 1: TERM begins its benefit-cost assessment by considering the benefits derived from all of TERM's proposed capital investment actions for a given agency-mode, including all identified rehabilitation, replacement, and expansion investments. If the total stream of benefits from these investments exceeds the costs, all assets for this agency-mode are assigned the same (passing) benefit-cost ratio. If not, the B/C module proceeds to Step 2.
- Step 2: Having "failed" the Step 1 B/C test, TERM repeats this B/C evaluation, but this time excludes all expansion investments. In effect, this test suggests that this agency-mode does not generate sufficient benefits to warrant expansion, but might generate enough benefits to warrant full reinvestment. If the agency-mode passes this test, all reinvestment actions are assigned the same, passing B/C ratio. Similarly, all expansion investments are assigned the same failing B/C ratio (as calculated in Step 1). If the agency-mode fails the Step 2 B/C test, the B/C module proceeds to Step 3.
- Step 3: The Step 3 B/C test provides a more realistic assessment of agency-mode benefits. Under this test, agency-mode benefits are assumed to exceed costs for at least some portion of that agency-mode's operations; hence, this portion of services is worth maintaining.

Investment Benefits

TERM's B/C module segments investment benefits into three groups of beneficiaries:

- Transit riders (user benefits),
- Transit operators, and
- Society.

Rider benefits: By far the largest individual source of investment benefits (roughly 86 percent of total benefits) accrue to transit riders. Moreover, as assessed by TERM, these benefits are measured as the difference in total trip cost between a trip made via the agency-mode under analysis versus the agency-mode user's next best alternative. The total trip cost includes both out-of-pocket costs (e.g., transit fare and station parking fee) and value of time costs (e.g., access time, wait time, and in-vehicle travel time).

Transit operator benefits: In general, the primary benefit to transit agencies of reinvestment in existing assets comes from the reduction in asset operations and maintenance costs. In addition to fewer asset repair requirements, this benefit includes reductions of in-service failures (technically also a benefit to riders) and the associated response costs of in-service failures (e.g., bus vehicle towing and substitution and bus for rail vehicle failures).

At present, none of these agency benefits is considered by TERM's B/C model. As noted above, little to no data are available to measure these cost savings. The available data relate primarily to fleet reinvestment and were not available at the time the B/C module was developed. FTA could incorporate some of these benefits in future versions of TERM.

Societal benefits: TERM assumes that investment in transit provides benefits to society by maintaining or expanding an alternative to travel by car. More specifically, reductions in vehicle miles traveled made possible by the existence or expansion of transit assets is assumed to generate benefits to society. Some of these benefits might include reductions in highway congestion, air and noise pollution, greenhouse gases, energy consumption, and automobile accidents. TERM's B/C module considers no societal benefits beyond those related to reducing vehicle miles traveled (hence, benefits such as improved access to work are not considered).

appendix D

Reimagining the C&P Report

Reimagining the C&P Report

Over the past 47 years, the C&P report series has provided an objective assessment of current system conditions and future investment needs. Its target audience includes the U.S. Congress, all levels of government, policy makers and analysts, academia, transportation associations, industry, news media, and the public. It raises public awareness of the physical conditions, operational performance, and financing mechanisms of highways, bridges, and transit systems, and promotes an understanding of the importance of these transportation investments.

The C&P report is a dynamic and evolving product, which has periodically undergone substantial overhauls and improvements. A good example is the introduction of benefit-cost analysis (BCA) to the process for estimating future investment needs through application of the Highway Economic Requirements System (HERS) introduced in the 1995 C&P Report; the Transit Economic Requirements Model (TERM) introduced in the 1997 C&P Report; and the National Bridge Investment Analysis System (NBIAS) introduced in the 2002 C&P Report. These models are presented and described in Appendices A, B, and C, respectively.

As discussed in the Introduction to Part I of this report, MAP-21 (Moving Ahead for Progress in the 21st Century Act) incorporated performance management principles into its requirements. States will set targets for several key performance measures and report on their progress in meeting these targets. This shift toward more performance-driven and outcome-based programs has direct and indirect implications for the C&P report. At the most basic level, the introduction of other performance reporting requirements in MAP-21 might necessitate some content changes to the C&P report, both to take advantage of newly available data and to avoid unnecessary duplication of information presented elsewhere. The shift in the processes that States and metropolitan planning organizations (MPOs) use for planning and performance management also has implications for assessing future transportation investment needs. State and local agencies are adopting more outcome-based approaches to investment decision-making, which has significant implications for the potential impacts of future investment on system performance and how these impacts are simulated. In addition, the data, analytical tools, and techniques developed to support the implementation of MAP-21 could yield new approaches that can be adapted to refine or replace HERS, NBIAS, and TERM.

With these issues in mind, the Federal Highway Administration (FHWA) initiated the *Reimagining the C&P Report in a Performance Management-Based World* effort in late 2012. Preliminary scoping work was conducted in 2013 to document who uses the C&P report, to assess the utility of the report to FHWA program offices in communicating key information, and to identify options for presenting information more effectively. This effort identified two areas of potential improvement to align better with performance measures: methodology and communication. Two major research projects were initiated in 2014, with the objectives of improving estimation methodologies to compute investment needs and enhancing communication approaches, respectively.

Methodology Improvement

Simulation modeling inherently involves compromises, as the desire for detailed, reliable predictions must be balanced against data collection burdens and computational tractability. The tools and methodologies currently used in the C&P report reflect several analytical shortcuts and simplifications introduced to accomplish the desired analysis with the available data and resources. Since the initial introduction of these tools, a new generation of analytical tools and models has been developed that provide advanced methodologies in asset management and performance management.

HERS, NBIAS, and TERM have been constantly revised and updated to incorporate newly developed data and tools. Building on this ongoing improvement effort, a research project is currently underway to scan and compare methods for assessing investment needs and to propose new and improved methods for more precise and comprehensive needs estimation in the C&P report. Several analytical frameworks are being explored to identify potential alternative methodologies and upgrades to the current BCA approach. This project includes a systematic review of performance management tools that States and local governments currently use. The goal is to identify practical approaches for improving the C&P report methodology in the future.

Evaluation of Alternative Methodologies

The first stage of this research effort involves evaluating alternative methodologies that could be used to replace or supplement the BCA-driven tools currently used in the C&P report. Two specific decision methodologies that will be reviewed are the multi-criteria decision method (MCDM) and value for money.

MCDM allows for consideration of performance objectives that are difficult to monetize, and therefore MCDM frequently includes some performance measures that are not limited to monetary terms or condition matrices. It is a flexible tool, enabling the evaluation of projects based on multiple performance measures such as environmental sustainability, livability, and safety. MCDM is a viable potential method for enhancing a revised C&P report that is better aligned with MAP-21 and strategic goals. Its application, however, hinges on the selection of appropriate performance measures and assignment of weight to each performance measure, which could be challenging for national investment analysis.

As defined in the Eddington Transport Study of the United Kingdom,[1] value for money is another methodology that measures wider economic and reliability benefits. It assesses the economic, environmental, social, distributional, and fiscal impacts of an investment based on both quantitative, monetized information and qualitative information at the project level. Although this approach helps guide the modeling of reliability and economic impacts, scaling the findings from individual projects to the national system and obtaining a strategic allocation of resources for infrastructure investment could be challenging.

Other methodologies that could be studied include impact analysis tools that attempt to estimate the economic impacts of highway investments on the overall economy. Alternative methodologies for evaluating indirect user benefits not currently captured in HERS or NBIAS also might be explored.

Identification of Alternatives for Refining Benefit-Cost Analysis

The next stage in this research effort involves identifying alternatives for refining the current BCA approach to align with performance management principles. Two specific options under review are the potential for integrating needs analysis of pedestrian and cycling infrastructure into the C&P process and the feasibility of integrating network analysis into the C&P highway needs assessment.

Local and regional stakeholders are increasingly demanding consideration of active transportation modes (i.e., pedestrian and cycling) in needs assessment. Data availability issues have hampered such efforts in the past, but significant advances in recent years could make this option more feasible.

Although HERS currently incorporates some limited procedures for estimating network effects, the system is fundamentally a highway segment-level evaluation tool. Potential alternatives are the adoption of a more corridor-focused analysis process or a complete network analysis. The NPMRDS (National Performance Management Research Data Set) discussed in Chapter 5 might prove useful in identifying existing corridor conditions and in calibrating forecasting procedures.

Other potential enhancements that could be explored include options for estimating needs specific to freight movements and the direct integration of operations treatments and assets within the core procedures for highway investment analysis.

Integration of Performance Management and Needs Estimation

The next stage of this research effort will involve identifying existing State- and local-level tools that incorporate performance management principles, possibly leading to additional future refinements to the C&P report analytical procedures. Later stages will involve combining these findings with those identified in the assessments of BCA refinements and alternative decision methodologies. This combination would enable a detailed evaluation and comparison of several comprehensive approaches to upgrade the current national needs estimation process. Once appropriate analytical frameworks are identified, new components could be added to HERS and NBIAS or a new generation of analytical tools could replace these models.

Enhanced Communication

Currently, the C&P report is issued in paper form and the entire report is posted online using standard Adobe Acrobat and HTML formats. The look of the C&P report, however, has remained largely unchanged, despite the wide adoption by FHWA offices and several other government agencies of enhanced communication tools for presenting complex data. Preliminary scoping work conducted in 2013 revealed several basic concerns about the current approach.

Although the C&P report contains useful information and serves as a valuable reference document, its sheer size creates some problems for users. Because writing and reviewing the document is a lengthy process, the report is often transmitted to Congress after newer data have been published elsewhere. Even when this is not the case, many of the data in the biennial report and many of the data sets upon which the report relies are updated annually, which means that readers must often look elsewhere to find the latest available data.

One option under consideration is to develop a more robust website to complement the paper report. Under this approach, some of the more detailed, supplementary analyses currently presented in the report could be migrated to the website, allowing the paper version to focus on key findings. Such an approach also would facilitate more frequent data updates than are currently possible for the C&P report.

A research effort is underway to explore alternatives for enhancing the current report, focusing on data visualization and an interactive Web-based design. The underlying goal is to facilitate ease of use by a wider audience of readers and enable the alignment of performance-based information in the C&P report with the information obtained from State and MPO performance management processes.

Data Visualization

Data visualization is the representation of data in a pictorial or graphical format. It is the easiest way for the brain to receive and process large amounts of information quickly and intuitively. As part of this research effort, alternatives are being explored to improve the communication of data on both paper and the Web through advanced data visualization tools and infographics. For the paper version of the C&P report, new static graphics could be developed to help visualize complex information on highways, bridges, and transit that is easier to understand at a glance. Contents of each chapter could be condensed into a format that is more accessible to the public, such as bullet points, at-a-glance boxes, and content optimization for print layout.

For the online version, selected contents could be presented through interactive data visualization to convey information from in-depth and complex analytics. For example, an online platform might support the use of more dynamic and interactive graphics, such as customized dashboards and charts as the underlying data change according to the user's unique needs. Through the intuitive interfaces, data visualization tools enable customized analytical views with flexibility and ease by multiple users with diverse demands.

Web-Based User Interface

As part of this research effort, discussions are ongoing about how to upgrade the Web page of the report to inform, attract, and retain visitors through new methods of electronic communication. The goal of any Web page improvements is to combine good information architecture with the art of expression to guide users to contents grouped into appropriate categories. A new digital publishing platform could integrate traditional format like PDF with many interactive elements such as embedded video/audio and interactive graphs. To attract and maintain the attention of an increasingly mobile audience, an upgraded Web page could use a responsive Web design to accommodate data exploration and communication across myriad devices, including touchscreen and mobile devices.

Recognizing the current shifts in media and technology preferences, a suite of communication methods could be used to improve user experience via a highly interactive Web-based platform for the C&P report. Such a platform could enable users to extract information relevant for their specific purpose and produce customized data and reports for distribution. Functions of the website ultimately could be substantially expanded to support requests like search information, zoom in and out on maps, and sort and filter databases in real time.

A critical part of developing an enhanced future C&P report website is ensuring that it complements existing online resources and potential new resources coming on line in response to the MAP-21 State and MPO performance reporting requirements. In many cases, providing links to information posted in other locations might be sufficient so that limiting the content of C&P Web page focuses mainly on elements unique and central to the report.

Moving Forward

Although FHWA began the particular research initiatives described in this appendix, the Federal Transit Administration (FTA) as a full partner in the development of the C&P report is closely involved in these efforts. FTA has initiated its own reviews regarding future analytical approaches and report presentation and content. As potential enhancements become more fully refined through the current research efforts, external outreach will be conducted to ensure that any changes to the report content and structure will improve its utility for the members of Congress and other key readers. Although the objectives of the report will remain unchanged, the goal of this effort ultimately is to provide a multimodal product with cutting-edge analytics that improves user experience.

---

[1] The Eddington Transport Study (2006). The case for action: Sir Rod Eddington's advice to Government. Available at http://webarchive.nationalarchives.gov.uk/20090104005813/http:/www.dft.gov.uk/about/strategy/transportstrategy/eddingtonstudy/.

www.ingramcontent.com/pod-product-compliance
Lightning Source LLC
LaVergne TN
LVHW070844210926
848482LV00016B/866
9780160937064